WITHDRAWN
UST
Libraries

Management Impacts of Information Technology:

Perspectives on Organizational Change and Growth

Edited by

Edward Szewczak
Canisius College

Coral Snodgrass
Canisius College

Mehdi Khosrowpour
Pennsylvania State University at Harrisburg

IDEA GROUP PUBLISHING
Harrisburg, Pennsylvania U.S.A.

Senior Editor: Mehdi Khosrowpour
Managing Editor: Jan Travers
Copy Editor: Karen Cullings
Printed: Jednota Press

Copyright © 1991 by Idea Group Publishing
4751 Lindle Road, Suite 116
Harrisburg, Pennsylvania 17111
(717) 939-7320

All rights reserved. No part of this book may be reproduced in any form or by any means, electronic or mechanical, including photocopying, without written permission from the publisher.

Printed in the United States of America

Library of Congress Card Catalog No: 90-082960

ISBN: 1-878289-08-X

HC
79
.I55
M35
1991

Management Impacts of Information Technology:
Perspectives in Organizational Change and Growth

TABLE OF CONTENTS

PART I
EVOLUTION AND EVALUATION OF INFORMATION TECHNOLOGY IN ORGANIZATIONAL SYSTEMS

335451

PART II
MANAGING INFORMATION TECHNOLOGY, MODELS AND DATABASES

PART III
INFORMATION TECHNOLOGY IMPACTS ON INDIVIDUALS AND GROUPS

PART IV
INFORMATION TECHNOLOGY IMPACTS ON JAPANESE AND AMERICAN ORGANIZATIONS

PART V
INFORMATION TECHNOLOGY AND ORGANIZATIONAL STRATEGY

Preface

Commenting on the information technology (IT) issues facing corporate management in the 1990s, Dixon & John (1989, p. 251) observe that the "senior IT executive is facing the need to manage a network of technology resources, supporting a network of uses determined by line management." The scope of this management challenge is rapidly expanding to encompass virtually all aspects of organizational activity.

IT has had a wide range of impacts on the management of organizations. As the technology improved from a technical vantage point, managers have witnessed changes in organizational life and ways of doing business that have transformed their thinking about what it means to manage effectively and efficiently. Various aspects of management have been influenced by IT in ways that were not always anticipated by the developers of the technology. In some respects IT has been a boon to organizational life, but in other respects IT has proven to be threatening to successful organizational functioning.

Managerial responses to IT have influenced the nature of managerial work and the quality of work life. The introduction of IT has influenced organizational strategy and strategy making, has altered organizational structures and cultures, and has reshaped organizational communication and learning as well as management thinking on organizational design. All aspects of organizational management — from manufacturing to R&D to human resources — have changed (in many cases for the better) in light of the evolution of IT.

Management Impacts of Information Technology: Perspectives on Organizational Change and Growth is a collection of scholarly and professional chapters which together provide differing viewpoints and critical assessments on how managers are responding to the challenges and opportunities presented by the proliferation of IT in modern society. Each chapter takes the reader through an interesting and important aspect of IT management by focusing on past, present and anticipated developments in IT. Discussions of solutions or remedies relevant to solving problems are provided for greater utilization and improved management of IT.

The authors of the chapters represent a wide range of management specializations — from organizational behavior to organization theory to strategic management to management information systems. These different specializations provide a forum for a debate wherein insights about future issues and trends and research opportunities are compared and contrasted. Taken together, the chapters provide a snapshot of the different perspectives which exist today in management thinking about the advantages and disadvantages of IT, and suggest a composite vision of where organizations may change and grow in response to IT capabilities.

We have organized the chapters into broad categories: the Evolution and Evaluation of IT in Organizational Systems, Managing IT, Models and Databases, IT Impacts on Individuals and Groups, IT Impacts on Japanese and American Organizations, IT and Organizational Strategy, and IT and Organizational Design. We believe the chapters have practical relevance for managers, but they may also serve as the basis for the education of future managers. This book may be utilized for an advanced undergraduate or graduate course in Information Technology Management or in Information Systems and Organizations — actually, any course in which a thoughtful approach to the issues and problems of IT management is adopted and encouraged.

Reference

Dixon, J., & John, D.A. (1989). Technology issues facing corporate management in the 1990s. MIS Quarterly, 13(3), 247-255.

Acknowledgements

Many people deserve our thanks for their contribution to this book. The chapter authors did a remarkable job, not only in putting together high quality chapters, but also in submitting the chapters in a timely fashion so that progress in compiling the book was greatly facilitated. Our institution was very supportive of our efforts to complete the book. In particular, we would like to thank Richard Shick, dean of the Wehle School of Business; Alan Duchan, chair of the MIS department; and Alan Weinstein, chair of the management/marketing department, all of Canisius College for their unwavering support throughout this project. Ernie Dishner, dean of faculty and Gayle Yaverbaum, acting director of the School of Business Administration, both of Penn State Harrisburg also deserve a special thank you for all their support. Finally, thanks are in order to Idea Group Publishing for their many efforts toward completing this project as planned, in particular, Jan Travers, managing editor, for her work in bringing this project together, Karen Culling for her fine copy editing, and Carol Kalbaugh and Victoria Cuscino for all their hard work.

Edward J. Szewczak
Coral R. Snodgrass
Mehdi Khosrowpour

Part I

Evolution and Evaluation of Information Technology in Organizational Systems

Chapter 1

Information System Capabilities and Organizational Applications: An Evolutionary Perspective

William I. Bullers, Jr. and Richard A. Reid
University of New Mexico

Managers require information to control daily operations and to support planning activities in their organizations (Davis & Olson, 1985). Information is a ubiquitous organizational resource used at every managerial level and by each functional area for managerial planning, organizing, and control (Brancheau & Wetherbe, 1987; Meltzer, 1981). In recent years, a rapidly changing competitive environment has led to an increase in the demand for accurate, timely, and relevant information, particularly of a strategic nature (Mockler, 1989; Porter, 1980). Service-oriented organizations need information to help them understand prevailing markets, to analyze competitive behavior, and to model customer-competitor interactions (Porter, 1980). In short, managers need information not only to manage their internal environment through a better understanding of their production, marketing, and financial capabilities, but also to provide assistance in predicting the behavior of those relevant aspects of their external environment that are not controllable.

This increased demand for information can be attributed to a variety of factors. Government deregulation in many industries, expanded international competition, shifts from an industrial-based to a service-based economy, and revolutionary technological developments reflect an increasingly turbulent environment. This situation has fostered significant changes in information requirements in every organization. Production, marketing, and financial decision-makers need readily-accessible and up-to-date information to develop cost-efficient and effective strategic, tactical, and operational plans. Without appropriate information, management performance is impaired.

Moreover, as organizational effectiveness deteriorates, a firm's very survival is threatened.

The role of computer-based information systems (IS) in organizational control and managerial planning varies by organization. For some organizations, the focus of IS is on helping managers design cost-efficient solutions to operational problems. For other organizations, however, IS support decision-making at the strategic level (Brancheau & Wetherbe, 1987; Cash & McLeod, 1985; Cash, McFarlan & McKenney, 1988; Mockler & Dologite, 1988), and thus, are recognized as critical components of the long-term strategic planning process (Leavitt & Whistler, 1958). More recently, progressive organizations have used information technology to help them determine and exploit their unique advantages or strengths (Bruns & McFarlan, 1987; Ives & Learmonth, 1984; McFarlan, 1984; Parsons, 1983). In such a role, IS are used to support the firm's strategic plans in order to gain a competitive advantage, generally by aiding the firm to become the low cost producer, differentiate its products, or to focus on a particular market niche (Porter, 1980, 1985).

In addition to facilitating organizational decision-making, information is also becoming an important marketable commodity in its own right. Information is reusable, portable, expandable, compressible, and shareable. When properly organized, information becomes a very valuable product, namely knowledge (Gross, 1988). Information supports not only interorganizational activities, but increasingly represents a core service, changing the operational parameters of many organizations in both the industrial and service sectors. IS now comprise the basic infrastructure for trade in goods and services. Sauvant (1986) argues that they are responsible for the redefining of operations in many transnational corporations, and contribute to increased international trade in such key business services as banking, insurance, accounting, engineering design, legal services, management, and consulting. Information as a commodity has become a valued economic resource of significant importance not only to the generating organization, but to its host country as well (Gassman, 1985; Sauvant, 1986; Yourdon, 1986).

Peters (1988) has stated that "excellent firms don't believe in excellence - only in constant improvement and constant change". The need for innovation, flexibility, and adaptability in the face of an increasingly dynamic organizational environment has fostered a significant change in IS requirements. Since humans have basic limitations in their information processing capabilities (Simon, 1965), there is increasing reliance on IS for efficiency in management control and effectiveness of managerial planning. Fortunately, changes in information technology have been commensurate with changes in the organization's environment. Computer capabilities and performance have improved by a factor of ten every several years (Yourdon, 1986). Of course, rapid technological developments in computers, telecommunications, and office automation also contribute to an increased rate of change within society as a whole (Drucker, 1985; Naisbitt, 1983; Toffler, 1980). Thus, while technological advances permit organizations to develop increasingly sophisticated IS, the environmental changes brought about by these advances compel organizations to continually improve their IS capabilities in order to remain competitive.

IS consist of integrated collections of (a) computer, telecommunication, and office automation hardware; (b) business application, data management, and operating system software programs; (c) data; (d) organizational policies and procedures; and (e) support personnel (Kroenke, 1981). This interrelated federation of components seeks to provide management with feedback information that promotes organizational control and supports managerial planning.

Four basic types of IS have evolved since the advent of computers: electronic data processing (EDP) systems, management information systems (MIS), decision support systems (DSS), and expert systems (ES). EDP systems, which were first introduced around 1951, tended to be transaction-based and control-oriented. They supported accounting and clerical operations such as payroll, accounts payable, sales order entry and invoicing, and inventory control (Aron, 1969; Davis & Olson, 1985). MIS, initially introduced in 1965, provided data storage, retrieval, and processing capabilities to aid routine, repetitive, managerial decision-making activities (Aron, 1969; Davis & Olson, 1985; McFarland, McKenney & Pyburn, 1983). MIS often filtered and condensed data that were generated by operational EDP systems. In contrast, DSS have provided flexible, user-directed modeling capabilities for non-routine planning and policy decisions since 1978 (Bonczek, Holsapple & Whinston, 1981; Keen & Scott Morton, 1978; Sprague & Carlson, 1982). Four years later in 1982, ES began to exploit artificial intelligence (AI) techniques to capture human expertise or managerial judgment in the form of problem-solving rules within a computer program (Hayes-Roth, Waterman & Lenat, 1983; Holsapple & Whinston, 1987). Although many organizations are in the process of shifting their focus from EDP and MIS to the development of DSS and ES (Yourdon, 1986), all four IS types continue to enhance an organization's capability for effective decision-making.

Decision-making in a changing information technology environment requires that managers understand the capabilities, limitations, and relationships between the various IS types, as well as the role that IS play within the larger organizational system. The purpose of this chapter is to promote this understanding through an examination of critical IS ideas and issues from an evolutionary perspective. First, several important concepts related to information, control, and decision-making are examined. Next, the evolutionary changes that have occurred in the functional capabilities of IS are analyzed.

Finally, some guidelines for managerial action in the future development of their organization's IS are presented.

Underlying Conceptual Frameworks

The growth in IS sophistication, use, and worth to the organization has occurred in stages (Gibson & Nolan, 1974; Nolan, 1979). Organizational evolution from initiation through maturation is influenced by both external and internal knowledge. The development of external knowledge is associated with progress in information technology and related disciplines such as operations research, organization behavior, and computer science. Often the new ideas and concepts require a lengthy gestation period before their adoption and implementation becomes technically and/or economically feasible. Internal knowledge is primarily experiential as managers, developers, and users learn through formal and informal experimentation with different IS products (Nolan, 1979). The interaction between external knowledge development and internal knowledge generation has promoted evolutionary changes in IS. Thus it is useful to review some conceptual frameworks that have influenced the evolution of the four IS types. Since the evolutionary nature of IS must be understood to benefically engage in IS planning, it is incumbent upon the organization's managers to systematically seek external knowledge, as well as to continue experimentation required in the development of an organization's internal body of knowledge about information technology. This dual approach to knowledgedevelopment will allow the manager to exploit IS for the benefit of the organization.

Important distinctions between data, information, and knowledge have been influenced by long-standing, formal research in the fields of information theory (Shannon & Weaver, 1949; Weiner, 1948) and information economics (McDonough, 1963). Data are raw facts, especially symbols, which represent objects, actions, and events (Davis & Olson, 1985). Symbols may be created, modified, copied, and eventually destroyed, but they do not have direct value to a decision-maker (Simon, 1981). Information is data that has been processed into a form that is meaningful to the recipient and is of value in decision-making (Davis & Olson, 1985). Information has various attributes, including time, form, place, and possession utility, along with relevance, accuracy, and verifiability. These attributes contribute directly to information's value in reducing uncertainty in a decision situation. Knowledge is the use of previously established object descriptions, relationships, and generalized procedures for processing the object descriptions and relationships that can be applied to a problem of interest in order to make a decision (Hayes-Roth, et al., 1983). Knowledge is essentially a set of rules that govern how information is interpreted and applied effectively for decision-making in a context-dependent situation. Management science has contributed various modeling perspectives and analytic techniques that have influenced the evolution of IS. The use of simulation as a decision-making aid (Law & Kelton, 1982), operations research models for decision optimization (Hillier & Lieberman, 1986), heuristic models based on human problem-solving techniques (Newell & Simon, 1972), and the general systems point of view (Churchman, 1968; Forrester, 1961; Simon, 1965) are all influential factors. Three frameworks associated with decision-making are of particular importance in the analyses of IS (Barros, 1981; Bonczek, et al., 1981; Forrester, 1961; Keen & Scott Morton, 1978). These frameworks include Simon's (1965) decision type, Anthony's (1965) hierarchical level of managerial activity, and Simon's (1965) idealized decision-making process.

Decision type ranges from structured to unstructured depending on whether the managerial processes under analysis can be prescribed comprehensively. Structured decisions refer to routine and repetitive decisions for which a problematic situation can be anticipated and appropriate conditional responses programmed. Unstructured decisions are those which cannot be preformulated because of problem uniqueness, complexity, and/or critical nature. Semi-structured decisions encompass the broad continuum between highly structured and unstructured problems. The amount of knowledge that must be applied to a problem, and hence the degree of expertise required to solve the problem, is highly dependent upon the decision type.

Hierarchical level of managerial activity corresponds to Anthony's (1965) widely accepted framework for classifying decisions according to their purpose in planning and control. This model identifies three levels of managerial activity, namely, strategic planning, management control, and operational control. Strategic planning is the process of identifying organizational objectives and the creation of policies required to achieve these objectives. Management control involves the determination, acquisition, and allocation of resources required to effectively and efficiently achieve the organizational objectives identified in strategic planning. Management control decisions tend to require managerial judgment based on an experienced, but subjective interpretation of the information available. Operational control seeks to assure that specific tasks are implemented effectively and efficiently. Even though operational control may not require as much judgment as managerial control, it may occasionally involve semi-structured or unstructured decision-making tasks which cannot be preprogrammed.

Simon's (1965) normative structuring of the decision-making process depicts the sequence of managerial activity required between recognition of a problem and implementation of a solution. This process can be partitioned into four stages: intelligence, design, choice, and implementation (Simon, 1965; Gerrity, 1971). Intelligence consists of monitoring an organization's environment and internal operations to identify problems or opportunities that call for action. Design involves developing alternative courses of action to solve problems or exploit opportunities identified in the intelligence phase. Choice is the process of evaluating alternative solutions and selecting a particular course of action. Implementation requires carrying out the selected course of action.

Four Types of Information System

Increasing requirements for organizational information processing, coupled with advances in information technology, have led to the development of four basic types of IS. Table 1 provides a summary of characteristics associated with each type. This table indicates the evolutionary nature of efforts to utilize the computer as a tool for enhancing the decision-making process.

Electronic Data Processing (EDP) Systems

EDP systems automate structured clerical tasks by providing basic operations for collecting, storing, manipulating, and distributing data (Phister, 1979). They process transactions, store and retrieve records, and generate periodic reports (Davis & Olson, 1985). Typical EDP applications include payroll, general ledger, and sales order entry.

The advent of EDP systems coincided with the genesis of early computers. EDP systems originated about 1951 with the delivery of a UNIVAC computer to the U.S. Bureau of the Census for use in connection with the 1950 census (Rosen, 1969). The UNIVAC was the first production-line computer designed specifically for commercial data processing instead of scientific or military use. Usage was relatively limited with only 36 machines in operation by 1955 and 85 active during the peak usage period in 1959. Limited application capabilities and significant operating expenses, including an average monthly rental of $25,000, restricted early EDP system usage (Phister, 1979). Thus while the the UNIVAC led to the development of prototype EDP systems, it did not engender widespread development.

IBM's first entries into the electronic computer field were scientific machines, the IBM 701 in 1953 (Hurd, 1983) and the IBM 650 in 1954 (Hurd, 1986). IBM initiated a generous educational grant program and universities seized the opportunity to acquire the 650 for use in research, development, and education. Fueled by accelerating interest in computers on college campuses, the IBM 650 was soon adopted for use by businesses. Subsequent EDP system development began in earnest.

IBM recognized that the commercial uses of computers would quickly exceed that of the military and scientific communities (Drucker, 1985). They

CONTRASTING DIMENSION	**Types of Computer-Based Information Systems**			
	EDP	MIS	DSS	ES
YEAR OF ORIGINATION	1951	1964	1971	1958
YEAR OF MAJOR ACCEPTANCE	1960	1968	1978	1982
PRIMARY FOCUS:				
Data Processing	X	X		
Information Processing	X	X		
Knowledge Processing				X
ORGANIZATIONAL LEVEL:				
Operational	X	X	X	X
Tactical/Managerial		X	X	X
Strategic			X	X
USER INTERFACE:				
Batch	X	X		
Interactive		X	X	X
PROBLEM TYPE ORIENTATION:				
Simple, Routine, Highly Structured	X	X	X	
Complex, Unique, Wide Variety, Semi-structured			X	X
Complex, Unique, Narrow Focus, Unstructured				X
AVAILABLE SUPPORT TOOLS:				
Database	X	X	X	X
Database Management System		X	X	X
Model Base (Software):				
Ad Hoc Inquiry/Retrieval Capabilities			X	X
Statistical Analysis/Operations Research Tools			X	X
Flexible (Programmable) Packages			X	X
Context Dependent Heuristics			X	X
General Problem Solving Structures				X

Table 1. A Comparison of Information Systems Functions and Capabilities

introduced the IBM 702 in 1955, and IBM 705 in 1956, both specifically intended for use in accounting, finance, sales, manufacturing, and general administration applications (Hurd, 1986). Other manufacturers including Burroughs, Control Data Corporation, Honeywell, National Cash Register (NCR), Philco, Radio Corporation of America (RCA), and Sperry-Rand (UNIVAC) also introduced commercial computers, providing further impetus to growth in EDP.

IBM soon became the dominant computer vendor in the commercial EDP field, particularly after the introduction of the transistorized IBM 1401 in 1959 (Rosen, 1969). The IBM 1401 could be leased for an average monthly rental of $2,500, permitting its installation by an increasing number of organizations. The number of general purpose computers used in the United States grew from about 240 in 1955, to 3,110 in 1959, and to 11,700 in 1963 (Phister, 1979). Of the nearly 12,000 computers in use in 1963, 45% were IBM 1401's, a market penetration that illustrates the impact of the 1401 on computing and early IS growth. Parallel developments in computer peripheral devices, including card readers, card punches, printers, magnetic tape, magnetic drum, and eventually disk, were motivated by organizational needs to efficiently process the voluminous data inherent in EDP applications.

In addition to hardware innovations of the late 1950's and early 1960's, software developments also contributed to the rising popularity of computers. High-level procedural programming languages, such as Algol-60, Fortran, and COBOL began to appear. COBOL (COmmon Business Oriented Language), initially defined in 1960 (Department of Defense, 1960) and regularly updated since then (Sammet & Garfunkel, 1985), provided EDP system programmers with a high-level language that was compatible across different machines, offered significant improvements in programming productivity over low-level assembly languages, and advanced the state of the art in practical data processing (Shneiderman, 1985). One testament to its importance in the development of EDP systems is found in its continued widespread use today, nearly 30 years after its introduction.

Spurred by continuing advances in computer hardware and software, most EDP systems were implemented solely to automate transaction processing applications. Few organizational behavior, decision-making theoretic, or management science perspectives were applied to EDP system development, since there was amplejustification for building these systems solely from an efficiency standpoint. Improved decision-making effectiveness may have been seen as a by-product in justifying some EDP system installations, but it was not considered to be a major factor in their development.

EDP systems typically deal with highly structured tasks at the operational control level in the organization. Yet in spite of the highly structured nature of many of these applications, the EDP system rarely controls the organizational system with which it interacts. The semantics of control theory require that a control system automatically execute the intelligence, design, choice, and implementation stages in the decision-making process without human intervention (Bullers & Whinston, 1983; Larson, McEntire & O'Reilly, 1984). EDP systems generally fail to accomplish all four stages, particularly from an implementation standpoint. Lacking a true control capability, many EDP systems might more appropriately be considered as highly structured, operationally-based, decision support systems. An inventory management system which identifies items that require a reorder (intelligence), produces a list of possible suppliers (design), and then prepares a purchase order to a particular supplier based on predefined criteria (choice) comes close to providing full-fledged automatic control. Yet unless the order is automatically sent without human review, the EDP system does not provide control. It merely supports the implementation stage of the decision-making process. Similarly, an airline reservation system supports the choice phase of a passenger's decision-making process by generating a list of possible flights (design), after recognition by the prospective passenger of the need to reserve a seat (intelligence). Actual booking of the seat (implementation) is carried out by the reservation clerk in conjunction with the EDP system. Many time-critical, embedded systems which serve to control a host industrial process, military weapon system, or scientific apparatus, do provide complete control, at least under normal operating conditions (Larson, et al., 1984). In the future, perhaps more administrative EDP systems will provide complete control, particularly when integrated with an ES component that could exercise judgment over the control function. The principal components of an EDP system usually include a set of master files containing application data, procedures for data collection and report distribution, and a library of programs for data manipulation. Adds, changes, and deletions to the master files are coded as transactions to be applied to the appropriate master files by the data manipulation programs. Reports are produced by interrelating the individual master files via other data processing programs. Recently designed EDP systems often take advantage of a data base management system (DBMS) to support more sophisticated data organization and manipulation than is generally available in a file environment. A commonplace processing sequence tends to include transaction input, edit, update, and report steps. In all EDP systems, the vast majority of the transaction processing is performed on a periodic basis by programs specifically designed and written to perform highly structured tasks.

The limited focus of EDP system applications from 1951 through 1964 on highly structured, operational level transaction processing was primarily due to technical limitations. The computer hardware simply did not permit either the breadth in scope or depth in complexity that was desired even at that time (Zachman, 1987). The conceptual and financial constraints imposed by computing technology prevented the development of IS beyond that of simple transaction processing and reporting. For example, the IBM 1401 computer system had only 16K bytes of internal memory. This limitation severely constrained the size and complexity of EDP programs that the 1401 could run. In contrast, a small personal computer today typically has at least a megabyte of internal memory, with up to 32Mbyte on larger machines. EDP application complexity was further constrained by the 1401's 6-bit byte size which permitted only 64 different data characters to be represented.

Early computer data storage devices further constrained EDP system sophistication. The IBM 1401 was primarily card and tape-oriented, supporting only sequential files in which data records are not individually accessible without reading the entire card deck or tape file. Tape drive technology in the mid-1950's permitted only 5 Mbytes of data per tape reel, with an increase to 15 Mbytes by the mid-1960's. Line printers provided 150 lines per minute (LPM) capability in 1953, and still only 600 LPM through the early 1960's. Disk drives, which permit individual records to be accessed from a data file, had been invented in the 1950's, but were small, slow, and expensive. Costing about $50,000, an early IBM 355 disk had a storage capacity of 12 Mbytes and took an average of 600 milliseconds (ms) to access an individual record. In 1960, the IBM 1405 disk was introduced for use with the popular IBM 1401 computer. The 1405 disk had a maximum unit capacity of 10 Mbytes, average access time of 625 ms, and cost "only" about $10,000 per unit.

Clearly the technical limitations of the early computing hardware constrained IS development and led to suboptimal approaches to automating functions within an organization. Significant increases in computing technology were needed if the sophistication of IS was to progress beyond rudimentary capabilities offered by the early EDP systems.

Management Information Systems

The impetus for the evolution to a more sophisticated type of IS was provided not so much by pure technological developments, as by a business decision. The key decision was embodied in a confidential, but landmark report issued by an IBM internal task force in 1961 (Galler, 1983). IBM assembled the Systems Programming Research and Development (SPREAD) task group in late 1961 to consider the feasibility and desirability of a family of computer systems to merge IBM's separate lines of scientific and commercial computers, and at the same time have both upward and downward compatibility. Within two months, this task force had recommended that IBM produce the IBM System/360 family of computers. IBM corporate management's decision to wholeheartedly embrace the SPREAD task group's recommendations and commit itself to a bold strategic development program was one of the most important decisions in the history of the computing industry (Galler, 1983).

Prior to 1962, IBM had produced seven families of computer systems - the 1400, 1620, 7030, 7040, 7070, 7080, and 7090 groups. These computers were not compatible with one another, and both IBM and external EDP users perceived a number of problems (Evans, 1986). With so many types of computer architectures, IBM was spending most of its development resources propagating the wide variety of processor units. Fewer resources were available to devote to the design of peripherals or development of better software products. The different architectures of the seven families made it impossible for IBM to develop optimized peripheral equipment, to standardize circuits in the face of rapid innovation in transistor technology, or to provide appropriate programming, operating system, and general customer support to all seven families. Users were limited by separate scientific and business products. Scientific users increasingly needed the alphanumeric and peripheral powers of business computers, and business EDP system users needed the logical and computational powers of scientific machines. User migration from one family architecture to another was difficult or impossible. Main memories were too small, and a new character structure was needed to allow lower-case and other special symbols.

The SPREAD Report outlined a family of central processing units, each of which could handle both scientific and business EDP applications. The processors were to have facilities for communications-processing and event-driven applications as well as common data formats and instruction sets. Standardized integrated circuits were to be used as a unifying technology. Main memory size was to be increased between 100 and 1,000-fold. An 8-bit character called the "byte" was selected for the basic alphanumeric data structure. The new character representation scheme would allow 256 different characters instead of the previous limit of 64, thus preparing the planned product line for most application areas, and for most countries in the world with different alphabets and symbols. Common operating systems, programming language compilers, and utility programs were to be developed for the entire family. To address the data and EDP transaction "throughput" problems brought about by limited peripheral devices, a standard peripheral-device interface was to be developed. Thus peripheral devices such as disks could be designed for the entire family and thus enjoy greater volume potential. Finally, user programs could be developed independent of individual processors. Given sufficient main memory on a given machine, and availability of peripheral devices required by a program, any processor in the family could run the program. The SPREAD Report identified definitive approaches to be undertaken to provide each of the proposed capabilities in order to substantially improve the capabilities of existing product lines. IBM's public commitment to the SPREAD strategic plan was embodied in the IBM System/360, announced on April 7, 1964. Six machines were initially released, the 360 Model

30, 40, 50, 60, 62, and 70, followed by the introduction of several additional systems over the next few years. The SPREAD plan eventually allowed IBM to commit substantial resources toward the development of more complete systems that focused on integrating peripherals, operating systems, communications, and new user applications. Both IBM and competing vendors that supplied System/360 compatible hardware and software were able to focus on developing capabilities and performance enhancements not previously possible in computer product-lines with differing architectures. This resulted in the development of powerful new peripherals, operating systems and supporting utilities, language compilers, terminals, high-volume application programs, and other complementary products and services whose development continues today. Almost single-handedly, the IBM System/360 provided the foundation for organizations to move beyond the capabilities offered by EDP systems, and begin moving toward the design of a true MIS. MIS have the capability to perform highly structured tasks and to provide limited managerial support for making semi-structured decisions. The major difference between MIS and EDP systems lies in the managerial orientation of its output and the degree of integration across an organization's functional areas (Ackoff, 1967). Whereas EDP system output is typically utilized by clerical personnel for operations management within a given functional area, MIS output is generally intended for middle managers and integrates data from multiple functional areas. The hardware advances in peripheral storage devices in the mid-1960's permitted the integration of production, accounting, marketing, finance, and personnel data beyond that typically found in EDP systems. As a result, both management control and operational control tasks could be supported with the MIS output. However, the early idea that MIS should be a single, completely integrated super-system was quickly shown to be an overly optimistic misconception (Deardon, 1972). Developers began to view MIS more appropriately as a loosely coupled collection of semi-independent, functional IS (Davis & Olson, 1985; Emery & Sprague, 1972). This view was based on emerging ideas of organizational processes which focused on the formal and informal structure of the organization, its standard operating procedures, and channels of communications between relatively independent, yet loosely integrated functional areas (Cyert & March, 1963).

Organizational requirements for increased amounts of managerial decision support served to direct MIS towards a more informational focus than the data orientation that was characteristic of EDP systems. MIS designers began to recognize that critical deficiencies under which many managers operated included not only a lack of relevant information, but also an overabundance of irrelevant information (Ackoff, 1967; Rappaport, 1968). As a result, management-by-exception principles were adopted in many MIS applications, with information filters and condensation techniques employed in order to help identify problematic conditions calling for decisions. This focus on the intelligence stage of decision-making distinguished MIS from EDP systems.

Typical MIS applications include departmental expense reports which contrast actual to planned expenditures, summaries comparing forecasted and actual sales revenues by product lines, manufacturing machine loading reports, periodic inventory investment summaries, and work-in-progress reports. These types of MIS reports illustrate significant decision support activities provided under the auspices of the intelligence stage in a manager's decision-making process. However, a budget variance report, for example, only identifies a situation requiring managerial attention. Frequently, support for the final three stages in the decision-making process (design, choice, and implementation) is not provided by the MIS. The standard operating procedures embodied in the MIS are not sufficiently developed to provide support in the design of alternative strategies to alleviate cost overruns, nor to assist in the evaluation and selection of an appropriate alternative, nor to help in the installation of appropriate corrective action. The incorporation of management science and statistical analysis techniques into MIS do tend to provide more direct support of the decision-making process. These analytical tools contribute valuable assistance in the choice stage. Optimal product mix from linear programming, manufacturing resource planning, assembly line balancing, and manufacturing process simulation illustrate modeling capabilities that can be embedded in MIS to identify preferred choices from among alternative actions.

Although early MIS components were the same as those found in traditional EDP systems, a greater emphasis in contemporary MIS is directed toward incorporating a DBMS. The use of a DBMS is necessary to support the complex data relationships required in the integrated context of MIS and to provide data management facilities for utilizing both centralized databases and data distributed throughout the organization. Besides providing data relatability, a DBMS is also useful in supporting special reporting requests and unique inquiry responses. MIS processing programs occasionally utilize statistical analysis techniques or management science models to summarize data or to create and assess alternative courses of action. In general, however, the MIS user can use these tools or models only in a predetermined way through the implementation of the organization's standard operating procedures in MIS programs. The models are typically not available for use in new situations or for analyses not previously anticipated.

Decision Support Systems (DSS)

Unlike the evolution from EDP systems to MIS which was spurred by technological developments, the evolution from MIS to DSS came about more as a result of change in philosophy of the role of IS in decision-making (Gorry & Scott Morton, 1971). In particular, the primary stimulus that led to the development of DSS as a distinct class of IS was the realization that a new conceptual framework was needed to more fully utilize the benefits associated with the previous partitioning of decision types, decision-making process stages, and managerial hierarchical levels. As a result, by 1978, IS developers had begun to personalize the design of DSS for individual organizations. This was achieved by incorporating a greater capacity to handle organizationally-specific, less well-structured decisions. These decision-making situations are, of course, more prevalent at the strategic planning and managerial control levels within a particular organization.

Although providing only a secondary impetus, technological progress in the 1970's was not inconsequential to the unfolding of DSS. The following developments all contributed to making interactive computer systems a reality: the design of time sharing systems, the increased reliability of computers, the introduction of user-oriented programming languages, the spread of video display devices, and the production of desk-top terminals. These technology-based advancements facilitated the maturation of DSS from a practical perspective (Keen & Scott Morton, 1978). More recently, technological developments including the microcomputer revolution, further developments in end-user computing tools, and continuing advances in telecommunications have greatly accelerated the development and use of DSS (Turban & Watkins, 1987).

In general, DSS provide an easy-to-use communication interface to promote direct managerial access to and interaction with both data and models through sophisticated data base and model base management systems (Bonczek, et al., 1981). Observed characteristics of a typical DSS include: (a) a definite emphasis on supporting the resolution of semi-structured or unstructured problems as frequently encountered by upper level managers; (b) the availability of combinations of models and analytical techniques with traditional data access and retrieval functions; (c) a focus on user-friendly features in an interactive mode; and (d) an orientation toward flexibility and adaptability to accommodate changes in environment and the decision-making approach of the individual manager (Sprague & Carlson, 1982). These characteristics are desirable for IS in all but the most stable and structured managerial settings.

A DSS can make significant contributions in assisting managers to carry out their responsibilities in all stages of the decision-making process. Typical types of assistance associated with the intelligence or problem recognition stage include: structuring, producing, and storing specially designed reports to monitor organizational performance; conducting descriptive statistical studies of data collected on operations using standard summary models; and signaling unique performance levels and then analyzing any results that warrant further investigation. DSS functions related to the second stage, the design of alternative courses of action, include activities such as: building mathematical models for any desired analysis, utilizing various forecasting techniques, and performing statistical analysis using standard or specially created models. Decision support activities related to the third stage, alternative evaluation and selection, involve assessment functions like performing simulations of alternative courses of actions, especially financial impact analysis; generating optimal alternatives through the use of standardized or customized operations research techniques; and conducting sensitivity analysis to check the resulting impact of changes in input variable values. The fourth stage, implementing a selected course of action, usually requires some human involvement in a typical DSS environment. Although an analysis of alternative implementation strategies can be supported by DSS-produced findings, until recently most of the decision activity supported by a typical DSS would not qualify for automatic control within the organization's control systems. However, recent developments in the field of AI and their application to the design and operation of expert systems may support automation of even the fourth stage (Holsapple & Whinston, 1987).

Expert Systems (ES)

For years organizations have desired intelligent IS components which would not simply respond to a manager's requests, but would aid the decision maker in structuring problems and understanding the environment in which decisions are made (King, 1973). The idea that computer programs could reason intelligently (McCarthy, 1960) and perform general problem solving to achieve some set of goals (Newell, Shaw & Simon, 1957; Newell & Simon, 1972) was pioneered in the late 1950's. Early ES were developed as research prototypes in the areas of medicine, chemistry, and speech recognition at major universities such as Carnegie-Mellon, Massachusetts Institute of Technology, and Stanford. However, it has only been within the last several years that ES techniques have progressed to the point where they are being applied more generally to real-world, business application problems.

Knowledge-based ES involve the application of AI techniques to solve difficult problems in narrowly defined problem areas. These systems are considered knowledge-based because they provide not only a data and model base as with DSS, but also a knowledge base that defines a variety of processes, strategies, and structures used to coordinate the decision-making process (Nilsson, 1980). ES use the knowledge base developed for a particular problem domain in combination with inference mechanisms to resolve problems in a manner that exhibits judgment comparable to human experts. The underlying premise is that a computer system that embodies knowledge can achieve high levels of performance in fields where expert decision makers often require extensive experience and/or years of special training and education.

Without a knowledge base, an IS is not able to synthesize the relevant facts and apply the appropriate rules or decision heuristics that are needed to address relatively unstructured problems. In many such cases, the information-processing requirements exceed the information-processing mechanisms available to the decision maker. The result is information overload (Schneider, 1987). The increasing necessity for managers to deal with problems of information overload has prompted interest in the development of knowledge-based ES. Only through intelligent filtering can an IS reduce the volume of detail transactions and data to be considered by decision makers (Cyert & March, 1963).

The basic characteristics of a knowledge-based ES include: (a) the effective performance of critical activities that are usually assigned to highly experienced and acknowledged experts; (b) a focus on narrowly defined decision-making strategies in contrast to more general problem solving frameworks; (c) the utilization of knowledge to resolve dilemmas or develop solutions through inference processes with conclusions being explicitly justified by expert systems; (d) the solution of problems associated with the following general activities - interpretation, prediction, diagnosis, debugging, design, planning, monitoring, repair, instruction, and control (Hayes-Roth, et al., 1983). The emphasis on knowledge processing, in contrast to data or information processing, is the major difference between ES and other types of IS.

To date, the major application of ES in organizations has been in the area of stand-alone operational control (Hayes-Roth, et al., 1983; I/S Analyzer, 1988; O'Leary & Turban, 1987). ES applications include software for computer system configuration, DNA molecule analysis, oil rig drilling, telephone cable maintenance, diesel-electric locomotive repair, and factory shop-floor scheduling (Harmon & King, 1985; Hayes-Roth, et al., 1983). Only recently have ES begun to focus on more common business applications (Holsapple & Whinston, 1987), such as decisions associated with management planning and control (Mockler, 1989) and strategic corporate planning (Mockler, 1989a, 1989b).

Strategic planning activities undertaken by more senior level executives are often complex, non-routine, and rather "messy" (Ackoff, 1986; Young, 1987). There is considerable need for complex inferential reasoning skills to extract and distill relevant information about noncontrollable factors that impact organizational performance and to specify criteria for key dimensions of success that are related to future endeavors (Mockler & Dologite, 1988). Strategic planning involves both quantitative and qualitative analysis of economic and financial data, market trends, and production capabilities. It is highly integrative in nature and requires judgments based on data and analyses drawn from all functional areas of the organization. Even under the best of circumstances, ES will probably never actually do corporate planning. However, ES hold immense promise for supporting strategic planning, particularly in the intelligence and design phases of the decision-making process.

An ES to support the intelligence phase of strategic planning might monitor an economic and demographic database concerning customers, competitors, financial markets, and government regulations to identify trends or shifts that might be exploited in a strategic manner. From another perspective, an ES could perform statistical quality control analysis to pinpoint not only operational control problems in a manufacturing process, but to serve as a catalyst for planning strategic changes in manufacturing facilities that might improve product quality. The design phase of strategic planning could be supported by an ES that would perform a contingency analysis to evaluate the competitive status of an organization. It might then explore alternative strategies to capitalize on strengths or eliminate weaknesses by determining which strategies are potentially the most fruitful. A number of generic corporate planning strategies have been developed for potential application in typical industry situations (Hamermesh, 1986; Mockler & Dologite, 1988; Mockler, 1989; Porter, 1980, 1985). For example, a firm seeking to gain a competitive advantage might adopt one of three potential strategies: (a) become the low cost producer of a good, (b) differentiate its product or service, or (c) focus on a particular market niche (Porter, 1980, 1985). An ES for strategic planning in this situation might weigh various factors such as competitors' size and number, the threat of new entrants, likelihood of substitute products and services, and the bargaining power of buyers and suppliers in order to evaluate previously formulated generic courses of competitive action that should be considered for adoption. Such an ES is unlikely to devise new courses of action itself, at least with the current level of ES technology, but could be of great assistance to a senior planner in thinking about alternative approaches to a strategic scenario.

Virtually any ES developed to support strategic planning will require access to data, information, and knowledge that is scattered throughout the organization. Unfortunately, many AI researchers did not foresee the need for such extensive integration of ES technology with traditional organizational databases. Thus, while ES and traditional EDP and MIS systems are being integrated in some cases, developing the required custom-made linkages can be the most challenging aspect of developing an ES (I/S Analyzer, 1988). This dilemma must be overcome if ES are to contribute extensively to improving organizational effectiveness.

The Evolution into an Integrated Information System

EDP systems seek to provide an accurate description of performance by summarizing large amounts of data. Their primary focus on data processing at the operational level for highly structured tasks limits their managerial support capabilities to the operational level. MIS evolved out of a need to increase the relevance of EDP system output for managerial personnel. The MIS focus on information processing at the tactical/managerial level of the organization increases their relevance to managerial support activities. However the orientation of MIS to highly structured and relatively routine problems limits their usefulness in the management decision-making process.

A modern integrated IS seeks to provide greater assistance for the management decision-making process than is possible with EDP and MIS processing. It incorporates computer-based support for solving managerial problems that have less structure, exist frequently at higher levels within an organization, and require more comprehensive and flexible tools for analysis. DSS utilize a variety of data and model base software to help managers analyze problems in an interactive operating environment that is user-friendly. The provision of such computer-based analysis tools directly to the managerial user allows the active utilization of a DSS in a wide variety of decision-making processes. This symbiosis of manager and system permits the DSS information output to be applied more successfully to the kinds of complex, semi-structured problems most often encountered by middle to senior level managers. ES represent an attempt to incorporate pre-established, application specific modeling structures into a comprehensive framework for analyzing difficult problem situations and synthesizing results into an effective decision-making or supporting process. In an integrative IS, embedded ES components apply model base software tools to the database to provide problem solving support for situations which are very complex, unique, and relatively unstructured to all but the most experienced manager. ES offers management the promise of knowledge processing support for semi-structured and unstructured problems at all levels of the organization.

From a managerial perspective, each of the four computerized systems has both benefits and limitations. EDP systems provide excellent data summarization capabilities, but the output often has limited applicability for decision-making. MIS provide relevant decision-making information and increase the perspective of managers through an integration of data from the organization's primary functional areas. However, MIS often lack the flexibility to support higher level, unstructured problems constantly encountered by mid-to-upper level managers. DSS provide the capability to handle decision-making situations that require additional flexibility by permitting user-directed access to a wide variety of data base and modeling tools. ES seek to synthesize the relevant components of knowledge to analyze problems and perform tasks typically delegated to experienced and competent human decision makers.

Implications for Organizational Change and Growth

User and IS managers alike need to promote the design of an integrated IS incorporating EDP, MIS, DSS, and ES components in order to provide comprehensive organizational control and managerial decision-making support. Of course, the development of a totally integrated, completely comprehensive IS to perform all managerial functions represents an ideal or unattainable goal. Still, it is a goal which should serve as a focus for the IS development effort. Managers should also recognize that inherent in this development process is the redefinition of decision-making responsibilities and management functions. The willingness of each manager to adapt to the changing environment is as important to the successful implementation of an integrated IS as is the level of technological sophistication which is attained (Storey, 1987). The IS evolution has been facilitated not only by hardware and software

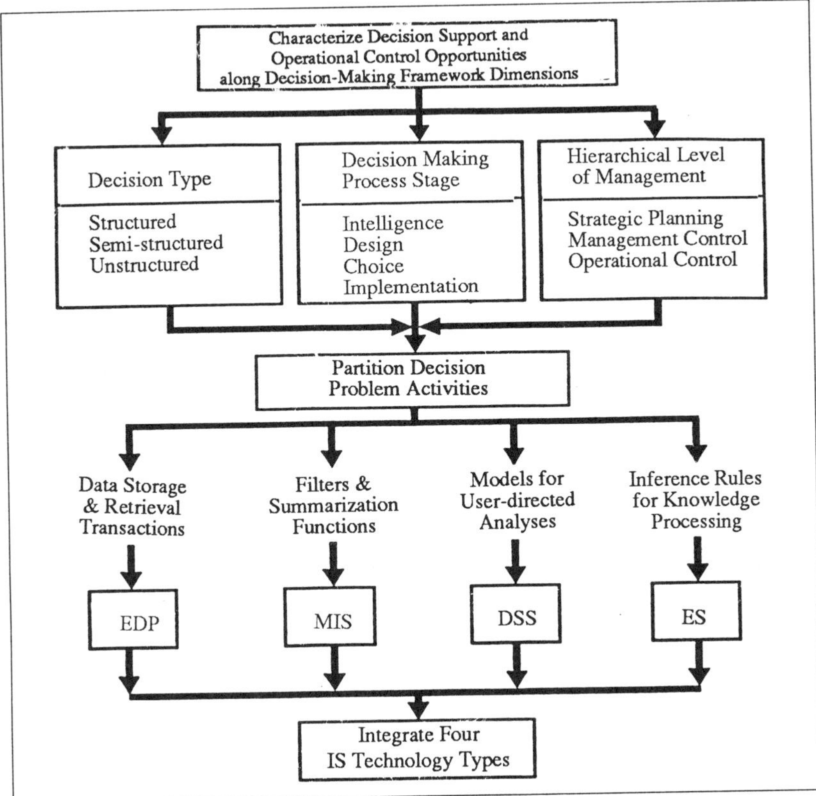

Figure 1. A Framework Toward Totally Integrated Information Systems

technological developments, but also by an expanded conceptualization of the managerial support function. The continuing evolution of the organization's IS reflects an enlargement of its domain of influence and scope of responsibility. These changes have occurred naturally as society has moved into a post-industrial era that is increasingly dominated by the service economy with its special emphasis on the processing and management of information.

A design framework to facilitate the development of an integrated IS is illustrated in Figure 1. User managers should first attempt to identify decision support and organizational control opportunities. Each manager's strategic planning process should include such an intelligence activity. Once appropriate opportunities have been identified, the user and IS managers together must determine the characteristics and attributes of each decision which is a candidate for IS implementation. IS design requires that each decision problem be partitioned into data storage and retrieval activities, filters and summarization functions for routine reporting and information processing, models for user-directed data retrieval and analysis, and inference mechanisms for knowledge processing. Partitioning of the problem serves as the blueprint for selection of the appropriate EDP, MIS, DSS, or ES information system technology. IS analysts, in consultation with the managerial users, can then design the appropriate IS components to support the necessary transaction processing, routine decision-making, ad hoc decision support capabilities, and knowledge processing requirements.

Actual development of the IS component systems should be undertaken in an evolutionary manner. This requires that prototype IS be implemented in which the appropriate EDP, MIS, DSS, and ES technology is applied to the decision problem subsets. The prototype systems should integrate not only the different organizational functions involved in the target decision applications, but also integrate the four IS technology types. As experience is gained through the use of the prototype systems in their particular decision contexts, more sophisticated prototypes can be developed. Thus the IS can evolve to more fully provide organizational control and more ably support the managerial decision-making process.

IS of the future will require continual integration and component replacement as rudimentary EDP, MIS, DSS, and ES subsystems are enhanced, and especially as new ES components are developed to perform additional control and decision-making tasks. Meanwhile EDP, MIS, and DSS components of the IS will continue to address those managerial problems that cannot be easily subsumed by ES components. Public and private sector organizations must actively exploit opportunities to improve their IS capabilities, particularly by extending DSS functions and integrating ES components. Managers must capitalize on every occasion to increase the comprehensiveness of the IS to improve control and decision-making at all organizational levels. The development of increased IS control over managerial activities will allow managers to spend their limited time and energy on the more challenging issues facing their organizations.

References

Ackoff, R. L. (1967). Management Misinformation Systems. *Management Science, 14*(4), B147-B156.

Ackoff, R. L. (1986). *Management in Small Doses.* New York: Wiley Interscience.

Anthony, R. N. (1965). *Planning and Control Systems: A Framework for Analysis.* Boston: Harvard Graduate School of Business Administration, Studies in Management Control.

Aron, J. D. (1969). Information Systems in Perspective. *Computing Surveys, 1*(4), 213-236.

Barros, O. (1981). Management Informations Systems Structure, Types and Integration. *Information Systems, 6*(4), 243-254.

Blumenthal, S. C. (1969). *MIS: A Framework for Planning and Development.* Englewood Cliffs, NJ: Prentice-Hall.

Bonczek, R. H. Holsapple, C. W., & Whinston, A. B. (1981). *Foundations of Decision Support Systems.* New York: Academic Press.

Brancheau, J. C., & Wetherbe, J. C. (1987). Key Issues in Information Systems Management. *MIS Quarterly, 11*(1), 23-45.

Bruns, W. J. Jr., & McFarlan, F. W. (1987). Information Technology Puts Power in Control Systems. *Harvard Business Review, 65*(5), 89-94.

Bullers, W. I. Jr., & Whinston, A. B. (1983). An Analysis of Information Processing Capabilities for Control and Decision Support. *Proceedings of the Decision Sciences Institute,* 188-191.

Cash, J. I. Jr., & McLeod, P. L. (1985). Managing the Introduction of Information Systems Technology in Strategically Dependent Companies. *Journal of Management Information Systems, 1*(4), 5-23.

Cash, J. I. Jr. McFarlan, F. W., & McKenney, J. L. (1988). *Corporate Information Systems Management: The Issues Facing Senior Executives* (2nd Ed.). Homewood, IL: Dow Jones-Irwin.

Churchman, C. W. (1968). *The Systems Approach.* New York:

Dell. Cyert, R. M., & March, J. G. (1963). *A Behavioral Theory of the Firm.* Englewood Cliffs, NJ: Prentice-Hall.

Davis, G. B., & Olson, M. H. (1985). *Management Information Systems: Conceptual Foundations, Structure, and Development* (2nd Ed.). New York: McGraw-Hill.

Deardon, J. MIS Is a Mirage. (1972). *Harvard Business Review, 50*(1), 90-99.

Department of Defense (1960). *COBOL, Initial Specifications for a Common Business Oriented Language.* Washington, DC: U.S. Government Printing Office #1960 0-552133.

Drucker, P. F. (1968). *The Age of Discontinuity.* New York: Harper & Row.

Drucker, P. F. (1985). *Innovation and Entrepreneurship: Practice and Principles.* New York: Harper & Row.

Emery J. C., & Sprague, C. R. (1972). MIS: Mirage or Misconception?. *Harvard Business Review, 50*(3), 22-23.

Evans, B. O. (1986). System/360: A Retrospective View. *Annals of the History of Computing, 8*(2), 155-179.

Forrester, J.W. (1961). *Industrial Dynamics.* Cambridge, MA: MIT Press.

Galler, B. A. (Ed.). (1983). Special Feature on the SPREAD Report: The Origin of the IBM System/360 Project. *Annals of the History of Computing, 5*(1).

Gassman, H. P. (1985). The Changing Nature of Transborder Data Flows. in Transborder Data Flows: *Proceedings of an OECD Conference*, Amsterdam: North-Holland, 7-14.

Gerrity, T. P., Jr. (1971). The Design of Man-Machine Decision Systems: An Application to Portfolio Management. *Sloan Management Review, 12*(2), 59-75.

Gibson, C. F., & Nolan, R. L. (1974). Managing the Four Stages of EDP Growth. *Harvard Business Review, 52*(1), 76-88.

Gorry, G. A., & Scott Morton, M. S. (1971). A Framework for Management Information Systems. *Sloan Management Review, 13*(1), 55-70.

Gross, A. C. (1988). The Information Vending Machine. *Business Horizons, 31*(1), 24-33.

Hamermesh, R. G. (1986). *Making Strategy Work.* New York: John Wiley & Sons.

Harmon, P., & King, D. (1985). *Expert Systems: Artificial Intelligence in Business.* New York: John Wiley, & Sons.

Hayes-Roth, F. Waterman, D. A., & Lenat, D. B. (1983). *Building Expert Systems.* Reading, MA: Addison-Wesley.

Hillier, F. S. Lieberman, G. J. (1986). *Introduction to Operations Research* (4th Ed.). Oakland, CA: Holden-Day, Inc.

Holsapple, C. W., & Whinston, A. B. (1987). *Business Expert Systems.* Homewood, IL: Irwin.

Hurd, C. C. (Ed.). (1983). Special Issue on the IBM 701. *Annals of the History of Computing, 5*(2).

Hurd, C. C. (Ed.). (1986). Special Issue on the IBM 650. *Annals of the History of Computing, 8*(1).

I/S Analyzer, (1988). *Trends in Artificial Intelligence. 26*(2), 3-16.

Ives, B., & Learmonth, G. P. (1984). The Information System as a Competitive Weapon. *Communications of the ACM, 27*(12), 1193-1201.

Keen, P. G. W., & Scott Morton, M. S. (1978). *Decision Support Systems: An Organizational Perspective.* Reading, MA: Addison-Wesley.

King, W. R. (1973). The Intelligent MIS - a management helper. *Business Horizons, 16*(5), 5-12.

Kroenke, D. (1981). *Business Computer Systems: An Introduction.* Santa Cruz, CA: Mitchell Pub. Co., 25-33.

Larson, R. E. McEntire, P. L., & O'Reilly, J. G. (1984). The Theory of Real-Time Control. In C. R. Vick (Ed.), *Handbook of Software Engineering*. New York: Van Nostrand Reinhold.

Law, A. M., & Kelton, W. D. (1982). *Simulation Modeling and Analysis*. New York: McGraw-Hill.

Leavitt, H. J., & Whisler, T. L. (1958). Management in the 1980's. *Harvard Business Review, 36*(6), 41-48.

McCarthy, J. (1960). Programs with Common Sense. *Proceedings of the Teddington Conference on the Mechanism of Thought Processes*, 1. London: Her Majesty's Stationery Office, 77-84, (reprinted in Minsky, M. (Ed.). Semantic Information Processing. Cambridge, MA: MIT Press).

McDonough, A. M. (1963). *Information Economics and Management Systems*. New York: McGraw-Hill.

McFarland, F. W. McKenney, J. L., & Pyburn, P. (1983). Information Archipelago: Charting the Course. *Harvard Business Review, 61*(1).

McFarlan, F. W. (1984). Information Technology Changes the Way You Compete. *Harvard Business Review, 62*(3), 98-103.

Meltzer, M. F. (1981). *Information: The Ultimate Management Resource*. New York: AMACOM.

Mockler, R. J., & Dologite, D. G. (1988). Developing Knowledge-based Systems for Strategic Corporate Planning. *Long Range Planning, 21*(1), 97-102.

Mockler, R. J. (1989). *Knowledge-Based Systems for Management Decisions*. Englewood Cliffs, NJ: Prentice-Hall.

Mockler, R. J. (1989). *Knowledge-Based Systems for Strategic Planning*. Englewood Cliffs, NJ: Prentice-Hall.

Naisbitt, J. (1983). Megatrends. New York: Warner Books.

Newell, A. Shaw, J. C., & Simon, H. A. (1957). Empirical Explorations of the Logic Theory Machine: A Case Study in Heuristics. *Proceedings of the 1957 Western Joint Computer Conference*, 218-230.

Newell, A., & Simon, H. A. (1972). *Human Problem Solving*. Englewood Cliffs, NJ: Prentice-Hall.

Nilsson, N. J. (1980). *Principles of Artificial Intelligence*. Palo Alto, CA: Tioga Pub. Co.

Nolan, R. L. (1979). Managing the Crises in Data Processing. *Harvard Business Review, 57*(2), 115-126.

O'Leary, D. E., & Turban, E. (1987). The Organizational Impact of Expert Systems. *Human Systems Management, 7*(1), 11-19.

Parsons, G. L. (1983). Information Technology: A New Competitive Weapon. *Sloan Management Review, 25*(1), 3-13.

Peters, T. (1988). Facing Up to the Need for a Management Revolution. *California Management Review, 30*(2), 7-38.

Phister, M. Jr. (1979). *Data Processing Technology and Economics* (2nd Ed.). Bedford, MA: Digital Press.

Porter, M. E. (1980). *Competitive Strategy: Techniques for Analyzing Industries and Competitors*. New York: The Free Press.

Porter, M. E. (1985). *Competitive Advantage: Creating and Sustaining Superior Performance*. New York: The Free Press.

Rappaport, A. (1968). Management Misinformation Systems - Another Perspective. *Management Science, 15*(4), B133-B136.

Rosen, S. (1969). Electronic Computers: A Historical Survey. *Computing Surveys, 1*(1), 7-36.

Sammet, J. E., & Garfunkel, J. (1985). Summary of Changes in COBOL, 1960- 1985. *Annals of the History of Computing, 7*(4), 342-347.

Sauvant, K. P. (1986). *International Transactions in Services: The Politics of Transborder Data Flows*. Boulder, CO: Westview Press.

Schneider, S. C. (1987). Information Overload: Causes and Consequences. *Human Systems Management, 7*(2), 143-153.

Shneiderman, B. (1985). The Relationship Between COBOL and Computer Science. *Annals of the History of Computing, 7*(4), 348-352.

Shannon, C. E., & Weaver, W. (1949). *A Mathematical Theory of Communication*. Urbana, IL: Univ. of Illinois Press.

Simon, H. A. (1965). *The New Science of Management Decision*. New York: Harper & Brothers.

Simon, H. A. (1981). *The Sciences of the Artificial* (2nd Ed.). Cambridge, MA: The MIT Press.

Sprague, R. H. Jr., & Carlson, E. (1982). *Building Effective Decision Support Systems*. Englewood Cliffs, NJ: Prentice-Hall.

Storey, J. (1987). The Management of New Office Technology: Choice, Control and Social Structure in the Insurance Industry. *Journal of Management Studies, 24*(1), 43-62.

Toffler, A. (1980). *The Third Wave*. New York: William Morrow and Co.

Turban, E., & Watkins, P. R. (1987). The Impacts of Emerging Management Support Systems. *Human Systems Management, 7*(1), 7-10.

Wiener, N. (1948). *Cybernetics or Control and Communication in the Animal and the Machine*. Cambridge, MA: MIT Press.

Young, L. F. (1987). A Systems Architecture for Supporting Senior Managers' Messy Tasks. *Information & Management, 13*, 85-94.

Yourdon, E. (1986). *Nations at Risk: The Impact of the Computer Revolution*. New York: Yourdon Press.

Zachman, J. A. (1987). A Framework for Information Systems Architecture. *IBM Systems Journal, 26*(3), 276-292.

Chapter 2

Evaluating the Impacts of Information Technology in Organizational Systems

Edward J. Szewczak
Canisius College

The importance of evaluating IT impacts on modern organizations and society cannot be overstated. Huff & Munro (1985, pp. 327-328) have identified IT as "the broad range of technologies involved in information processing and handling, such as computer hardware, software, telecommunications and office information, and includes such 'technologies' as new systems development methodologies." As so defined, IT serves as the technological basis for the design and implementation of computer-based information systems whose purpose is the production of useful information in support of management functions. Two broad classes of such systems are management information systems (MIS) and decision support systems (DSS). IT impacts are usually exhibited in organizations and society through the development and use of MIS and DSS.

A MIS is an integrated, user-machine system for providing information to support operations, management, analysis and decision-making functions in an organization. A DSS is a kind of MIS which is used in planning, analyzing alternatives, and trial and error search for solutions, and is usually operated through terminal-based interactive dialogs with users (Davis & Olson, 1985).

A MIS/DSS is said to be "effective" if it produces the desired result for which it was developed. Being effective implies doing the "right" thing. This differs from MIS/DSS "efficiency" which refers to the use of the system

Support for this research was provided by a Wehle School of Business Faculty Summer Research Grant.

resources to achieve the desired results. Being efficient implies the system is operating the "right" way (Davis & Olson, 1985).

MIS/DSS evaluation has generally been considered an integral part of MIS/DSS development (Davis & Olson, 1985; Keen, 1981). The large investment in hardware, software and personnel which organizations must make to develop and maintain various MIS/DSS, as well as in the organizational mechanisms required to manage them, dictates the need to assess whether the MIS/DSS are actually doing the jobs for which they were designed. Not surprisingly, this need is particularly felt during economic hard times, when cost-cutting and belt-tightening compel organizations to closely evaluate the worth of projects requiring large capital outlays (King, 1975).

Evaluating MIS/DSS is a fundamental issue in the management of IS. Dickson & Wetherbe (1985, p. 164) identify six reasons for evaluating MIS/DSS:

> (1) Senior executives want to evaluate MIS/DSS because organizations devote a large amount of resources to this activity.
> (2) In many organizations there is a vague feeling that the benefits received from MIS/DSS are not worth the level of expenditure for the systems.
> (3) Usually the MIS/DSS function has not been subjected to the same standards that have been applied to other organizational units, and it is time to attempt a formal measurement of the MIS/DSS function.
> (4) Senior management may feel that the organization is at risk should the computing function experience failure such as computer crime or misuse of computer resources.
> (5) Management needs a basis for comparing the performance of MIS/DSS over time in order to perform corrective action to improve the systems.
> (6) It is very useful to accurately assess the current state of the MIS/DSS before embarking on new systems planning activity.

Though the importance of evaluating MIS/DSS effectiveness is relatively easy to identify, the task of MIS/DSS evaluation is very complex due to a number of factors related to the scope of the MIS/DSS concept itself. Since an organization's environment usually includes external as well as internal components, the outputs of MIS/DSS may have various impacts which directly affect not only the organization itself but also society at large. Further, since systems are often designed to be interrelated, it is not always clear how to delineate which system (or systems) is being evaluated. The current impetus to network various systems (such as office automation systems, factory automation systems, accounting information systems) compounds the evaluation problem greatly.

In addition, various general measures of MIS/DSS effectiveness have been identified, including decision quality, user satisfaction, changes in user behavior, and system utilization. Assuming a clear identification of a system for evaluation purposes, it is not always clear which evaluation measures should be applied to which system to arrive at an adequate determination of its effectiveness impacts. For example, a determination of decision quality may be appropriate for a marketing DSS, but be inappropriate for an office automation system designed for word processing.

This paper reviews various MIS/DSS evaluation frameworks and approaches which have been proposed in the literature. Based on this survey of research, critical issues relevant to MIS/DSS evaluation are identified and related in a comprehensive research framework which attempts to integrate the various aspects of IT impacts. Future directions for research using this comprehensive framework are suggested.

MIS/DSS Evaluation Frameworks

Various MIS/DSS evaluation frameworks have been proposed. Each framework tends to reflect one or more basic viewpoints on what aspects of MIS/DSS evaluation are most important. Three basic viewpoints have been identified (Dickson & Wetherbe, 1985): economic (the net contribution of MIS/DSS to the organization), managerial (the extent to which "good" practice is being followed), and accounting (the minimization of risk or exposure of the organization to inadequacies in MIS/DSS performance).

The five evaluation frameworks which have had the most influence among MIS/DSS researchers are discussed below.

Marschak (1968) framework. The Marschak framework addresses the economics of inquiring, communicating and deciding (Figure 1). According to this framework, data about events are gathered and communicated to a decision maker in the form of a message, which he/she uses to decide which action is to be taken. The action taken produces a result which has some value.

Prior to the decision being made, some decision must have been made previously about the form of the three interrelated services — inquiring, communicating, deciding — which provide the basis of the MIS. The person making this decision — the "leader" — may or may not be the same person making the result-relevant decision. The leader must consider the cost of the combined services (the MIS) in relation to the expected value of the result. Clearly, the cost of the MIS cannot exceed the expected value of the result if the decision is to be economically optimal. On the other hand, the MIS itself may contribute to the realization of a result with greater value than would have

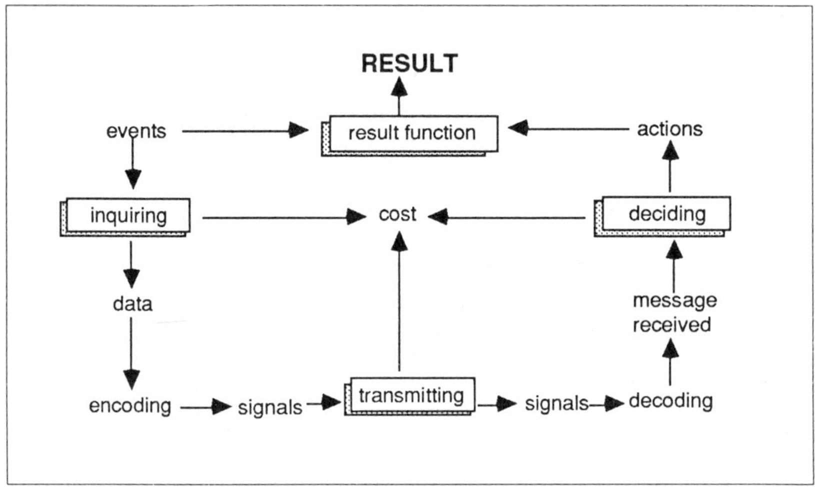

Figure 1. Marschak (1968) Framework

been achieved with another MIS or no MIS at all.

The Marschak framework graphically depicts the parameters of the evaluation issue. In effect, the next four frameworks expand on this (essentially economic) framework to emphasize different aspects of MIS/DSS development and use in organizations.

King & Rodriguez (1978) framework. The King & Rodriguez framework addresses the integral role of MIS evaluation throughout the systems development life cycle. According to this framework, MIS evaluation should be conducted at various states of the systems development to detect systems problems and solve them constructively before continuing further, i.e., not merely in the initial need assessment phase and/or when the system has been completed. This normative recommendation differs from current evaluation practice. It has been found that the reason for post-implementation evaluation of systems is project closure and not project improvement (Kumar, 1990).

The King & Rodriguez framework recognizes that decision performance (ala the Marschak framework) is only one measure of MIS effectiveness. Also relevant to MIS evaluation is the assessment of user attitudes toward the system, user value perceptions (beliefs) about the system, and actual information usage. The framework explicitly encourages comparisons on the basis of the evaluation measures. For example, does use of the system result in a more systematic assessment of choice situations, in the greater use of more relevant information, or in the use of previously unused decision models?

Hamilton & Chervany (1981a) framework. The Hamilton & Chervany framework addresses both effectiveness-oriented and efficiency-oriented perspectives of MIS evaluation. MIS effectiveness may be assessed at three levels of objectives: the information provided (Level 1), the use of information and its effect on user processes and performance (Level 2), and the effect of the MIS on organizational performance within its environment (Level 3). Associated with each objectives level are unique objectives and performance measures. For example, the Level 1 objective of improving the quality of information content may be assessed by looking at performance measures appropriate to data evaluation, such as their accuracy and reliability. In contrast, the Level 2 objective of improving the decision making process may be assessed by considering such performance measures as the comprehensiveness of the analysis and the quantification of action consequences.

In the framework, the support provided by MIS personnel during MIS operations is explicitly seen as impacting Level 2 objectives, which in turn impact Level 3 objectives. Thus, the quality of the information provided and the support services provided by MIS staff must be assessed together to arrive at an assessment in terms of Levels 2 and 3 objectives. The linkages between the three levels have received some empirical support in the finding that there is a positive correlation between perceived organizational performance and the perceived importance of particular aspects of MIS, including the information attributes of conventional systems (Miller & Doyle, 1987).

MIS efficiency may also be assessed at four levels of objectives: the IS (Level 0), the resources needed to provide the IS (Level 1), the production capability of the resources (Level 2), and the level of investment in the resources (Level 3). Associated with each objective are unique objectives and performance measures. For example, the Level 0 objective of providing technical IS quality may be assessed by judging compliance to systems development standards for program design, database design, testing, etc. In contrast, the Level 1 objective of efficient resource consumption during development and operation of the IS may be assessed by considering budget variances.

Keen (1981) framework. The Keen framework addresses DSS evaluation, i.e., the evaluation of systems which are developed to support decision making in non-routine task situations. Such task situations involve frequent ad hoc analysis, fast access to data, generation of non-standard reports and the addressing of "what-if" questions. Rather than focus on traditional cost/benefit analysis, "value analysis" is conducted in two steps:

> (1) Establish value first using a prototype system, then test if the expected cost is acceptable.
> (2) For the full system, establish expected cost first, then test if the expected benefits are acceptable.

The Keen framework advocates a systematic methodology which

focuses on four basic issues.

(1) Value and cost. The framework emphasizes that value and cost be kept separate by focusing on the operational benefits of the DSS to the decision maker first, then keeping costs within acceptable levels to the decision maker. Once a prototype is built, the decision maker can experiment with it to learn more about the capabilities of the DSS. Determination of value and cost may change in light of experimentation.

(2) Simplicity and robustness. A determination is made about expanding the prototype into a full system by first considering cost, then judging what threshold of value must be obtained to justify the cost. Since the value of the DSS has been established during the prototype stage, the decision maker does not have to provide precise estimates of uncertain, qualitative future variables relating to system use. Rather, a judgment is made about how much of an investment should be made in the DSS project to achieve its perceived expected value.

(3) Reducing uncertainty and risk. The methodology reduces uncertainty and risk by reducing the initial investment in the DSS project, and delay between the approval of the project and the delivery of the prototype DSS.

(4) Innovation. The DSS development effort is viewed as an R&D project. As such, the prototype system is built at a cost below the capital investment level of a full system. If the prototype system is viewed as having little value, the project can be written off.

Ives & Learmonth (1984) framework. The Ives & Learmonth framework addresses the value of an IS as a competitive weapon. Building on the work of Porter (1980) and McFarlan & McKenney (1983), the Ives & Learmonth framework identifies and categorizes an application of IS technology in terms of the degree to which it contributes to a supplier's competitiveness in all phases of its customer relationships. Using the concept of the customer's resource life cycle (CRLC), the Ives & Learmonth framework can be used to evaluate whether an application of IT to any of the CRLC stages serves to build and maintain the loyalty of a supplier's customer base, thus enhancing the supplier's overall business strategy. For example, does the application serve to determine where customers will buy a resource and/or to ensure that the resource meets specifications?

The Ives & Learmonth framework involves certain key assumptions relating to the customer-supplier relationship. As Ives & Learmonth (1984, p. 1197) observe:

• Type of system (MIS, DSS)
• Scope of system/subsystems
• Systematic evaluation
• Levels of MIS/DSS effectiveness
• MIS/DSS and the organization
• MIS/DSS and the world: one-way versus two-way emphasis
• Goals of evaluation
•Various effectiveness measures

Table 1. MIS/DSS Evaluation Issues

If the supplier can assist the customer in managing this life cycle, the supplier may be able to differentiate itself from its competitors, usually on the basis of enhanced customer service or, in some cases, by introducing direct cost savings. In the process, the supplier will typically introduce switching costs (costs that customers must bear if they wish to switch to another supplier) for its customer base.

In fact, the supplier may choose to provide the customer with IT to develop and install support systems that the customer would not otherwise be willing or able to afford.

The five frameworks are noteworthy not only for their individual contributions but also for their differences. Table 1 presents a list of fundamental MIS/DSS evaluation issues which are identifiable from an examination of the five frameworks. In section 5 below, a comprehensive framework is presented which captures these fundamental MIS/DSS evaluation issues together with various evaluation approaches which have been discussed in the literature and relates them to the issue of IT impacts.

Type of system. Various types of MIS/DSS are addressed in the frameworks. MIS and DSS are primary foci, but it is clear that other system types may also be identified, such as telecommunications systems, office automation systems, and expert systems. One might argue that MIS and DSS are the two broadest categories of systems, and that other system types are simply subcategories of the two broadest categories. For example, Snitkin & King (1986) approached the evaluation of personal DSS, a subset of the DSS category. Expert systems may be evaluated as having certain characteristics which are similar to DSS as well as some which are dissimilar (Gashnig, Klahr, Pople, Shortliffe & Terry, 1983; Turban & Watkins, 1986; Liebowitz, 1986).

Scope of system/subsystems. The scope of the system evaluated may include the whole system or various subsystems of a larger system. For example, in the Marschak framework, three interrelated subsystems are identified (inquiring, transmitting, deciding), each of which may be evaluated by itself or in relation to other subsystems. The type of system and scope of system are closely related. For example, in the evaluation of MIS, the transmission component may be transparent to a manager whose primary concern is decision making, particularly when it is functioning well. However, local area networking and telecommunications systems may have negative as well as positive impacts on managerial work (Szewczak & Gardner, 1989), and these various impacts may influence a manager's perception of the MIS as a whole, or of a particular subsystem related to the transmitting subsystem.

Systematic evaluation. The systems development life cycle (SDLC) may be divided up in different but essentially equivalent ways. For example, Powers, Adams & Mills (1984, p. 50) identify five phases (and associated activities) of the SDLC: investigation phase (initial investigation, feasibility study), analysis and general design phase (existing system review, new system requirements, new system design, implementation and installation planning), detailed design and implementation phase (technical design, test specifications and planning, programming and testing, user training, system test), installation phase (file conversion, system installation), and review phase (development recap, post-implementation review). Evaluation takes place between various stages or steps of the development process, which may allow pre-post comparisons of key criteria. Systematic evaluation is closely linked with control of the development process insofar as the results of evaluation may indicate that remedial action must be taken to ensure successful development and perhaps even specific courses of action themselves.

Levels of MIS/DSS effectiveness and efficiency. MIS/DSS effectiveness and efficiency may be evaluated in terms of the information provided by the MIS/DSS (information product), the use of the information for organizational work, the level of service provided to information users by the MIS staff (information service), the organizational resources consumed to provide the information and service, and the impact of the MIS/DSS on the organization, especially on organizational performance. The levels of MIS/DSS effectiveness and efficiency are related in various ways and to different degrees.

MIS/DSS and the organization. The MIS/DSS may be evaluated in terms of how it impacts the internal activities of an organization, for example, enhancement of organizational efficiency, strengthening of managers' decision making power, or improvement of the quality of working life (Whistler, 1970; Attewell & Rule, 1984). The MIS/DSS may also be evaluated in terms of how it is impacted by the organization, for example, the effect of the organization's business strategy on the MIS/DSS (Davis, Dhar, King & Teng, 1984).

MIS/DSS and the world. The MIS/DSS may be evaluated in terms of (1) how it is affected by the world (which may be loosely defined as comprising the totality of the various environmental influences — both direct and indirect — on the organization), (2) how the MIS/DSS affects the world, or (3) how the MIS/DSS both affects and is affected by the world. (1) and (2) reflect a one-way emphasis, while (3) reflects a two-way emphasis, of the interaction of the MIS/DSS and the world.

Goals of evaluation. One basic reason for evaluating MIS/DSS is to arrive at an understanding of the degree to which the systems are achieving the objectives originally set down for them. Another reason for evaluation is to identify systems problems and provide creative solutions to them so as to improve systems performance. Forward looking managers will evaluate systems to determine whether additional investment in IT is called for to further improve existing systems or to develop new systems whose purpose will be to achieve corporate objectives.

Various effectiveness measures. Various effectiveness measures can be used to evaluate MIS/DSS effectiveness. These measures exhibit general characteristics which can be identified as unidimensional versus multidimensional, objective versus subjective, and quantitative versus qualitative. For example, a measure such as cash flow may be considered unidimensional (if it is considered the only measure of MIS/DSS effectiveness), objective (since it can be measured precisely and no rational person would disagree with the measurement), and quantitative (since it can be expressed as a number). It has been reported that, in actual practice, a set of effectiveness measures is often used to evaluate MIS/DSS (Singleton, McLean & Altman, 1988). The categories of effectiveness measures and methods for measuring them are discussed in the next section.

Measures of MIS/DSS Effectiveness

Measures of MIS/DSS effectiveness may be used to determine IT impacts. These measures fall into two broad categories: general measures and performance measures. Various methods of measurement can be used to assess these categories.

General measures. Though a number of MIS/DSS effectiveness measures have been identified in the literature under various names, five general measures seem to subsume all of them: decision effectiveness/quality, utilization, user satisfaction, changes in user behavior/attitudes, and organizational performance. These measures have been identified from a number of studies which review the MIS/DSS evaluation literature (Dickson, Senn & Chervany, 1977; King & Rodriguez, 1978; Zmud, 1979; Hamilton & Chervany, 1981a; Ives, Olson & Baroudi, 1983; Ives & Olson, 1984; Trice & Treacy, 1988). See Table 2.

Performance measures. Performance measures represent more specific operationalizations of the general effectiveness measures. Many performance measures have been used in the literature on MIS/DSS evaluation; no attempt will be made here to identify them all. However, as an example, decision effectiveness/quality has been measured in terms of decision cost, decision profit, decision time, and decision confidence (Dickson, Senn & Chervany, 1977).

Methods of measurement. Various methods of measurement have been used to evaluate MIS/DSS. These include dollar estimates, expert evaluation, historical records, n-item Likert-type scales, n-item semantic differential scales, n-item bipolar scales, interviews, and n-item difference scales, among others (Ives & Olson, 1984). There is a close relationship between the effectiveness measure(s) used and the method(s) of measurement. For example, if cash flow were the effectiveness measure used, then dollar estimates would be an appropriate method of measurement to adopt whereas semantic differential scales would not.

MIS/DSS Evaluation Approaches

The various effectiveness measures are used in conjunction with evaluation approaches. Table 3 lists the various approaches that have either

General Effectiveness Measures	Sources
Decision effectiveness/quality	Dickson, Senn & Chervany,1977; King & Rodriguez, 1978; Zmud, 1979; Ives & Olson, 1984
Utilization	Dickson, Senn & Chervany, 1977; King & Rodriguez, 1978; Zmud, 1979; Ives & Olson, 1984; Trice & Treacy, 1988
User satisfaction	King & Rodriguez, 1978; Zmud, 1979; Hamilton & Chervany, 1981a; Ives, Olson & Baroudi, 1983; Ives & Olson, 1984
Changes in use behavior/attitudes	King & Rodriguez, 1978; Ives & Olson, 1984
Organizational performance	Hamilton & Chervany, 1981a

Table 2. Sources of General Effectiveness Measures

been taken to, or advised for, the evaluation of MIS/DSS.

Cost/benefit analysis involves the identification of factors contributing to system costs and benefits, and measuring, calculating, and comparing cost/benefit values. Usually two or more system alternatives are analyzed. Various quantitative methods are employed in the calculation of cost/benefit values, such as payback, net present value and internal rate of return. However, measuring costs and benefits often presents different kinds of problems. Costs may be categorized and quantified without excessive trouble (King & Schrems, 1978). However, benefits are classified into three categories: tangible monetary benefits, tangible nonmonetary benefits, and intangible benefits (Ahituv & Neumann, 1982). Whereas costs can be compared directly to tangible monetary benefits, they can be compared only indirectly to tangible nonmonetary benefits (for example, improving the chances of saving a life). Intangible benefits defy quantification, generally speaking, and so comparison to costs becomes problematic. This problem directly impacts the evaluation of MIS/DSS effectiveness, since many of the benefits avowed to result from MIS/DSS are intangible, such as improved decision making (Chervany & Dickson, 1970) and improved organizational communication (Montgomery & Benbasat, 1983).

Multiattribute analysis evaluates MIS/DSS by identifying and combining various system attribute values into some overall effectiveness measure. Methods of combination vary and may include additive, multiplicative, or more complex forms of multiattribute value functions (Dyer & Sarin, 1979). Linear valuation models are employed where the weights represent the relative importance of the attribute values. The attribute values vary for different systems, and may be objective as in the case of utility functions (Ahituv, 1980) or subjective as in the case of user reported satisfaction (King & Epstein, 1983). (Strictly speaking, cost/benefit analysis is a kind of multiattribute analysis

Multiattribute analysis
Cost/benefit analysis
Value analysis
Simulation analysis
Goal programming analysis
Critical success factors analysis
Attitudinal response analysis
Utilization analysis
Informal analysis

Table 3. MIS/DSS Evaluation Approaches

since the utility of a system may be measured in terms of system costs and system benefits.) Multilinear valuation models may also be employed (Kleijnen, 1980). Where linear valuation models are not possible or impractical, lexicographical ordering, the efficient frontier, and statistical analysis may be used (Ahituv & Neumann, 1982).

The variety of methods of measurement used and the general lack of consistent results with respect to MIS/DSS evaluation has led to the development of a proposed standard instrument to measure "user information satisfaction" (Ives, Olson & Baroudi, 1983). Based on the work of Bailey & Pearson (1983), and reportedly developed according to accepted psychometric theory, the user information satisfaction instrument is intended to measure MIS effectiveness through the collection and analysis of system user subjective responses to questions using the semantic differential technique. Both a long-form measure and a short-form measure have been developed (Ives, Olson & Baroudi, 1983; Baroudi & Orlikowski, 1988). The user information satisfaction instrument has been used to evaluate large-scale transaction processing systems, but has not been tested in a DSS environment (Baroudi & Orlikowski, 1988). However, the reliability of the instrument has been challenged recently (Galletta & Lederer, 1989).

Value analysis is an alternative to traditional cost/benefit analysis which emphasizes the value of a DSS project rather than its cost, focuses on identifying and interpreting intangible benefits, and is usually applicable to a single investment proposal (Money, Tromp & Wegner, 1988). Experience with developing DSS suggests that there is no common cost/benefit recipe that can be applied successfully to all DSS, since the degree to which supported decisions are structured, the level of managerial activity, the level of uncertainty, and the source of information used are factors which have differing impacts during the phases of the decision making process (Pieptea & Anderson, 1987). It has been suggested that decision support and office automation systems be evaluated in the aggregate, rather than individually, in terms of overall productivity improvements attributable to a portfolio of systems (Gremillion & Pyburn, 1985).

Simulation is a method of representing the general characteristics of MIS/DSS within a dynamic model of a firm, and of evaluating MIS/DSS in terms of the economic benefits of improved total performance attributable to better information provided by MIS/DSS (Bonini, 1963; Boyd & Krasnow, 1963). The approach assumes that the intangible benefits of MIS are linked to the tangible economic benefits of performance, and that the provision of improved information will translate into improved performance (e.g. writing more accurate purchase orders based on current inventory figures). This approach has been criticized as being too general to be useful to MIS development specialists (Chervany & Dickson, 1970).

Goal programming, a mathematical programming algorithm from the operations research/management science literature, has been proposed as a technique for evaluating MIS by establishing a causal relationship between user multiple goal attainment and system activity, and iteratively examining different sets of targets in light of a set of constraints (Chandler, 1982). The approach assumes that all goals are quantifiable for purposes of comparison.

The *critical success factors* (CSF) method was proposed by Rockart (1979) as a way of assisting managers to define their significant information needs. It has been used as a methodology in MIS planning and control (Munro & Wheeler, 1980; Shank, Boynton & Zmud, 1985) and in determining the information needs of top MIS managers (Martin, 1983). The CSF method usually involves two basic steps: interview the manager, and review the results. Questionnaires may also assist the data collection effort. The CSF method has been used to perform the evaluation of the information function of MIS by comparing the information needs of a management team involved in planning and control to assess the usefulness and availability of critical information (Bergeron & Begin, 1989).

Various *attitudinal* approaches to MIS/DSS evaluation have been reviewed in Ives, Olson & Baroudi (1983) and Goodhue (1988). Attitudinal approaches generally stress the futility of trying to evaluate MIS/DSS using objective measures, and the need to examine the subjective attitudinal responses of users in assessing the value of their MIS/DSS. These responses may be influenced by the social group to which an individual belongs which may affect how the MIS/DSS is interpreted and used (Robertson, 1989). Data may be analyzed about user attitudes toward the effect of the system on a manager's job performance, on interpersonal relations and communication, on changes in organizational structures, etc. (King & Rodriguez, 1978).

Utilization was identified as a general effectiveness measure in Table 2. However, it has been observed that both actual usage measures and perceived effectiveness measures (user satisfaction) must be considered together when evaluating MIS/DSS effectiveness, since neither measure category alone is sufficient (Srinivasan, 1985). It has been argued that utilization should be viewed as a factor affecting system performance rather than as an indicator of performance itself (Trice & Treacy, 1988). Utilization analysis can indicate flaws in system design (such as requiring users to expend more time than necessary on a task, or having features which prompt users to utilize the system in nonproductive ways) as well as design successes (such as improving the decision making process by decreasing the length of time required to make decisions) (Hamilton & Chervany, 1981a; Trice & Treacy, 1988).

Informal approaches to MIS/DSS evaluation may also be used, such as talking to users during the normal work day (Halbrecht, 1977; Hamilton & Chervany, 1981a). Informal approaches may be used alone or in conjunction with the more formal approaches discussed above.

Comprehensive Framework and Future Directions for Research

The fundamental issues relevant to MIS/DSS evaluation addressed in the five frameworks and the various MIS/DSS evaluation approaches may be subsumed by a comprehensive framework for MIS/DSS evaluation (Figure 2). The various elements of the comprehensive framework have been related in a manner which is intuitively appealing but which is as yet empirically unverified.

The critical role of evaluator viewpoints in MIS/DSS evaluation has been identified in Hamilton & Chervany (1981b). Different functional groups may be involved in system implementation — users, MIS/DSS development personnel, internal audit personnel, and top management. These groups may disagree about assessments of MIS/DSS effectiveness since evaluations tend to be subjective, and influenced by perceptions of systems objectives and by experiences with system performance in accomplishing organizational objectives. Hamilton & Chervany (1981b) argue for the integration of multiple viewpoints into evaluation approaches (see Dickson & Wetherbe (1985, Chapter 7) for suggestions on conducting an overall assessment of the MIS function), and advocate participatory approaches to system evaluation to facilitate communication of evaluative information between the functional groups. The role of evaluator viewpoints cannot be underestimated in MIS/DSS evaluation. Though evaluator viewpoints are not explicitly depicted as an element in the comprehensive framework, they impact the various decisions relevant to each of the framework's main elements.

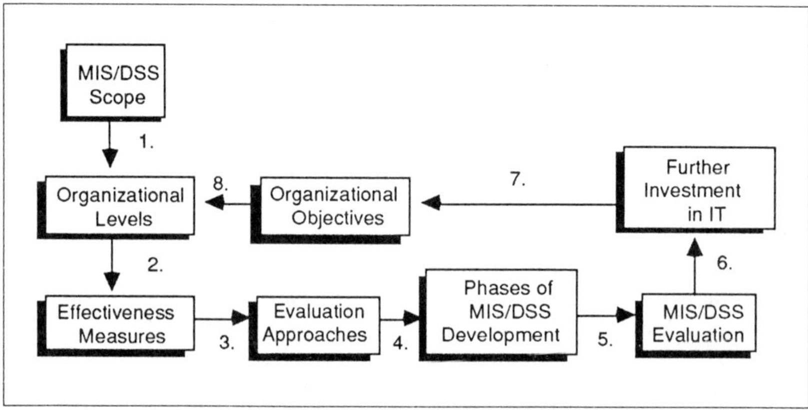

Figure 2. Comprehensive MIS/DSS Evaluation Framework

The scope of the MIS/DSS to be evaluated is directly related to the various organizational levels of MIS/DSS effectiveness (relation 1 in Figure 2). Information product, information use, information service, and organizational performance (measured as the accomplishment of organizational objectives) are all related to the scope of the system as well as the task or tasks for which the system is designed. If the system to be evaluated is part of a larger system (i.e., a subsystem), then some decision must be made as to the limits or boundaries of the subsystem. This decision may reflect an evaluator viewpoint, even in cases where system/subsystem scope is formally defined in system development documentation such as data flow diagrams, since the system boundaries may have been decided on by the system developer(s) and may not be perceived as such by members of the other three functional groups involved in system evaluation.

MIS/DSS scope is also related to the impact of the MIS/DSS on the organization and the world. Depending on how the limits or boundaries of the system are perceived by the evaluator, the system will be viewed as having an internal impact on (or as being impacted by) the organization and/or as having an external impact on (or as being impacted by) the world.

The various effectiveness measures are related to organizational levels (relation 2) and evaluator viewpoints insofar as these two elements impact the decision as to what measures are most appropriate for the evaluation. The choice of effectiveness measures is directly related to the selection of evaluation approaches (relation 3), since the type(s) of effectiveness measure (unidimensional versus multidimensional, objective versus subjective, quantitative versus qualitative) will suggest the type(s) of evaluation approaches. Evaluator viewpoints also play a role in the choice of evaluation approaches in the case where more than one approach appears appropriate and/or acceptable for arriving at an evaluation. For example, an approach such as goal programming may require mathematical sophistication beyond the abilities or training of the evaluator.

The chosen evaluation approaches may be applied at various phases of the MIS/DSS development process (relation 4). Which evaluation approaches are applied at which phases will be determined by the evaluator, who in turn may be constrained by the characteristics of the evaluation approaches themselves. For example, cost/benefit analysis may be appropriate during the investigation phase for determining system feasibility but inappropriate during the detailed design and implementation phase.

The results of applying the evaluation approaches are reviewed by the evaluator to arrive at the MIS/DSS evaluation (relation 5). The results of the evaluation may be used to arrive at a decision as to whether to invest further in IT (relation 6) to accomplish organizational objectives (relation 7). (This decision may be the output of a strategic planning system. The difficulties in assessing the accomplishment of organizational objectives are discussed in King (1983).) Organizational objectives help to shape evaluator viewpoints on what is significant in the areas of information product, information use, information service, and organizational performance (relation 8).

The comprehensive MIS/DSS evaluation framework has an intuitive basis in previous research but is in need of empirical validation. Although various research studies have reported empirical results concerning the use of evaluation approaches in field and laboratory study contexts, where usually the researcher determined the approaches to be studied, very little empirical evidence has been reported on actual organizational practice in the area of MIS/DSS evaluation. Observers of MIS/DSS evaluation practice often lament the apparent lack of organizational interest in the activity. Yet little evidence exists (with the exception of personal observation) to either support or refute this view (Hamilton & Chervany, 1981a,b).

Table 4 lists a number of research questions about the organizational practice of evaluating MIS/DSS which are in need of empirical investigation. Providing answers to these related questions appears to represent a fundamental empirical task for MIS/DSS researchers:

(1) Do organizations evaluate their MIS/DSS?

Some organizations may systematically evaluate their MIS/DSS as a matter of organizational policy. Other organizations may choose not to evaluate their MIS/DSS, while still other organizations may evaluate some

Do organizations evaluate their MIS/DSS?
Which functional groups are involved in evaluating MIS/DSS effectiveness in organizations?
What approaches do organizations use to evaluate the effectiveness of their MIS/DSS?
At what phases of the system development process do organizations apply evaluation approaches to assess MIS/DSS effectiveness?
Are organizations satisfied with the results of using the approaches for evaluating MIS/DSS effectiveness?
Do organizations use the results of evaluating the effectiveness of their MIS/DSS to justify further investment in IT?

Table 4. MIS/DSS Evaluation Research Questions

MIS/DSS but not other MIS/DSS.

The absence of an organizational policy may explain why some organizations do not evaluate their MIS/DSS, but other factors may also be at work. There may be an organizational policy which provides guidelines for MIS/DSS evaluation but which is not adhered to for various reasons including lack of time, lack of trained personnel for conducting the evaluation, and management who do not perceive adequate benefits from evaluation (Kumar, 1990). Organizations may judge that there are no useful approaches for determining MIS/DSS effectiveness (see question 4 below). Or organizations may conclude that the pace of technological change makes it practically impossible to conduct a careful evaluation of MIS/DSS effectiveness.

The issue of the scope of MIS/DSS may have a bearing on why organizations may evaluate some MIS/DSS but not other MIS/DSS. It may be that MIS/DSS which are perceived as having well-defined boundaries are better candidates for evaluation. If this is true, then networking various MIS/DSS may contribute to an organizational unwillingness to evaluate MIS/DSS while at the same time producing (in some intuitive but unmeasured sense) more effective organizational systems. Perhaps the development of clear, explicit system documentation detailing the scope of MIS/DSS which is communicated to all functional groups involved in system implementation is one approach to managing the MIS/DSS scope issue. Requesting feedback from all functional groups to reach a consensus may be one way of avoiding differing perceptions of system scope.

(2) Which functional groups are involved in evaluating MIS/DSS effectiveness in organizations?

Users, MIS/DSS development personnel, internal audit personnel, and top management have been identified as the functional groups which may be involved in evaluating MIS/DSS (Hamilton & Chervany, 1981b). Data are needed to determine which groups are actually involved in the evaluation of MIS/DSS effectiveness specifically in organizational practice. Also to be determined is the degree and quality of communication between the groups, and whether one or more groups are assigned primary responsibility for making the final evaluative judgment on the effectiveness of MIS/DSS.

(3) What approaches do organizations use to evaluate the effectiveness of their MIS/DSS?

Of the variety of MIS/DSS approaches identified (Table 3), some are well-known while others are of relatively recent origin. Research is needed to determine the reasons for organizational adoption of specific approaches.

Organizations may adopt a single approach or a set of approaches to evaluate MIS/DSS. Which approach(es) are adopted may be a function of past experience, familiarity with an approach, or evaluator bias toward an approach.

The close relationship between evaluation approaches and effectiveness measures may also impact the adoption of evaluation approaches. Which effectiveness measures are viewed as most important or revealing to the evaluator may influence the adoption of evaluation approaches. In this case, organizational reality may reflect a problem in search of a solution. On the other hand, it is also possible that bias toward the use (or neglect) of an approach may dictate which effectiveness measures are assessed as most important. In this case, organizational reality may reflect a solution in search of a problem. An approach requiring some advanced intellectual skill (for example, mathematical sophistication) may fall into the neglected approach category.

> (4) At what phases of the system development process do organizations apply evaluation approaches to assess MIS/DSS effectiveness?

Evaluator viewpoints about the appropriate times to evaluate systems will affect the decision as to when to evaluate. Organizational past experience may play a large role in determining at what phases of the system development process organizations apply various evaluation approaches.

The effectiveness measures used and approaches adopted may also influence the choice of application phases. It may well be that approaches which measure the subjective beliefs and attitudes of MIS/DSS users are best suited to evaluating MIS/DSS during the various developmental phases, while approaches which measure objective criteria are best suited for early and later developmental phases.

Reasons for evaluation are a critical factor in choosing when to evaluate. Evaluation may be performed in order to creatively improve systems development, or it may be performed ritualistically to bring closure to a systems development effort (Kumar, 1990).

> (5) Are organizations satisfied with the results of using the approaches for evaluating MIS/DSS effectiveness?

This question is closely related to question (1). From an efficiency perspective, organizations may find that certain evaluation approaches are better suited than others for conducting an evaluation of MIS/DSS effectiveness. From an effectiveness perspective, organizations may find that the outputs of certain evaluation approaches are easier to interpret and use in making a determination of MIS/DSS effectiveness than the outputs of other approaches. The degree of satisfaction with the results of using the approaches may influence the organizational decision to continue the practice of evaluating the effectiveness of their MIS/DSS.

> (6) Do organizations use the results of evaluating the effectiveness of their MIS/DSS to justify further investment in IT?

Organizations may have many reasons for making further investment in IT, including increased information processing demands on the organization and the threat of industry competition. The decision to invest in further IT may be the output of an organization's strategic plan for its MIS/DSS (King, 1978). A determination of the effectiveness of existing MIS/DSS may have a direct bearing on the decision as to the level of further investment as well as to the type of IT to acquire. Even a negative evaluation may result in a decision to invest further in IT for a number of reasons, not the least of which is the observation of other organization's successful implementation and use of IT.

The answers to these questions should interest both practitioners and researchers. Practitioners will learn how their organizations compare to other organizations in terms of commitment to MIS/DSS evaluation, of effectiveness measures and evaluation approaches adopted, and of the relative importance of evaluator viewpoints. In addition to this information about MIS/DSS evaluation practice, researchers will get valuable feedback on which evaluation approaches are viewed as most useful by practitioners. Such feedback will be helpful in the further improvement and/or promotion of the approaches to make them more relevant and appealing to practitioners. Further, the answers may have relevance to the evaluation of the use of IT in non-business organizations. For example, educational institutions at both the high school and college level may benefit from learning which evaluation approaches are utilized to advantage in business, and apply this knowledge to the measurement of needs and resources for providing effective IT in higher education (cf. Mosmann, 1976).

The comprehensive MIS/DSS evaluation framework provides direction for further empirical research. Of critical importance is the study of the reasons for differing evaluator viewpoints with a view toward providing a behavioral/social explanation of the reasons for the differences. In addition, the impact of evaluator viewpoints on the success of the MIS/DSS development process itself may also be approached systematically. One key issue in this regard is whether one or more functional groups have a greater influence on the evaluation of MIS/DSS effectiveness during different phases of the MIS/DSS development process. Mathematical approaches such as the analytic hierarchy process (Saaty, 1980), which seeks to address the prioritization of relevant issues, may be of help in managing the issue of differing evaluator viewpoints, since people disagree about the significance of various IS characteristics according to their respective interests (Liebowitz, 1986). The comprehensive framework identifies the critical areas that may be targeted for focused empirical study of these issues.

Summary

Evaluating MIS/DSS has been argued to be a critical activity in the organizational management of IS. Various theoretical frameworks and practical approaches have been proposed in the literature for addressing the evaluation of MIS/DSS. This paper has attempted to synthesize the fundamental issues in these frameworks and approaches into a comprehensive framework for further empirical research on actual organization practice. A better understanding of actual practice in this area will help researchers to develop better tools to assist IS managers to develop more effective MIS/DSS using IT and to analyze the possibly reciprocal impacts of systems on the organization and the world.

References

Ahituv, N. (1980). A systematic approach toward assessing the value of an information system. *MIS Quarterly, 4*(4), December, 61-75.

Ahituv, N., & Neumann, S. (1982). *Principles of information systems for management.* Dubuque, IA: Brown.

Attewell, P., & Rule, J. (1984). Computing and organizations: What we know and what we don't know. *Communications of the ACM, 27*(12), December, 1184-1192.

Bailey, J.E., & Pearson, S.W. (1983). Development of a tool for measuring and analyzing computer user satisfaction. *Management Science, 29*(5), May, 530-545.

Baroudi, J.J., & Orlikowski, W.J. (1988). A short-form measure of user information satisfaction: A psychometric evaluation and notes on use. *Journal of Management Information Systems, 4*(4), Spring, 44-59.

Bergeron, F., & Begin, C. (1989). The use of critical success factors in evaluation of information systems. *Journal of Management Information Systems, 5*(4), Spring, 111-124.

Bonini, C.P. (1963). *Simulation of information and decision systems in the firm.* Englewood Cliffs, NJ: Prentice-Hall.

Boyd, D.F., & Krasnow, H.S. (1963). Economic evaluation of management information systems. *IBM Systems Journal, 2*(1), March, 2-23.

Chandler, J.S. (1982). A multiple criteria approach for evaluating information systems. *MIS Quarterly, 6*(1), March, 61-74.

Chervany, N.L., & Dickson, G.W. (1970). Economic evaluation of management information systems: An analytical framework. *Decision Sciences, 1*(3/4), 296-308.

Davis, G.B., & Olson, M.H. (1985). *Management information systems*. New York, NY: McGraw-Hill.

Davis, J.G., Dhar, V., King, W.R., & Teng, J. (1984). The impact of the organization on the computer. *Business Forum, 9*(3), Fall, 40-44.

Dickson, G.W., Senn, J.A., & Chervany, N.L. (1977). Research in management information systems: The Minnesota experiments. *Management Science, 23*(9), May, 913-923.

Dickson, G.W., & Wetherbe, J.C. (1985). *The management of information systems*. New York, NY: McGraw-Hill.

Dyer, J.S., & Sarin, R.K. (1979). Measurable multiattribute value functions. *Operations Research, 27*(4), July-August, 810-822.

Galletta, D.F., & Lederer, A.L. (1989). Some cautions on the measurement of user information satisfaction. *Decision Sciences, 20*(3), Summer, 419-438.

Gashnig, J., Klahr, P., Pople, H., Shortliffe, E., & Terry, A. (1983). Evaluation of expert systems: Issues and case studies. In Hayes-Roth, R., Waterman, D.A., & Lenat, D.B. (Eds.), *Building expert systems*. Reading, MA: Addison-Wesley.

Goodhue, D. (1988). I/S attitudes: Toward theoretical and definitional clarity. *Data Base, 19*(3/4), Fall/Winter, 6-15.

Gremillion, L.L., & Pyburn, P.J. (1985). Justifying decision support and office automation systems. *Journal of Management Information Systems, 2*(1), Summer, 5-17.

Halbrecht, H.Z. (1977). Interview with R.E. McDonald. *MIS Quarterly, 1*(2), June, 7-11.

Hamilton, S., & Chervany, N.L. (1981a). Evaluating information system effectiveness — Part 1: Comparing evaluation approaches. *MIS Quarterly, 5*(3), September, 55-69.

Hamilton, S., & Chervany, N.L. (1981b). Evaluating information system effectiveness — Part II: Comparing evaluator viewpoints. *MIS Quarterly, 5*(4), December, 79-86.

Huff, S.L., & Munro, M.C. (1985). Information technology assessment and adoption: A field study. *MIS Quarterly, 9*(4), December, 327-339.

Ives, B., & Learmonth, G.P. (1984). The information system as a competitive weapon. *Communications of the ACM, 27*(12), December, 1193-1201.

Ives, B., & Olson, M.H. (1984). User involvement and MIS success: A review of research. *Management Science, 30*(5), May, 586-603.

Ives, B., Olson, M.H., & Baroudi, J.J. (1983). The measurement of user information satisfaction. *Communications of the ACM, 26*(10), October, 785-793.

Keen, P.G.W. (1981). Value analysis: Justifying decision support systems. *MIS Quarterly, 5*(1), March, 1-15.

King, J.L., & Schrems, E.L. (1978). Cost-benefit analysis in information systems development and operation. *ACM Computing Surveys, 10*(1), March, 19-34.

King, W.R. (1975). *Marketing management information systems*. New York, NY: Petrocelli/ Charter.

King, W.R. (1978). Strategic planning for management information systems. *MIS Quarterly, 2*(1), March, 27-37.

King, W.R. (1983). Evaluating strategic planning systems. *Strategic Management Journal, 4*(3), July-September, 263-277.

King, W.R., & Epstein, B.J. (1983). Assessing information system value: An experimental study. *Decision Sciences, 14*(1), January, 34-45.

King, W.R., & Rodriguez, J.I. (1978). Evaluating management information systems. *MIS Quarterly, 2*(3), September, 43-51.

Kleijnen. J.P.C. (1980). *Computers and profits: Quantifying financial benefits of information*. Reading, MA: Addison-Wesley.

Kumar, K. (1990). Post implementation evaluation of computer-based information systems: Current practices. *Communications of the ACM, 33*(2), February, 203-212.

Liebowitz, J. (1986). Useful approach for evaluating expert systems. *Expert Systems, 3*(2), April, 86-96.

Marschak, J. (1968). Economics of inquiring, communicating, deciding. *American Economic Review, 58*(2), May, 1-18.

Martin, E.W. (1983). Information needs of top MIS managers. *MIS Quarterly, 7*(3), September, 1-11.

McFarlan, F.W., & McKenney, J.L. (1983). *Corporate information systems management*. Homewood, IL: Irwin.

Miller, J., & Doyle, B.A. (1987). Measuring the effectiveness of computer-based information systems in the financial services sector. *MIS Quarterly, 11*(1), March, 107-124.

Money, A., Tromp, D., & Wegner, T. (1988). The quantification of decision support benefits within the context of value analysis. *MIS Quarterly, 12*(2), June, 223-236.

Montgomery, I., & Benbasat, I. (1983). Cost/benefit analysis of computer based message systems. *MIS Quarterly, 7*(1), March, 1-14.

Mosmann, C. (1976). *Evaluating instructional computing*. Irving, CA: University of California Press.

Munro, M.C., & Wheeler, B.R. (1980). Planning, critical success factors, and management's information requirements. *MIS Quarterly, 4*(4), December, 27-38.

Pieptea, D.R., & Anderson, E. (1987). Price and value of decision support systems. *MIS Quarterly, 11*(4), December, 515-527.

Porter, M. (1980). *Competitive strategy*. New York, NY: Free Press.

Powers, M.J., Adams, D.R., & Mills, H.D. (1984). *Computer information systems development: Analysis and design*. Cincinnati, OH: South-Western.

Robertson, D.C. (1989). Social determinants of information system usage. *Journal of Management Information Systems, 5*(4), Spring, 55-71.

Rockart, J.F. (1979). Executives define their own data needs. *Harvard Business Review, 57*(2), March-April, 81-93.

Saaty, T.L. (1980). *The analytic hierarchy process*. New York, NY: McGraw-Hill.

Shank, M.E., Boynton, A.C., & Zmud, R.W. (1985). Critical success factor analysis as a methodology for MIS planning. *MIS Quarterly, 9*(2), June, 121-129.

Singleton, J.P., McLean, E.R., & Altman, E.N. (1988). Measuring information systems performance: Experience with the management by results system at Security Pacific Bank. *MIS Quarterly, 12*(2), June, 325-337.

Snitkin, S.R., & King, W.R. (1986). Determinants of the effectiveness of personal decision support systems. *Information & Management, 10*(2), February, 83-89.

Srinivasan, A. (1985). Alternative measures of system effectiveness: Associations and implications. *MIS Quarterly, 9*(3), September, 243-253.

Szewczak, E.J., & Gardner, W.L. (1989). Social and organizational impact of local and telecommunications systems — Open questions. *Information Resources Management Journal, 2*(1), Winter, 14-25.

Trice, A.W., & Treacy, M.E. (1988). Utilization as a dependent variable in MIS research. *Data Base, 19*(3/4), Fall/Winter, 33-41.

Turban, E., & Watkins, P.R. (1986). Integrating expert systems and decision support systems. *MIS Quarterly, 10*(2), June, 121-136.

Whistler, T.L. (1970). *The impact of the computer on the organization*. New York, NY: Praeger.

Zmud, R. (1979). Individual differences and MIS success: A review of the empirical literature. *Management Science, 25*(10), October, 966-979.

Chapter 3

Negative Consequences of Information Technology

Joseph Williams
Colorado State University

Most of the chapters in this book applaud and encourage the development of information technology (IT). Indeed, the substantive body of literature in this field is replete with references to the bold promises and spectacular success stories of information technology. Peter Drucker (1989) summarizes a prevailing attitude when he argues that computerized IT will benevolently revolutionize the organizational structure of American industry. As he suggests, simple economics will dictate the development and extensive use of IT.

Emerging amidst the optimistic reports and rhetoric is a trickle of troubling reports of instances where IT has been unable to fulfill its promises. More than just stories about technology that failed to perform, these accounts describe situations in which successfully implemented technology has produced chillingly negative consequences.

There are, then, two dark sides to the development and implementation of IT. First, there is the ever-present problem of performance failure. Performance failure occurs whenever the technology does not deliver substantially what was promised. These failures can take place in one or more of the development, implementation, or use phases of the technology. Generally speaking, this class of failures is referred to as systems failures. Second, however, is the problem of negative consequences which arise from the successful implementation of IT. In these cases, the technology may deliver what was promised, but the success of the technology causes problems for the firm or its industry.

This chapter presents an overview of the negative consequences of IT. No issue is taken with the benefits of and comparative advantages to computerized IT. However, the road to high-tech is not paved with gold, and many of its potholes can seriously damage an unsuspecting or unprepared organization. In other words, this chapter confronts the problems rather than the potential of IT.

For reasons of clarity, negative consequences of systems failures are hereafter referred to as "negative impacts"; the term "negative consequences" is reserved for problems which follow the successful implementation of IT systems. After some preliminary discussion of the context in which the issue has developed, the chapter proceeds by investigating in turn negative impacts and then negative consequences. Because the subject of negative impacts has been cohesively examined in other contexts, it will not receive the systematic attention given the problem of negative consequences. However, a model for assessing, classifying, and ranking negative impacts will be presented.

Background

Computerized IT has been available to businesses in some form for nearly forty years. The formalized study of IT from a managerial perspective (as opposed to an engineering perspective) is relatively new, commencing about twenty years ago. During the last decade, the advent of the personal computer, combined with significant price breakthroughs in processing power and memory capacity, has created an explosive growth in IT penetration into the work environment. Concurrently, academic and professional interest in IT literature has soared.

These have functionally been the "Go-Go" years of IT. Most of what has been written on the subject is extremely positive, exhibiting an almost evangelical fervor in exhorting the benefits of the technology. Such a posture is hardly surprising. IT can be revolutionary, and its "zealots" have been quick to proselytize the uninformed by regaling them with tales of efficiency, cost-reductions, and competitive advantage. Fascination with IT and its potential has spawned an industry of technophiles who trade on the promises of each breakthrough, regardless of the development of any practical applications. For example, hundreds of articles and books have been written on artificial intelligence (AI) even though the technology as yet has failed to deliver its promised benefits. Nonetheless, interest in AI continues to grow.

Of course, not all the news has been good. Engineers and systems professionals cannot help but notice that at times hardware broke down, operating systems failed to operate, and software refused to run properly. Moreover, information systems departments have found it increasingly difficult to accomplish their tasks in the face of rising end-user expectations

(Konsynski, 1985). Unfortunately, the pervasiveness of IT failures has never been adequately documented. In part, companies are reluctant to advertise their IT investment failures for competitive and image reasons. On the other hand, scholars and other IT professionals generally make their reputations researching or presenting successes, not failures. In combination, these two phenomena have historically provided little opportunity or incentive to investigate the negative sides of IT.

Although there has been considerable positive press given IT, a few studies have been conducted to investigate the phenomena of IT systems failure. Lucas's (1975) early work demonstrated that there are complex relationships among technology, organizational behavior, environmental exigencies, and individual characteristics. Lucas postulated that absence of or difficulty with any factor in this constellation could cause the failure of a IT system. However, he further noted that his experiences led him to conclude that most IT failures were caused by problems in organizational behavior (OB). Sadly, quantifiable numbers were not presented to support that conclusion.

Essentially, Lucas argued that end-users were responsible for most IT systems failures. Either the end-users were unable to understand, process, or use the information generated by the IT, or they were unhappy in some way with the system itself. In the former instance, the IT system functioned properly but the end-user was a poor consumer of the system. In the latter instance, communication between the end-user and the IT system's developers broke down or otherwise failed to permit the appropriate IT to be developed. In either event, the organization and the behaviors engendered by it were seen as being at the root of most IT systems failures.

The OB-driven explanation for IT systems failures became and remains a pervasive theme in the information systems literature. Turner (1982) and Markus (1984), among many others, have confirmed the problems organizational resistance poses to successful IT systems implementations. Bostrum and Heinen (1977) and Langefors (1978) wholly embraced the OB-driven view. However, most studies of IT systems failures have been poorly articulated (Lyytinen, 1988) and biased toward some OB-driven explanation. Although McFarlan and McKenney (1983) have noted that the sources of IT systems failure are diverse, complex, and certainly not restricted to organizational or individual behavioral problems, the literature until very recently has been dominated by the one viewpoint.

It is easy to understand why studies of IT systems failures have tended to produce results fingering the organization or end-users as culprits. First, most studies have been of IT systems which were either unsuccessfully implemented or improperly used. That these systems had reached the implementation stage suggests that most if not all of the technological challenges of the IT system had been overcome. By fait accompli, failures after this point almost had to be of an organizational or individual nature. Second, the germane literature tends to focus on managerial issues involving IT. Almost by definition, then, other sources of failure, including the technology itself, have not been promising prospects for investigation. Obviously most researchers are cognizant that many other sources are responsible for IT systems failure; researchers have merely restricted most of their efforts to systems with managerial problems or solutions.

Many IT systems are designed, implemented, and used properly and they perform according to original specifications. Nevertheless, some of these systems create profound problems. For example, Williams and Nelson (1990) report that decision support systems (DSSs) successfully installed at several large oil companies have inhibited—not facilitated—executive decision making. Similarly, Grant et al. (1988) report that successfully implemented computerized performance monitoring and control systems (CPMCSs) actually led to degradation in customer service. Both sets of findings relate curiously ironic and certainly undesirable consequences.

It can be argued that these negative consequences are little more than a perverse set of IT systems failures. After all, both studies suggest that the problems arose from either an oversight in the systems analysis phase or a training mistake in the implementation phase of systems development. However, unlike systems failures, these systems were not scrapped, replaced, or substantially modified. Instead, they required an adjustment to the IT system by the organization. The oil companies studied opted to use their DSSs for organizational control (in effect, treating their systems as expert systems). Firms using CPMCSs had to develop non-quantitative measures to ensure "fair" evaluations of employees. In effect, these IT systems are not failures, they are successes with problems.

Some relatively recent studies have addressed the negative consequences of seemingly successful IT systems. In these studies, the IT system functions acceptably (or better), but negative consequences for the business enterprise derive nonetheless from the system. Markus (1984) investigated the problems of negative consequences from the perspective of their impact on organizations. Other authors (e.g., Vitale, 1986; Barton and Bobst, 1988) have inspected the phenomena in terms of strategic and operational risks. On the whole, however, the problem of negative consequences remains relatively unexamined.

Historically, the focus of research has not provided a clear picture of the negative consequences which derive from either the successful or the unsuccessful implementation of IT. The following sections fine-tune that focus for further discussion of the issues involved.

The Negative Impacts of IT Systems Failures

The model proposed here for examining the negative impacts of IT systems failures is rather modest. As explained below, systems failures are classified as having negative investment or organizational impacts. The criticality of these impacts is measured on a grid ranging from insignificant impacts to impacts which result in the organization's total demise. The model is not intended to be exhaustive; rather, it is designed for discussion purposes.

"Failure" has a variety of meanings in the context of computerized IT. Myers (1979) defined "failure" as the occurrence of an error, weighted by its seriousness. This definition demarcates "error" from "failure" in terms of consequences for the organization. While the orientation is appropriate, the definition lacks specificity. Failures potentially can be correctable, fatal for the IT system, fatal for the organization, or some combination or variation thereof. The definition also does not distinguish when a failure occurred (e.g., during development or during operation). In short, the term is vague.

The technical press popularly uses "failure" in one of two contexts: (1) to describe IT systems which have been abandoned because they never performed acceptably; and (2) to describe IT systems which have suddenly ceased to performed acceptably (e.g., "crashes"). The first context can be otherwise characterized as development failure, while the second is better described as an operations failure. While these definitions lack the elegance of Myers', they more precisely provide a nexus by which to link the failure activity, risk, and consequence. Thus, the popular definitions are adopted here.

To illustrate the distinction between development and operation failures, consider the recently reported problem with Lockheed Aeronautical Systems Corp.'s executive information system (Scheier, 1990b). Lockheed's system is used by executives to track production scheduling, sales negotiations, and financial performance. The system was installed in 1978 and had performed acceptably through 1989. Therefore, the system was a development success (or, more precisely, it was not a development failure). However, in 1989 it was discovered that the system had not been designed to collect and report a critical piece of production data (i.e., the weight of an aircraft). Although it had never been a problem before, this time the omission would cost the company approximately $300 million. The design error resulted in an operational failure of a system that was otherwise performing acceptably.

Operations failure, because it occurs in a system that had previously been identified as successfully implemented, is discussed in the next section as a negative consequence.

Development failures can occur at any time in the systems development life cycle (SDLC). Traditional IT systems literature posits a series of phases—corresponding to the systems development life cycle model—through which new systems usually must move as they are created (Wetherbe, 1988). The number and sequence of these phases vary among authors and practitioners, but integral to all models are design, development, and implementation activities (Senn, 1989). Although it is subject to some debate (Swanson, 1988), user acceptance of a system is usually an implementation activity or treated as a separate phase of the SDLC.

Most researchers (e.g., Strassman, 1985) treat the problem of development failure as having two components: investment risk and organizational risk. Investment risks derive from the direct costs associated with the IT project. Organizational risks are those costs which result from the development failure; these include damage to customer relations and inability of the organization to properly function. Regardless of when or where a development failure occurs, negative impacts will include investment costs and may generate organizational costs.

Mehring and Gutterman (1990) report a development failure at Amoco (U.K.) that illustrates a negative impact primarily limited to investment costs. Amoco U.K. had a good manual system to support supply and distribution planning, but anticipated expansion would eventually strain the existing system. Amoco U.K. decided to build an IT system to perform the manual tasks, with the costs of developing the system covered by savings in supply and distribution costs during the first year of the system's use. The system was built, installed, and functioned properly, but its users decided not to use it. Thus, the system was a development failure.

The negative impacts of this development failure were limited primarily to the costs of developing the system. Resulting impacts on the organization were negligible. As a result of the development failure, Amoco U.K. must still manually perform supply and distribution planning, but it was doing so before the system was installed. Of course, some residual impacts may have been generated for the business unit suffering the investment loss and for the IS group responsible for the failed system, but these residual impacts would be localized.

A recent development failure with negative organizational impacts was reported by Schatz (1989). The Department of Defense decided in 1982 to modernize the World Wide Military Command and Control System (WWMCCS) it uses for monitoring and planning military activities. The development project was given to the Electronic Systems Division of the U.S. Air Force Systems Command. After seven years and $395 million, the project was certified a complete development failure. As a result, the Air Force lost all control over the project, which was ultimately budgeted at $2.3 billion. In addition, the Air Force and the Department of Defense lost credibility with Congress, who responded to WWMCCS's development problems with severe budget cuts. Worse, military preparedness has suffered.

The WWMCCS fiasco illustrates that development failures have negative impacts that are not limited to investment losses. Not only did the Air Force lose control of a substantial budget item, it lost its ability to ensure the system met its specific needs. Instead, a new system will be developed by the Department of Defense that will be subject to lobbying by the various branches of the armed forces. The impact on military preparedness thankfully remains intangible.

In 1981, Bankamerica Corp. embarked on an ambitious program to attract and retain bank customers by attempting to automate a vast array of its services (Rifken and Betts, 1988). Gross investment in the various projects approached nearly $5 billion over seven years. Several major development failures have been held responsible for accelerating the plunge taken by the bank as it fell from being the largest bank in the world to a more humble 29th place.

Typical of Bankamerica's problems were those experienced in the development of its Masternet institutional trust accounting system (Ludlum, 1988). Masternet was designed to support the bank's activities as trustee for institutional trusts such as pension funds. The system was to track the performance of money managers, settle trades, collect investment income, and make payments to beneficiaries. Built around a platform supplied externally, after five years of development Masternet went on-line in 1987. Masternet was a disaster; it failed to maintain current data and was beset by hardware and communication glitches. After losing some customers who were disenchanted with the system, Bankamerica eventually scrapped Masternet, taking an $80 million loss on the project and losing customers who moved an estimated $4 billion in assets to other financial institutions. The Chief Information Officer and the executive responsible for the system were both fired. Staff reductions and poor morale were subsequently reported. Most of the asset management functions were stripped from the corporate headquarters and given to a subsidiary, but some of the more complex accounts had to be transferred to an external source.

The organizational impacts of the Masternet fiasco cannot be underestimated. The investment loss pales in comparison to the organizational costs incurred by the project's failure. Bankamerica's self-inflicted credibility wounds cast aspersions on its ability to compete. Coupled with severe financial problems created by shaky foreign and domestic loan portfolios, the bank now has something of a bumbling and hapless image.

The use of external system developers does not immunize an organization from the problems of negative impacts. The State of New Jersey contracted with a major accounting firm's consulting unit to modernize the computer operations of the state's Division of Motor Vehicles (Rifken and Betts, 1988). The accounting firm "professionally mismanaged" the implementation, resulting in registration errors involving a million vehicles. The development failure cost the state $6.5 million. While a subsequent investigation hung blame for the failure on the external developers, the snafu did nothing to detract from the state government's reputation for incompetent administration.

Over the years, there have been many undocumented allusions to organizations suffering ruin because the negative impacts of a development failure were too severe to survive. Such a scenario is not difficult to imagine if a firm has invested a substantial percentage of its capital in the IT project or has staked the firm's marketing success on the operability of the technology. Nor is it difficult to imagine why none of these failures are documented: the principles in a failed business rarely publicize their mistakes. The lack of a documented incident does not detract from the very real possibility that a IT development failure could have fatal results for the entire organization.

A preliminary model for classifying the negative impacts of development failures emerges from the preceding examples. Investment risk from a IT project can be plotted on a Likert scale that measures the proportionate impact on the firm from the project's failure. Organizational risk can be similarly measured. Investment impacts and organizational impacts are usually intertwined when a IT development project fails, so the two scales are superimposed, creating a four-quadrant diagram (Figure 1). Although the model

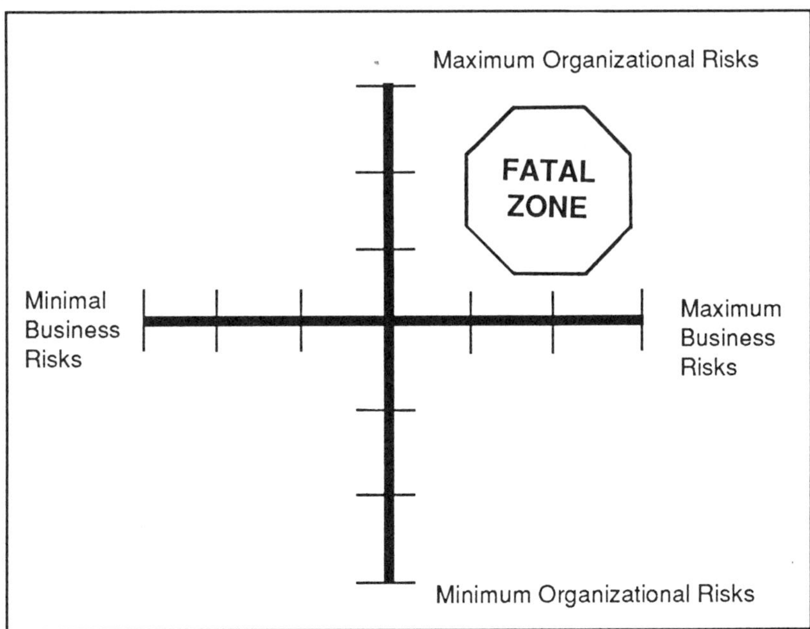

Figure 1. Four Quadrant Diagram

currently lacks an exhaustive validation, it can be used preliminarily to roughly present and compare the relative risks of IT development projects.

Negative impacts from investment costs derive from the risks associated with losing the IT project investment. The minimum point on the scale represents an insignificant capital loss problem, while the maximum point represents a situation in which the capital loss is fatal to the firm. Intermediate points represent varying degrees of financial hardship. The organizational scale operates similarly. The maximum point of this scale represents a fatal situation for the firm in which, for example, failure of the IT project causes critical personnel to leave or cripples a key marketing effort.

Use of this model requires that a business enterprise tailor each scale to their firm, identify and assess the potential negative investment and organizational impacts of the IT project, and plot the results. Projects plotting in the south-west quadrant would present the least risk to the firm, while those in the north-east quadrant would seriously jeopardize the firm if the project failed. Combinations of negative investment and organizational impacts can be used to determine threshold numbers that identify particularly risky projects (Edgeman, 1983).

At a medium-sized U.S. oil production company, for example, investment impacts are given twice the weight of organizational impacts because that particular organization lacks financial resiliency. Any IT project with a negative impact rating of ten [(2 * investment rating) + organizational rating] requires executive committee approval. In the case of Bankamerica, the negative investment impacts from failure of the Masternet project were minimal. However, negative organizational impacts were severe. Businesses of this size rarely embark on IT development projects that pose potentially significant negative investment impacts. The Masternet fiasco suggests that threshold numbers for these types of firms should be heavily weighted by the organizational impacts rating.

The model presented here for assessing the negative impacts of IT development failures is not designed to be comprehensive. Rather, it is designed to be a simple tool for addressing the problem of negative impacts. Most business organizations exhibit the zealotry and optimism for computerized IT displayed by computer professionals. Use of a negative impacts model will force companies to understand and assess the ramifications of development failures. However, the model does not address the problem of negative consequences deriving from successful development projects.

Negative Consequences

The problem of negative consequences is best introduced through an analogy. Consider the first-time purchase of an automobile by a wage-earner domiciled within walking distance of the place of employment. Initially, of course, there are concerns about the technical performance of the vehicle. Assume, however, that the automobile works splendidly. Over time, relying upon the dependability of the vehicle, the wage-earner moves to the suburbs and joins the happy ranks of those who commute to work.

Now consider the worst things that could happen to the wage-earner because of that automobile. To begin with, the vehicle could suddenly and unexpectantly cease to function, thereby stranding the wage-earner. The vehicle could develop an undiagnosible glitch (e.g., it won't start when the weather is above 100 degrees temperature, but no one can determine why or how to fix it). Some nefarious character could pour sugar into the gas tank (ruining the engine) or even steal the automobile. Other members of the household could demand to use the vehicle, thereby depriving the wage-earner of access. The automobile could crash, causing injury to the wage-earner (the forces responsible for the crash are unimportant at this juncture). The automobile could become so alluring to the wage-earner that the latter soon begins to divert investment funds to accessories such as car phones, sound systems, and the like. The wage-earner could so integrate the automobile into the job environment that continual investment in new and more expensive automobiles is necessary to maintain the job environment. And there are problems of repairs, taxes, insurance, operating expenses, and pollution.

Several of these consequences of automobile ownership are clearly negative. Others are negative depending upon one's viewpoint. Still others are negative only under certain conditions. For example, the temperature glitch is unimportant if the wage-earner lives in a temperate climate or, if the wage-earner lives in Texas, unless the condition occurs at an inconvenient time. Importantly, all of the consequences described occurred when the automobile was (or continues) functioning properly.

IT is not unlike the automobile in that successful systems nonetheless can generate negative consequences for the organization. As in the analogy, some of these consequences are clearly negative, some are negative depending upon one's viewpoint, and others are negative only under certain circumstances. And all are consequences generated by IT systems that can be characterized as successfully implemented.

An examination of published accounts of negative consequences resulting from successful IT implementations yields problems which fall somewhat neatly into four categories: (1) vulnerability; (2) encumbrances of operations; (3) social issues; and (4) unintended consequences. In brief, vulnerability results from damage to a IT system. Encumbrances of operations involve the business issues and costs of having and using an IT system, with social issues being the phenomenological counterpart. Unintended consequences arise when a IT system produces results considerably different from what was expected. As noted above, not all incidents assigned to each category will be unanimously appraised as undesirable, but each will exhibit distinctly negative characteristics. The balance of this section will in turn examine and discuss each category.

Vulnerability

Dependency on any resource generates vulnerability (Tate, 1988). Recent natural disasters in San Francisco and the U.S. southeastern seaboard, together with moments of near hysteria generated by the media over computer viruses, have underscored the vulnerability of IT systems to destruction and corruption from external agents and disgruntled employees. One of the most frustrating axioms of computerized IT is that the more successful and pervasive the system, the more difficult it is to protect and the more important that protection becomes. Vulnerability is a business risk that is the natural consequence of relying on IT systems because it is virtually impossible to completely eliminate the problems of computer crime, disaster, and system penetration.

Computer crime encompasses a broad range of activities including fraud, theft of computer equipment, theft of computer services, industrial espionage, and sabotage. Among many other stories receiving extensive media coverage are the Brooklyn College grade-changing conspiracy (Norman, 1983), Security Pacific National Bank's $7 million electronic funds transfer embezzlement, and Volkswagen's $259 million "cooked books" accounting fraud (Keefe, 1988). Hammer (1986) reported an incident involving a public employee (a financial analyst with minimal computer skills) who changed the computer's access codes and conveniently "forgot" them, denying the agency access to millions of its own dollars. The perpetrator then started a "Guess the Password" contest in the local press by providing daily clues; ultimately, the agency had to break into their own computer system. The pervasiveness of these and other villainous activities has given birth to a billion dollar industry in computer security (Keefe, 1988).

Computer crime and its security counterpart should not be confused with the challenge of data integrity. In general, the latter concerns difficulties which arise from simple errors and omissions in the data life cycle (Fisher, 1984). Data integrity is fundamentally an accounting problem (Perry, 1981), and will be dealt with later under the heading of operations burden. Computer crime has accounting consequences, but is primarily a security problem. Also not covered under this heading are hackers and computer viruses. Although both normally involve criminal activities, they present special problems and are treated separately.

There are three kinds of losses which result from the perpetration of a computer crime (TABLE 1). Norman (1983) and Bequai (1983) constructed

Loss of Asset	Exclusive Use	Denial of Rights
Money	Customer lists	Harrassment
Software	Designs/patents	Privacy
Hardware	Research data	Blackmail
Computing services		
Information		

Table 1. Losses Resulting from Computer Crime

different categories, but the list of criminal activities are similar. Losses of money due to fraud and embezzlement generate the most publicity, but software piracy, thefts of customer lists and proprietary data, and concerns about individual privacy, particularly with regard to electronic mail (Higgins, 1990), have recently generated great concern. It has been estimated that 44% of computer crimes involve monetary theft, 16% damage to software, 10% theft of information, and 10% theft of computer services, with the remainder split amongst the other activities (Bidgoli and Azarmsa, 1989).

As many as 2,000 cases of computer crime have been documented (Keefe, 1988). Bequai (1983) estimated in 1983 that losses to computer crime might exceed $1 billion annually in the United States; the typical incident currently nets over $1,000,000 for the perpetrator (Nash, 1989). However, because only an estimated 10-15% of computer crimes are ever reported (Bidgoli and Azarmsa, 1989), pervasiveness and severity of the problem is truly unknown.

Total security from criminal activities is not economically viable, so some security failures must be expected. The optimal security strategy should endeavor to minimize the sum of security expenditures, expected losses, and insurance. According to Murray (1988), this policy of prevention, absorption, and indemnification should make a net contribution to profits if properly employed so long as the losses it prevents or mitigates are greater than the required expenditures. Deterring the computer criminal is seen as ineffective because computer crime is so difficult to prove that prosecution may be considerably more expensive than the loss (Carroll, 1977). As few as 1 in 20,000 computer criminals ever goes to jail, and there are no international laws relating to computer crime (Nash, 1989).

Any IT system is vulnerable to computer crime. Industry experts suggest that most computer crimes are caused by insiders (Schwartz, 1990), although there is as yet little documented evidence to support this conclusion. IT systems are convenient targets for computer crime because the computer is relied upon for financial transactions and because information is a valuable corporate resource. The IT systems of banks and other financial institutions are particularly vulnerable to computer crime (Price, Cotner, and Dickson, 1989).

Each day in the United States, $1 trillion is electronically transmitted amongst financial institutions, and each transaction is vulnerable to tampering (Schwartz, 1990).

The "World Series" earthquake reiterated the vulnerability of IT systems to disasters. The October 17th earthquake caused mainframes and minicomputers to roll around and PCs to fly off their desktops (Bozman, 1989). Disruption of computing services was widespread, and damage to systems resulted not only from the quake but from flooding, fires, and power spikes that were the quake's aftermath. In a different incident, a wind storm ripped the roof off of the computer services subsidiary of a large financial institution, allowing rain to pour down on its data center for several hours (Hamilton, 1990). The center was out of operation for two days, and physical damage to the computing equipment was estimated at more than $5 million. Disaster need not strike the IT system directly to disrupt services, as the Hinsdale fire demonstrated. In 1988, an Illinois Bell local telephone switching office was nearly destroyed by fire, causing near-total disruption of telecommunications routing for hundreds of corporations. Motorola, for example, was left without 35 domestic data circuits, four international circuits, in-coming 800 service, international toll-calling ability, and incoming-call service (Wiexel, 1989).

IT systems disasters can originate from flooding, earthquakes, fire, power failure, sabotage, wind, and many other causes (Rohde and Haskett, 1990). Although prevention strategies can minimize disaster risks, they can never completely protect a IT system (Toigo, 1989). Thus, any operating system is vulnerable to disaster. Consequences of a disaster include a reduction or complete loss of basic business functions, loss of revenue, increases in operating costs, and reduced competitive edge (Christensen and Schkade, 1989).

Toigo (1989) estimates that the average company has a one percent chance of experiencing a IT systems disaster and that 93% of disasters affecting the corporate data processing center will be fatal for the firm. A business suffering a disaster lasting longer than 10 days will rarely fully recover, and over half of these firms are out of business within five years. Costs to recover can be staggering. World Life Insurance Company, after suffering a fire that completely destroyed its data center, nearly went bankrupt during the 50 days it took to recover (Kolodziej, 1988).

The proliferation of computer viruses has raised interest and concerns regarding the vulnerability of IT systems to penetration (Wilkes, 1990). Penetration can occur in one or both of two ways: by hacker or by virus. While hackers are a problem primarily for distributed systems and for systems permitting external access (e.g., by modem), viruses can strike any system vulnerable to hackers and any system where data and/or programs are shared. Some hackers, such as Robert T. Morris, are also progenitors of deadly viruses

(Littman, 1990).

Hacking has been described by the U.S. Attorney's office as the white-collar crime of the 1990s (Alexander, 1990d). Hackers are individuals who, through unauthorized means, penetrate the security of IT systems for a variety of different reasons. One prevalent theory explains the motivation for hacking as arising from the intellectually-stimulating technological challenge of "beating" the network security system. Atkins (1985) more sternly characterizes the hacker as a "Jesse James at the Terminal." Regardless of their depiction, hackers pose a serious problem for any operating IT system because, at a minimum, they steal computer services. At worst, they could steal millions of dollars and/or destroy the entire system.

Fitzgerald (1990) cites an example of a band of hackers who penetrated the messaging system of a wholesale grocery firm in Los Angeles. After penetrating the system, the hackers reprogrammed 200 voice mailboxes and used the victim's free 800 number to run a prostitution ring and pass drug information. The hackers also stole MasterCard, Visa, and American Express numbers via the system. The system's internal help programs assisted the hackers in most of their activities.

Hacking has become increasingly more difficult to defense. In response to the problem, the U.S. Secret Service recently used 150 agents in a national dragnet ("Operation Sundevil") to crack down on 28 hackers in 14 cities (Alexander, 1990d). The raids netted 42 computers and more than 23,000 floppy diskettes allegedly used in criminal activities responsible for more than $50 million in damages. Agents have also seized a computer game (GURPS Cyberpunk) that is reputed to teach players how to become computer hackers (Alexander, 1990c). However, civil libertarians are battling the raids on constitutional grounds (Alexander, 1990e), so it is unclear how the matter will be resolved legally. To date, few hackers have been sent to prison (Alexander, 1990b).

In November, 1988, Robert T. Morris, a university student, hacked his way into the Internet network in order to plant a program specifically designed to disrupt the system, an umbrella for three national communications networks (Littman, 1990). Morris's program brought down over 6,000 workstations across the country. Damage by Morris's program is estimated to be as high as $186 million (Payne and Raiborn, 1989).

Morris's program was a worm, a powerful kind of computer virus. A computer virus is a segment of self-replicating program code that attaches itself to programs or other executable system components (McAfee, 1989b). If well-designed, the virus will not show any external manifestations. The virus can have a variety of effects, depending on the designer, but most are more irritating than malicious or destructive (Payne and Raiborn, 1989). Worms, such as the one created by Morris, seek out unused computer space and use the available capacity to propagate itself until the system's memory is exhausted. As happened with Internet, an undetected worm can shut down a network. A trojan horse is a program that contains destructive code designed to sabotage a system, disguise a virus, or permit the perpetration of fraud. A logic bomb is a program that contains code to activate a worm or virus once some specified condition (e.g., date) is met.

Viruses often have colorful names, but the damage they cause to IT systems is not amusing. The "Dark Avenger" virus struck a software development company's administrative system, wiping out 400 files and crashing three local area networks in a matter of hours ("Dark Avenger," 1989). The "Jerusalem" virus struck Andersen Consulting's Manila, Philippines offices, costing the firm a week of access time to its PCs (Scheier, 1989). The same virus nearly interrupted the 1990 U.S. census (Alexander, 1990a). Some virus scares are overblown (e.g., the "Datacrime" virus), but most will cause real damage. Preventative measures are available, but most are built to react to known viruses, worms, and trojan horses. As yet, legal recourse against virus designers, assuming the individuals can be identified, is minimal (Payne and Raiborn, 1989). Antivirus legislation has been proposed (Betts, 1989), but progress is not expected quickly.

One proposed solution to limiting the proliferation of computer "bugs" has been to adopt a corporate policy prohibiting the use of public-domain and shareware software programs (McAfee, 1989a). Many of these programs are available from electronic bulletin boards that are of unverifiable integrity. However, shareware and public-domain software are a central part of many software libraries ("Shareware," 1990) and a source of grass-roots innovation and creativity. Many corporations are using PC shareware as a less expensive alternative to mainstream programs. Ernst & Young, the accounting and consulting firm, uses 10 shareware programs, including 140,000 copies of one program (Scheier, 1990a).

What can be done to eliminate the vulnerability of an IT system? According to computer security expert Robert Morris (father of hacker Robert Morris), it is easy to have a secure computer system: you merely disconnect all dial-up connections, put all the machines in a shielded room, and post a guard at the door (Littman, 1990). Of course, this measure makes IT tools rather cumbersome and severely limits their usefulness.

Vulnerability has become a critical corporate issue and has stirred considerable international debate (Tate, 1988). The American Federation of Information Processing Societies (AFIPS) has concluded that corporate America is at risk because of its IT dependence. Similarly, professional organizations in Norway and Sweden have reported the seriousness posed by the vulnerability of social structures and activities to IT systems interference.

Encumbrances of Operation

Reconsider the automobile analogy. At a minimum, operation of the vehicle requires an ongoing investment in supplies (e.g., gasoline) and maintenance. During times of maintenance or when the automobile is not running properly, the driver is not only deprived of its use but must normally rely on specialists for the vehicle's repairs. These are predictable burdens of automobile ownership, but they are often overlooked or underestimated costs until they are incurred. Similar encumbrances accrue to the operation of any IT system.

There are many obvious and hidden costs to the operation of any IT system (Strassman, 1985). One ever-present issue is whether an IT system's performance is worth its ongoing investment costs (Morris, 1989; Hammer, 1988). However, the problem of investment viability is too broad to be meaningfully addressed in this context. Upon dissection into its major components, however, the problem yields the interesting issues of data integrity, system downtime, technological dependence, and power politics. All are costs which usually derive as negative consequences of IT ownership.

Data integrity is one of the critical stumbling blocks to any IT system. The majority of IT systems problems arise from simple errors and omissions in the data life cycle (Fisher, 1984). Consequently, organizations invest considerable resources in developing and operating data controls (Perry, 1981). These controls function to ensure that data is properly gathered, processed, and stored throughout the entire information system (Nash, 1989). The distribution of computing, the advent of the personal computer, and increases in on-line programming have made it difficult to keep information resources under control (Dickson, et al, 1984). While the issue of data controls has apparently declined in importance for executives (Brancheau and Wetherbe, 1987), it remains a vital problem for accountants (Nash, 1989).

The cost to ensure that all data is proper and correct and to correct every subsequent error would be exorbitant for most organizations. To determine an appropriate level of expenditure on controls, most researchers advocate a cost-benefit analysis that balances the costs of controls with the cost of possible errors (Fisher, 1984). However, legislative and strategic imperatives require many business enterprises, particularly banks, to have controls in place to assure particularly accurate data quality (Nash, 1989).

A management concern similar to data integrity is that of system downtime. Even the best IT systems experience time losses due to equipment, software, or operator failure. On average, the real cost of downtime, which includes lost business, administrative costs, repair expenses, lost productivity, and outside support costs, is thought to be about five percent of company revenues (Schwartz, 1990). It is estimated that each hour of downtime costs American Airlines approximately $34,000 in booking fees alone.

Disasters, computer criminals, and failing technology have previously been identified as causes for IT system downtime. However, the major cause of downtime is employee error. Daly (1989) reports that 55% of all unplanned downtime results from accidents and errors by employees. To minimize people as a cause of downtime, IT systems are being designed to automate monitoring, error-checking, diagnostics, and recovery operations (Wysocki and Young, 1990). However, automation increases the risks of vulnerability and technological dependence, so there are important tradeoffs to consider.

Technological dependence can be both a physical and a psychological addiction. Warner (1987) discovered that a common first response to management problems is to "throw" a computer solution at them, often with disastrous results. For example, John Deere and Company rushed billions into factory automation in order to develop a competitive production system. The poorly conceived program resulted in chaos, massive losses, and a complete retrofitting of the production process. When investment becomes a knee-jerk reaction to management problems, IT is a competitive burden, not a strategic weapon.

Some industries have invested so heavily in IT that their very existence is dependent on their computers (Barton and Bobst, 1988). Downtime of even a few days could seriously damage banking, airline, and financial services companies. Fault-tolerant computing, which involves fail-proof designs that use intelligent systems software, redundant hardware, and parallel processing, have been developed to guarantee system availability (Wysocki and Young, 1990). However, these systems, even when imperative, are very expensive.

There are two other faces to technological dependence. One, involving the technophile, has been alluded to earlier; this form of dependency requires investment in the latest technology regardless of real need. Executive information systems, for example, have been rushed into operation before their need was established; many of these systems have subsequently languished (Watson, 1990).

While technophile dependency is often costly, it is less insidious than its counterpart, technological complacency. In this form, organizations grow complacent with their existing technology. Although investment is made to maintain or perhaps incrementally improve the systems, complacent firms are reluctant to decommission or replace them, regardless of real need. The City of Oakland, California, for example, had resisted the implementation of local area networks, in part because the existing system was adequate and in part because the investment was seen as too costly (Scheier, 1990d). A subsequent earthquake severed data communications and disrupted the entire system. Only after the disaster did the City Council realize the need for the new technology. Hindsight is not uncommon for firms suffering from technological complacency. When their competitors develop technology to considerable strategic advantage, technological complacency can be fatal.

The deployment of computerized IT can also have profound impacts on an organization's structure and behavior. Thurow ("An Unvarnished View...," 1989) notes that computerization of the office has had important sociological ramifications because the deployment of IT interferes with organizational power arrangements. In fact, white-collar worker productivity in the United States has declined recently, despite the massive investment in computerization. According to Thurow, the "office" needs to be reinvented before productivity benefits from IT are realized, yet office politics make it difficult to change traditional administrative structures. At present, technology and administrative science are out of synch.

Researchers have long recognized that deployment of IT is fraught with consequences for the power and political structure of an organization (Markus, 1983). Originally, IT was so complex that its deployment had to be supported by a centralized group of professionals. This arrangement led to the insulation and elevation of the IS organization within the typical business (Foster and Flynn, 1984); in most organizations, the centralized IS function became nonresponsive to its constituents. Conflicts arose between IT managers and end-users over control of IT and information itself. Vast IS department empires were created and, in some cases, became resolutely despotic. Some IS departments were so out of touch with the needs of their constituents that they created monstrous, self-serving systems that the business could not or would not use (Janus, 1988). Almost in self-defense, IS constituents lobbied for decentralization, distributed computing, and end-user computing in order to break the organizational monopoly of IT (Sprague, 1987).

Moving from mainframe applications to minis and micros has resulted in downsizing of the IT function (Alexander, 1988). As a result, the influence of the MIS department has diminished. In fact, Dearden (1987) predicts that downsizing will ultimately lead to the demise of the IS organization. Recently, companies such as the health-care giant Baxter International have restructured their MIS departments and made large cuts in IS staff. Other firms have resorted to out-sourcing. Quotron Systems, Inc., for example, has slashed its MIS staff and farmed out most of its IT operations to external companies (Scheier, 1990c). Recent cuts by some businesses have been so severe that IT professionals recently have wondered aloud whether Dearden is right and this is the decade to dump data-processing departments (Currid, 1990).

Social Issues

A myriad of social issues have developed from and have accompanied the rise of IT. Initially, there were concerns that computers would replace workers. While it is true that some workers were displaced by computers, their numbers were compensated for by increases in staff to work with and maintain the computers. As noted above, businesses have yet to realize real productivity gains from the introduction of IT, so the worker replacement issue has, for the moment, melted away. However, concerns regarding privacy, legal ramifications of IT, and health issues raised by exposure to IT persist.

One pervasive theme in the early technology literature was the concern for privacy (Turn, 1978). In this context, privacy is a term that refers to the social rights of individuals vis-a-vis the collection, processing, storage, dissemination, and use of personal data about them. The automation of record-keeping established linkages between databases and data-gathering organizations that allowed tremendous volumes of information to be collected and correlated about any individual. Not only does the availability of detailed data present a potential threat to individual rights, but problems abound regarding the accuracy and fairness of the collected data.

The privacy fear has not abated (Mason, 1986). A recent public opinion poll shows that almost 80% of U.S. consumers are concerned about threats to their privacy posed by IT (Betts, 1990). Nearly half of the poll's respondents agreed that technology is almost out of control. Government regulation of information databases has been proposed in the form of a federal privacy oversight board. Another oversight board has been proposed to regulate Caller ID telephone services. In addition to the threat of regulation, court rulings make it clear that legal liability will be imposed on information managers who are not diligent in preventing information abuse or protecting information privacy (Westermeier, 1979).

IT is dogged by a pack of other legal issues as well. Software piracy by employees and by business organizations is estimated to cost vendors about $1.5 billion a year ("Avast, Mateys!...", 1990). To combat the problem, the Software Publishers Association is conducting corporate audits to determine whether companies are using pirated software. Of the 27 companies it audited in the first quarter of 1990, all but one were found with pirated software. The settlements negotiated with these companies ranged from $20,000 to $50,000 each. Careless companies could face lawsuits and severe penalties if they do not police themselves diligently.

Other aspects of IT which expose companies to legal liability include on-line libel and the more serious problem of the loss of legal documentation. On-line libel hit Stanford University as an issue when an on-line joke bulletin board was shut down by the university's legal counsel because many of the jokes were racist, anti-Semitic, and sexist (Fitzgerald, 1989). A subsequent uproar (the action was alleged to be a suppression of academic freedom) caused the bulletin board to be reinstated and no court case was brought. A more serious episode occurred in San Jose, where police have charged two men with using a bulletin board in a bizarre plot to find a boy who would be kidnapped, raped, tortured, and killed in a "snuff" film. Who is liable if such illicit activities are orchestrated using a corporate or private bulletin board? At present, the matter is unclear, but legal exposure should be considered as a potential problem.

An intriguing legal consequence of the office computer is that hard disk files have supplanted photocopies and carbon copies that are retained for documentation of letters, contracts, and other important documents. Most people incorrectly assume that as long as the document is not changed, it can be printed out at any time in the future as a "carbon copy" of the original (Waegemann, 1988). Unfortunately, because computer files can be easily changed without leaving a trace of the alteration, the data stored on disk usually may not be accepted as a true record by a court. Consequently, an organization without proper paper backups may risk losing in business disputes. There currently are no commercial word-processing packages that meet the legal criteria that define a true document, so the storage of formal documents on disks without paper backups is a potential legal time bomb.

Ironically, of course, the computer was touted as a way to create a "paperless" office environment.

Documentation of health risks from exposure to IT is on the rise. Brodeur (1990) presents research and case studies indicating that video display terminals (VDTs) cause birth defects, cell mutations, and cancer. Brodeur's work directly disputes AT&T's Bell Laboratory report that VDTs are safe, although his evidence of a national coverup to suppress information about the health risks is questionable.

The problem of "technostress" has attracted attention from researchers in many disciplines (Koenenn, 1990). Data processors have reported tension, paranoia, overstimulation, fatigue, sagging libido, high anxiety, headaches, psychic numbing, and low self-esteem related to their jobs and from the introduction of new IT. Low productivity and potential legal liability may be the consequences if technostress is not mitigated.

In addition to enterprise-wide issues of health, safety, legal liability, and privacy, scholars have raised concerns over the computerization of society and its impacts on the behavior and socialization of people (Koenenn, 1990). Psychologists have noted that high-tech environments are not meeting people's needs for personal contact (witness the phenomena that people sitting next to each other in an office nonetheless communicate by electronic mail). Similarly, the high tech environment is accused of replacing a naturally leisurely "human sense" of time with demands for instantaneous service, instantaneous feedback, and impersonal contacts. The ultimate disposition of these and other social issues is yet to be decided, but they currently pose problems for which planning and evaluation would be prudent.

Unintended Consequences

Vitale (1986) exposed an intriguing problem with information systems: namely, technical success can be catastrophic to the business organization. Vitale used Porter's (1980) "five forces" of strategic opportunities as a mirror in which to present his own case that information systems used for competitive advantage may have negative strategic consequences. To wit, instead of using the five as lenses through which to find opportunity, Vitale employed them to identify opportunities which, once found, were fumbled, or went awry.

The first area of opportunity involves changing the basis of competition. Vitale observed that once a given technology is used to gain competitive advantage, use of that technology by all players in that industry may become obligatory for continued competitive viability. For example, American Airlines' success with its SABRE reservations system enabled the company to gain a substantial competitive advantage by allowing it to become the dominant service provider to travel agents. In order to stay competitive, other airlines were forced to develop their own technology (e.g., United Airlines) or to buy into someone else's system. However, an organization that introduces such a technology had better be prepared to maintain its investment. Otherwise, larger and better funded competitors will move in, adopt and improve the technology, and turn the pioneering company's initial IT success into a competitive failure. Vitale noted that if the other side has bigger and better guns, it might be best not to start a gunfight.

The second area of opportunity involves the creation and maintenance of entry barriers to keep out potential competitors. In industries where IT is essentially the price of entry to the competition, Vitale observed that the industry itself may become vulnerable to outsiders with underutilized IT resources. This situation might occur whenever IT is the major vehicle for producing, selling, distributing, or servicing a product. This problem did arise for a small but growing long-distance telephone company faced with considerable investment in additional capacity. After noting that several large national hotel chains had unused telecommunications capacity in their systems, the telephone company tried to lease the excess. The idea, although innovative and very cost-effective, was quickly dropped after one particularly large hotel chain began making noises about creating its own telephone company subsidiary.

The third area of opportunity addressed by Vitale entails increasing the costs of switching to a competitor. Using again the American Airlines example, its IT reservations systems' success (combined with United Airlines') has been so complete that competitors have filed lawsuits against and demanded federal investigation of both companies. Thus, strategic IT systems which are too successful may lead to unwanted legislative or regulatory intervention as well as potential legal liability.

The fourth area of opportunity lies in strengthening or altering the balance of power with customers and suppliers. The danger here lies in a company giving its customers or suppliers the tools or position to dictate terms or bypass that company entirely. Federal Express made a considerable investment to market facsimile services to its customers under the product name ZAP Mail. Customers have been quick the embrace facsimile technology, but discovered they did not require Federal Express as a middleman. Functionally, Federal Express helped create the market, then found itself shut out as customers dealt instead directly with facsimile suppliers. Conversely, companies with substantial IT investments who rely too heavily on a single vendor may find themselves unable to negotiate favorably or, worse yet, locked into old technology if the vendor fails to keep pace with the marketplace.

Vitale illustrated the fifth area of opportunity, new product development, by pointing out that IT can be used to develop new products that, in the end, do not generate revenues. A New York City bank developed a product that linked all of the databases supporting its various banking products. This allowed the bank to consolidate all of its customers' accounts into a single statement. As no other bank of similar size offered this product, it was considered a competitive advantage. However, customers were able to more efficiently manage their banking affairs by switching their funds into a more productive array of accounts which, while laudable, actually reduced bank fee income and increased the bank's interest expense. The bank quickly modified the system to discourage switching.

Vitale's work is limited by design to negative consequences of strategic information systems. Other researchers have noted similar negative consequences accruing to IT used to support company operations. Williams and Nelson (1990), for example, have found that oil companies have built technologically successful decision support systems that actually stifle decision-maker behavior. While some might argue that these systems are failures because their results are at cross-purposes with their design, the companies have retained and continue to develop the systems because they feel the benefits of being able to formalize and control decision making throughout the organization far outweigh the costs of any chilling effects on decision behavior. However, to counter a situation they see as unduly restrictive, decision-makers have begun to "cheat" on the decision support systems by manipulating the variables that they input into the model. Ultimately, chaos may result.

Negative consequences of successful IT systems are only now being identified and investigated. Nevertheless, companies deploying IT systems need to prepare to deal with all the ramifications of IT. Sometimes, solving one problem creates two more.

Conclusion

IT has been and will continue to be a boon to business, and nothing in this chapter should distract from its fabulous potential. However, IT does not come without a price. More than just measurable costs, the true price of IT must include the negative impacts and consequences of its deployment. How much risk is involved if the technology fails? And if it works, how risky is success? This chapter has not offered a rigorous program to assess and manage these risks. Instead, an initial step has been taken by identifying that the risks exist, the form they take, and the negative impacts and consequences that have resulted for organizations caught unprepared.

References

Alexander, Michael. (1988, November 28). "Downsizing Threatens MIS Influence." *Computer-World, 22* (48): 10.

Alexander, Michael. (1990a, February 5). "U.S. Recalls Virus- Laden Census Disks." *Computer-World, 24* (6): 1, 6.

Alexander, Michael. (1990b, May 7). "Two Hackers Found Guilty in Voice-Mail Crime Schemes." *ComputerWorld, 24* (19): 16.

Alexander, Michael. (1990c, May 7). "Agents Confiscate Fantasy Handbook." *ComputerWorld, 24* (19) 16.

Alexander, Michael. (1990d, May 14). "Secret Service Busts Alleged Crime Ring." *Computer-World, 24* (X): 128.

Alexander, Michael. (1990e, June 25). "Hacker Raids Stir Up Battle Over Constitutional Rights." *ComputerWorld, 24* (26): 1, 6.

"An Unvarnished View of IS Efficacy." (1989, November 13). *ComputerWorld, 23* (46): 87-91.

"Avast, Mateys! Prepare to be Boarded!" (1990). *CIO, 3* (8): 14.

Atkins, William. (1985). "Jesse James at the Terminal." *Harvard Business Review, 63* (4): 82-87.

Barton, Ron, and Bobst, Richard. (1988). "How to Manage the Risks of Technology." *Journal of Business Strategy, 9* (6): 4-7.

Bequai, August. (1983). *How to Prevent Computer Crime*. New York: John Wiley & Sons.

Betts, Mitch. (1989, July 24). "Antivirus Legislation Proposed." *ComputerWorld, 23* (30): 100.

Betts, Mitch. (1990, June 18). "Consumers Fear Threat to Privacy." *ComputerWorld, 24* (25): 4.

Bidgoli, Hossein, and Azarmsa, Reza. (1989). "Computer Security: New Managerial Concern for the 1980's and Beyond." *Journal of Systems Management, 40* (10): 21-27.

Bostrom, Robert P., and Heinen, J. Steven. (1977). "MIS Problems and Failures: A Socio-Technical Perspective." *MIS Quarterly, 1* (3): 17-31.

Bozman, Jean S. (1989, October 30). "Fault-y Tales: Mainframes Rock 'n Roll." *ComputerWorld, 23* (44): 6.

Brancheau, James C., and Wetherbe, James C. (1987). "Key Issues in Information Systems Management." *MIS Quarterly, 11* (1): 23-45.

Brodeur, Paul. (1990). *Currents of Death.* New York: Simon & Shuster.

Carroll, John M. (1977). *Computer Security.* Los Angeles: Security World Publishing Co.

Christensen, Steven R., and Schkade, Lawrence L. (1989, March 13). "Surveying the Aftermath." *ComputerWorld, 23* (7): 82-83.

Currid, Cheryl. (1990, March 26). "IS Reorganization May Be Alternative to Out-Sourcing." *PC Week, 7* (12): 111.

Daly, James. (1989, February 20). "No Excuse for Unpreparedness." *ComputerWorld, 23* (7): 25, 28.

"Dark Avenger Wreaks Havoc at Software Firm." (1989, December 25). *PC Week, 6* (52): 4.

Dearden, John. (1987). "The Withering Away of the IS Organization." *Sloan Management Review, 28* (4): 87-91.

Dickson, Gary W., Leitheiser, Robert L., Wetherbe, James C., and Nechis, Mal. (1984). "Key Information Systems Issues for the 1980's. *MIS Quarterly, 8* (3): 135-159.

Drucker, Peter F. (1989). *The New Realities.* New York: Harper & Row.

Edgeman, Rick L. (1983). *Optimal and Quasi-Optimal Sequences of Decisions.* Ann Arbor, MI: University Microfilms Intl.

Fisher, Royal P. (1984). *Information Systems Security.* Englewood Cliffs, NJ: Prentice-Hall, Inc.

Fitzgerald, Jerry. (1990). Business Data Communications, 3d. New York: John Wiley & Sons; citing (1988) "Voice Mail User Held Up by Hackers." *Network World, 5* (37): 1.

Fitzgerald, Mark. (1989, September 2). "Coming to Grips With On-line Libel." *Editor & Publisher, 122* (5): 34, 42. Editor & Publisher,

Foster, Lawrence W., and Flynn, David M. (1984). "Management Information Technology: Its Effects on Organizational Form and Function." *MIS Quarterly, 8* (4): 229-236.

Grant, R.A., Higgins, C.A., and Irving, R.H. (1988). "Computerized Performance Monitors: Are They Costing You Customers?" *Sloan Management Review, 29* (3): 39-45.

Hamilton, Rosemary. (1990, February 19). "Picking up the Bytes and Pieces After a Wind Storm." *ComputerWorld, 24* (8): 6.

Hammer, Carl. (1986, October 1). "Computer Security and the Human Interface." *Datamation, 32*

(19): Advertising Supplement.

Hammer, Carl. (1988). "Is Today's Office Receiving Full Value From its Computers?" *Information & Management. 15* (1988): 15-23.

Higgins, Steve. (1990, April 9). "Emergency cc:Mail Upgrade Combats Security Breach." *PC Week, 7* (14): 1, 6.

Janus, Susan. (1988, December 26). "Managers Face a Crisis in Systems Development." *PC Week, 5* (52): 1, 8.

Keefe, Patricia. (1988, April 6). "It Can't Happen Here...." *ComputerWorld, 22* (14A): 13-16.

Koenenn, Connie. (1990, May 23). "The Curse of the Cursor." *Albuquerque Journal*, p. B1, B3.

Kolodziej, Stan. (1988, April 6). "Beyond the Blue: Recovery Services Aren't Just for IBM Anymore." *ComputerWorld, 22* (14): 19-22.

Konsynski, Benn F. (1985). "Advances in Information Systems Design." *Journal of Management Information Systems, 1* (3): 5-32.

Langefors, Borje. (1978). "Discussion of MIS Problems and Failures: A Socio-Technical Perspective." *MIS Quarterly, 2* (2): 55-62.

Littman, Jonathan. (1990, June). "The Shockwave Rider." *PC Computing, 3* (6): 142-159.

Lucas, Henry C. (1975). *Why Information Systems Fail.* New York: Columbia University Press.

Ludlum, David A. (1988, February 1). "$80M MIS Disaster." *ComputerWorld, 22* (5): 1, 10.

Lyytinen, Kalle. (1988). "Expectation Failure Concept and Systems Analysts' View of Information Systems Failures: Results of an Exploratory Study." *Information and Management, 14* (1988): 45-56.

Markus, M. Lynn. (1983). "Power, Politics, and MIS Implementation." *Communications of the ACM, 26* (6): 430-444.

Markus, M. Lynn. (1984). *Systems in Organizations: Bugs & Features.* Boston: Pitman Publishing.

Mason, Richard O. (1986). "Four Ethical Issues of the Information Age." *MIS Quarterly, 10* (1): 5-12.

McAfee, John. (1989a, February 13). "Managing the Virus Threat" *ComputerWorld, 23* (6): 89-96.

McAfee, John. (1989b, February 15). "The Virus Cure." *Datamation, 35* (4): 29-40.

McFarlan, F.W. and McKenney, James L. (1983). *Corporate Information Systems Management.* New York: Richard D. Irwin.

Mehring, Joyce S. and Gutterman, Milton M. (1990). "Supply and Distribution Planning Support for Amoco (U.K.) Limited." *Interfaces*, Forthcoming Summer 1991.

Morris, Linda. (1989). "Useful Tool or 40-Pound Paperweight?" *Training & Development Journal, 43* (9): 80-84.

Murray, William. (1988, April 6). "How Much is Enough? Expert Says Security Efforts Should Pay, Not Cost." *ComputerWorld, 22* (14A) 30.

Myers, Glenford J. (1979). *The Art of Software Testing*. New York: Wiley-Interscience.

Nash, John F. (1989). *Accounting Information Systems*. New York: PWS-Kent Publishing Co.

Norman, Adrian R.D. (1983). *Computer Insecurity*. New York: Chapman and Hall.

Payne, Dinah and Raiborn, Cecily A. (1989). "Computer Viruses: What Accountants Should Know." *Accounting Today CPE Program*, (Lesson 89-6).

Perry, William E. (1981). *Computer Control and Security*. New York: Wiley-Interscience.

Porter, Michael E. (1980). *Competitive Strategy*. New York: The Free Press.

Price, Leon R., Cotner, John S., and Dickson, Warren L. (1989). "Computer Fraud in Commercial Banks: Management's Perception of Risk." *Journal of Systems Management, 40* (10): 28-33.

Rifken, Glenn, and Betts, Mitch. (1988, March 14). "Strategic Systems Plans Gone Awry." *ComputerWorld, 22* (11): 1, 104.

Rohde, Renate, and Haskette, Jim. (1990). "Disaster Recovery Planning for Academic Computing Centers." *Communications of the ACM, 33* (6): 652-657.

Schatz, William. (1989, September 1). "The Pentagon's Botched Mission." *Datamation, 35* (17): 22-26.

Scheier, Robert L. (1989, November 6). "Virus Victims Outline Basic Steps for Prevention." *PC Week, 6* (44): 103, 107.

Scheier, Robert L. (1990a, January 22). "Corporate Users Say Yes to Shareware." *PC Week, 7* (3): 120.

Scheier, Robert L. (1990b, January 29). "Lockheed Learns Costly Lesson in Systems Design." *PC Week, 7* (4): 1,6.

Scheier, Robert L. (1990c, March 26). "Stock-Quote Firm Cuts Costs by Cutting IS." *PC Week, 7* (12): 111, 114.

Scheier, Robert L. (1990d, June 11). "City Finds Fault with Outdated PC Strategy." *PC Week, 7* (23): 115-116.

Schwartz, Melvin. (1990). "Computer Security: Planning to Protect the Corporate Asset." *Journal of Business Strategy, 11* (1): 38-41.

Senn, James A. (1989). *Analysis and Design of Information Systems*. New York: McGraw Hill Publishing Company.

"Shareware: A Vital Part of the PC Industry." (1990, February 12). *PC Week, 7* (6): 12.

Sprague, Ralph H. (1987). "New Directions in Information Systems Management." *Information Technlogy, 1* (3): 31-38.

Strassman, Paul A. (1985, February 1). "The Real Cost of OA." *Datamation, 31* (3): 82-94.

Swanson, E. Burton. (1988). *Information System Implementation.* Homewood, IL: Richard D. Irwin.

Tate, Paul. (1988, April 15). "Risk: The Third Factor." *Datamation, 34* (8): 59-64.

Toigo, Jon W. (1989). *Disaster Recovery Planning*. Englewood Cliffs, NJ: Yourdon Press.

Turn, Rein. (1978). "Privacy Protection in Record-Keeping." *Information & Management, 1* (1978): 187-197.

Turner, Jon A. (1982). "Observations on the Use of Behavioral Models in Information Systems Research and Practice." *Information & Management, 5* (1982): 207-213.

Vitale, Michael R. (1986). "The Growing Risks of Information Systems Success." *MIS Quarterly, 10* (4): 327-334.

Waegemann, C. Peter. (1988, May 3). "Hard-Disk Archives Won't Guarantee Legal Validity of Business Documents." *PC Week, 5* (22): 45.

Warner, Timothy N. (1987). "Information Technology as a Competitive Burden." *Sloan Management Review, 29* (1): 55-61.

Watson, Hugh. (1990, June 25). "Avoiding Hidden EIS Pitfalls." *ComputerWorld, 24* (25): 87-91.

Weixel, Suzanne. (1989, March 13). "Survivors Know the Value of Preparedness." *ComputerWorld, 23* (11): 75-81.

Wetherbe, James C. (1988). *Systems Analysis and Design* (3rd). St. Paul, MN: West Publishing Company.

Wilkes, Maurice V. (1990). "Computer Security in the Business World." *Communications of the ACM, 33* (4): 399-401.

Williams, Joseph and Nelson, James A. (1990, March 15). "Striking Oil in Decision Support." *Datamation, 36* (12): 83-86.

Wysocki, Robert K., and Young, James. (1990). *Information Systems.* New York: John Wiley & Sons.

Part II

Managing Information Technology, Models and Databases

Chapter 4

Model Management:
Concepts, Challenges, and Opportunities

Clyde W. Holsapple
University of Kentucky

Linda Ellis Johnson
Western Kentucky University

Ramakrishnan Pakath
University of Kentucky

The decade of the '90s may be regarded as heralding the beginning of a "modeling revolution" in many organizations. To a substantial extent, this revolution is being facilitated by a proliferation in the availability of increasingly powerful hardware and modeling software to organizational participants. These tools enable the construction, maintenance, and solution of sophisticated models of complex decision problems.

When properly exploited, such modeling facilities could positively impact organizational decision tasks by enabling participants to make more effective decisions. Further, the astounding speeds of modern computing devices, coupled with advances in the computational sophistication of model solvers, could render the decision making process more efficient as well.

While these factors indicate the existence of a wealth of decision modeling and analysis opportunities to organizations, numerous issues still tend to inhibit the complete exploitation of this wealth. In addition, organizations may find it necessary to control modeling activities of organizational participants as well. The focus of this chapter is on examining existing technological opportunities, highlighting inhibiting issues that need resolution, and drawing attention to emerging concepts and technology vis-a-vis model management for organizational decision support and problem solving. Before doing so, we present a brief review of historic milestones in organizational computing that forms the basis for our subsequent discussions.

During the 1960s and 1970s, information technology revolutionized business computing and managerial decision making by introducing data base management systems (DBMSs) to the business world. The impact of this revolution could be measured, in terms of the significant shift in emphasis from traditional, file-oriented, electronic data processing (EDP) systems to management information systems (MISs). During this period, the receipt of timely information was viewed as the key to good decision making as opposed to efficient transaction processing. The role of the MIS was to generate descriptive reports based largely on transactional data contained in the organizational data base. The tasks of building, maintaining, and administering MISs were entrusted to the EDP staffs of organizations.

It soon became apparent, however, that good decisions are not merely a result of timely information alone. Managers, the end-users of MISs, were often dissatisfied with the contents and structures of the generated reports. Further, these individuals were highly dependent on technical EDP staff for ad hoc information needs. Consequently, MISs soon came to be regarded as rigid, monolithic, inflexible report generators with limited decision support capabilities.

Subsequent attention focused on systems that could overcome these drawbacks. The new class of support systems were called decision support systems (DSSs) (Sprague and Watson, 1975 and Bonczek, Holsapple, and Whinston, 1981). Cited as being paramount among its numerous features is the ability of a DSS to facilitate the creation, maintenance, and manipulation of decision models (in addition to traditional data manipulation) to facilitate effective decision making. The term "model management" is a collective reference to all such model-related activities.

Academic research in the development and use of models first appeared in the early 1970s (Gershefski, 1970; Naylor and Schauland, 1976; Little, 1970; Will, 1975). Several early studies noted that existing modeling knowledge was virtually not being utilized by a significant proportion of practicing managers due to real or perceived difficulties in model formulation, solution, output analysis, and interpretation. Early model management research focused on resolving these issues on an ad hoc basis, without regard to developing comprehensive software systems for managing "banks" of models. Recent research in model management arose as a component of research conducted in the area of DSSs. Because the goals of the model management systems (MMSs) approach to managing models is essentially similar to that of the DBMS approach for data management, it is not surprising that development of the recent model management literature itself closely parallels that of data management. In addition, model management research certainly has implications for DSSs developers.

Broadly, research in model management is concerned primarily with two issues, model representation and model processing. To better understand these issues it will be necessary to first (i) define what exactly we mean by the term "model", and (ii) examine MMSs in the context of a generic framework for DSSs. These topics are covered in the next section of this chapter. Then, *A Perspective on Model Management Technology* contains a discussion on the capabilities of commercial model management technology and identifies where this technology falls short of performance expectations. This is followed by the section, *Research in Model Management*, where we present an analysis of model management theory and present a representative review of theoretical research. Viewed in conjunction with the previous section, this examination highlights both the existing chasm and the emerging bridge between theory and practice. Finally, we focus on future trends in model management research and technology and their potential impact on organizational decision support.

What is Model Management?

Definition of Concepts

Table 1 provides a list of some "definitions" of the term "model" found in the model management literature. It is not readily apparent what each of these definitions means. What is apparent is that different authors use the term "model" to mean different things. It is, therefore, useful to identify a common basis that underlies different usages of the term.

Author/Reference	Definition
Elam, Henderson and Miller [15:98]	A common characteristic of all decision makers is the use of a "model" as a basis to gather data, analyze this data and eventually make a choice. These models may be intuitive or externalized.
Konsynski and Dolk [26:187, 27:131]	Models are representations we find useful in our decision making processes.
Blanning [29:141]	A model is viewed as a (virtual) relation with input and output attributes, just as a file in relational data management is viewed as a relation with key and content attributes.

Table 1. Definition of a Model

Broadly speaking, a model is an abstract representation of reality. The extent of abstraction can vary, resulting in a continuum that ranges from less abstract models to more abstract ones. This continuum may be further subdivided into regions corresponding to various classes of models, based largely on the computer representation schemes utilized (see, e.g. Sprague and Carlson, 1982).

Sprague and Carlson (1982) identify three distinct model representation schemes that translate, in effect, into three different interpretations for the term model. Thus, a model may be viewed as a subroutine or procedure, as a set of data, or as a set of modeling statements. In a recent survey of MMS/DSS literature, Holsapple and Whinston (1988) show that different usages of the term essentially reduce to one of the three interpretations suggested by Sprague and Carlson. We examine each of these interpretations in greater detail below.

First, a model may be an executable procedure. This is the most common interpretation and derives from the traditional approach of storing algorithmic procedures as computer program subroutines. One manifestation of this interpretation occurs in the context of executing a particular mathematical programming procedure, such as the simplex method for the solution of linear programs.

However, the interpretation is broad enough to allow us to consider procedural "modules" (i.e., relatively simple models) which can be linked together to build a larger procedural model (Hwang, 1985). Thus, individual formulas in a set can be considered as procedures. When considered collectively, the set of formulas constitute a relatively more complex procedure, such as a "net income model". As another example, Katz and Miller (1977) construct an interactive model system which "provided guidance to decision makers on the relative benefits and costs of alternative mitigation and recovery policies for natural hazards." This system consists of several models (stored as subroutines) that calculate values for different attributes of interest.

A second interpretation of the term model allows a model to be represented as a set of input data. Thus, in the context of mathematical programs (for example, a linear program), a common approach is to structure the input values as data matrices (for example, in mathematical programming system [MPS] format) that are then fed to solvers. Here, the solvers are mathematical programming procedures.

In essence, the "models as procedures" approach presupposes the existence of structured input data whereas the "models as data" approach assumes the existence of suitable algorithmic procedures. With such assumptions, model management would appear to emphasize either "procedure management" or "data management," one at the expense of the other. However, regardless of such differences in emphasis, it is clear that a computational "procedure" is always executed with the aid of the requisite input "data". An interpretation that permits the emphasis of either data or procedure management or both, as necessary, would constitute a more comprehensive view of models and of model management. Researchers who view a model as a statement of a problem subscribe to this comprehensive view.

In the "models as statements" approach, the focus is on specifying a problem as a set of statements called "modeling statements". Konsynski (1982) refers to the set of statements as a set of "equations" that must be solved. For instance, a linear programming procedure of a problem uses a set of equations that describe the objective and the various constraints of the problem. The extent of procedurality of these statements could vary from highly procedural to highly non-procedural. This latitude extends to the specifications of both the necessary computations as well as the appropriate input data. Researchers in this area seek to develop languages that facilitate the creation of such modeling statements.

As shown in Figure 1, a discussion of the representation and processing of models can involve each of the three interpretations discussed above. Each of these approaches places a slightly different perspective on the study of model management.

Along with these viewpoints, it would also be beneficial to have a setting in which models exist. Two possible model settings relevant to this work are presented in Figure 2. Models may be examined from the perspective of an organizational setting or a business computing setting.

That is, we could discuss models and their management in the organizational setting of managerial functions (i.e., planning, organizing, controlling), managerial roles (i.e., interpersonal, decisional, and informational), organizational level (i.e., strategic, tactical, and operational), and organizational functions (e.g., marketing, finance, production). Holsapple and Whinston (in press) provide an introduction to the differences between these

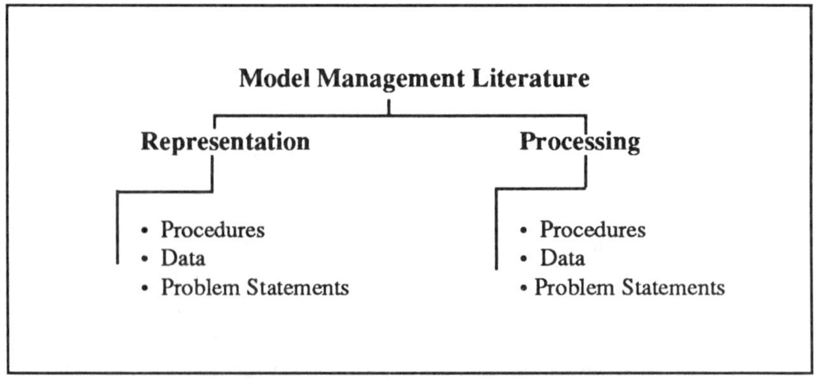

Figure 1. The Model Management Focus

- **Organizational**
  - Managerial Functions
  - Managerial Roles
  - Organizational Level
  - Organization Function
- **Business Computing**
  - Electronic Data Processing
  - Management Information Systems
  - Decision Support Systems

Figure 2. Model Settings

settings and provide a basis for thinking about models in each of these settings.

For example, it is certainly possible to consider the management of models within the realm of organizational functions such as marketing, finance, and accounting and how each of these functions might deal with managing procedures, data, and/or problem statements. Along these lines, Applegate, Konsynski, and Nunamaker (1986) have utilized their own framework to design a model management system for supporting the planning function in an organization.

On the other hand, "business computing" is the generic label placed on the field of study encompassing MISs, DSSs, and expert systems (ESs). Procedural models are pervasive in both MISs and DSSs, and certainly conceivable in ESs. The management of data and problem statements are also within the realm of discussion of each of these types of systems.

In this chapter, we examine model management largely from the more comprehensive models-as-statements perspective and from within a business computing setting. We do this only because business computing pervades virtually all of the organizational settings discussed above. While this is a limitation, exhaustive consideration of model management from the perspective of all of the organizational settings is beyond the scope of this chapter.

Model Management and Decision Support Systems

Within the context of business computing, we focus largely on DSSs in order to concisely define the role of models in this setting. We believe that there is no loss of generality with this approach, as we view MISs and ESs as very special cases of the class of systems called DSSs.

DSSs are computerized systems which support a decision maker in any decision making task. We will utilize the generic framework developed

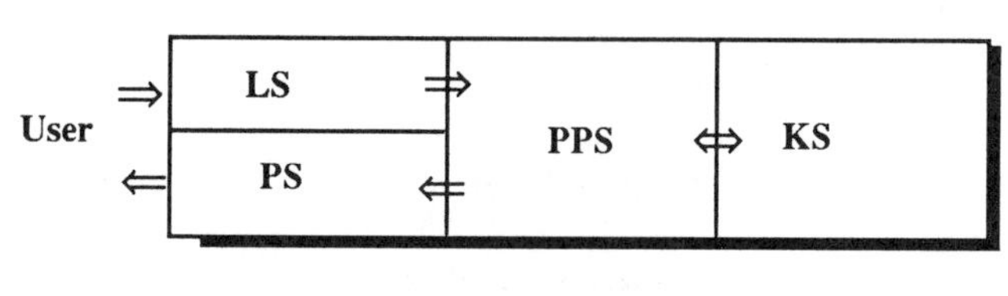

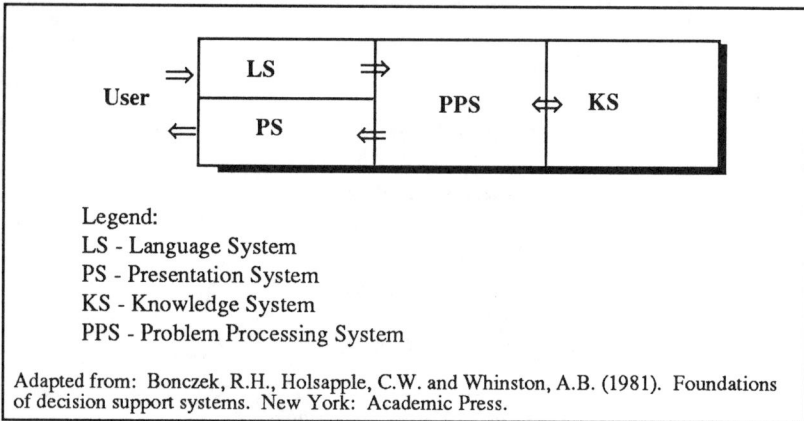

Legend:
LS - Language System
PS - Presentation System
KS - Knowledge System
PPS - Problem Processing System

Adapted from: Bonczek, R.H., Holsapple, C.W. and Whinston, A.B. (1981). Foundations of decision support systems. New York: Academic Press.

Figure 3. Generic Framework for DSS

by Bonczek, Holsapple, and Whinston to present an overview of the model management literature as relates to each of the components of their framework. The generic framework contains four components: a language system (LS), a presentation system (PS), a knowledge system (KS), and a problem processing system (PPS). For a comprehensive explanation of this framework consult Bonczek, Holsapple, Whinston, 1981; Holsapple and Whinston, in press. This framework is shown in Figure 3.

The LS, PS, and KS are representation systems, while the PPS is a software system. The KS can contain many different types of knowledge. Six types of knowledge have been identified. Three primary types of knowledge contained in the KS are descriptive knowledge (i. e. data), procedural knowledge (i.e. models), and reasoning knowledge (i. e. rules). Other frameworks (Sprague and Carlson, 1982) developed for DSSs can be viewed as special cases of the generic framework. An indepth discussion of this can be found in Holsapple and Whinston, 1988.

Frameworks for DSS development sought to incorporate the interaction of data and procedures in supporting a decision maker's tasks. Recognizing that managers often deal with non-routine, non-repetitive, ill-structured problems and decisions, a DSS can be distinguished from MISs which have as their goal supporting routine repetitive structured decisions and problems. It is in support of these ill-structured decisions and problems that procedures and data could interact to lend decision support.

The key to understanding the nature of this support is the distinction between structured decisions and structured problems and "unstructured" decisions and "unstructured" problems. That distinction is — that in order for a DSS to aid managers in making "unstructured" decisions or solving "unstructured" problems, structure must first be found for some components of the task.

That is, either the DSS or DSS user must seek structure for some component(s) of the problem or decision. Marsden and Pingry (in press) have identified five types of unstructuredness relating to the development of mathematical programming procedures. These are fundamental unstructure (white noise), representation unstructure, information unstructure, stochastic unstructure and complexity unstructure. Fundamental unstructure is defined as a problem for which no structure can be found. The remainder of their categorization could result in a decision or problem becoming structured if the appropriate type of knowledge can be obtained. That is, the manager (or possibly the DSS) must first subdivide an "unstructured" decision or problem into several structured components which can be solved. Both the task of structuring and then the resulting interaction is what model management researchers are interested in studying. In particular, researchers are interested in building software systems to support these tasks.

With the generic framework in mind, and a recognition of the differences between structure and unstructured as characteristics of a decision or problem, we will now proceed with a survey of the model management literature as it relates to each of the components of a DSS. Both Holsapple and Whinston (1988) and Blanning (1989) will provide supplementary survey literature which we will not review within this chapter. Rather, it is the intent of this review to clearly understand the role of model management within a DSS context and provide an overview of recent literature not yet surveyed.

The generic framework defines a LS as the set of all possible messages that can be given by the user to the DSS. The PS is the set of all possible responses the DSS can make to the user. In this context, Binbasioglu and Jarke (1986) questioned how general a DSS tool should be in order to support the model formulation task. They propose that the KS should contain both syntactic knowledge and semantic knowledge which is domain specific. In this light, they question how general a PPS should be in order to support the model formulation task. Within this line of questioning, the topic of automatic model building arises. Hwang (1985) has also addressed the issue of automatic model building by presenting a survey of research work, primarily in artificial intelligence, along these lines. He views the task of model building as being twofold. First, how can a decision model be formulated? Secondly, how should procedures be selected for a model base? Liang and Jones (1988) have addressed the issue of automatic modeling as well. They suggest a graph-based framework for model integration.

Brennan and Elam (1986) have identified a need to construct a "model" of the problem situation. They suggest that model management researchers have been preoccupied with "the solution component of modeling systems to the neglect of the model definition or generation component and the report generator or analysis component". These observations fit into the language system and the presentation system of the generic framework.

Brown (1987) has proposed the development of an interface for stochastic problems where data can be incomplete, uncertain, and subjective. He develops a relative entropy model management system in order to handle the data problems existing in stochastic environments. In this system, he sought to incorporate both subjective and objective data where inconsistent information could be accepted by the system from the user.

The PPS can be defined as the software system which acts on the contents of the LS, PS, and KS. Liang (1985, 1988) proposes a model management system consisting of a model utilization subsystem, a modeling subsystem, and an inference engine, where the inference engine is driven by a graph-based framework. Jarke and Radermacher (1988) suggest that intelligence is the ability to model reality and draw conclusions. In order to facilitate the ability of the model management system to model, they propose that there must be an "abstract representation of models (we will call this the logical view) and there must be one or more modeling paradigms." Specifically, they discuss the architecture for a "normative model management" system. This system contains an object-oriented DBMS kernel; an integrated data, algorithm and model administration system; and a knowledge-based system for manipulation of model bases.

d'Alessandro, Dallas Mora, and DeSantis (1989) focus on the design and architecture of the PPS when they write:

> "To begin with, the user should be assumed to have concrete but unstructured problems. This means that the system and not the user is in charge of structuring and formalizing the problem in question."

The knowledge system is defined as all knowledge which can be drawn upon by the PPS in responding to a user's request. Dolk (1986) recognizes that there is a duality between data and models and that perhaps there has been a preoccupation by information systems professionals and operations research/management science professionals to treat the two separately. He proposes a broad perspective of model management to include what he terms "data-as- models".

While we have viewed model management in this subsection primarily in the context of DSSs, other means of thinking about models and how models might be managed are possible. In particular, a logical view of model management which has received attention in the literature is the work on relational model management. A logical view can be defined as the logical structure which exists between models and the natures of their interrelationships.

A Perspective on Model Management Technology

A Framework for Assessing Current Technology

Thus far we have focused on theoretical concepts underlying model management for decision support and problem solving. We next examine a descriptive taxonomy developed by Bonczek, Holsapple and Whinston (1979, 1980) that enables us to evaluate representative MMS "tools" and MMS "generators" with regard to their capabilities and limitations. The basis for the taxonomy is a more detailed consideration of the "models as statements" paradigm, with the emphasis on languages that facilitate such modeling statements.

Decision makers utilize a specific MMS to perform numerical computations that result in the generation of desired output data. Models themselves require data pertaining to input parameter values for successful execution. Bonczek, Holsapple and Whinston (1979) observe that, in essence, two classes of languages are involved, referred to as "Languages to Direct Retrieval" (LDR) and "Languages to Direct Computation" (LDC). In essence, model specification and execution entails the use of some member of the class LDC as well as a member of the class LDR.

Each of these language classes may, in turn, be divided into three broad subclasses based on the degree of procedurality involved. Thus, the class LDC consists of languages where,

> (1) The necessary computational steps are explicitly specified. The user acts as model builder and encoder.
>
> (2) A particular model (that performs the necessary computations) is invoked by specifying only its name. Several such "specific" models are predefined and stored in the MMS, for use with various applications. A language processor translates the invocation request into an explicit computational procedure. No modeling knowledge on the part of the user is required.
>
> (3) Statements concerning what information is required are provided to the MMS. The system takes over the tasks of model logic formulation, program code generation, and execution. The user is not concerned with knowledge about model construction or model invocation.

Correspondingly, the class LDR consists of language subclasses that facilitate input data retrieval via any of the following approaches:

> (a) The data retrieval procedure is explicitly specified. The user must have knowledge of physical or logical data organization.
>
> (b) An appropriate report generator (that retrieves the required data) is invoked by specifying its name. Several such "specific" report generators, pertaining to various applications, are predefined and stored in the MMS. A language processor translates the invocation request into an explicit data retrieval procedure. It is the language processor and not the user, that is concerned with data organization.
>
> (c) Statements concerning what data are required are provided to the MMS. The MMS performs the tasks of retrieval logic formulation, program code generation, and execution. The user is not concerned with knowledge about data organization or knowledge about report invocation.

With reference to the subclasses in LDC and LDR, observe that we have not explicitly specified the entity "using" the language. This entity could be a human decision maker or yet another model. The latter is the case when one model depends on the outputs of another model for its own input data and consequently, assumes the role of a model user. Our concern, however, is only with a human user of the MMS and this is our implication when we use this label. However, recently there has been a trend in MMS research to incorporate AI techniques into such systems. It is entirely conceivable that the '90s will see the emergence of systems where models themselves behave as versatile users of the aforementioned languages for computation and retrieval.

In essence, there are nine different language combinations that facilitate model representation and processing. For ease of exposition, we label the resultant categories of languages as A, B, ..., I, corresponding to the combinations <1, a>, <1, b>, ..., <3, c>, respectively, considered in increasing lexicographic order.

Some Representative Tools and Generators

In this subsection, we utilize the above taxonomic scheme as a framework for examining how representative MMS tools and generators "fit" the categories of the scheme in terms of their computational and retrieval language capabilities. In doing so, we focus only on well known, commercially available products. Discussions on experimental tools and generators, that are still in the research and development stage and not as yet freely available in the

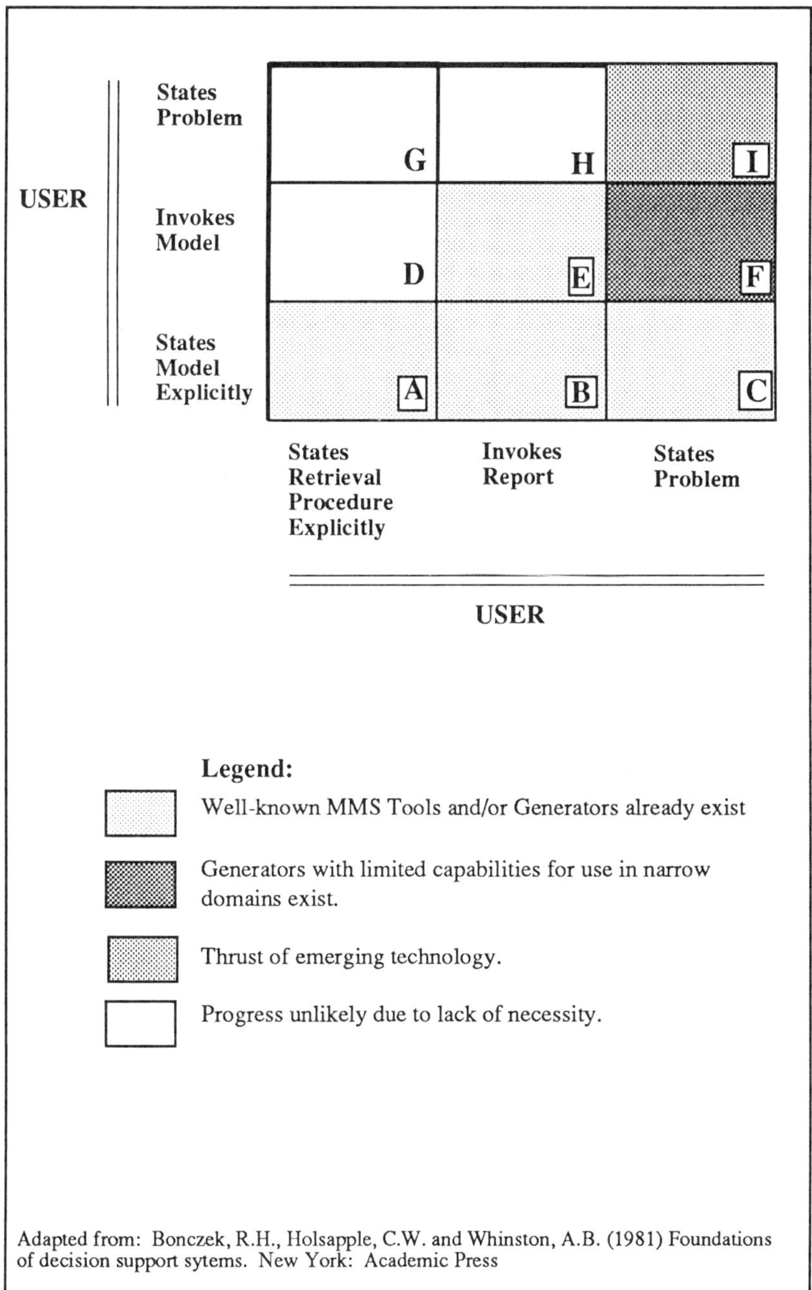

Figure 4. Classification Scheme for Viewing Representative MMSs

marketplace, are deferred to a subsequent section. Often an "exact" fit is impossible as some products span several categories. In such cases, we suffice by providing a "best" fit approximation.

Categories A, B, and C. MMSs that belong to categories A, B, and C require the explicit specification of computational procedures but differ in the extent of procedurality of data retrieval specifications. From the perspective of directing computations, essentially any MMS that a user constructs from scratch, using a commercially available programming language (i.e., an MMS software tool) or the language provided by an appropriate software package (i.e., an MMS generator) fits this category. Traditional programming languages include BASIC, FORTRAN, COBOL, PL/I, Pascal, C, and the like as well as machine and assembler languages. Apart from such general-purpose languages, specialized languages are also widely available. For example, in the context of simulation modeling we have languages such as SLAM, GPSS, and SIMSCRIPT. Likewise, in the context of expert support systems, the user may direct computations using rule-oriented languages such as OPS-5 and Prolog. As examples of MMS generators in these classes, we have spreadsheet software such as Lotus 1-2-3 or VisiCalc and rule-based software such as GURU. Each of these generators provides the user with a specific language for directing computations.

Many of these languages also allow user-directed access to subroutine libraries for invoking frequently used procedural modules. For example, we have the IMSL library for accessing mathematical and statistical FORTRAN subroutines, the C Standard Library for accessing precompiled C functions, and user-defined macros in the context of spreadsheets. Observe that, although a computational program may almost entirely be made up of such subroutine or function invocations, the approach is still procedural as the user must decide (and encode) what subroutines are invoked and the sequence in which they must be invoked.

With regard to data retrieval, virtually all of the aforementioned MMS tools and generators possess language constructs that permit data retrieval through traditional file manipulation techniques. This is one approach to handling data retrieval for category A systems. Alternately, in some contexts, the data may be extracted from an integrated data base with relatively less effort. In this event, the user must utilize the data manipulation language (DML) of the associated data base management system (DBMS) along with the host language (i.e., the language used for directing computations). That is, DML commands are invoked as needed, via explicit host language calls on data manipulation procedures provided by a DBMS. For example, in IBM's hierarchic DBMS product IMS, a user can incorporate, as part of applications written in COBOL, PL/I, and System/370 Assembler Language, calls to data retrieval procedures in its data manipulation language, DL/I.

For a category B MMS, data retrieval is via report invocation as discussed in the preceding subsection. Thus, when specifying the inputs for a specific (user-defined) computational procedure, the user merely invokes the appropriate generators from within the host language computational program. It is possible that the generator permits conditional retrieval, where the conditions are passed on as arguments to the report generation subroutine. Thus, the same generator could generate different subsets of data. This is the only distinction between category A and category B systems.

This approach presumes complete knowledge about the anticipated data-retrieval needs of users. In general, the method is feasible only as long as the number of (predefined) generators required are few and these requirements remain relatively stable over time. That is, (i) all existing computational procedures within the system derive their inputs from some subset of the set of preprogrammed generators available and (ii) few additional subroutines need be created as the user enhances the computational capabilities of the system over time. Thus, despite their apparent convenience, category B systems are clearly rigid and inflexible in relation to systems in category A.

The data retrieval approach in category C systems seeks to capture the best of the desirable features of the approaches in category A and B systems. Here, a user merely states what data is desired. There is no dependence on preprogrammed subroutines and neither is there a need for explicit user effort in encoding retrieval procedures. This is only possible if the system allows for a declarative interface with the user. For example, query language commands for data retrieval may be embedded within a host language program that directs computations. A precompiler provided with the DBMS translates the query language commands into host language calls to DML procedures. Thus, the INGRES relational DBMS allows one to use EQUEL (Embedded Query Language) statements within FORTRAN, BASIC, COBOL, Pascal, and C language applications.

Many simulation packages (e.g. SLAM) fall in category C. The user must define the simulation procedure explicitly, but input data is typically generated by the system once the user specifies the kinds of data (typically random numbers or random variates) required (e.g. the arrival process is Poisson distributed with known mean). As a very sophisticated case of this class, the interaction could also occur via a natural language. For instance, the Knowledgeman 2 software package provides users with K-Chat, a natural language interface.

In general, particularly with the types of MMS development tools described, a technically skilled decision maker could create MMSs that fit well into any of the categories, A through C (perhaps with the assistance of technical staff personnel).

Categories D, E, and F. Tools and generators belonging to categories

D, E, and F, differ from their counterparts in categories A through C in that computations are carried out through model invocations only. Model invocation in these contexts is very much like report invocation for category B products. The MMS maintains a set of such invocable, preprogrammed models based on anticipated user needs.

Ideally, the set of models would constitute what is called a "model bank". A model bank contains both relatively simple, general purpose procedures (e.g., random number generation, matrix manipulation, and so on) as well as more complex procedures (e.g., regression analysis, inventory control, linear programming, etc.) that may be used on a standalone basis or may be integrated as desired.

In practice, however, it is often the case that commercially available generators are highly biased towards certain classes of models. For instance, software packages such as MPSX, IMSL, and the more recent OSL, for use in mainframe and supercomputing environments, facilitate the invocation of the closely-knit family of linear, integer and nonlinear programming procedures. Some microcomputer products like GAMS and LINDO also exhibit this bias. However, the situation is somewhat better from a microcomputing perspective where more general products like STORM and QSB+ are available that seek to accommodate a larger variety of mathematical programming and optimization procedures as well as forecasting, regression and statistical analysis routines. Regrettably, comprehensive micro-based products are also extremely limited in terms of the sizes of the problems they can handle. In both mainframe and microcomputing environments, the products themselves do little to facilitate model integration.

Model integration is only possible through an approach called "chaining" whereby the outputs of one model are readily accessible, if necessary, by another model in the system. The products just discussed treat each model in isolation, having its own input requirements and output capabilities. In general, most micro-based products facilitate inputs through special languages that permit both formatted and format-free interactive data entry and editing. Most mainframe products are fairly inflexible in this regard. In both environments, input data is typically stored in isolated data files. Likewise, considerable flexibility in the structure and content of output reports is also typically provided in a microcomputing environment (while this is generally not true with mainframe products). Unfortunately, this flexibility is tuned more toward user satisfaction than toward facilitating model integration.

In summation, once the necessary input data files have been created, a user can typically invoke a particular model and have the procedure execute for a particular set of input data (i.e., named input file). From the perspective of our taxonomic scheme, the commercial products of the kind just discussed would largely fall in category E. As with conditional data retrieval for category

B products, a limited kind of conditional model execution is also possible by simply specifying alternate input files for the same model.

To the best of our knowledge, there are no commercial products that lie in category D, where preprogrammed models are invoked for execution and these require the aid of user-specified, prescriptive input data retrieval procedures. However, a user may choose to create his/her own prescriptive procedures for a category E product to improve upon it (for instance, to integrate a set of stand-alone models) and, thus, create a category D product.

Likewise, we are not aware of any well-known category F products where data retrieval is via descriptive user-defined procedures. However, as with category D products, it is conceivable that a user and/or technical staff could move a product from category E to category F (for instance, by building a data base query language interface for the product). Some commercial products, defined for very narrow application domains, may be regarded as MMS generators belonging to category F. For example, SIMFACTORY and COMNET II.5 are two such packages that allow one to simulate respectively, factory and telecommunications network operations. In SIMFACTORY, for instance, once a factory's layout and operations have been described to the system (in a fairly non-procedural way), the user merely invokes the resultant model by name and can execute it with a variety of input data by only specifying the type of data (i.e., random numbers or variates) required. The drawback with such products is their narrow domains and the consequent lack of a comprehensive set of models in the system.

Categories G, H, and I. In terms of directing computations, products in categories G through I are an "improvement" on the preceding categories in that a user merely specifies what output information is required. Thereafter the system takes over the tasks of problem recognition, algorithm (model) construction, encoding and solution. As in the earlier cases, the difference between category G, H, and I products lies in the ways in which data retrieval is accomplished.

We are unaware of any well-known products currently available in the market that lie in these categories. However, as mentioned in subsection A, we believe that in the '90s and beyond, organizations will begin to taste the fruits of the application of AI research to such model management problems. We note that any conceivable system with the intelligence to design and execute its own computational procedure will also exhibit similar competence with regard to data retrieval (such capabilities already exist, as discussed in the cases of category C and F products). As such, in our opinion, categories G and H are essentially superfluous and emerging products will largely belong to category I.

Research in Model Management

The Status of Current Research

By way of introduction, we stated that the research in model management revolved around the representation and processing of models. We have discussed that there are three very different definitions for the term model. In the previous section, we examined current products available. In this section, we will present a sampling of the theoretical research being conducted in model management. It is important to present this research because it may well result in new commercial products available to users of MMSs. In particular, we will examine the development of logical views for representing data, procedures, and problem statements.

These logical views have their origins in the DBMS literature, where five data models are prevalent. These include: the CODASYL, hierarchical, shallow network, relational, and post-relational data models.

Stohr and Tanniru (1980) utilized the CODASYL data model to incorporate the handling of data input necessary for procedure execution. Another logical view called the relational model management view has sought to draw on the well-established field of relational data base management in developing a means of viewing models.

The goal of relational model management is to provide a framework in which a software system, specifically, a relational model management system, could be developed. In order to draw an analogy from relational database concepts, it is necessary to discuss the organization of the model bank (a collection of procedures/models), the relational completeness of model query languages, and implementation. Blanning (1987, 1985, 1984, 1983, 1982, 1981) has developed a detailed discussion of each of these three issues which we will briefly describe.

To begin a discussion of a logical view of representing procedures, consider the goal of utilizing a procedure. Typically, it involves inputs which are processed by the procedure and result in an output. Relational model management, as defined by Blanning, proposes several definitions necessary for our discussion.

> A model is viewed as a "(virtual) relation with input and output attributes, just as a file in relational data management is viewed as a relation with key and content attributes". (Blanning, November, 1983) Note: This can be written as M=<I,O> where M is the model and I and O are input and output attributes respectively.

> A model bank may be viewed "as a set of (virtual) relations with input

and output attributes and functional dependencies between them, just as a database may be viewed as a set of relations with key and content attributes". (Blanning, June, 1983)

A model set is "a collection of models (which may consist of only one model) used to respond to a specific user query". [Similarly, a model set is a subset of the model bank. (Blanning, June, 1983)

A model management system "is a software system that insulates its users from the physical aspects of model bank organization and processing, just as a database management system insulates its users from the physical aspects of database organization and processing". (Blanning, June, 1983)

In order to discuss how a model bank should be organized, it would be beneficial to think about what, if any, processing problems could arise from the way in which models are organized within the bank. Database management has identified anomalies which can occur as a result of the way in which data is added, deleted, or changed. The recognition of these update anomalies (sometimes called storage anomalies) in database management resulted in the identification of normal forms which attempt to reduce or eliminate these anomalies. Because the inputs and outputs to models are dynamic, storage anomalies do not exist in model management. However, Blanning (June, 1983) has identified three processing anomalies which arise as a result of model processing and need to be addressed with respect to model bank organization. He calls these input, search, and output anomalies. The identification of these processing anomalies resulted in a development of normal forms (alpha, beta, and gamma respectively) for relational model management.

Relational completeness of database languages is based upon the five primitive relational algebra operations of union, difference, product, selection, projection and three derived operations of join, intersection, and division. In order to draw the analogy from relational database languages, it is necessary to develop a relational algebra for model management and then show that it is relationally complete with respect to the calculus. It has been shown that the first-order predicate calculus is relationally complete with respect to relational algebra (Blanning, 1984).

Blanning (1984) has identified the three algebraic operations for relational model management as execution, optimization, and sensitivity analysis, which are relationally complete with respect to first order predicate calculus. Similar to data management, languages for model management can be categorized into two components: a model description language (MDL) and a model specification set language (MSSL).

A third issue in the design of relational model management systems revolves around implementation. Implementation as it relates to database management includes topics such as security, integrity, query optimization, and the implementation of joins. With the exception of the implementation of joins, very little research work has been conducted in model management analogous to security, integrity, and query optimization. It has been shown that the join problem does not exist in relational model management because functional dependencies are causal, not assigned dependencies as previously discussed above. (Blanning, 1985) The implementation of joins in relational model management are transparent to the user because only one procedure can produce an output which becomes input to another procedure in a model set. That is, while several different procedures might be utilized to calculate a specific output, only one can be selected as input to another procedure. For example, there may be several different ways to calculate net income; but if net income is input to a finance model, only one number can be selected.

Query optimization in model management differs from database management in that the latter uses a single criteria (usually time minimization) approach for query optimization. Model management would most likely utilize multiple criteria in selecting a path through the model bank to arrive at a response. Multiple criteria query optimization for model management could include criteria such as time, cost, and reliability (Johnson and Pakath, 1990). Essentially, this view proposes that more than one model set can respond to same user query and the characteristics of that model set differ not only based on the specification set utilized, but also by the criteria for query optimization. The question facing the user of the model bank is which model set should be selected given either known parameters or stochastic parameters for the criteria.

While the theory of relational model management has not yet resulted in a MMS implementation, it provides a means of representing models which is consistent with a widely utilized data model. Other logical views for representing models include the entity relationship approach proposed by Blanning (1986) and the "data-as-models" approach proposed by Dolk (1986). Current research continues to focus on the need to integrate both data input and models to support decision making in the context of DSSs.

In addition, research in the area of artificial intelligence may result in additional commercial products for use by organizational participants. It is to these possible future developments that we now turn our attention.

Future Research Trends and Emerging Technology

Research in model management has historically followed three streams. As we discussed in the first section of this chapter, research in model management was initiated by MS/OR academicians who were concerned about the disuse of mathematical programming procedures developed to assist managers in problem solving and decision making. With the development of DSSs, DSS researchers saw MMSs as a potential tool to be utilized as a PPS or as part of the PPS. In fact, the main content of this chapter has revolved around MMSs in the context of DSSs. However, there is a third stream of research which is providing fuel to the development of MMSs, both now and in the future, and that field of research is artificial intelligence.

There are two primary areas of research in artificial intelligence that may result in emerging future technology with relevance to MMS. These include the development of representation schemes and processing techniques. One such technique that is currently receiving substantial research attention is object-oriented model representation and processing. Object-orientation enables one to take an aggregrate view of the contents of a model bank. This is in contrast with the traditional algorithmic and programming approaches of viewing each model and its inputs and outputs in isolation. The object-orientation paradigm encourages the user to view an interrelated set of models and data as a single unit for conceptualization and processing ease. This approach will provide organizational participants with the ability to construct and manipulate complex, aggregate models of a realistic decision problem. An example of a MMSs tool based on the object-oriented paradigm is the Simula (simualation) language and Small Talk, a traditional programming language.

Another stream of emerging research is the integration of symbolic data manipulation with traditional means of data manipulation. Mathematica is an example of a symbolic manipulation software package that permits non-numeric manipulation of calculus expressions. Such a tool has definite uses at the operational control level in many engineering decision making contexts.

Conclusions

Models can provide organizations with valuable resources which have a need to be both exploited and controlled. Similar to the problems that organizations encountered several decades ago in managing data, managing models also presents a challenge and an opportunity to both researchers and the organization. The proliferation and widespread use of software packages which allow for easy procedure formulation and execution (such as spreadsheets and statistical packages) present organizations with a need to address this issue.

This chapter has presented a framework for viewing existing technology available to organizations and a discussion of research work done toward this end. In particular, it has been noted that DSSs researchers are interested in research work being conducted in model management. While we recognize that DSSs are not limited to managing data and models in response to problem solving and decision making, the ability to assist the DSS user in model formulation, analysis, and interpretation are certainly worthwhile goals of MMSs.

Future research in model management will continue to primarily revolve around both the representation and processing of procedures, data, and problem statements. There are organizational issues with respect to each of these to be addressed as well. This could include both an investigation of the security and integrity of models being developed as a result of proliferation and the availability of DSSs tools available in organizations for managers.

References

Applegate, L.M., Konsynski, B.R., & Nunamaker, J.R. (1986). Model management systems: Design for decision support. *Decision Support Systems, 2*, 81-91.

Binbrasrough, M., & Jarke, M. (1986). Domain specific DSS tools for knowledge-based model building. *Decision Support Systems, 2*, 213-223.

Blanning, R.W. (1987). A relational theory of model management. In C.W. Holsapple & A.B. Whinston (Eds.), *Decision support systems: Theory and applications* (pp. 19-53). Berlin: Springer-Verlag.

Blanning, R.W. (1989). *Model management systems: An overview.* (Working paper 89-23), Vanderbilt University, Owen Graduate School of Management.

Blanning, R.W. (1985). A relational framework for join implementation in model management systems. *Decision Support Systems, 1*, 69-82.

Blanning, R.W. (1985). A relational framework for assertion management. *Decision Support Systems, 1*, 167-172.

Blanning, R.W. (1983, November). TQL: A model query language based on the domain relational calculus. *Proceedings of the IEEE Workshop on Languages for Automation* (pp. 141-146).

Blanning, R.W. (1983, June). Issues in the design of relational model management systems. *Proceedings of the National Computer Conference* (pp. 395-401).

Blanning, R.W. (1984). Language design for relational model management. In S.K. Chang (Ed.), *Management and office information systems* (pp. 217-235). New York: Plenum.

Blanning, R.W. (1981, June). Model structure and user interface in decision support system. *DSS-81 Transactions* (pp. 1-7).

Blanning, R.W. (1981). Model-based and data-based planning systems. *Omega, 9*, 163-168.

Blanning, R.W. (1986). An entity-relationship approach to model management. *Decision Support Systems, 2*, 65-72.

Blanning, R.W. (1982). Data management and model management: A relational synthesis. *Proceedings of the 20th Annual Southeast Regional ACM Conference*, 139-147.

Blanning R.W. (1982, June). A relational framework for model management in decision support systems. *DSS-82 Transactions* (pp. 16-28).

Brennan, J.J., & Elam, J.J. (1986). Understanding and validating results in model-based decision support systems. *Decision Support Systems, 2*, 49-54.

Brown, D.E. (1987). An information theoretic interface for a stochastic model management system. *Decision Support Systems, 3*, 73-81.

Bonczek, R.H., Holsapple, C.W., & Whinston, A.B. (1982). Specification of modeling knowledge in decision support systems. In H. G. Sol (Ed.), *Processes and tools for decision support*, Amsterdam: North-Holland.

Bonczek, R.H., Holsapple, C.W., & Whinston, A.B. (1981). A generalized decision support system using predicate calculus and network data base management. *Operations Research, 29*, 263-281.

Bonczek, R.H., Holsapple, C.W., & Whinston, A.B. (1981). The evolution from MIS to DSS: Extensions of data management to model management. In M.J. Gintzberg, W. Reitman, & E. A. Stohr (Eds.), *Decision Support Systems*,

Bonczek, R.H., Holsapple, C.W., & Whinston, A.B. (1980). The evolving roles of models in decision support systems. *Decision Sciences, 11*, 337-356.

Bonczek, R.H., Holsapple, C.W., & Whinston, A.B. (1981). *Foundations of decision support systems*. New York: Academic Press.

Bonczek, R.H., & Whinston, A.B. (1977). A generalized mapping language for network data structures. *Information Systems, 2*(4), 171-185.

Bonczek, R.H., Holsapple, C.W., & Whinston, A.B. (1979). Computer-based support of organizational decision making. *Decision Sciences, 10*, 268-291.

Bonczek, R.H., Whinston, A.B., & Holsapple, C.W. (1979). The integration of network data base management and problem resolution. *Information Systems, 4*(2), 143-154.

d'Alessandro, P., Dalla Mora, M., & DeSantis, E. (1989). Issues in design and architecture of advanced dynamic model management for decision support systems. *Decision Support Systems, 5*, 365-377.

Dolk, D.R. (1986). Data as models: An approach to implementing model management. *Decision Support Systems, 2*, 73-80.

Elam, J. (1979). Model management systems: A framework for development. (Working Paper 79-02-04). University of Pennsylvania, Department of Decision Sciences.

Elam, J.J., & Konsynski, B. (1987). Using artifical intelligence techniques to enhance the capabilities of model management systems. *Decision Sciences, 18,* 487-501.

Elam J.J., Henderson, J.C., & Miller L.W. (1980). *Model management systems: An approach to decision support in complex organizations.* International Conference on Information Systems (pp. 98-110). Philadelphia.

Gershefski, G.W. (1970). Corporate models—The state of the art. *Management Science, 16*(6), B303-B312.

Gorry, G.M., & Scott Morton, M.S. (1971). A framework for management information systems. *Sloan Management Review, 13*(1), 55-70.

Holsapple, C.W., & Whinston, A.B. (in press). *Decision Support Systems*

Holsapple, C.W., & Whinston, A.B. (1988). *The information jungle.* Homewood, IL: Dow Jones-Irwin.

Holsapple, C.W., & Whinston, A.B. (1988). Model management issues and directions. *Kentucky Initiative for Knowledge Management.* Paper No. 7

Hwang, S. (1985, April). Automatic model building systems: A survey. *DSS-85 Transaction* (pp. 22-310), San Francisco.

Jarke, M., & Radermacher, F.J. (1988). The AI potential of model management and its central role in decision support. *Decision Support Systems, 4,* 387-404.

Johnson, L.E., & Pakath, R. (1990). *Model management and the query optimization problem.* Working paper. University of Kentucky.

Katz, N., & Miller, L. (1977). *An interactive modeling system.* (Working Paper 77-09-02). University of Pennsylvania, Department of Decision Sciences.

Konsynski, B.R. (1981). On the structure of a generalized model management system. *Proceedings of the 14th Hawaii International Conference on System Science, 1,* 630-638.

Konsynski, B.R., & Dolk, D. (1982, June). Knowledge abstractions in model management. *DSS-82 Transactions* (pp. 187-202).

Konsynski, B.R. (1982). Model management in decision support systems. In A.B. Whinston (Ed.), *Data base management and decision support systems.* Boston: Reidel.

Liang, T., & Jones, C.W. (1988). Meta design considerations in developing model management systems. *Decision Sciences, 19,* 72-92.

Liang, T. (1988). Development of a knowledge based model management system. *Operations Research, 31,* 849-863.

Liang, T.P. (1985). Integrating model management and data management in decision support systems. *Decision Support Systems, 1*(3), 221-232.

Little, J.D.C. (1970). Models and managers: The concept of a decision calculus. *Management Science, 16,* 466-485.

Marsden, J.R., & Pingry, D.E. (in press). A theory of DSS design evaluation. *Decision Support Systems.*

Naylor, T.H., & Schauland, H. (1976). A survey of users of corporate planning models. *Management Science, 22*, 927-937.

Sprague, R.H., & Watson, H.J. (1975). MIS concepts part I. *Journal of Systems Management, 26*(1), 34-37.

Sprague, R.H., & Watson, H.J. (1975). MIS concepts part II. *Journal of Systems Management, 26*(2), 35-40.

Sprague, R.H., & Carlson, E.D. (1982). *Building effective decision support systems*. Englewood Cliffs, NJ: Prentice Hall.

Sprague, R.H., & Watson, H.J. (1975). Model management systems in MIS. *Proceedings of Decision Sciences Institute* Atlanta, Georgia. 213-215.

Stohr, E.A., & Tanniru, M.R. (1980). A database for operation research models. *International Journal of Policy Analysis and Information Systems, 4*(1), 105-121.

Will, H.J. (1975). Model management systems. In Grochla & Szyperski (Eds.), *Information systems and organization structure*. Berlin: Gruyter.

Chapter 5

Management Implications of the Object-Oriented Paradigm

John E. Gessford
California State University, Long Beach

The use of microcomputers in business and governmental organizations has developed rapidly in the last decade and microcomputers seem likely to become even more ubiquitous in the future. The single-user PC is giving way to PC LANS (Local Area Network Systems) on which messaging systems, departmental calendering systems and shared databases can exist. As capabilities to handle voice and image data are added to the PC , the telephone system may be merged into the LANS, creating a system that opens new possibilities for higher productivity and closer cooperation between people.

The fact that greater productivity and cooperation is possible does not necessarily mean that it will happen, however. Leadership and managerial planning and control actions (including training and budgetary provisions) are required to take advantage of the means for enhanced productivity and cooperation that information technology makes possible. As an example of this distinction between a capability and its use, we all know of at least one software package that was purchased with the best of intentions, but never used because the investment of professional time required to attain mastery was not made.

Managerial prerequisites to effective use of information systems is the broad issue addressed in this chapter. What policies, organizational arrangements, systems and managerial actions are required for an organization to make full use of information technology? Just as certain steps must be taken to successfully introduce a new product or bring a construction project to completion, so a systematic approach to applying new information technology should lead to effective use. The question is, What is this systematic approach?

In this chapter, we consider the question of managerial prerequisites with respect to a particular type of information technology, namely, database management. We take for granted the effective use of computer/communications hardware and systems software. Word processing and other office automation systems are treated only to the extent that they involve the accessing of databases. The proliferation of database systems in organizations is making it increasingly worthwhile to find ways to coordinate and share these independently developed and operated data resources.

The limited use being made of the departmental LANS that exist in most companies is evidence that a knowledge of the managerial prerequisites for full use of these systems is lacking. Word processing, spreadsheet processing, electronic messaging, calendaring, and querying copies of certain application databases down-loaded from a mainframe are the principal uses being made of these general purpose computing systems. In most cases, there is only limited access to basic business data about the products, customers, employees, facilities, sales and inventories of the organization.

In this chapter, we show that the limited access to business data is a result of the way information systems are defined and developed. They are defined and developed in terms of specific business applications instead of as a set of component parts of a corporate information resource. The idea of treating the data of an organization as a single resource has been recognized for many years. But, the commonly accepted approach to developing systems, the System Development Life Cycle (SDLC), has prevented its realization.

In this chapter, the object-oriented paradigm, which has so far mainly been used at the programming level to write highly interactive software, is shown to be an approach to planning and developing information systems that lends itself to developing coordinated components of a corporate information resource in a decentralized fashion. It is shown that application of this paradigm to coordinated, organization-wide systems development requires changes in management policies, organizational arrangements, systems and actions. Only as these changes in management are made can the paradigm be successfully applied and truly integrated systems be developed.

Background

In most organizations, there are "line" and "staff" functions in the sense that some functions serve to achieve the basic purposes of the business while other functions support these "line functions." Although, the line functions depend on the basic purposes, it is common for finance, marketing and production (or operations) to be considered "line functions." Accounting and personnel are commonly viewed as staff functions. The accounting function serves the finance function and provides cost and revenue data to the managers of all functions. The personnel function assists in meeting the staffing need of all managers and in administering the employee policies and programs of the organization.

The information systems function of an organization is usually considered a staff function. It develops, and has traditionally operated, the information systems required by all functions of the organization. In a financial services business, such as banking, information systems may be seen as a line function, but even in this case, information processing can as easily be viewed as a service to the branch banking, commercial loan, and other functional units of the bank.

Staff Relationships

It is obvious that the relationships between a staff function and the other functions that it serves should be very important to those in the staff function. These relationships should be as important to the staff function as are the customer relationships of an organization to that organization. The justification for the existence of the staff function can be found in these relationships.

The relationship between a staff function and the other functions that it serves involves certain "objects." In the case of accounting, the transaction, the report and the account are three basic types of objects involved. The Accounting Department defines its services in terms of these objects and it is generally considered to be responsible for the consistent use of these objects throughout the organization, and for the reports issued to outside parties.

In the case of personnel, position, person and benefit are three basic types of "objects" involved in its relationships with the other functions that it serves. Personnel defines these objects in concert with the managers involved. The services personnel provides can also be defined in terms of them, and certain other objects. Personnel is generally considered to be responsible for the consistent treatment of these types of objects throughout the organization.

The information systems function (I/S department) has traditionally seen itself as dealing with relationships that involve data processing systems and the transactions and reports processed by such systems. These, then, are the types of "objects" with which I/S management has been concerned. I/S has traditionally been considered responsible for the consistent treatment of these objects throughout the organization.

Erosion of Information System Department Relationships

Several fundamental changes, however, have been eroding the authority of I/S over the data processing systems of an organization. One that has already been mentioned is the advent of the microprocessor and desktop computer. Another is the growing importance of information systems to the competitiveness of an organization. A third is the large increase in complexity of interactive systems compared to traditional batch processing systems. The effects of these changes have been noted by various authors.

With respect to the first change, Wysocki and Young (1990) observe:

> As computers and software become less expensive, more varied, sophisticated, and functional, user skills and computer literacy have improved. In many cases this has allowed users to acquire systems without having to follow traditional IS procedures or gain the approval of an IS department. The result, in some organizations, is a central IS department surrounded by many unrelated pockets or islands of technology. (p. 13)

Concerning the effect of the growing importance of information systems to organizational combativeness and its effect on IS department relationships with other functions, Joe Izzo (1987) quotes from a company meeting on business objectives, "What do you consider the greatest block in achieving the long-range business plan? was the question a large corporation's senior management posed to each of its business units. Every manager gave the same answer: Data processing." (p. 21) The natural tendency in such a situation is to give each business unit manager his or her own I/S staff and facilities.

With respect to the increase in complexity of interactive systems, McNurlin and Sprague (1989) observe,

> In the early 1980's, a major shift began in the way applications were being built—away from the use of third generation languages, such as COBOL and PL/1 and toward the use of fourth generation languages (4GLs). The shift was also away from the traditional system development life cycle, where the attempt is made to define system requirements fully before design and construction begin. (p. 256)

The myriad interfacing details that human factor engineering raises when interactive systems are being developed, made the third generation language approach totally inadequate for the on-line, interactive systems demanded by system users in the 1980's. The move to 4GLs and the concentration on sophisticated user interfaces has tended to shift responsibility for system development from I/S to the functional units using the systems.

Some Prescribed Remedies

The effect of these centrifugal forces is to create multiple information systems in an organization that are incompatible. Incompatibilities in hardware and operating systems lead to larger expenditures for training, maintenance and backup facilities than compatible systems entail, all other things being equal. Incompatibilities in application programs and data structures lead to cumbersome and expensive ways of sharing files and redundant and conflicting data collection, storage and processing activities.

The question is, how to counteract the centrifugal forces that promote many independent information systems. An intelligent response is needed that does not cause other problems in the process of opposing the development of many uncoordinated systems. A number of strategies have been proposed and some of them have been tried. So far, a clear winner has not emerged. To give a perspective on the state of affairs, three strategies are briefly discussed. These are major currents of thought on managing the future of the information systems function.

Business Systems Planning

In the 1970's, IBM introduced BSP (Business Systems Planning) as a means of ensuring that an organization develops integrated information systems. For a recent description see (IBM, 1984). Since the original BSP methodology, many variations on it have been developed by other organizations.

The basic strategy in BSP is to obtain top management support for a set of integrated systems, integrated with respect to hardware, software and data. In effect, the result of BSP is a description of one very large information system which is to be implemented over a period of years.

BSP is a brute force approach to counteracting the centrifugal forces. It calls for a special project in which all of the information system needs of the organization are identified in a top-down fashion. Planned systems are then defined in terms of data classes and business processes. These are compared to the existing systems and a plan is formulated to build and revise systems so that the identified needs are satisfied in an orderly fashion, and the resulting systems are integrated with respect to data and computer-communication system usage.

According to a study of database management in 30 large corporations conducted by Professor Rockart and his students at the Sloan School of Management, very few organizations have found the BSP approach viable (Goodhue, Quillard, & Rockart, 1987). The BSP project takes too long and yet does not produce a clear enough description of system requirements to achieve the coordination and integration of systems that is its basic justification.

In 1986, Dan Appleton, an information systems consultant, defined a "very large project" in an article in Datamation as one that is never completed. His conclusion is based primarily on experience. Business is simply too dynamic to hold still for years while very large projects are undertaken to build very large systems.

Information Engineering

Although BSP has not proven to be the solution to the problems of I/S management, it contains the idea of treating data as an important resource to be defined and managed. In 1982, James Martin proposed a modification of BSP in Strategic Data-Planning Methodologies that featured a more detailed modeling of the data requirements of the organization. He argued that the types of data used by an organization were more stable than the ways in which the data was used and therefore attempts to coordinate and integrate application systems should concentrate on the data rather than the application systems per se. Martin's information engineering utilizes the entity-relationship approach to modeling data requirements first formulated by Peter Chen, (Chen, 1976).

Unfortunately, Martin did not abandon any of the other aspects of BSP so the resulting methodology was even more ambitious and detailed than BSP. To cope with the mass of detail, Martin developed the concept of a "system encyclopedia," a database that describes the data and information systems of the organization. But, information engineering as defined by Martin does not gracefully overcome the "very large system" problem of BSP. Like BSP, it attempts to used brute force to build and maintain a very large system for the organization.

Develop Systems for Flat Organizations

A number of thought leaders counsel more foresight in defining the the information system needs of modern organizations. Richard Nolan in a speech before the Society for Information Management in Los Angeles on October 5, 1989, complained that neither I/S managers nor other corporate executives have thought through the implications of doing business in the information age in which the vast majority of all employees are "information workers." In the Nolan, Norton Company publication, Stage by Stage, Nolan, Pollack and Ware give their attempt to foresee future implication in an article entitled, "Toward the Design of Network Organizations."

Peter Drucker in *The New Realities* (1989) describes the emerging form of relationship between workers as follows.

> In pharmaceuticals, in telecommunications, in papermaking, the traditional sequence of research, development, manufacturing, and

> marketing is being replaced by synchrony: specialists from all these functions work together as a team, from the inception of research to a product's establishment in the market. And this is likely to become the model generally for the performance of knowledge work." (p. 211)

In "McNurlin and Sprague, (1989)" views of Peter Keen, a leading information systems consultant, are described that echo the same theme of Nolan and Drucker:

> Organizations can be simplified by increasing direct contact between people, Keen believes. Information technology can help by reducing the need for intermediaries, thereby flattening organizational hierarchies. The technology also can help organize information for easier access. This is particularly needed for document-based information, where complex organizational structures have emerged just to process documents. Management should shift its attention from using information technology for competitive purposes to using it to redesign their organizations to operate more simply, says Keen. (p. 36)

These observations show the need to rethink the information systems function in information-age organizations. Even if the very large system described by BSP could be built, it would not meet the evolving need of the information-age organization if it failed to foresee the flexible, "synchronous" type of organizational structures described by Drucker.

However, the observations by Nolan, Drucker and Keen on what is needed in the way of information systems are more in the nature of problem definitions than solutions. They advise us to take an approach to information systems design and management that takes into account the organizational effects information systems can have. But, they do not define that approach.

Our purpose in this chapter is to address the issue of how to approach the development of information systems in a dynamic organization in which the vast majority of the workers are sophisticated specialists who work with different types of information. In one form or another, almost everyone in the organization is primarily engaged in collecting, analyzing, organizing, communicating and otherwise using information to accomplish a business function. In such an organization, information systems are too important to be left to experts, just as nuclear weapon usage policy is too important to be left to the military.

We investigate the proposition that I/S needs to change the "objects" on which its relationships with other functions in the organization are based. Instead of basing these relationships on data processing systems, I/S should base them on all objects about which the organization needs information.

A New View of Information Systems

Nothing is more difficult than to detect the limitations of a relationship or viewpoint once it has been accepted. The limits are taken for granted. As Marshall McLuhan put it, "I don't know who discovered water, but it wasn't a fish." Unfortunately, the familiar, conventional viewpoint does not appreciate the novel idea that only makes sense from a totally different perspective. Alan C. Kay, the originator of the object-oriented viewpoint, in speaking to Stanford engineering students about the future, said, "In some sense our ability to open the future will depend not on how well we learn anymore but on how well we are able to unlearn. Can you imagine a course at Stanford on unlearning? That would be revolutionary." (1989, p. 6.)

The way we think about information systems (alias "management information systems," "decision support systems," "data processing," etc.) is encrusted with some basic concepts that need to be unlearned. One is the concept of what computers do. Another is the concept of what an application system is. Only as we open to question our notions about these basic aspects of information systems will we be able to see how the object-oriented approach applies to information systems planning.

What Computers Do

Conventional wisdom holds that the role of computers in business is to process data. Virtually all texts on the subject describe the computer as a machine that executes three logically related functions: input, process, output (IPO). Data is entered, the computer processes it according to instructions, and the results are output. If this description is true, it is only because it describes the paradigm of the programmer/analysts who designed the software. But, the basic nature of programming does not prohibit other models of the computer's function.

In the object-oriented view of the computer, it is a simulator of objects, not a processor of data. Simulated objects can be created, displayed, changed, made to interact, stored and retrieved, sent to other computers, printed on paper. By working with the computer as a simulator of objects, one turns it into a medium of communication rather than a transformer of data. It becomes a medium of expression. In 1977, Alan Kay described its advantages as a communications medium in the following passage,

> "...unlike conventional mediums, which are passive in the sense that marks on paper, paint on canvas and television images do not change in response to viewer's wishes, the computer medium is active: it can

respond to queries and experiments and even engage the user in a two-way conversation." (p. 231)

Computers simulate objects by executing procedures, or "methods" as they are sometimes called. These procedures can even be described as "input-process-output procedures." But there is a difference between asking an object to do something and invoking a procedure. The difference lies in the constancy of the response. When a procedure is invoked, we tend to think of the result as always being the same. A procedure is like a mathematical function: for a given value of the independent variable, the value of the dependent variable is always the same. But, the response of an object to a request depends on the "state" of the object. This state is defined by the variables (attributes) of the object, and the state may change as objects interact.

As an example, consider a message that one person might send to another via a computer system. Suppose message objects can respond to requests to create, send, display, and delete themselves. If User A issues commands to create and send a message to User B, then when User B issues a request to display messages, the message from A will be displayed. On the other hand, if User A does not issue the create and send commands, then when User B asks for new messages to display themselves, there will be none from A. Thus, the response to the User B display request depends on the state of the message objects being simulated.

Alternative System Concepts

So, to assert that a computer system simulates objects is not to deny that it executes procedures. Simulation is accomplished by procedure executions. However, the object simulation viewpoint leads to defining and organizing procedures in a very different way. The object-oriented approach is to organize procedures according to the data they use. In contrast, the traditional procedural approach organizes data according to the procedures that use it.

This difference fundamentally alters the way computer applications should be defined. From the object-oriented viewpoint, an "application" should be defined in terms of the objects it involves instead of in terms of the procedures executed or business functions supported. To define an application system in terms of procedures and then define the software logic in terms of objects is to "switch horses in the middle of the stream." As is shown in a subsequent example, it unnecessarily complicates the system design process; it leads to poor object definitions, and it creates application systems that redundantly store the same data.

When we think of a *system* in terms of objects, we identify objects that have some relationship or interaction with each other in the system. For example, in describing a system of traffic, we include the vehicles, streets, signal lights and so forth in a particular locality. We would not include vehicles in Los Angeles and Chicago in the same traffic system description.

The object-oriented approach to defining a business information system, then, should involve identifying objects that have some relationship or interaction with each other in the system. An inventory system, for example, would involve stock items, stock locations, stock replenishments, and so forth. A customer invoicing system would involve customers, sales, stock items, etc.

If all the other objects that a given object interacts with are not in one application system, then that object must be included in one or more other *system* definitions. This is exactly what happens when systems are defined in terms of procedures rather than in terms of objects. The result is object redundancy, and incomplete (and therefore potentially misleading) simulations of a given object in any one application system.

For object-oriented systems planning, the set of all of the objects about which the business requires information needs to be identified. Specific applications can then be defined in terms of these objects and the information about them that the application requires. The information requirements for a given object by all applications can then be provided for in the behavioral simulation capabilities created for that object.

We need a term to designate the set of related objects, related in the sense that one or more business functions needs to simulate them and their interactions. Calling such a set an "application" or a "system" could cause semantic confusion because of the established meanings of these terms. To identify this set of related objects is not to fully define the objects and their relationships; therefore we should not call it a "database." For these reasons and others that will become clear later, we call this set of related objects a Business-wide Information Structure (BwIS).

To explore the implications of the object-oriented viewpoint for systems planning in more detail, let us analyze an example using two different planning methods. One is the object-oriented approach to system planning and the other is a more traditional, structured analysis approach. To avoid obscuring the differences by complexity, a very simple example is used. It is the problem of designing an Electronic Mail System (EMS). Only a small number of objects and procedures are involved, yet there are a number of design alternatives.

It is possible to devise a system planning method that results in an object-oriented design of the programming but the system planning method itself is not object-oriented. It may be called an object-oriented approach to system planning but it is really an adaptation of the structured analysis methodology to the demands of an object-oriented programming philosophy.

To show the effects of trying to combine a traditional structured analysis approach with an object-oriented implementation of the software, we first present a combined process-object analysis of the EMS system as given by Seidewitz and Stark in a 1987 paper. This is followed by a purely object-oriented approach to the same problem. The differences between the two approaches are then reviewed.

A Combined Process-Object Approach to System Planning

The approach described by Seidewitz and Stark begins with a process-oriented description of the EMS "business" as a whole. Their overall characterization of the system is summarized in the "context diagram," shown in Figure 1. The circle labeled EMS represents all of the processes of the system and the two squares represent external data stores (also called external entities). The system includes procedures that allow users to send, read and respond to messages. It also contains programs for listing the system users and adding to, or deleting users from, that list.

The second step in the process-object approach is to represent the system by a more detailed data flow diagram. In Figure 2 this data flow diagram is shown with the EMS procedure now exploded into five procedures and three data store (plus the message queues data store which is not show). In a more complex system, further decomposition of the processes would be necessary; however, for the EMS system this data flow diagram is sufficient for this second step.

The third step is abstraction analysis. This is where the transition is

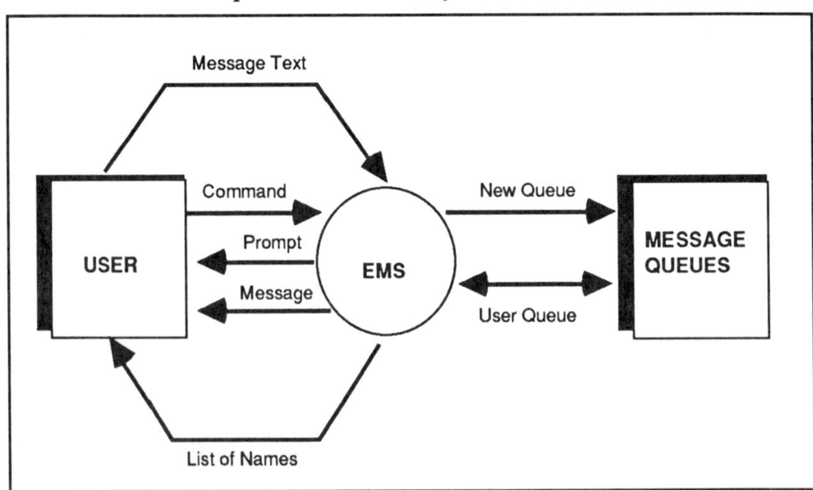

Figure 1. EMS Context Diagram

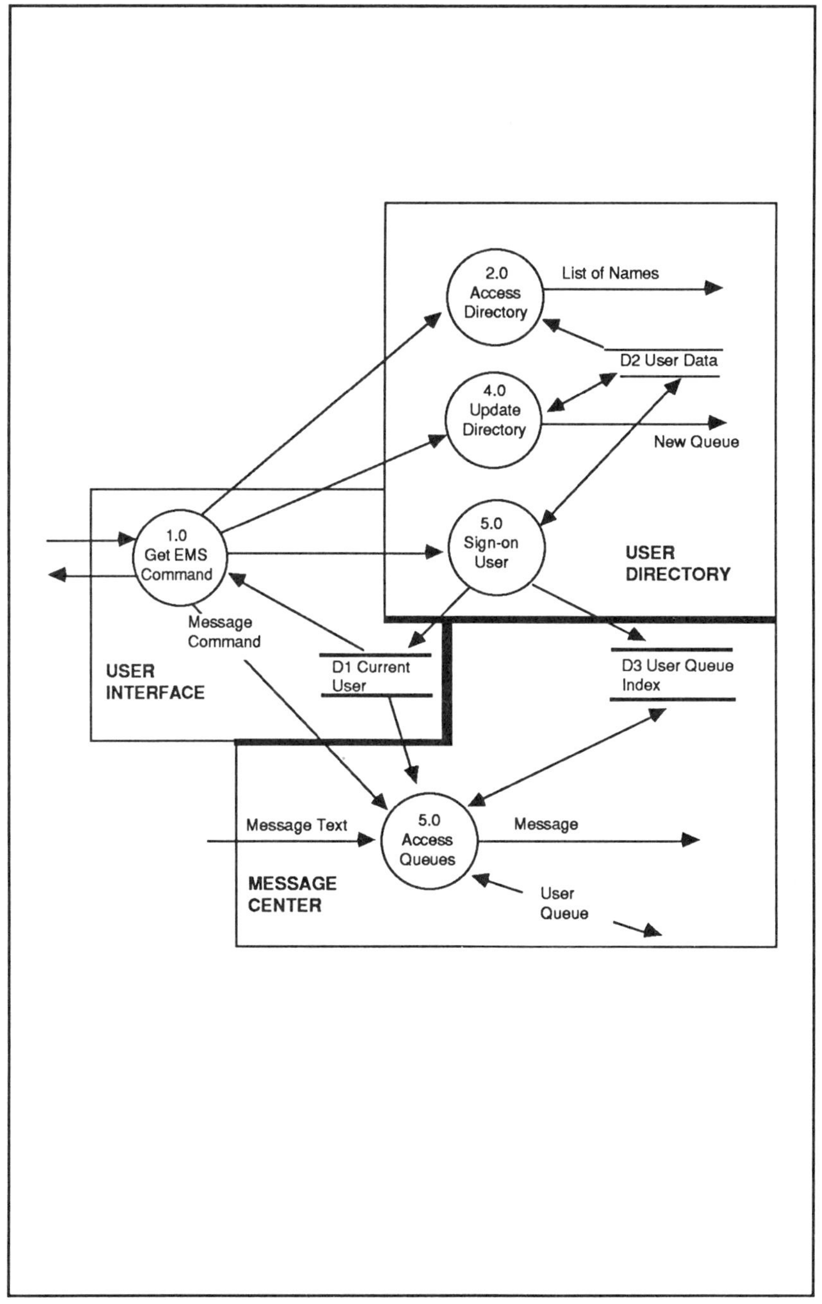

Figure 2. EMS Level 0 Data Flow Diagram

made from a structured description of the system to an object-oriented description. The data flow diagram is analyzed to identify system "entities." An entity is defined to be a set of processes and data stores accessed by them. A "central entity" is first identified, then other entities are identified by following the data flow lines emanating from the central entity. In the EMS case, the central entity was determined to be the USER INTERFACE because "EMS is a system serving a person sitting at a terminal sending and receiving messages." In Figure 2, the dashed rectangle that contains the "Get EMS Command" process and the "D1 CurrentUser" data store symbolizes the USER INTERFACE entity.

Two other entities are identified in Figure 2. One is the USER DIRECTORY which includes three processes and the User Data store. The other is the MESSAGE CENTER which includes the "Access Queues" process and the "User Queue Index" data store. In general, the analysis of the flow diagram continues until all processes and data stores are associated with an entity.

The fourth step is to construct an entity graph of the system. This contains the entities identified by abstraction analysis plus a "most senior entity" and the two external data stores from the context diagram (Figure 1). The lines between entities in the entity graph shown in Figure 3 represent the flow of control from one entity to another. If there are no arrowheads, the exact nature of the interaction has not yet been determined. The "most senior entity" (EMS in Figure 3) represents the system as a whole and is the entity to which control is given by the User.

The fifth step is object identification. This is done by altering the entity graph (Figure 3) in ways that achieve a balance between design simplicity and object abstraction and also clarify the control hierarchy. For example, combining the USER DIRECTORY and MESSAGE CENTER entities into one object would simplify the overall design but the combined "object" would have little coherence as an abstraction. For this reason they are not combined in the object diagram shown in Figure 4. On the other hand, the EMS and USER INTERFACE entities of Figure 3 are combined in Figure 4 and called the USER INTERFACE object. The USER INTERFACE object serves as the control center of the system once the user issues the "run" command.

The sixth step is to define the variables and operations (program modules) for each object in the system. Because the entities were defined in terms of sets of processes and data stores in Step 3, this is a matter of consolidating and clarifying the sets previously defined. Figure 5 shows a possible result of Step 6 for the EMS system.

The "Update Directory" process of Figure 2 is decomposed in Figure 5 to "Add user" and "Delete user." Also, the "Access Queues" process has been decomposed to "Receive/store" and "Retrieve/send." The variables shown in

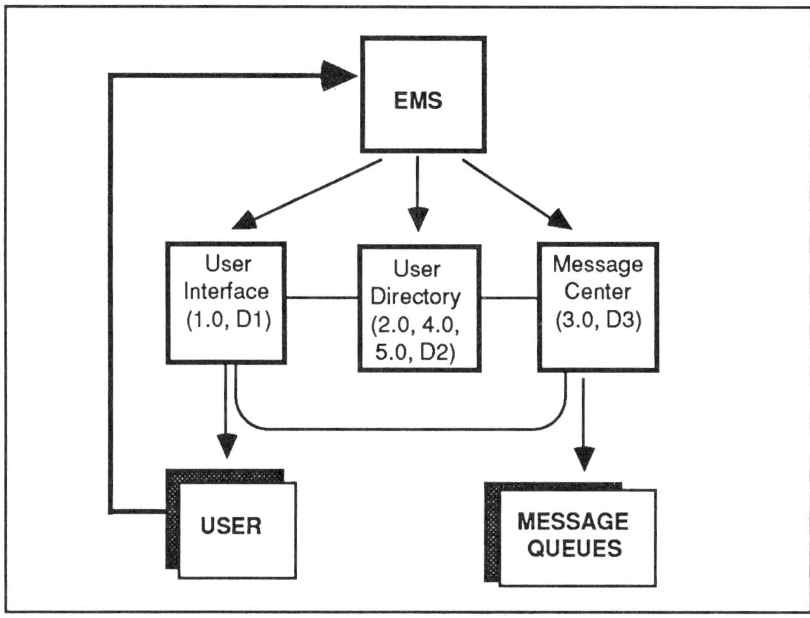

Figure 3. EMS Entity Graph

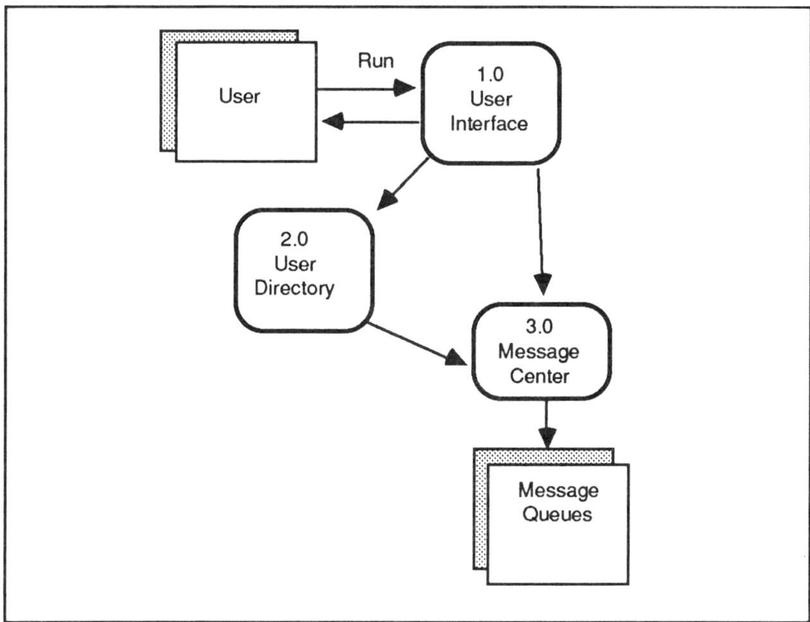

Figure 4. EMS Object Diagram

OBJECT	VARIABLE	OPERATION
User Interface	Current user ID	Identify user
	User node	Display
		Add user
		Delete user
		List users
		Send message
		Read message
		Edit message
User Directory	User ID	Sign-on
	User name	Add user
	Password	Delete user
		List users
Message Center	User index	Receive/store
	Text	Retrieve/send
	Sender ID	
	Receiver ID	
	Time received	

Figure 5. Objects in Process-Object System

Figure 5 are straightforward interpretations of the references to "data" to Figure 2 except for those of the Message Center. In this case User Index is taken directly from Figure 2 and the remaining four variables describe an item in a Message Queue, which the Message Center operations must handle.

Step 7 is to translate the description of the system in Figure 5 into the language of an object-oriented programming system. This constitutes the end of system design and the beginning of coding. Thus, consideration of the example can terminate at this point. The terms and syntax of the program module descriptions will depend on the programming system used. It is an easy translation if the programming system is truly object-oriented.

A Purely Object-Oriented Approach to System Planning

The purely object-oriented approach begins by defining the purpose of the business in terms of the functions it performs. Our example EMS enterprise is so simple that this definition can be given as a sentence: The purpose of EMS is to enable users to send and receive messages. For a more complex business, a functional model is needed to define the purpose. The development of such models is a separate subject, discussed in (Gessford, 1990) and (Martin, 1982).

The second step is to characterize the information needed to accomplish the purpose of the business in terms of entity types and their relationships. In the case of EMS, it is obvious from the statement of purpose that information about Users and Messages is required. In a more complex case, the information requirements are not obvious and must be obtained from experts on the functions of the business.

Two relationships between the User and Message entity types are important for the EMS system. One is *send* and the other is a *receive* relationship. The *send* relationship identifies the sender of a message. The *receive* relationship defines the message queue for each user. These entity types and relationships are depicted in by an Entity-Relationship diagram in Figure 6. "Crows feet" at the end of a connecting line mean that many occurrences of the entity type can be involved in an instance of the relationship.

The third step in the pure object-oriented approach is to define the objects of the system in terms of their variables and operations. This is done in Figure 7. Three objects are defined. The first two come directly from the User and Message entity types directly. The third one, labeled "Receive," simulates the message queue for each user. It implements the *receive* relationship of Figure 6.

Comparison of the Two Approaches

The process-object (PO) and purely object-oriented (POO) approaches to system design are clearly different and they yield different system designs. Although they both happen to result in three objects in the EMS example, they do not identify the same three objects. The User object of the POO approach combines the User Interface and User Directory objects of the PO approach. The PO approach fails to incorporate the pointers structures of the POO approach because it treats the message queue as an entity that is external to the system.

Before discussing the EMS example, it was asserted that the PO approach had three disadvantages: (1) it unnecessarily complicates the system design process; (2) it leads to poor object definitions, and (3) it creates application systems that redundantly store the same data. Consider how the EMS example illustrates each of these disadvantages.

More Complex Approach

The PO approach has six steps whereas the POO approach has only three. Step 2 of the PO approach is complex and results in a very detailed system description early in the design process. The detailed flow diagrams are then referred to, but not fully used, in subsequent steps.

Poor Object Definitions

The objects identified by the PO approach lack coherence. They

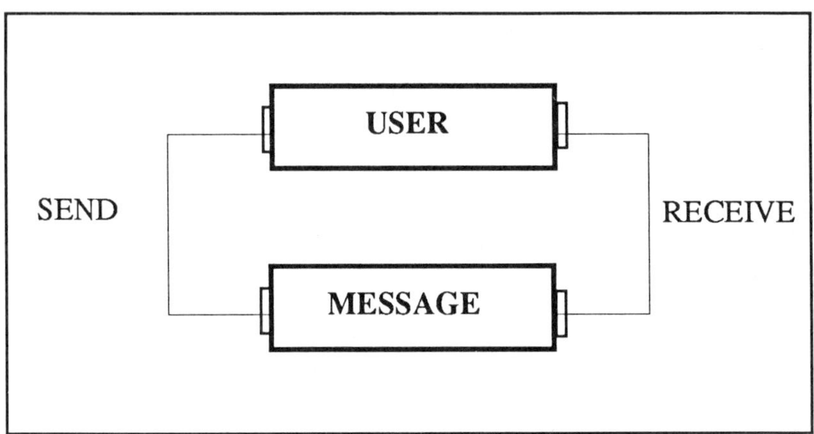

Figure 6. Entity-Relationship Diagram for EMS Database

OBJECT	VARIABLE	OPERATION
User	User ID	Sign-on
	Current node	Display menu
	Password	Add user
	User name	Delete user
	Message sent pointer	List users
	Message received pointer	Send message
		Read message
		Edit message
Message	Text	Receive/store
	Next sent pointer	Retrieve/send
	Sender pointer	Link to sender
	First receiver pointer	Link to receivers
	Time received	Delete
Receive	User ID	Store next received
	Message ID	Store next sent
	Next received pointer	Retrieve next rec'd
	Next receiver pointer	Retrieve next sent
		Delete next received
		Delete next sent

Figure 7. Objects in Pure Object-Oriented System

reflect the process-oriented viewpoint that defined them. "User Interface" is simply a collection of processes; it has no identity as an object. The same is true of "User Directory." "Message Center" also connotes a group of processes more than an object.

Redundant Storage

The distinction between User Interface and User Directory in the PO design would lead to some redundancy in user data. Also, the PO designed system would store a separate copy of a message for each receiver of the message. The POO system would store only one copy of the message and use pointers to link it to its receivers. (At least this is one interpretation of the Seidewitz/Stark description of their EMS design, which is actually ambiguous as to how the "message queues" entity is organized.)

Another source of redundancy that is not illustrated by the EMS example is the replication of an object in more than one application. To illustrate such redundancy a more complex "enterprise" for which the PO approach leads to more than one application would need to be analyzed.

Object-Oriented Systems Management

The object-oriented perspective on computers and their application to business has implications for managing the entire system development process. If the way an "application" is defined needs to change then it should not be surprising that the whole approach to managing the development of "applications" needs to be revised.

Instead of viewing data as something that exists within an application, we need to view an application as something that exists within a concept of data. The conceptual framework of this "database" is the BwIS (Business-wide Information Structure). At a minimum, the BwIS is the set of all data object classes of interest to the business. For example, Figure 7 shows the BwIS for an artificially simple business, the "EIS business."

Some Object-Oriented Terms

Unfortunately, it seems necessary to use some unfamiliar terms in discussing the object-oriented approach. In the definition of BwIS just given, the term "data object class" was used. It should be understood that a class of objects is a certain type of object, such as the class of all customers or the class of all products (of a certain kind). Using this definition, we can redefine an "application" to be a set of behavior capabilities that objects of certain classes possess.

The term "data object" is used to distinguish objects whose existence through time needs to be simulated by storing their definition and "state" on a permanent storage device, such as a disk. This is in contrast to an object such as the cursor appearing on the monitor of a PC which only needs to exist for the duration of PC work session. (A data object differs slightly from the concept of a "persistent object" as defined in the object-oriented literature because concurrent simulations of a data object for multiple system users can occur and must be coordinated. Whereas, a persistent object, such as a "window," is commonly considered to be unique to a single user and simulated by only one user at a time.)

The fact that multiple applications need to use the same data, such as data about customers, is provided for in the object-oriented approach by the fact that objects of a class can possess multiple behavior capabilities. Printing a customer address label can be one behavior of a customer object, for example, and displaying account receivables data another behavior. There is no limit on the number and variety of behaviors that can be given to an object class.

Information Systems Management Responsibilities

In the object-oriented approach, the primary responsibility of the I/S function shifts from data processing systems to object classes. Instead of providing data processing systems for users, I/S is responsible for providing object classes, an object class being a capability to simulate the behavior of objects of the class. I/S is responsible for defining object classes, delegating ownership responsibilities for object classes, and setting standards for object class behavior implementations (programs) and the hardware environment.

The first impression may be that this is just another way of describing traditional I/S management responsibilities. To simulate the behavior of an object is to execute a program that causes the computer to provide (or capture) information about that object. So, isn't this just another way to say that I/S is responsible for printing the payroll?

The difference is in the level of control. Instead of exercising control at the level of application systems, it is exercised at the higher level of object classes. An object class concerns all the information about a certain type of entity that is relevant to the business. Different users can have different subsets of this information for different purposes but it is managed as a single quantum of data to be shared by all who have a need to know. This centralized management of each object class makes it possible for all functions with a "need to know" to have access to all the information available about an object class.

How does managing object classes differ from managing databases? The difference lies in the management of object behavior. Data management and database management functions exist in many organizations and there is general agreement on the difference between these two functions. In object-oriented terms, the data management function is concerned with the definition of object state variables and object class ownership issues. The database administration function is concerned with certain fundamental behaviors of data objects, such as the way data is stored on disk and data access alternatives, and with other details of state variable definitions. But neither of these functions commonly has responsibility for defining the entire repertoire of data object class behaviors. This is the big change in the way responsibility is structured in the object-oriented approach. It leads to a radical shift in the way applications are defined and in the work assignments given to programmers. Data administration and database administration logically become subfunctions of an information resource management function.

The Business-wide Information Structure Planning Project

The basis for managing data object classes, which is the management problem of interest in this chapter, is a BwIS (Business-wide Information Structure). As already stated, this structure defines the set of all data objects that are related in the sense that they define the information that the application systems of the business have in common. Figure 6 is an entity-relationship diagram representation of the BwIS for the EIS example discussed in the previous section.

The initial definition of the BwIS for an organization is best done as a project. At this stage, there are no standard BwIS and the structure must be developed from scratch. As the object-oriented approach to system management develops, however, standard BwIS's seem likely.

Prospects for Standard Business-wide Information Structures

The fact that the value of information is relative to the goals of the organization means that any BwIS is only valid to the extent that it describes objects relevant to the goals. To the extent that different organizations have different goals, each organization can be expected to require a unique BwIS. Conversely, to the extent that different organizations have the same goals and objectives, the same BwIS should be useable.

This suggests that once a valid BwIS has been developed for an organization in a certain type of business or social activity, it should be applicable, with minor adjustments, to all organizations of the same type. Such a standard BwIS could at least serve as a starting point for developing the BwIS suited to the particular organization.

Project Overview

The project to develop a BwIS should use a more structured way to identify the entity types and relationships that form the basis for data object identification than is the case in BSP (as defined in [IBM, 1984]). But the approach can be the same in the sense of beginning with a definition of data requirements at the top level of an organization and then systematically identifying the requirements of subfunctions.

The methodology for this top-down approach to BwIS definition has two parts. One is functional analysis; the other is information decomposition. The concept of functional analysis is explained in Chapter 2 of (Gessford, 1990) and elsewhere. Functional analysis defines the organization functionally and results in a "road map" that is useful in the information decomposition process.

Information requirements decomposition is explained in Chapter 5 of (Gessford, 1990). It constitutes a structured approach to refining entity type and relationship concepts to the degree required to clearly define the information needs of the organization in terms of data objects. Applying these concepts in a top-down fashion to the functional breakdown developed in the other part of the methodology results in an "information structure" that defines the information needs of the business.

A final step in developing the BwIS is the integration of the information structures developed for each functional area. This is called "view integration." It is a process of identifying entity types and relationships that occur in more than one view or information structure. These are the redundancies that can be eliminated by taking an object-oriented approach to systems planning and development. The process of identifying these "overlaps" in information structures and integrating the structures is described in Chapter 12 of (Gessford, 1990).

It is important that the project be completed as quickly as possible. It is an overhead activity that will be seen as an unproductive burden if it does not progress according to schedule. It is reasonable to expect the project to be completed within six months regardless of the size of the organization. The main work of the project is done by "user experts" and the number of such experts should be proportional to the size of the organization. Thus, a large organization can complete the project in approximately the same time as a small one, if it devotes sufficient manpower to it.

A six-step approach to the project is shown in Figure 8. Each of these steps can be completed within one month if the project is properly managed. Delegation of responsibility by the steering committee is instrumental to maintaining the pace.

The following description of the six steps shown in Figure 8 summarizes this overview of the project. Space does not permit a more complete discussion of the many management and technical issues that arise in the process of executing these steps. The IRDS mentioned in Step 5 is the Information Resource Dictionary System used to manage the inputs from the many end-user experts involved in the project. This metadatabase is called an IRDS by the American National Standards Institute, a Repository System by IBM, and a System Encyclopedia by James Martin and others.

Organize for the project. The most important action in this step is the selection of a project leader. However, this step is also intended to include initial staffing decisions by that project leader, setting up a project office, and preparing for orientation and training activities in the next step.

Start project. The project is started by a series of executive orientation and project planning meetings. The Chief Executive defines the reasons for the project and its objectives. Then the project scope is defined by means of a review of the functions and existing information systems of the business and discussion about which should be included in the project. This provides a basis for defining who should be included on the project steering committee so that all functional areas are adequately represented. Starting the project includes organizing the steering committee and educating the members of it in functional breakdown analysis, critical success factor analysis, and entity-relationship analysis.

Defining upper-level data requirements. The steering committee, with staff help, develops a definition of the upper-levels of the functional breakdown and the data requirements at these levels. Interviews are conducted with each executive on the committee to identify the critical success factors for that executive and the data useful in measuring those factors.

Functional breakdown. Training of end-user experts, appointed by the steering committee, in functional breakdown analysis is conducted. The upper-level functions are broken down so that the elemental activities involved in each are identified. These breakdowns are entered into the system encyclopedia and reviewed and approved by the steering committee.

Database design. Training in E-R analysis is provided for the user experts. The data requirements of each subfunction are defined by the appropriate user expert. These requirements are entered into the IRDS and the user views are consolidated to create the BwIS.

Information resource management/Data administration. The BwIS is documented, and policies and procedures for its implementation in future system development projects are formulated, approved and promulgated. A data administration staff is organized to maintain, formulate and execute policies and procedures to maintain the BwIS on a continuing basis.

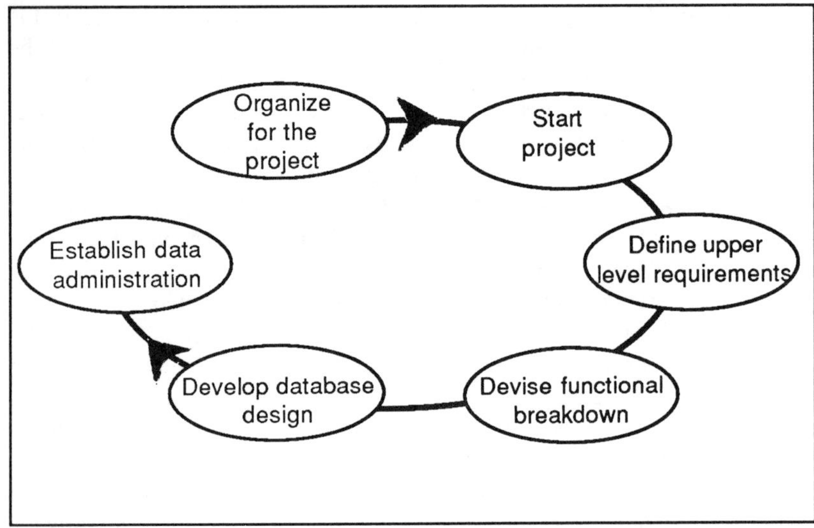

Figure 8. Business-Wide Information Structure Planning Project

Managing Information Systems Development

The SDLC (System Development Life Cycle) is the traditional paradigm for managing the development of information systems. Given that an application is to be defined in terms of what it does to support, or execute, a business function, the SDLC is a very straightforward, common sense description of steps to be taken in developing an application system.

Figure 9 depicts the SDLC as embedded in a larger system development concept. The traditional SDLC is shown in a large rectangle as a seven-step cycle that begins with the System Investigation and Planning Step and ends with Maintenance. The SDLC macro-step interacts with another macro-step labeled "Implement BwIS" in Figure 9.

The top box in Figure 9 concerns the definition of the Business-wide Information Structure (BwIS), the set of entity types and their relationships, about which a "business" needs information. This set is viewed as a single conceptual system. (This does not imply a single physical system.) Figure 6 shows the BwIS for the EMS example in the form of an entity-relationship diagram. The project that develops this set definition is described in the previous subsection.

Establish Development Policy

The second box in Figure 9 concerns the implementation of the BwIS.

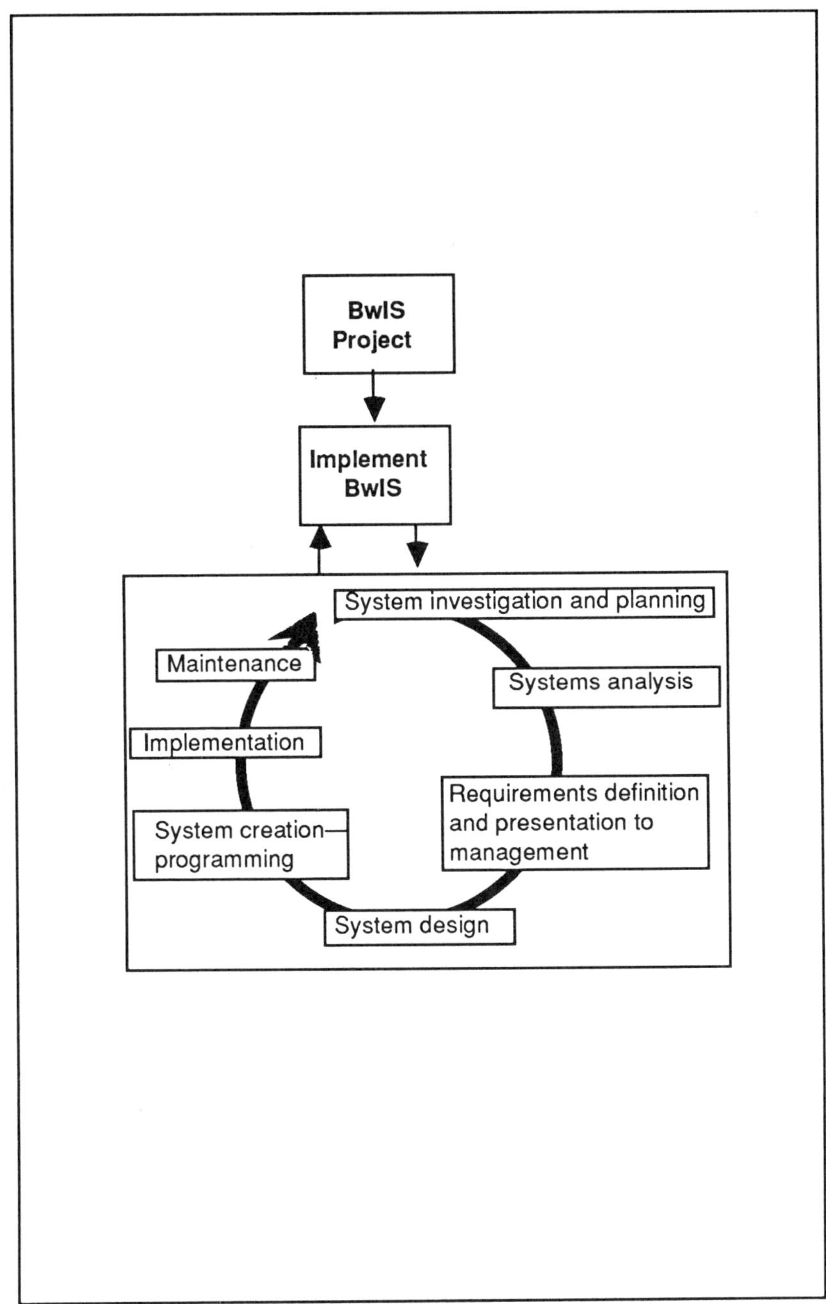

Figure 9. An Object-Oriented SDLC

It should be interpreted to include both basic policy formulation and ongoing management of the development processes. An alternative title for the box could be "Establish Systems Development Environment."

The strategy in making the basic policy decisions should be to start with the application software acquisition issues and work back toward the operating system and hardware procurement issues. The strategy requires an upper-level management that is sophisticated enough to resist the temptation to focus on hardware decisions first and then let subordinates worry about the software.

This is not the place to consider the policy decisions in any detail, however, the sequence can be indicated. The first policy decision concerns the alternative of purchasing standard software packages. One important criterion for this decision should be how accurately the database of the standard software represents an implementation of the BwIS. If suitable standard application software is not available, the alternative is customized software. Within this alternative there are in-house and outsourcing alternatives. An important criterion for these decisions is the quality and quantity of the software development talent each alternative offers. Before this can be evaluated for an alternative, particularly an in-house alternative, the CASE (Computer-Aided Software Engineering) tool set that would be used needs to be identified. It is in this selection of CASE tools that limits on the degree of object-orientedness in the implemented systems are set. The present generation of CASE tools leaves much to be desired in this respect.

The ongoing management of development processes involves data administration and interacting with system development project staff in the definition of objects and their repertoire of behavioral capabilities. It needs to be carried out on a continuing basis by what can be called an "object administration" function. The function is an adaptation to the object-oriented database environment of what has been traditionally known as database administration. Classes and superclasses of objects must be defined in a way that satisfies the needs of business functions for object simulation capabilities.

Improvements in the System Development Life Cycle

The existence of a BwIS does not mean that all the software for an entire business needs to be created in one development project. On the contrary, software can still be developed to support specific subfunctions of the business in a modular fashion. We will refer to the software for a subfunction as an "application system." But, it should be remembered that the applications are no longer independent. All the applications for the subfunctions of a given business simulate a common set of objects. Any one application is a partial simulation from the viewpoint of the business as a whole, but it is presumably sufficient for purposes of the business subfunction that it serves.

Thus, the SDLC macro in Figure 9 should be interpreted as a step that is replicated for each subfunction application that is developed. Each replication is an application development project that results in a partial implementation of the BwIS. And each project can have interactions with the Object Administration Function that results in evolution of the BwIS. Some of these evolutionary steps can affect existing applications of other subfunctions of the same business. Such side-effects should, of course, be minimized.

With this understanding of the SDLC macro in Figure 9, let us consider the changes in the micro-steps of the SDLC that the existence of a BwIS (and the associated business functional model) make possible. It turns out that significant simplification and standardization opportunities exist for each of the seven system planning steps. The business-wide functional model provides guidelines for planning and designing individual systems. So, for systems that involve database processing, significant improvements can be made in the way the seven steps of the SDLC are performed.

It should be noted that the improvements are made within the framework of the SDLC. It is a case of re-thinking the best way to perform the seven steps, not a matter of changing the basic sequence of steps.

System Investigation and Planning

The first step in the SDLC, system investigation and planning, is a survey of a computer application possibility in which a manager is interested. It is an initial cycle in a dialogue that should take place between those who will use the computer system and those who will develop it. The users of the potential system describe the computer tool capabilities they could profitably use and the system developer estimates the cost and time to develop.

Facilitate communication. Educating system analysts in the intricacies of the business functions to which the users want to apply the computer is often not a trivial task. In most organizations, these functions are not defined and documented in a way that makes it possible to give the analyst a definition of the relevant subfunctions which can be studied prior to giving an estimate of cost and development time for the application. Rather, a time-consuming sequence of meetings, memos and observations are commonly required to convey the system application requirements to the developer.

If a BwIS has been developed, as described in a previous section, then a functional breakdown, or model, of the enterprise is available. This can be used by the developer to quickly get a fix on the scope and details of the subfunctions involved in the application. Thus, a reduction in the time required for system developers to learn about the business functions to be supported by an application is one important impact that the object-oriented approach to systems planning can have.

Identify programs. There are other impacts that are equally or more important. One is the ability to quickly count and categorize the computer programs needed for an application. In the traditional SDLC, the definition of computer programs is not considered until Step 4 or 5. The existence of a BwIS makes it possible to identify the computer programs required at the first step, assuming an object-oriented approach. This makes a more accurate estimate of development cost and time possible.

The existence of the BwIS makes it possible to identify most of the objects that the application will simulate even before the procedures for doing so are defined. User experts in the subfunction originally defined the entity types and relationships about which the subfunction requires data during the Develop BwIS step of Figure 9. Thus, the relevant entity types and relationships are known and the relevant objects can be deduced directly from them as is illustrated in Figure 7 which is derived directly from Figure 6.

Because each computer program is associated with a specific object in the object-oriented approach, identifying the computer programs that will be required is a matter of determining the programs required for each object. This determination can be systematized by classifying the objects derived from the BwIS into two types, permanent and temporary.

The distinction between permanent and temporary entity types (or many-to-many relationships) is analogous to the distinction between a noun and a verb. A noun can either serve as a subject or object of a simple sentence. A verb denotes the action (or existence) described by the sentence. For example, in the sentence, "The product is delivered to the customer," there are two nouns and one verbal phrase.

Entity types that either take actions or are affected by actions (nouns) are permanent entity types. "Customer" is a permanent entity type that requests quotations, gives orders, and takes delivery, for example. "Product/Service" is a permanent entity type that is quoted, sold, shipped, and invoiced. Entity types that represent events or actions are temporary entity types. "Delivery" is a temporary entity type. "Purchase Order" and "Customer Order" are temporary entity types.

"Day" and "Year" are temporary entity types because they refer to something that happens. On the other hand, "Age Interval" is a permanent entity type. It does not refer to a specific time that comes to pass but rather to a measure of time. If the entity type "Day" is used to refer to a unit of time rather than to a specific day then it also should be classified as static.

Finally, it should be noted that the concept of an event as applied to business includes more than raw action. Orders and decisions are events. As a matter of fact, four types of events can be distinguished which have a sequential relationship to one another with respect to the duration of their impact: (1) actions, (2) orders to take actions, (3) decisions on plans for future actions, and (4) strategic decisions on goals. In the case of order type events, the order as well as the action ordered are temporary entity types.

In addition to permanent and temporary objects, there are programs that merely access data but make no changes in it. These are commonly known as report generator or query processing programs. When an object-oriented approach is taken, the formatted reports and special query responses required become additional objects to be included in the system in addition to data objects which are based on the BwIS requirements.

The programs required to simulate permanent objects are highly predictable. This is because the procedures that these "master file" programs must be able to carry out are easily enumerated. There are only four of them:

(1) Add a permanent object,
(2) Modify an existing object,
(3) Delete an existing object,
(4) List existing objects.

In general, then, four programs are needed for each permanent entity type in the BwIS. Existential dependency relationships between entity types do not invalidate this rule although they complicate the relationships between the programs. (One entity type is existentially dependent on another, if the dependent one can only have occurrences when a corresponding occurrence of the other entity type exists. An example of two permanent entity types that are existentially dependent is Customer and Customer Location.)

The programs required for temporary entity types are not as easily defined as master file programs. In addition to the four procedures listed previously for permanent entity types, these "processing programs" for temporary objects can involve combinations of adding an occurrence of one temporary object and changing the state of other permanent (and temporary) objects. In addition, temporary objects require programs that do archival updating. These may work with many occurrences of many entity types removing them from disk storage and placing them in some form of less accessible and lower cost storage.

Estimating the number of processing programs for an application is not too difficult, however. Most applications require less than a half-dozen processing programs. The programs are larger and more complex than the master file programs but they are not as numerous. In general, one processing program is required for each set consisting of a temporary entity type and the other temporary entity types that are existentially dependent on it. (Existential dependencies are more common between temporary entity types than they are between permanent entity types.)

System planning. The third impact of an object-oriented approach on the first step of the SDLC is a different approach to envisioning the system.

Instead of relying upon only a brief investigation of how the relevant subfunctions are now carried out, the planners can think in terms of how the database can most efficiently be populated and accessed to serve the basic purposes of the functional managers. This tends to free thought from the pattern of the present way of performing the subfunctions. For example, the fact that there is one database, which has already been defined in terms of the entities and relationships involved, and this one database can be shared simultaneously by all who carry out the subfunctions, makes it easier to detect parallel processing opportunities, where before a serial approach to processing an document was necessary because of a limitation in the availability of the document.

It is not expected that in this initial step a final system design will be developed. Nevertheless, the fact that the number of programs required by the application can be estimated means that the initial vision of the system is more concrete and detailed than is the case if the planners have only the process-oriented description of the application that a description of the functions to be supported tends to create. This concreteness about the software makes possible better estimates of the computer-communication system requirements of the application and of the time it will take to develop the system.

Systems Analysis. The analysis of the existing systems that will be affected if a new computer system is developed is much more detailed in the systems analysis step than in the system investigation and planning step. The existing forms and reports must be scrutinized. Timing requirements in performing various subfunctions must be defined. The use of existing data needs to be assessed. If there is data not currently available that would improve performance, the cost and utility of making it available should be estimated. The sequence of events in the current systems, and current and projected transaction volumes are defined and documented.

The existence of a BwIS and the program analysis described above can have the effect of making systems analysis more focussed and less laborious. In this environment, systems analysis should really have only two objectives. One is to identify all of the attributes (state variables) that each entity type (object class) in the conceptual database (BwIS) must have for this application. The other is to glean all of the insights that the existing system can reveal about the capabilities and features that the processing programs should have.

Exhaustive flow diagramming of the existing systems, particularly if they are manual systems, is a questionable use of resources. An analysis of those parts of the systems concerned with transaction processing and a careful analysis of all items on each form and report in the existing systems is a better use staff resources. The purpose in both of these analyses is to sift what the new system should incorporate of the old from all that should be left behind by the new system.

The attributes detected by this analysis that may be relevant to the new system should either already exist in the BwIS, or be added to it. The BwIS is assumed to be stored in an Information Resource Dictionary System, IRDS. IBM's announced Repository is an example of an IRDS that should be suitable for this purpose.

New entity types and relationships, about which data is required, can be detected by systems analysis. Particularly, in the cases of systems to support strategic and planning decisions, required additions to the BwIS are likely to come to the attention of the analysts. At this systems analysis stage it is not appropriate to make a final decision to include the new entity types in the BwIS because the new system has not yet been approved for development. However, they should be recorded in the IRDS for possible incorporation in Step 4 (System Design).

Requirements Definition and Presentation to Management

This third step is a key milestone in the development of a system. It is the point at which a definition of what the system will and will not do is formulated and a project to develop the system is proposed by the system developer to the system user. The project proposal normally defines the resources required for the development and sets a schedule.

The existence of the BwIS makes it easier for the two parties to communicate and understand one another at this stage. The functional model can be used to state more clearly which subfunctions will be supported by the proposed system and in what ways. The entity types and relationships in the BwIS that will be populated and/or used by the proposed system can be specified to clarify the data that must be entered by the users of the system and the data that they will obtain from the system.

System Design

This fourth step in the SDLC is the one in which the application programs are defined in detail. It is the one in which structured system design techniques are traditionally used. The result of this step should be software specifications that can be given to a programmer who performs the next step in the process. To serve this purpose, the specifications need to be quite detailed.

Prototyping. In working with users on the details of a design, one important guideline is to minimize the level of abstraction used. Prototyping is recommended by many as the best way to present alternatives to users in a way that they can understand. If a prototyping approach is taken, the BwIS can be used to obtain the descriptions of relations used in the relational database of the prototyping system. Most prototyping systems are built on a relational database management system and include the capability to easily create screen layouts by referencing attributes of the relations in the database. Thus, the business-wide conceptual model can be an important resource used by the prototyping system.

Access control policy. While prototyping is clearly an effective way to perfect screen designs and user interactions with the system, there are other aspects of the design for which it is not adequate. One of these is database access controls. Access to data that various classes of users should have is more efficiently reviewed with management by referring to a table of user types versus data types. An example of such a table is shown in Figure 10. In this Figure, objects are grouped by major business areas, such as Accounting, Travel Expenses, and Sales Orders. But, for each of the five types of objects shown, a list of specific entity types and relationships in the BwIS should exist. These lists are the link between the general access policy shown in the table and its implementation in physical databases. The existence of an BwIS makes it possible to coordinate access policy across all physical databases.

Design decisions. The main impact of the object-oriented approach on the design step, however, does not concern prototyping or access controls. It relates to the way the design issues are formulated, the whole approach to system design. Rather than approaching the design task as an unconstrained procedures modeling exercise, the object-oriented approach starts with the master file and processing programs (and query programs) identified in Step 1 as given. The only open design questions concern the screen designs and processing logic for these programs.

The main issues in the design of master file programs concern formats, list capabilities, and the extent of the help screens to be provided. The procedures to be executed by these programs are simple enough to be predefined (add, modify, delete and list occurrences of a permanent entity type). These programs can be automatically generated, with the programmer supplying only the parameters that vary from one program to another. (One exception is special validity checks that must sometimes be added.)

The logic of the design of a processing program is made easier to develop and communicate if the temporary entity type that the program manages (and its dependent entity types) is made the focus of attention. The temporary entity type and its dependent types can be made the basis for modeling the processing as a hierarchical structure. Omitting the details of this analysis, we can summarize the result by observing that the programs for the entity types at the lowest level in the hierarchy turn out to be similar to master file programs. The only difference is that they are called from the program of another object (invoice or cash receipts, for example) rather than from a menu of user options. Each program obtains the data for a certain dynamic object

OBJECT TYPES	Sales Staff	Inside Sales	Marketing Mgr.	Materials Staff	Materials Mgr.	Accounting & Controls Staff	General Management
Accounting		S:R/U	S:R	S:R/U	S:R	R/U	R
Travel Expenses	S:R		S:R		S:R	R/U	R
Sales Orders	S:R	R/U	R	S:R/U	R	R	R
Purchasing & Inventory	S:R	S:R	S:R	R/U	R	S:R/U	R
Products	S:R/U	R/U	R	S:R/U	R	S:R	R

S: = selected occurrences
R = retrieval access to data
U = ability to update and add data

Figure 10. Object-Oriented Database Access Control Policy

class and manages the storage and accessing of that type of object in the database (in the same way as a master file program).

In the design stage, the standard queries to be selected from a system menu may also need to be defined. This is best done by prototyping. A query can be formulated using SQL or a similar language in a matter of minutes by an experienced database administrator. Queries suggested by users can be run using sample data so that the user can confirm that the result is useful and in an appropriate format.

Remaining Steps in the SDLC

The remaining steps in the System Development Life Cycle are Programming, Implementation and Maintenance. Because they are beyond the scope of systems planning they will not be discussed here. It should be noted, however, that the standardization of program types described in the considera-tion of Step 1 has beneficial effects on these later steps.

In the programming step, the standardized program types facilitate the development of code generators. For example, the source code for programs that execute each of the four procedures for "permanent objects" can be developed with all references to the permanent data "parameterized." The program only needs to define these parameters for a specific object for the code generator to create an appropriate program.

This method of generating programs has advantages for implementa-tion and maintenance as well as for the programming step in the SDLC. It assures that all programs have the same formats and behave in a similar manner. This has many benefits for both the training and maintenance programming steps.

Conclusion

The object-oriented approach can give an I/S function a new basis for its relationships with the rest of the organization in which it exists. It can create relationships in which the I/S function is seen as facilitating the development of quality information systems rather than being the source of delays and obstacles to progress in computer usage.

The object-oriented viewpoint is not compatible with the traditional view of data processing. Attempts to mix the two by taking a process-oriented approach to system planning and an object-oriented approach to programming make systems planning unnecessarily complex. They also prevent I/S from changing its perceived role in the organization from that of application builder to that of systems coordinator.

The need for a BwIS (which IBM's announced Repository system can handle) both complicates and simplifies information systems development. It complicates it by adding a Business-wide Information Structure Planning Project at the beginning of systems development and an object class administration function on a permanent basis. It simplifies it by creating a framework that allows virtually all steps within the SDLC to be simplified and standardized.

These effects on the SDLC have very important implications for the design and development of CASE (Computer-Aided Software Engineering) tools. Unfortunately, the existing generation of CASE tools is based on the traditional approach to the SDLC steps. As the possibilities for simplification and standardization become better understood, a new generation of more powerful CASE tools seems likely.

References

Appleton, D. (1986). Very Large Projects. *Datamation*. January 15, 1986.

Chen, P. (1976). The entity-relationship model—Toward a unified view of data. *ACM-TODS. 5*, 9-36.

Gessford, J.E. (1990). *Business-wide Database Planning*. New York: John Wiley & Sons.

Goodhue, D.L., Quillard, J.A., & Rockart, J.F. (1987). *Managing the data resource: A contingency perspective*. Cambridge, MA: Center for Information Systems Research, Sloan School of Management (CISR WP No. 150).

International Business Machines (1984). *Business systems planning: Information systems planning guide*, IBM (order from your local IBM sales office), GE20-0527-4.

Izzo, J.E. (1987). *The embattled fortress*. San Francisco: Jossey-Bass.

Kay, A.C. (1977). Microelectronics and the Personal Computer. *Scientific American*. 237, 3

Kay, A.C. (1989). Predicting the Future. *Stanford engineering* (Fall Issue). Stanford, California: Stanford University Press.

Martin. J. (1982). *Strategic data-planning methodologies*. Englewood Cliffs, N.J.: Prentice-Hall.

McNurlin, B.C., & Sprague, R.H. (1989). *Information systems management in practice* (2nd ed.). Englewood Cliffs: Prentice-Hall.

Nolan, R.L., Pollack, A.J., & Ware, J.P. (1989). Toward the design of network organizations. *Stage by Stage. 9*, 1-12

Seidewitz, E. & Stark M. (1987). Towards a General Object-Oriented Software Development Methodology. *Tutorial: Object-Oriented Computing*, (Vol. 2). New York: Computer Society Press.

Wysocki, R.K., & Young, J. (1990). *Information systems: Management principles in action*. New York: John Wiley.

Part III
Information Technology Impacts on Individuals and Groups

Chapter 6

Researching the Impact of Computer Technology in the Workplace:

A Psychological Perspective

Laura L. Koppes
Wanda A. Trahan
E. Alan Hartman
Baron Perlman
David J. Nealon

University of Wisconsin Oshkosh

> ...information technology ...the convergence of several streams of technical developments, including micro-electronics, computer science, telecommunications, software engineering, and system analysis. It is a technology that dramatically increases the ability to record, store, analyze, and transmit information in ways that permit flexibility, accuracy, immediacy, geographic independence, volume and complexity. (Zuboff, 1988, p. 414)

This definition describes the wide range of activities included under the label "computer technology." The pervasiveness of computer technology also is apparent in the number of people touched by it. The past decade has experienced a rapid growth of computer technology in organizations. In fact, there are few individuals in the United States who are not affected by computer technology. Almost a decade ago it was estimated (hard data about usage are nonexistent) that between 40% and 50% of all American workers used electronic terminal equipment at some 38 million workstations (Giuliano, 1982). More recently Zuboff (1988) estimated that 65% of all professional, managerial, technical and administrative workers depend upon computer-based technologies. It is predicted that by the year 2000 computer terminals will be as commonplace as telephones or office desks (Office of Technology Assessment [OTA], 1985). Given computer technology's obvious extensive presence in the work environment, understanding its impact on people, workgroups and organizations is critical.

Computer technology's effects on people, workgroups and organizations can manifest themselves in a variety of ways. For example, profits of organizations with advanced computer technologies may be greater than those of comparable organizations with less advanced technology. The productivity of work groups with more computers may be greater than that of workgroups with fewer computers. Furthermore, workers who have computer hardware and software on their desks may be more efficient than workers who do not have access to this technology. However, understanding *how* computer technology affects organizations or workgroups requires more than simply comparing organizations or workgroups. For example, if we find differences between "computer literate workgroups" and "computer nonliterate workgroups," we don't know how or why these differences occurred. If we find that production rates of computer intensive groups are greater than those of less intensive groups, we don't know if this difference is due to each individual producing more, which may increase the workgroup's overall performance, or if it is due to greater efficiencies in how individuals work together in computer intensive groups. Because this linkage between group level performance and individual level performance is not specified, we don't know "why" the computer technology is related to greater productivity. Knowing why there are differences would provide a greater understanding of the effects of computer technology on people.

As stated above, computer technology's impact can occur at the individual, workgroup and/or organizational level. The other chapters in this book focus on the effect of computer technology on either workgroups or organizations; this chapter explores the possible effects of computer technology on *people* in organizations and what they actually do and feel. We are not concerned with the physical characteristics of the technology any more than we would have been interested in the physical characteristics of the shovel, tractor, or telephone as technologies of earlier times. These characteristics are important, but *our* interest lies in studying the impact of computer technology on people and *how* they work, not on the technology itself. Therefore, we present a non-exhaustive literature review to develop a research strategy to answer the question "Does computer technology make a difference in what people do and how well they do it?"

Studying the effects of computer technology in organizations requires developing a theory for the individual, workgroup and organization. Since computer technology's impact may be different at each of these levels, a theory would help determine the appropriate information to obtain for the level(s) of analysis desired. If a theory is not developed, it is possible that the data obtained will not answer questions of interest to management and/or academics. To assist in developing a theory and a research strategy to study the impact of computer technology on individuals at work, we review literature, present a taxonomy of relevant variables, describe research decisions, offer recommendations, and identify the most important research questions.

Background

To help identify a theory to study the effect of computer technology on people, it is useful to review the history of machine technology (the assembly line) which was the last pervasive technology to be introduced (Huber, 1984). Parallels may be found between machine technology and computer technology.

Frederick W. Taylor (1911), in the early part of the 20th century, was interested in making workers more efficient by focusing on redesigning their jobs. He introduced time and motion studies, which frequently led to simplifying jobs. He also believed that workers were motivated only by economic rewards; their job attitudes were not considered important.

In 1924, a series of studies were conducted at the Hawthorne plant of the Western Electric Co. (Mayo, Roethlisberger, & Dickson, 1939). The studies, similar to Taylor's studies, were concerned with workers' productivity and efficiency. The researchers found that the performance of both experimental and control groups increased. Although these studies have been criticized (Bramel & Friend, 1981) for a variety of shortcomings (e.g., poor methodology), two important conclusions were made: (1) workers' job attitudes and their perceptions of job characteristics are important in understanding the effects of pay, work hours, physical environment, etc. on performance; and (2) workers react not only to formal rules and procedures of the organization, but also to norms and pressures established by informal work groups. These studies stimulated researchers to address issues such as workers' attitudes, perceptions, and emotions. Thus, research on machine technology had progressed from concerns about efficiency of the technology to concerns about how the technology affected people (i.e., the "Human Relations" movement).

Research on implementing computer technology has focused mainly on either user acceptance or the technology's efficiencies, not on the impact of the technology on individuals (Franz & Robey, 1986; Galletta & Lederer, 1989; Kiesler, Siegel, & McGuire, 1984). At this point, research on computer technology appears to be in transition between the "Taylor" and the "Human Relations" eras. Several people have speculated about the impact of this computer technology on people (e.g., Landy, Rastegary, & Motowidlo, 1987; Weick, 1985; Zuboff, 1988), but few empirical studies have been conducted (e.g., Kiesler, Siegel, & McGuire, 1984). Frese (cited in Frese, Ulich, & Dzida, 1987) reviewed 300 published works on human-computer interactions, but these studies address very specific issues such as the mental model and cognitive architecture of the computer user. Edwards (1989) reviewed the effects of computer aided manufacturing while others have reviewed studies that focus on physiology and performance (e.g., Smith, cited in Frese et al., 1987). However, empirical research on psychological issues of individuals is almost nonexistent.

It has been assumed that computer technology has positive effects in the work setting (Drucker, 1985); however, we do not know its true impact on people and work. This chapter provides a theoretical framework to examine the impact of computer technology from a psychological perspective at the individual level. In developing this framework it should be recognized that computer technology has characteristics of two current research streams within "Organizational Behavior." It has the potential to affect the entire organization (i.e., organization development), and it does so by affecting individual jobs (i.e., job design) in the organization. To better understand how to study the impact of computer technology, these two areas of literature are reviewed and some research on the impact of computer technology is presented. These reviews will then serve as a basis for developing a taxonomy of variables to be studied.

Organization Development

As stated, the Hawthorne studies revealed the importance of workers' attitudes and perceptions; this emphasis evolved into the concept of organization development and change. Organization development (OD) and change is defined as a planned intervention based on behavioral science knowledge, which is implemented to improve the organization (French & Bell, 1984). Its goal is to make the organization more effective by focusing on its different levels (i.e., individual, group, and organization).

Many different interventions are included in organization development and change, all emphasizing employees and their attitudes. For example, quality circles are comprised of employees from a work area who meet regularly to identify, analyze, and solve quality related problems (Griffin, 1988). Team building stresses problem-solving skills and interpersonal relations to improve the effectiveness of work teams (Dyer, 1977). Management by objectives (MBO) (Szilagyi & Wallace, 1983) also has been implemented to affect workers' attitudes, motivation and performance. Other OD interventions include survey feedback, process consultation, managerial grid, and laboratory training. All these interventions have been implemented to enhance the effectiveness of the organization by being concerned with the employees. (See French & Bell, 1984, for further discussion on OD interventions).

Studies attempting to evaluate the effectiveness of these interventions offer differing conclusions. The focus of this research has revolved around evaluating organizational change. Does change occur and if so, is it successful? Armenakis, Bedeian, and Pond (1983) described three phases of research on OD evaluation and offered suggestions for additional research. Furthermore, Terpstra (1981) and Randolph (1982) examined research on evaluating OD interventions, focusing on methodology. Although these researchers emphasized methodological issues, a consistent theme throughout the studies of intervention is the lack of a theoretical framework guiding both the intervention and its evaluation (French & Bell, 1984; Kahn, 1974). Computer technology is used to increase an organization's effectiveness and has the potential to affect employees' performance as well as their reactions to their jobs and the organization. Therefore, information on organization development and change, particularly the literature on OD evaluations, can aid in developing a framework and strategy to study computer technology's impact on individuals.

Job Design

Job design is a second research area in organizational behavior that may be useful in developing a theoretical framework for the impact of computer technology. Early changes in job design focused on making jobs simple and standardized; subsequently, the emphasis shifted to making jobs broader and more complex. Making jobs broader required workers to have more skills, and increasing complexity made jobs more challenging. Thus, these broader and more complex jobs helped employees feel a sense of accomplishment, satisfaction, and motivation to work harder. Three methods of this job design approach were frequently implemented: job enrichment, job enlargement, and job rotation. These methods appeared successful for many years (Hackman & Lawler, 1971; Walker & Guest, 1952). However, as its success was reported (Maher & Oberbagh, cited in Saal & Knight, 1988), researchers began to identify cases in which this approach was not successful (Brief & Aldag, 1975; Robey, 1974; Wanous, 1974). One plausible explanation was the lack of theory to guide how to redesign the jobs. A theoretical perspective would integrate these disparate results and provide an understanding of why they occurred.

Job Characteristics Model. Partially to develop such a theoretical approach, Turner and Lawrence (1965) generated a list of job attributes that would lead to favorable outcomes of satisfaction, motivation, and performance (e.g., autonomy, variety of tasks, worker responsibility). Hackman and Oldham (1976) elaborated their model and developed the most widely studied theory of job attributes, the job characteristics model. This model helps focus the research on computer technology's impact by identifying job characteristics that have been shown to be related to how individuals feel about their work.

The job characteristics model is based on Maslow's (1943) need theory which states that individuals have various psychological needs (physical, safety, belongingness, self-esteem, self-actualization). Hackman and Oldham (1976) believed that certain task attributes would satisfy the higher order needs. They identified five key dimensions or characteristics of jobs: skill variety, task identity, task significance, autonomy, and task feedback. Descriptions of each characteristic follow.

Skill variety refers to the number of different skills required to complete the task successfully. For example, being a Data Processing Manager requires several skills such as planning, organizing, decision making, and supervising; therefore, the job has high skill variety. In contrast, a Data Entry Clerk often is responsible only for keying data, and has no responsibility for quality control or input screen creation. In this case only the skill to key data is used; thus the job has low skill variety.

Task identity is defined as the degree to which the worker performs a complete, identifiable piece of work starting with raw materials and finishing with the final product. Again, a job with low task identity is a Data Entry Clerk who is responsible for one part of the job. A job with high task identity is a Systems Programmer who is responsible for developing entire computer systems.

Task significance refers to the impact that a person's job has on others in and out of the organization. An example of a job high in task significance is a Mainframe Repairperson; the success the person has in solving a hardware problem will eventually affect every member of the organization. Even though no job has a total lack of task significance, an example of one lower on this characteristic is a Custodian. One would hope employees could do their jobs whether or not there is dust on the floor.

Autonomy is defined as the freedom to make decisions, independent of supervision, about how to perform the job. An example of a job high in autonomy is an outside salesperson who makes sales calls and identifies ways to make sales. An example of a job low in autonomy is a person who packages PC's for shipment; written specifications for packaging are provided.

Finally, *task feedback* refers to the information an employee receives about effectiveness from doing his/her job. An example of a job high in task feedback is a Software Trainer. A trainer can tell immediately whether or not he/she is effective by observing participants. An example of a job low in task feedback is a Software Designer. People who design software do not know if they are effective until the product has been marketed and sold.

Hackman and Oldham (1976) theorized that these key characteristics are positively related to outcomes such as worker attitudes and behaviors; specifically they have an effect through critical psychological states. These states are "Experienced Meaningfulness of Work," "Experienced Responsibility for Work Outcomes," and "Knowledge of Results." Skill variety, task identity, and task significance affect "Experienced Meaningfulness of Work." The degree of autonomy affects "Experienced Responsibility for Outcomes of Work." The degree of task feedback affects the "Knowledge of Results." According to Hackman and Oldham (1975), four general outcomes may be affected by the critical psychological states: internal work motivation, work performance, satisfaction with work, and withdrawal (absenteeism and turnover).

In addition, Hackman and Oldham (1976) believed that these job characteristics are designed to satisfy certain higher order needs (i.e., Maslow's self actualization and esteem needs). This means that the model may not hold for those individuals who are at lower order needs. Thus, they proposed that growth need (higher order needs) strength (GNS) would moderate the relationship between the job characteristics and outcomes such as satisfaction. Only those individuals high in GNS would be affected by the job characteristics through their psychological states.

Implementing computer technology affects the nature of work; thus the job design theory of Hackman and Oldham (1976) provides an initial framework to study computer technology's effects. Before our framework is presented, it is necessary to provide a brief review of research on the impact of computer technology.

Impact of Computer Technology

Descriptions of three studies that represent three different approaches to examine the effects of computer technology follow. The first study (Kiesler, Siegel, & McGuire, 1984) used a laboratory experiment to examine the effect of computer communications on decision making. The second study (Rice & Contractor, 1990) explored the impact of computer technology in one organization by comparing measures taken before implementing the technology with measures taken after implementing it. The third study (Franz & Robey, 1986) examined perceived usefulness of computer technology across organizations.

Kiesler, Siegel, and McGuire (1984) had three-person groups reach consensus in one of three situations: (1) face to face, (2) using the computer with communications not being anonymous, and (3) using the computer with communications being anonymous. They found that computer mediated groups took longer to reach consensus and communicated less. However, there was greater equality of participation in computer mediated groups, and individuals were more willing to change their minds. This laboratory study allowed the researchers to gain information about specific phenomena. Even so, a theory to understand why the results were obtained was not presented.

Rice and Contractor (1990) utilized a research strategy proposed by

Golembiewski, Billingsley, and Yaeger (1976, see section on Data Collection Strategies for more information on this approach) to study the effects of organization development techniques. Rice and Contractor (1990) studied the impact of introducing desktop computers and a complete computer information management system in a federal office building. They obtained data both pre- and post-introduction of the system. Of interest was the impact of the technology. Workers reported being more effective on most measures, but it was found that this was because workers changed their beliefs about how effective they were before the technology was introduced. Also, the new technology changed how workers conceptualized work. For example, workers initially viewed electronic mail as an information management activity; after the technology was introduced they viewed it as a communication activity.

These results are of interest for several reasons. The study is one of the first to utilize an OD research methodology to study the impact of computer technology. The results show how individual workers will change how they "think" about work rather than just measuring short term productivity. Finally, the study points out the importance of studying the effect of computer technology at the individual worker level. Despite these important findings, the study stands alone without a framework or theory either to interpret the results or to direct the next study.

Franz and Robey (1986) employed a cross sectional research strategy to identify organizational and computer technology effects on perceived usefulness of the technology. Using data from 118 managers in 34 organizations, they found that organizational variables (e.g., size, decentralization) were not related to level of involvement of managers in computer technology, but they were related to perceived usefulness. The longer the MIS department had been in existence, the less useful the technology was perceived. Overall, they found that organizational variables played a significant role in manager's perceptions of technology's usefulness. Again, the importance of individuals and their perceptions is indicated by the obtained results. But, how and why individuals are affected by technology cannot be systematically explored without a theoretical framework.

Summary

All three literature reviews (OD, Job Design, and Impact of Computer Technology) lead to the same conclusion: an overall framework is needed to study how and why computers affect individuals within organizations. This framework should include variables related to decision making, organizational structure, characteristics of jobs, and outcomes. Following is a description of a taxonomy of variables that can provide a framework for studying computer technology's impact on people.

Taxonomy of Relevant Variables

Assessing the impact of computer technology on people at work requires a model identifying potential variables to measure. As the literature reviews demonstrate, little is known about the effects of computer technology on people's work from a psychological perspective. Some researchers have shown that computer technology does impact the organization structure, work roles, job requirements, decision making, and interactions/relationships. However, little empirical work investigates the nature of these effects. Much research focuses on specific issues such as outcomes (e.g., profits, speed, number of claim forms processed, tons of paper produced) (Zuboff, 1988). Some information on the specific types of tasks that can be performed (Broadbent, 1990) is available. However, this research attempts to maximize the effectiveness of the computer human interface. Its objective is efficiency, not describing technology's effects (including but not limited to effectiveness) on humans.

Because there is limited empirically based information about the effects of computer technology on people, it is not possible to construct a "model" of this process. However, based on literature reviews of computer technology, job design, and organization development, a taxonomy of variables that may be related to the effects of computer technology must be developed. Variables in the taxonomy come from three sources. Some variables were proposed by theorists (e.g., Zuboff, 1988) as being important for understanding computer technology and people. Others were studied by researchers who have empirically demonstrated effects on people (e.g., Hackman & Oldham, 1976). The remainder originated from interviews with individuals who described the effects of computer technology on their jobs and work groups. The taxonomy is presented in Figure 1.

Level 1 - Contextual and Computer Technology Variables

The taxonomy is divided into three levels, with the top row of variables describing the context (organization and the workgroup) and characteristics of computer technology. In this taxonomy, these variables are treated as "given" or "situational" variables in which people work. It is expected that computer technology will affect people differently in different types of organizations and workgroups. For example, the impact of computer technology on workers at McDonald's may be very different than on workers at Ford Motor Company. Similarly, the impact on a work group with few employees and a skilled manager is likely to be very different than on a work group with many employees and an unskilled manager.

Contextual variables include characteristics of the organization's environment when implementing computer technology. Rafaeli (1986) and

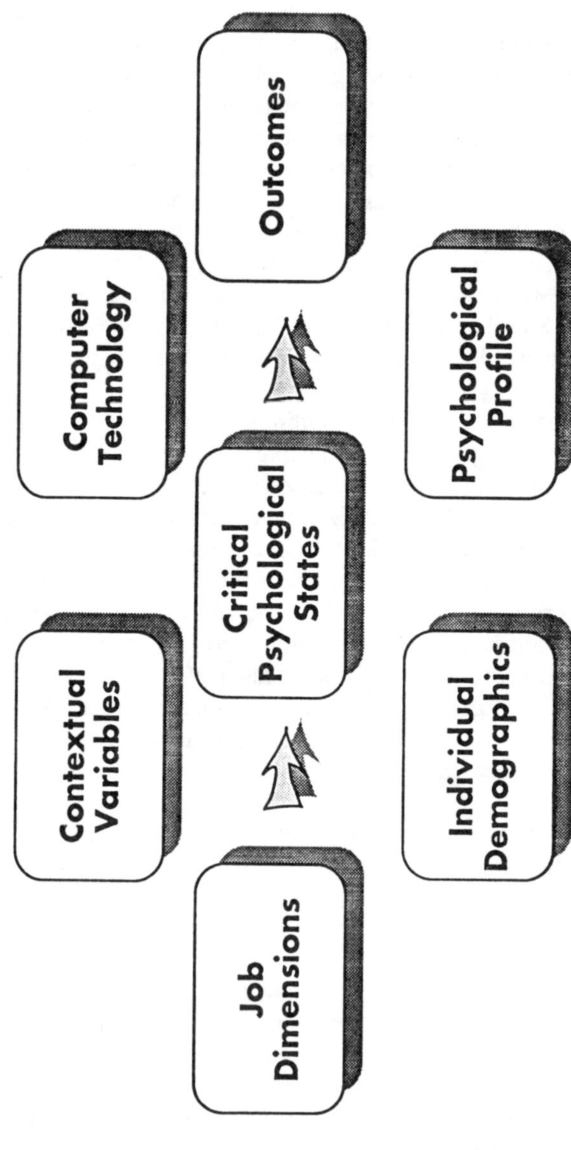

Figure 1. A taxonomy of the impact of computer technology in the workplace: A psychological perspective

Organization:
- Type of Industry/Niche
- Gross Sales
- Profits
- Return on Investment (ROI)
- Number of Employees
- Number of Locations
- Stage of Development (e.g., growth)
- Culture/Climate
- Training Opportunities
- Slack Resources
- Organization Structure
- Corporate Strategy
- Span of Control
- Reward System
- Communication Processes

Work Group:
- Number of Workers
- Leadership Style of Manager
- Computer Expertise of Manager
- Attitudes of Coworkers
- Homogeneity of Work Group

Table 1. Contextual Variables

Blomberg (1987) recommended that support, specifically opportunities to use the computer and access to knowledgeable and helpful colleagues, is needed to successfully implement computer technology. Metz (1986) identified the importance of culture, specifically participative decision making, to effectively implement technology. Klein and Hall (1988) concluded that five changes must take place in the organization when implementing technology: (1) training and counseling employees who lack skills in using the technology, (2) redesigning jobs to increase effectiveness of humans and machines, (3) increasing worker participation in selecting and implementing new technologies, (4) redesigning organization structure to create flatter organizational hierarchies, and (5) increasing the use of experts who design, program, and maintain the new automated equipment. Examples of organization and workgroup contextual variables are presented in Table 1.

It also is expected that the impact of computer technology will depend on the characteristics of the technology. Does the organization utilize state-of-the-art technology, or does it use older generations of technology? Is the technology stand-alone PC type, or is it a mainframe with local area networks? Majchrzak and Klein (1987) differentiate between two types of computer automated technologies. Programmable manufacturing automation includes equipment and software such as robots, computer assisted design, computer assisted manufacturing, and computer integrated manufacturing. Office automation includes technology such as microcomputers, word processors, facsimile machines, optical scanners and electronic mail. It is expected that all of these types may influence how technology impacts people at work. Table 2 presents variables related to computer technology.

Level 2 - Job Dimensions, Critical Psychological States, and Outcomes

The second level of variables consists of specific job characteristics and how they affect people at work. These are labeled "Job Dimensions," "Critical Psychological States," and "Outcomes." The first two categories were identified by Hackman and Oldham (1976) and have been extensively researched (see section on Job Design for detailed descriptions). They reflect how specific aspects of work influence individuals and what individuals do about them. Weber (1988) identified several job characteristics affected by computer technology. Others also have speculated about the effects on job content, decision making, organization structure and autonomy (Czaja, 1987; Olson & Lucas, 1982; Stout, 1981). These characteristics fit into the categories outlined by Hackman and Oldham (1976).

Some research has shown the importance of social relationships on the job, work group cohesiveness, group norms and productivity. For example, research has demonstrated that work group norms for productivity and work group cohesiveness (extent of forces on individuals to remain in the group) are related to group productivity (Keller, 1986). Groups that have goals that are compatible with the organization's goals and are cohesive will be more productive than groups that are not cohesive. However, if the group's goals are not compatible with the organization's goals, then more cohesive groups will be *less* productive than low cohesive groups. Based on this research, it might be expected that the effect of computer technology on core job dimensions will vary as a function of the group's goals and cohesiveness.

Newness of the Technology
Level of Integration
Length of Use in Organization
Uses of the Technology
Level of Expertise in the Organization
Who Controls Access
Level and Ease of Access
Types of Technology

Table 2. Computer Technology

One aspect of the job that may be affected by computer technology is the social interactions, but the direction of any effects is in question. Zuboff's (1988) research led her to conclude that computer technology dehumanizes and worsens the relationships of workers. Also, Landy et al. (1987) speculated that computer technology can significantly reduce the degree of social interactions. On the other hand, interviews we conducted in a University setting indicated that computer technology had enhanced the quality of relationships for many employees. (The computer technology in the large organizations Zuboff and Landy et al. studied may be different from the technology employed in the University, which perhaps is one source of these differences.) We need to study the time computer technology takes away from or adds to peoples' ability to relate with each other, the failing or enhanced quality of these interactions, and the nature of these relationships (social or work related, people defending decisions or collegial, etc.).

Job dimensions influence outcomes through critical psychological states described earlier (i.e., meaningfulness of work, responsibility, knowledge of results). For example, jobs providing little freedom to make decisions result in individuals accepting less responsibility and having little motivation or satisfaction with work. Various outcomes have been identified in the computer technology literature including motivation, work performance, satisfaction, affective reactions, organization effectiveness, stress, absenteeism and turnover (Czaja, 1987; Majchrzak & Klein, 1987; Stout, 1981). Internal work motivation refers to the degree to which the individual feels a sense of accomplishment and pride from performing the job. Work performance includes quality, quantity, effectiveness and efficiency. Satisfaction is defined as one's attitude toward his/her work. Stress is the self-perception that the task or job demands exceed the individual's resources. Absenteeism and turnover describe the withdrawal from work.

The question of interest is how computer technology affects "Job Dimensions," "Critical Psychological States," and "Outcomes." Tables 3 through 5 present the variables in these three categories.

Skill Variety (number of different skills used)
Task Identity (degree to which person completes whole product)
Task Significance (importance of job to organization)
Autonomy (freedom to make decisions regarding job)
Feedback (amount of information about job performance)
Interactions with Others

Table 3. Job Dimensions

Meaningfulness of Work
(Skill Variety + Task Identity + Task Significance)
Responsibility (Autonomy)
Knowledge of Results (Feedback)

Figure 4. Critical Psychological States

Work Motivation
Satisfaction
Quality of Work
Quantity of Work
Creative Solutions to Problems
Stay or Leave Company (Intentional and Actual)
Stress
Absenteeism
Tardiness
Satisfaction with Interactions with Others
Psychological/Physical Problems

Figure 5. Individual Level Outcomes

Level 3 - Individual Demographics and Psychological Profile

The bottom tier of the taxonomy depicts what the individual brings to the organization. The first category refers to personal demographics that are clearly verifiable such as age, gender, training received, and degrees received. The second category is likely to be more important but contains variables much less visible and verifiable relating to a person's psychological "profile." It is expected that these demographics and psychological profile variables will influence the relationships among job dimensions, critical psychological states and outcomes (e.g., some individuals prefer not to have to make decisions at work; therefore, lack of autonomy and lower job responsibility will actually cause these people to be satisfied instead of dissatisfied).

Some psychological profile variables are directly related to computer technology and an individual's attitude toward it. For example, some people may be "resistors"; they feel threatened by a computer and either refuse to learn or limit their learning on how to use the technology. On the other hand, some people are "acceptors"; they look favorably upon computer technology and are willing to learn and use it. In addition, locus of control (i.e. attributing events to either person or fate) can affect individuals' attitudes and performance.

Position in Organization
Gender
Age
Tenure in Organization
Education
Computer Training
Computer Expertise

Table 6. Demographics

General:	
	Growth Need Strength (degree a person needs to learn)
	Locus of Control (is success a result of person or fate?)
Computer Technology:	
	Resistor/Acceptor (resists or accepts change)
	Navigator/Procedural (degree a person experiments)

Table 7. Psychological Profile

Coovert and Goldstein (1980) found that persons with internal locus of control had more favorable attitudes towards computers than those with external locus of control.

Another dichotomous psychological profile variable refers to how one uses the technology. "Navigators" like to explore with the technology; they attempt to identify various ways of solving problems and manipulating the technology. On the other hand, "Procedural" users follow a specific set of steps and will not deviate from them. When they have a problem, they call an expert. Instead of exploring, they only want to know how to solve a specific problem.

Tables 6 and 7 present the variables included in the categories of "Individual Demographics" and "Psychological Profile."

As one can see, we have expanded the job characteristics model in a variety of ways to examine the impact of computer technology on work in its broadest sense. This taxonomy points to several important research issues that deserve elaboration.

Research Issues

Definition of Computer Technology

To study the effects of computer technology, it is important to have a clear definition of the concept of "computer technology." Technology refers to anything from "job routineness to the hardness of raw materials" (Rousseau,

1983, p. 230). Technology can be either an object, such as a computer, or the process of a flow of information within an organization. Hulin and Roznowski (cited in Goodman, Sproull, & Associates, 1990) define technology as "the physical combined with the intellectual or knowledge processes by which materials in some form are transformed into outputs used by another organization or subsystem within the same organization" (p. 47). Similarly, Law (cited in Goodman et al., 1990) defines technology as "a family of methods for associating and channeling other entities and forces, both human and nonhuman. It is a method, one method, for the conduct of heterogeneous engineering, for the construction of a relatively stable system of related bits and pieces with emergent properties in a hostile or indifferent environment" (p. 115). Berniker (1987) believes "Technology refers to a body of knowledge about the means by which we work on the world, our arts and our methods. Essentially, it is knowledge about the cause and effect relations of our actions... Technology is knowledge that can be studied, codified, and taught to others" (p. 10).

In Rousseau's (1983) review of 31 studies of technology, it was found that technology was defined most frequently as technical complexity, the transformation of input into output, or the operations performed on an object. These definitions reveal the distinction between technology as information and the technical system as a specific subsystem of the technology (Weick, 1990). Thus, technology refers to the information needed by a technical system. Computer technology, then, is information required by a computer-based technical system.

Even though computer technology falls under this umbrella of technology, and a simple theoretical definition of computer technology is easily derived from the technology literature, identifying a workable definition may be more difficult. To begin with, in the paucity of research on the impact of computer technology, a definition is rarely provided by researchers. Most researchers assume that the managers and/or employees know what computer technology is. For example, in Weber's (1988) study of computer technology's impact on job characteristics, a definition was not included in the article, but employees were asked to assess the impact of computer technology on their jobs.

When definitions of computer technology are provided, they vary widely in content. For example, one can conceptualize computer technology as automating assembly lines, utilizing personal computers, implementing the telephone, or using computer terminals attached to a mainframe.

As stated earlier, Majchrzak and Klein (1987) differentiated between two types of computer technologies: programmable manufacturing automation and office automation. Programmable manufacturing automation encompasses technologies that vary in size, technical complexity, and impact. Office automation refers to the variety of computerized technologies currently available to assist in the effective procurement, storage, analysis, retrieval, and communication of information.

Furthermore, in interviewing a variety of workers (managers to secretaries), we found that individuals defined computer technology differently. In response to the question "What does computer technology mean to you?" one individual replied, "When I think of computer technology, I think of it in broad terms, for example, the automation of assembly lines." Another person gave a more specific definition, "I define computer technology as hardware and software for personal computers."

It is apparent that even though we can theoretically define computer technology, the reality is that each potential subject of computer technology research defines it differently. These definitions seem to be a function of individuals' exposure to, knowledge of, and use of computer technology as well as their overall frame of reference. To conduct research on the impact of computer technology, it is necessary to identify a definition of computer technology that is relevant, meaningful, and useful to the people being studied.

Methodology

Introducing computer technology is a form of organization development; thus, reviewing organization development intervention studies may provide insights into methodological problems encountered when conducting this type of research (e.g., Porras & Berg, 1978; Terpstra, 1981; White & Mitchell, 1976). Three main problem areas have been addressed: (1) design of the study, (2) sampling, and (3) data collection.

Design. In designing an investigation within an organization, the choice between a laboratory or a field study often is made implicitly in favor of the latter. Even so, studying computer technology provides opportunities to utilize laboratory methodologies in which variables are directly manipulated and controlled. For example, studying fatigue resulting from different combinations of computer equipment, or how individuals would use different types of information provided by a computer system could be done most easily through a laboratory study. Because of the control of variables, stronger causal inferences could be made about the results. However, the conditions of a laboratory setting are less realistic; therefore, the generalizability of the findings may be lessened.

Field studies take place in a natural setting, such as an organization. Using field studies usually increases the generalizability of results. However, because of having less control over variables, it is difficult for researchers to establish causal relationships. Field studies have been the methodology of choice by researchers in organization development.

In addition to choosing either a laboratory or a field investigation, the design of the study should be considered. Campbell and Stanley (1963) discussed three main experimental designs: pre-experimental designs, true experimental designs, and quasi-experimental designs. Pre-experimental designs use one group. No control group is included, for example, when describing the effects of computer technology on one organization (Kiesler, Siegel, & McGuire, 1984; Rice & Contractor, 1990). Second, a true experimental design randomly assigns subjects to experimental and control groups. An example is a study where some managers received negative information about the success of implementing computer technology and some managers did not receive any information about computer technology (e.g., Galletta & Lederer, 1989).

While true experimental designs control more for threats to internal and external validity (i.e., history, selection, etc.), they are difficult to implement and often not possible in an organizational setting (Nicholas, 1979). For example, it is not always possible to randomly assign workers to different groups. Therefore, Campbell and Stanley (1963) suggested using a quasi-experimental design.

Quasi-experimental designs measure the effects of some "treatment" on one group over time or on groups who have not been randomly assigned. An example is comparing the effect of computer technology on existing work groups (or organizations) (e.g., Franz & Robey, 1986) that differ from each other in many ways. Campbell and Stanley (1963) pointed out that this leaves this type of study open to several threats to internal validity. For example, individuals divided themselves into the groups; therefore the groups may differ initially (selection). There are several other possible threats, and the reader is encouraged to refer to Campbell and Stanley (1963) for more information about threats to internal and external validity and for a complete taxonomy of quasi-experimental research designs.

Another research design approach is a longitudinal design, which allows the researcher to look at trends over time but does not include treatment effects (Armenakis et al., 1983; Porras & Berg, 1978; White & Mitchell, 1976). Many times the full impact of an intervention, such as implementing computer technology in a work setting, does not occur quickly; therefore, a longitudinal design would give the researcher and the organization the opportunity to learn about its effects at various points in time. It also may provide information about optimal length of time required. However, threats to internal validity (selection, history and regression to the mean) and external validity (e.g., multiple treatment effects may occur because many things are done in the organization over a period of time and isolating the effect of one is not possible) exist for this design.

Careful consideration must be given to the research design. The options should be closely examined and carefully selected, keeping in mind the purpose of the research.

Sampling. A second methodological issue is one of who to include in the research. Should everyone who has been affected by computer technology be included; or should we sample a portion of those organizational members? Do we include users as well as nonusers of technology? If we choose to sample workers, then it would be preferable to use random sampling, where each person has an equal chance of being selected. However, random sampling is not always possible in organizations, since often management prefers to choose those workers or work groups who will participate (Nicholas, 1979).

Another possibility is to use stratified sampling where first the population to be studied is divided into levels, then subjects are sampled from each of these levels (Stone, 1978). For example, in studying the impact of computer technology, a researcher might choose to divide organizational members according to their level or position in the organization. Following this, subjects would be sampled from each level. The advantage of stratified samples is that all levels of the organization are proportionally represented, which allows the researcher to determine if there is a differential impact of the technology at various organizational levels. Key informants or workers who are considered successful in the organization also may be used. It can be difficult to develop measures of effectiveness, so using key informants can be useful because they already have been identified as "effective" by the organization.

Finally, some type of control or comparison group is needed to determine the impact of computer technology. Porras and Berg's (1978) evaluation of organizational change research concluded that control and comparison groups are seldom used in these investigations. Often this is because management wants the intervention applied to everyone and does not want to "hold out" certain workers or work groups. Additionally, if control or comparison groups are used, often they are from different departments or are at different levels in the organization. This implies that differences other than the technology may exist. Nonusers of computer technology may be possible for a comparison group; however, the problems with matching also apply here. Although a group does not use the technology, we cannot be certain that its responses will not be influenced by the indirect effects of technology on the group members and their jobs.

Selecting subjects for experimental and control or comparison groups in organizational studies is an important consideration. When studying the effects of technology in an organization, it is preferable to sample members from as many levels in the organization as possible.

Data Collection. With any type of organizational change, measures should be taken before and after the intervention to determine the impact on workers (Nicholas, 1979; White & Mitchell, 1976). As Rice and Contractor

(1990) have demonstrated, measuring this impact also applies to studying computer technology; discovering that workers changed how they conceptualized work would not have happened if only simple post measures were used.

To study the impact of computer technology, we need to know about workers' productivity, satisfaction, etc. both prior to and following the implementation of computer technology. However, since most organizations already use some form of computer technology, it is difficult to get information about pre-computer days. Historical records may or may not be available, and these records may not contain all the information needed (i.e., worker perceptions). Therefore, using retrospective data may be helpful. While this may sound like an ideal solution, there are concerns with it. First, organizations, jobs, and people continually change; therefore, it is difficult to be certain that the changes measured before and after implementing technology are actually due to the technology itself and not to other organizational changes (White & Mitchell, 1976). Second, as mentioned earlier, people have different definitions and ideas of technology. When asked about conditions prior to computer technology, some may refer to personal computers while others may refer to earlier forms of technology. Therefore, retrospective data as a surrogate for "pre" data should not be used without careful planning, consideration, and pilot testing with debriefing.

Even with adequate pre and post measures, simply comparing these measures may not be sufficient to really assess the impact of computer technology on people. In 1976, Golembiewski, Billingsley, and Yaeger proposed that three types of change may occur as a result of organization development interventions. The first type of change, labeled alpha, is defined as a real change. People in the organization evaluate the organization as being better after the intervention than before it. An example of an alpha change is when on a pre-measure of "communications" the average score is three on a five point scale, and on a post measure the average score is four.

The second type of change, beta, is when workers recalibrate the response scale, indicating that workers' standards have changed since the intervention. The intervention may provide the workers with a new standard of comparison, which changes how they define the ends of the continuum. For example, prior to change, a "five" on the "communications" scale ("five" being the maximum) may mean to workers that their supervisor talks with them directly at least once every work day. After the intervention, a "five" may mean that their supervisor talks with them several times a day.

Finally, gamma change is represented by a redefinition of what is being measured. Prior to the intervention, workers might define "communications" as a method by which management obtains information as part of a control process. If, after the intervention, workers define "communications" as using electronic mail to communicate with coworkers, as Rice and Contractor (1990) found in their study, then gamma change would have occurred.

While pre and post measures may be used to detect alpha and beta change, some researchers have argued for using retrospective measures in conjunction with pre and post measures to identify gamma change (Terborg, Howard, & Maxwell, 1980). This means that when measuring the impact of computer technology, the ideal research strategy involves three separate measures; one taken prior to, and two taken after introducing technology.

Measurement of Variables

The final research issue to be addressed is measurement of variables. Many methods of collecting data have been used (e.g., questionnaires, interviews, and "hard" data such as productivity records). One major problem is that no standardized measures exist for the specific purpose of investigating the impact of computer technology in the workplace. Standardized instruments may be used to measure some aspects of the employee and the job; for other aspects either standardized instruments will need to be modified or homemade instruments developed.

Our taxonomy has theoretically defined the variables one would place in a model; but to test the model, we need to operationally define and measure these variables. The central issue is objectivity versus subjectivity. A common criticism of psychological research is its frequent use of self-report measures, which are subject to several biases. We understand the importance of objective measures; however, we found from our interviews and literature search that objective measures do not always exist! For example, two contextual variables of interest are the organization's climate and its culture. Both of these are defined as employees' perceptions, with climate being their perceptions of the organization's characteristics and culture being the meaning they attach to events in the organization. In measuring these, we must rely on either self-perceptions or third party observations. Organizational climate has been fairly well defined and several instruments are available (James & Jones, 1980). However, culture has only recently been used as a variable in organizational research, and we could find no comprehensive instruments that have been validated.

Hackman and Oldham (1975) designed the Job Diagnostic Survey (JDS) to measure the five core job characteristics and critical psychological states. Although Hackman and Oldham (1975) described the reliability and validity of the JDS, other researchers have not replicated their findings. In addition, the survey relies on employee self-report for the dimensions.

Several outcome variables lack objective measures as well. For example, from our interviews we found nine definitions of effectiveness. Each individual might rate the effectiveness of the organization, but results would be meaningless because of the different definitions of effectiveness. In fact in our interviews, some people struggled when identifying how they know they are effective. One person said, "We know we are effective when people are doing what they are supposed to be doing." Another person said the feedback received from students indicates to him how the department is functioning. Some organizational units did have objective measures. For example, one person commented that the number of students graduating and gaining employment is a measure of effectiveness.

Other outcomes of interest include emotional responses to and attitudes towards computer technology, which are difficult to measure objectively. For example, measures of job satisfaction, by necessity, are self-report questionnaires. The two most commonly used measures of satisfaction are the Minnesota Satisfaction Questionnaire (MSQ) (Weiss, Dawis, England, & Lofquist, 1967) and the Job Diagnostic Index (Smith, Kendall, & Hulin, 1969). While each instrument has been extensively researched, the short version of the MSQ contains only 20 items, and may be easier to include in a larger questionnaire.

Furthermore, objective outcome variables (e.g., leaving the organization, absenteeism, productivity), while often easy to quantify, pose a problem for the researcher. Such data are maintained by the organization; however, to tie these data to individual responses on a questionnaire requires that anonymity cannot be provided to the respondents. This causes concern about the validity of the data obtained. A respondent's "intention to leave" has been shown to be related to actual turnover of employees (Hartman & Perlman, 1986; Mobley, Griffeth, Hand, & Meglino, 1979) and has been used as a surrogate for actual turnover data.

In conclusion, since we could find no comprehensive measures of the impact of computer technology on people, most measures used in investigations will have to be either developed or modified from existing standardized measures. If these measures are used to investigate conditions prior to implementing computers, the researcher needs to present clear definitions of computer technology to give subjects a common frame of reference when answering questions.

Summary

Any researcher investigating the influence and impact of computer technology on people is entering a realm with many currents, at times difficult to read and navigate. We can make some recommendations for the issues we have raised. These are not meant to be all encompassing — research in the area of the impact of computer technology on people is in its infancy. Most importantly, it is necessary to think clearly. Researchers must thoroughly define what they want to know. Clarity of thought can be facilitated by realizing that the questions one wants answered drive the variables to be defined and measured, the methodology to be adopted, and the instruments or measures to be utilized. In addition, developing and utilizing a taxonomy or model brings to life the ideas of the researcher and forces him/her to be explicit about what is being studied. A model is simply a way of describing the domains under study, the variables within each domain, and the relationships of these variables. In this chapter, we presented a taxonomy, not a model, because we are unsure of the relationship of the variables within and between the domains.

With clear thinking, research on the impact of computer technology can be conducted successfully. First, computer technology must be defined. It is important that the definition includes a detailed description of the organization and computers (systems) being researched. Unless researchers describe the computer technology clearly and in detail, those who synthesize the research data in the years ahead will be unable to categorize the past research and will have difficulty drawing conclusions from the research findings.

Second, decisions about methodology must be made. As stated earlier, organizational research can be difficult and presents several problem areas including design, sampling, and data collection. Since we are interested in the impact of computer technology on people in the workplace, quasi-experimental and longitudinal designs are appropriate. However, using laboratory studies should not be ignored by those researchers interested in a tightly controlled investigation of specific effects. For sampling, it is suggested that members from as many levels as possible in the organization be included in the research. To collect the data, several measures should be taken prior to and after introducing the technology over a period of time.

Finally, with clear thinking, the variables can be theoretically and operationally defined; thus measures can be developed and selected. Although objective measures are preferable, subjective measures may have a place in this research. Existing standardized measures should be closely examined and considered. If new measures need to be developed, expertise and experience in instrument design is imperative.

To reiterate, before the research is conducted, the researcher needs to know the question(s) to be answered. Therefore, in the next section, we offer several research questions that will facilitate the research on the impact of computer technology on people from a psychological perspective.

Research Questions

The overriding research questions concern variables in the taxonomy and their relationship to the impact of computer technology in the work place. This includes questions such as: Which variables demonstrate the strongest relationship with outcome measures? Which variables are not significantly related to technology's impact on the work place? What are the causal relationships in the taxonomy (i.e., where should causal arrows be placed and in which direction)?

The following sets of specific questions will help form a research agenda for studying the impact of computer technology on the work place. These questions either are a direct result of the taxonomy or are logical extensions from the taxonomy; however, they are not exhaustive. They are divided into three categories: impact questions, context questions, and methodology questions.

Impact Questions:

- How much initial and ongoing training is necessary for organizational members to feel comfortable with computer technology?
- Does computer technology impact workers' job attitudes and their perceptions of job dimensions (job characteristics)?
- Can people be grouped based on their psychological profile variables, and do these groups react differently to computer technology?
- Which types of organizational members use computers and which do not? Why?
- Does computer technology simplify and standardize jobs?
- What is the impact of computer technology on skill variety, task identity, task significance, autonomy, and task feedback (Hackman & Oldham's five job dimensions)?
- How does computer technology affect, alter or enhance worker relationships?
- In what basic ways does using computer technology change the experience (meaningfulness, responsibility, feedback, participation in decision making) of work?
- How long after introducing computer technology does its impact on work and outcomes become apparent?
- How long after introducing computer technology does its impact on work and outcomes reach maximum effect?
- Is the length of time to impact and maximum impact different for different outcomes and various work characteristics?

Context Questions:

- How does the impact of computer technology on the psychology of work differ, given different types of contextual variables (different industries, training for individual workers, etc.)?
- How many computer technology types is it important to study? Are there similarities between different computer technologies so that research in this field generalizes from one type of computer technology to another?

- Does organizational culture/climate affect the way computer technology impacts on people or work?

Methodology Questions:

- What types of research designs "work best" in this arena of research?
- How useful are laboratory studies in this area of research?
- Can norms be developed for some of the variables presented in our taxonomy? Can standardized instruments be developed to use in this research area?

As stated, these questions are not all encompassing. However, we provided them to stimulate research on the impact of computer technology in the work place from a psychological perspective.

Future Trends

Experts predict the labor market may experience a shortage of workers throughout the next decade because there will be fewer younger people entering the job market than in the past (Offermann & Gowing, 1990). The labor force will be increasing at a slower rate than at any time since the 1930's (Rauch, 1989). In addition, the demographics of the workers will be changing. We will observe an influx of women and minority group members (i.e., blacks and hispanics). The skill level of the entry-level workers is of concern, especially in computer technology. Organizations can no longer be selective, retain the skilled and eliminate the uneducated (Offermann & Gowing, 1990). Organizations now find it difficult to recruit entry-level workers and spend close to 210 billion dollars for on-the-job training.

Furthermore, the next decade will continue to observe a change in the nature of work. New jobs will generally require higher education and skill levels than those that were made obsolete (Miller, 1989). In addition, U.S. companies are concerned about being competitive and maintaining organizational productivity. To be competitive and productive, it will probably be necessary to invest in state-of-the-art computer technology.

With the decrease in the number of workers, the change in demographics, and the increase in computer technology, it will be extremely important for companies to retain their employees. These employees will probably be unskilled, but will be required to do complex jobs.

Therefore, we must be concerned about the psychological effects of computer technology. It has been said that computer technology is as major and influential a technology as the telephone and typewriter in its impact on organizational functioning, productivity, and innovation. As is true of any major technology or technological system, the influence of computer technology is widespread, disparate, and complex; and it influences different people in different ways. As mentioned, these effects may not always be as positive as we once speculated. Careful decisions about implementing, managing and evaluating computer technology must be made.

Conclusion

To summarize, we have presented a variety of issues that need to be considered for researching the impact of computer technology in the work place from a psychological perspective. Because the empirical research is limited, examining other literature (e.g., organization development and change) and identifying a theory or model (e.g., job characteristics model) is necessary for guiding the researcher. In addition, the researcher needs to define computer technology, make methodological decisions and determine ways to measure the variables of interest. Once these issues are addressed, the next step is the gathering of data to answer the basic question, "What is the impact of computer technology on individuals?" This includes examining changes in jobs and attitudes about jobs. As we begin to understand the impact at the individual level, we then can explore the impact at the group and organization levels.

References

Armenakis, A., Bedeian, A., & Pond, S. (1983). Research issues in OD evaluation: Past, present, and future. *Academy of Management Review, 8,* 320-328.

Berniker, E. (1987). *Understanding technical systems.* Paper presented at symposium on Management Training Programs: Implications of New Technologies, Geneva, Switzerland.

Blomberg, J. (1987). Social interaction and office communication: Effects on user's evaluation of new technologies. In R. Kraut (Ed.), *Technology and the transformation of white-collar work* (pp. 195-210). Hillsdale, NJ: Erlbaum.

Bramel, D., & Friend, R. (1981). Hawthorne, the myth of the docile worker, and class bias in psychology. *American Psychologist, 36,* 867-878.

Brief, A.P., & Aldag, R.J. (1975). Employee reactions to job characteristics: A constructive replication. *Journal of Applied Psychology, 60,* 182-186.

Broadbent, D. (1990). Two books on human-computer interaction point up the general problems of applying cognitive psychology to human-artificial systems. *Psychological Science, 1,* 235-239.

Campbell, D.T., & Stanley, J.C. (1963). *Experimental and quasi-experimental designs for research.* Chicago: Rand McNally.

Coovert, M.D., & Goldstein, M. (1980). Locus of control as a predictor of users' attitude toward computers. *Psychological Reports, 47,* 1167-1173.

Czaja, S.J. (1987). Human factors in office automation. In G. Salvendy (Ed.), *Handbook of human factors* (pp. 1587-1616). New York: Wiley.

Drucker, P.F. (1985). *Innovation and entrepreneurship: Practices and principles.* New York: Harper and Row Publishers.

Dyer, W.G. (1977). *Team building: Issues and alternatives.* Reading: Addison-Wesley.

Edwards, J.R. (1989). Computer aided manufacturing and worker well-being: A review of research. *Behaviour and Information Technology, 8*, 157-174.

Franz, C.R., & Robey, D. (1986). Organizational context, user involvement, and the usefulness of information systems. *Decision Sciences, 17*, 329-356.

French, W.L., & Bell, C.H. (1984). *Organization development: Behavior science interventions for organization improvement* (3rd ed.). Englewood Cliffs: Prentice Hall.

Frese, M., Ulich, E., & Dzida, W. (1987). *Psychological issues of human-computer interaction in the work place.* North Holland: Elsevier Science Publishers.

Galletta, D.F., & Lederer, A.L. (1989). Some cautions on the measurement of user satisfaction. *Decision Sciences, 20*, 419-438.

Giuliano, V.E. (1982). The mechanization of office work. *Scientific American, 24*, 149-165.

Golembiewski, R.T., Billingsley, K., & Yeager, S. (1976). Measuring change and persistence in human affairs: Types of change generated by OD designs. *Journal of Applied Behavioral Science, 12*, 133-157.

Goodman, P.S., Sproull, L.S., & Associates (1990). *Technology and organizations.* San Francisco: Jossey-Bass, Inc.

Griffin, R.W. (1988). Consequences of quality circles in an industrial setting: A longtitudinal assessment. *Academy of Management Journal, 31*, 338-358.

Hackman, J.R., & Lawler, E.E., III (1971). Employee reactions to job characteristics. *Journal of Applied Psychology, 55*, 259- 286.

Hackman, J.R., & Oldham, G.R. (1975). Development of the job diagnostic survey. *Journal of Applied Psychology, 60*, 159-170.

Hackman, J.R., & Oldham, G.R. (1976). Motivation through the design of work: Test of a theory. *Organizational Behavior and Human Performance, 16*, 250-279.

Hartman, E.A., & Perlman, B. (1986, August). *Prediction of individual turnover versus organizational turnover rates.* Paper presented at the Annual Meeting of the Academy of Management, Chicago, IL.

Huber, G.P. (1984). The nature and design of post-industrial organizations. *Management Science, 30*, 928-951.

James, L.R. & Jones, A.P. (1980). Perceived job characteristics and job satisfaction: An examination of reciprocal causation. *Personnel Psychology, 33*, 97-135.

Kahn, R.L. (1974). Organization development: Some problems and proposals. *Journal of Applied Behavioral Science, 10*, 485-502.

Keller, R.T. (1986). Predictors of the performance of project groups in R&D organizations. *Academy of Managment Journal, 29*, 715-726.

Kiesler, S., Siegel, J., & McGuire, T.W. (1984). Social psychological aspects of computer-mediated communication. *American Psychologist, 39*, 1123-1134.

Klein, K.J., & Hall, R.J. (1988). Innovations in human resources management: Strategies for the future. In J. Hage (Ed.), *Futures of organizations: Innovating to adopt strategy and human resources to rapid technological change* (pp. 147-162). Lexington, MA: Lexington Books.

Landy, F.J., Rastegary, H., & Motowidlo, S. (1987). Human-computer interactions in the work place: Psychosocial aspects of VDT use. In M. Frese, E. Ulich, & W. Dzida (Eds.), *Psychological issues of human-computer interaction in the work place* (pp. 3-22). North Holland: Elsevier Science Publishers.

Majchrzak, A., & Klein, K.J. (1987). Things are always more complicated than you think: An open systems approach to the organizational effects of computer-automated technology. *Journal of Business and Psychology, 2*, 27-49.

Maslow, A.H. (1943). A theory of motivation. *Psychological Review, 50*, 370-396.

Mayo, E., Roethlisberger, F.J., & Dickson, W.J. (1939). *Management and the worker*. Cambridge: Harvard University Press.

Metz, E. (1986). Managing change towards a leading edge information culture. *Organizational Dynamics, 15*, 28-40.

Miller, K. (1989). *Retraining the American workforce*. Reading: Addison-Wesley.

Mobley, W.H., Griffeth, R.W., Hand, H.H., & Meglino, B.H. (1979). Review and conceptual analysis of the employee turnover process. *Psychological Bulletin, 86*, 493-522.

Nicholas, J.M. (1979). Evaluation research in organizational change interventions: Considerations and some suggestions. *Journal of Applied Behavioral Science, 15*, 23-40.

Offerman, L.R., & Gowing, M.K. (1990). Organizations of the future: Changes and challenges. *American Psychologist, 45*, 95-108.

Office of Technology Assessment, U.S. Congress (1985, December). *Automation of America's offices* (Publication No. OTA-CIT-287). Washington DC: U.S. Government Printing Office.

Olson, M.H., & Lucas, H.C. (1982). The impact of office automation on the organization: Some implications for research and practice. *Communication of the ACM, 25*, 838-847.

Porras, J.I., & Berg, P.O. (1978). Evaluation methodology in organization development: An analysis and critique. *Journal of Applied Behavioral Science, 14*, 151-173.

Rafaeli, A. (1986). Employee attitudes towards working with computers. *Journal of Occupational Behavior, 7,* 89-106.

Randolph, W.A. (1982). Planned organizational change and its measurement. *Personnel Psychology, 35,* 117-139.

Rauch, J. (1989). Kids as capital. *The Atlantic,* 56-61.

Rice, R.E., & Contractor, N.S. (1990). Conceptualizing effects of office information systems: A methodology and application for the study of alpha, beta, and gamma changes. *Decision Sciences, 21,* 301-317.

Robey, D. (1974). Task design, work values, and worker response: An experimental test. *Organizational Behavior and Human Performance, 12,* 264-273.

Rousseau, D.M. (1983). Technology in organizations: A constructive review and analytic framework. In S.E. Seashore, E.E. Lawler III, P. Mirvis, & C. Camman (Eds.), *Assessing organizational change: A guide to methods, measures, and practices* (pp. 229-255).

Saal, F.E. & Knight, P.A. (1988). *Industrial/Organizational psychology: Science and practice.* CA: Brooks/Cole.

Smith, P.C., Kendall, L.M., & Hulin, C.L. (1969). *Measurement of satisfaction in work and retirement.* Chicago: Rand McNally.

Stone, E.F. (1978). *Research methods in organizational behavior.* Santa Monica: Goodyear Publishing Company.

Stout, E. (1981). The human factor in productivity: The next frontier in the office. *Journal of Micrographics, 14,* 25-31.

Szilagyi, A.D., Jr., & Wallace, M.J., Jr. (1983). *Organizational behavior and performance* (3rd ed.). Glenview: Scott Foresman.

Taylor, F.W. (1911). *The principles of scientific management.* New York: Harper and Brothers.

Terborg, J.R., Howard, G.S., & Maxwell, S.E. (1980). Evaluating planned organizational change: A method for assessing alpha, beta, and gamma change. *Academy of Management Review, 5,* 109-121.

Terpstra, D.E. (1981). Relationship between methodological rigor and reported outcomes in organization development evaluation research. *Journal of Applied Psychology, 66,* 541-543.

Turner, A.N., & Lawrence, P.R. (1965). *Industrial jobs and the worker: An investigation of response to task attributes.* Boston: Division of Research, Graduate School of Business Administration, Harvard University.

Walker, C.R., & Guest, R.H. (1952). *The man on the assembly line.* Cambridge: Harvard University Press.

Wanous, J.P. (1974). Individual differences and reactions to job characteristics. *Journal of Applied Psychology, 59,* 616-622.

Weber, R. (1988). Computer technology and jobs: An impact assessment model. *Management of Computing, 31*, 68-77.

Weick, K.E. (1985). Cosmos vs. chaos: Sense and nonsense in electronic contexts. *Organizational Dynamics*, 51-65.

Weick, K.E. (1990). Technology as equivoque: Sensemaking in new technologies. In P.S. Goodman, L.S. Sproull, & Associates (Eds.), *Technology and organizations* (pp. 1-44). San Francisco, CA: Jossey-Bass Inc.

Weiss, D.J., Dawis, R.V., England, G.W., & Lofquist, H. (1967). *Manual for the Minnesota satisfaction questionnaire*. Minneapolis, Minnesota: University of Minnesota Press Industrial Relations Center, Work Adjustment Project.

White, S.E., & Mitchell, T.R. (1976). Organization development: A review of research content and research design. *Academy of Management Review, 1*, 57-73.

Zuboff, S. (1988). *In the age of the smart machine: The future of work and power*. New York: Basic Books, Inc.

We would like to thank the following people for their interest in the effects of computer technology on people and their support to continue our research with the next step — data collection: Richard Holtzman, Assistant Vice Chancellor, Information Systems and Technology, University of Wisconsin Oshkosh, and Mark Haslanger, Director, The Richard W. Koehn Institute for Information Systems and Automation. A special thank you goes to Paul Ansfield, Professor of Psychology, University of Wisconsin Oshkosh, who was involved in the conceptualization of the ideas found in this chapter and the empirical research forthcoming.

Chapter 7

Computer-Mediated Communications in Organizational Settings:
A Self-Presentational Perspective

William L. Gardner III
University of Mississippi

Joy Van Eck Peluchette
Southern Illinois University at Carbondale

Over the past decade, an ever increasing number of organizations have discovered a new type of communications medium: computer-mediated communications systems (CMCS) (Hiltz & Turoff, 1985). The variou manifestations of this medium — such as electronic mail systems (EMS), computer conferencing (CC), and computer bulletin boards (CBB) — employ computers and telecommunications networks to facilitate two-way communication among two or more persons more rapidly and efficiently than could be achieved by telephone, the postal service, or during face-to-face meetings (Culnan & Markus, 1987; Finholt & Sproull, 1990; Huber, 1990; Kiesler, 1986; Sproull & Kiesler, 1986; Rice, 1984). Some types of electronic media, such as CC, allow people to interact simultaneously (synchronously), or at different times (asynchronously) by leaving messages for recipients. Others are primarily asynchronous (e.g. EMS, CBB). An added benefit of a CMCS is the capability to generate an automatic written transcript of any exchange (Culnan & Markus,1987).

Potential advantages of CMCS for managers include more voluminous and timely information for better decision-making, the capability to send messages over long distances rapidly and efficiently, increased coordination of activities, enhanced documentation capabilities, and the ability to monitor individual performance more accurately (Culnan & Markus, 1987; Gardner &

Schermerhorn, 1988; Szewczak & Gardner, 1989). In light of their enormous potential, CMCS have been described by some as a new form o enhanced communication (Verity & Lewis, 1987; Zuboff, 1987). Because its social ramifications are not fully understood, however, this conclusion appears to be premature (Gardner & Schermerhorn, 1988; Huber, 1990; Kiesler, 1986; Szewczak & Gardner, 1989). Moreover, because many aspects of human relations rely on information that are not present with CMCS (e.g., visual cues and nonverbal behavior), one must question whether the assumptions made for other media still apply (Culnan & Markus, 1987). While certain aspects of communication might indeed be enhanced, an increasing number of researchers are uncovering "hidden consequences" of this technology (Kiesler, 1986).

Although most research on CMCS has focused on technical issues related to connectivity (Estrin & Cheney, 1986; Rowe, 1988; Thompson, 1987), a steady stream of studies into the social and organizational consequences of this medium has emerged over the years. One body of literature has concentrated on group processes, in terms of participation, consensus, efficiency, and group choice (Kiesler, Siegel, & McGuire, 1984; McGuire, Kiesler, & Siegel, 1986; Siegel, Dubrovsky, Kiesler, & McGuire, 1986), and group decision support systems (DeSanctis & Gallupe, 1987; Kraemer & King, 1988;Nunamaker, Applegate, & Konsynski, 1987). Other researchers have focused on the behavioral implications of particular networking systems, such as the Electronic Information Exchange System (EIES) (Kerr & Hiltz,1982;Hiltz, 1984; Hiltz & Turoff, 1981; Rice, 1982; Rice & Case, 1983) and EMS(Finholt & Sproull, 1990; Sproull & Kiesler, 1986; Steinfield, 1986). Finally, considerable theoretical attention has been devoted to the implications of CMCS for managerial work (Gardner & Schermerhorn, 1988: Szewczak& Gardner, 1989), and organizational design (Child, 1988; Foster & Flynn,1984; Keen, 1987; Huber, 1990; Huber & McDaniel, 1986).

Social Consequences of Electronic Media: Contradictory Findings

As research into the dynamics of CMCS accumulates, some unexpected and seemingly contradictory findings have emerged. For example, researchers have obtained discrepant findings with regards to the relative amounts of task-oriented versus socioemotional messages sent via CMCS. Hiltz andTuroff (1978) found that the proportion of socioemotional content is much lower (14 percent) in CMCS groups, than it is in face-to-face groups (33 percent). One explanation for this finding is that this medium is perceived by users to be highly impersonal because it is low in "social presence" (Short,Williams, & Christie, 1976). Support for this position is provided by research which indicates that CMCS are considered by users to be inappropriate for communication tasks involving interpersonal conflict (e.g., bargaining) or relations (e.g., getting to know people) (Rice & Case, 1983). However, contradictory results were obtained in a study by Rice and Love(1987), revealing that nearly 30 percent of the messages sent using a nationwide public computer conferencing system were socioemotional in content. Furthermore, Kerr and Hiltz (1982) concluded from a series of studies that CC systems can be successful at increasing friendships and people's sense of personal interaction. Thus, there appears to be considerable variability in the amount of socioemotional content found in CMCS across social settings.

Empirical research into CMCS has also produced contradictory results with regards to users' self-presentations. Several studies have shown that users of CMCS become extremely uninhibited in their behavior, presumably because the low social presence of this medium causes it to be seen as impersonal (Finholt & Sproull, 1990; Kiesler, 1986; Siegel etal., 1986; Sproull & Kiesler, 1986). Indeed, Sproull and Kiesler (1986) discovered that EMS users often engage in a practice known as "flaming," a term coined by computer hackers. Flaming involves name calling, profanity, a heavy use of exclamation points or capitalization, making personal remarks to others, or "the expression of strong and inflammatory opinions to others electronically" (Siegel et al., 1986, p. 161). In other cases, however, CMCS users appear to be much more concerned about the image they portray. Kerr andHiltz (1982) concluded that CC can support self-presentation and emotional subtleties. Similarly, Trevino, Lengel and Daft (1987) found that 15 percent of the reasons managers provided for using EMS were symbolic in nature. Finholt and Sproull (1990) likewise found that, in a recent field study of an EMS, some members of electronic work groups use this medium for symbolic purposes and "to display their knowledge before superiors and peers" (p. 56). Clearly, the EMS in this organization served as a vehicle for self-promotion.

Chapter Objective

From the above discussion, it is apparent that CMCS can support a wider range of behaviors than many theorists had expected (Short et al., 1976). Unfortunately, this diversity of behavior is not sufficiently explained by current theory in this area. Indeed, Steinfield and Fulk (1987) note, that "despite this impressive accumulation of research findings, there has been little synthesis, integration, and development of theoretical explanations" (p. 479). This lack of theory is especially pronounced with regards to micro-level analyses of CMCS, since very little attention has been focused on the cognitive, motivational, and affective processes that influence user behavior (Sproull & Kiesler, 1986; Trevino et al., 1987).

One theoretical perspective which appears to possess considerable potential for reconciling the contradictory findings described above is self- presentation theory (Baumeister, 1982, 1989; Gardner & Martinko,1988b; Schlenker, 1980, 1985; Tedeschi, 1981). Accordingly, the purpose of this chapter is to use this theory to provide a micro-level analysis of the self- presentational behaviors people exhibit when they communicate via computer networks. Toward this end, we have divided this chapter into four major sections. In the first section, we present a general model of the self- presentation process. The second applies this model to advance a series of propositions about the ways in which people use CMCS to present themselves to others. Pertinent research and theory from the information technology (IT) and self-presentation literatures will be provided in support of these propositions. In the third major section, we examine some of the potential benefits and liabilities of using CMCS as a medium for self-presentation. The fourth discusses the need for managers to consider both technical and social issues in designing CMCS for organizations. The chapter concludes by identifying future research opportunities in this area.

A Model of Self-Presentation

The model of self-presentation, depicted in Figure 1, is a refinement of an earlier model of the impression management process developed by Gardner and Martinko (1988b). The revised model differs from the previous model in two key respects. The first arises from its focus on self-presentation as opposed to impression management. Although these terms are often used interchangeably in the social psychology literature, they are not identical (Schlenker, 1980). Impression management "refers to the process by which individuals attempt to control the impressions others form of them" (Leary & Kowalski,1990, p. 34). The distinction between impression management and self-presentation is made clear by Schneider (1981).

> Self-presentation may be defined as the manipulation of information about the self by the actor. Obviously self-presentation is a close cousin of impression management,but they are different. Impressions can be managed by means (e.g.,third party conveying of information) other than self-presentation, and presentations may be used for goals (e.g., information seeking) other than impression management (p. 25).

The second key difference between the two models is that the current model explicitly includes the communication media and message as components of self- presentation. An understanding of these components is crucial to our

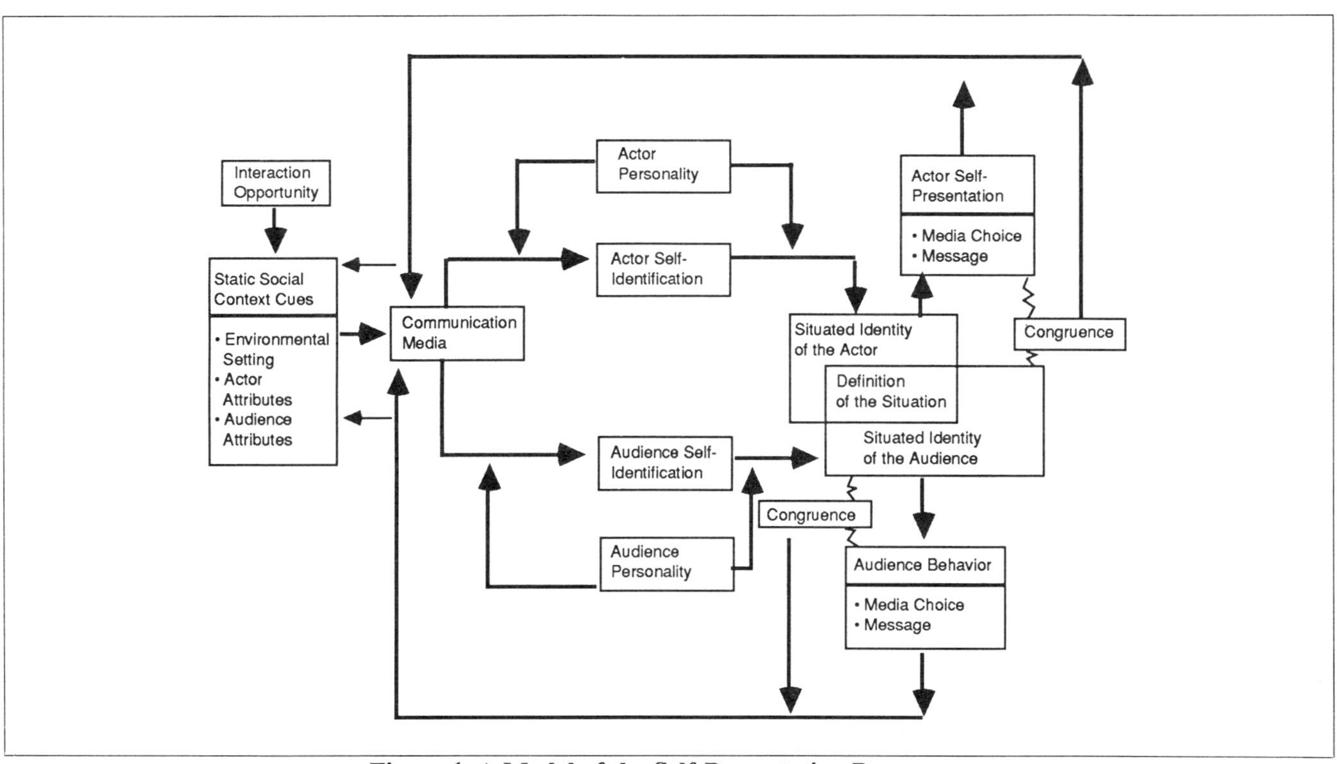

Figure 1. A Model of the Self-Presentation Process

analysis of the reasons and ways in which people present themselves using CMCS.

The primary conceptual foundation for both models is provided by Erving Goffman's (1959) classic work, *The Presentation of the Self in Everyday Life*, in which he advanced a dramaturgical perspective of social relations. Goffman views individuals as "actors" who perform in varied"settings" before different "audiences." Although he recognized that people jointly occupy the roles of actor and audience in ongoing social exchanges, he distinguished between them for purposes of conceptual clarity. Another key concept advanced by Goffman is the "definition of the situation," which involves the perceptions the interaction participants' hold of the situation and their roles in it. Each of these components of his perspective areincorporated into the self-presentation model that is illustrated in Figure 1.

The model begins with an opportunity for interaction between an actor and an audience. Self-presentation cannot occur if the actor is not given a chance to convey a desired image to the audience. When an interaction opportunity arises, the characteristics of the actor and audience combine with environmental stimuli to provide static social context cues. Dynamic cues are also generated by the ongoing behaviors of both parties. These social context cues are then selectively perceived and interpreted by both the actor and audience through a self-identification process, whereby various cognitive, motivational, and affective processes interact to produce their situated identities. This entire process is mediated by face-to-face, telephone, CMCS, written, numeric, or other forms of communication media.

As Figure 1 indicates, the personality characteristics of self-monitoring ability, needs for approval, social anxiety and public self-consciousness account for differences in the ways in which people define the situation. The area in the model where the situated identities of the actor andaudience overlap reflects agreement about the roles of each party. That is, this overlap represents the interaction participants' mutually agreed upon definition of the situation. While there is typically considerable agreement in their situated identities, there may also be notable discrepancies.

Using their situated identity as a guide, actors select from their repertoire of self-presentational behaviors those which are deemed to be most consistent with their interaction goals and situational requirements. Key aspects of this selection decision include choosing the medium and message for self-presentation. The degree of congruence between the audience's situational definition and the actor's behavior has a major impact on the audience's impression of the actor. Congruent behavior typically leads to favorable impressions and desirable responses by the audience. In contrast, perceptions of incongruence may lead to undesirable impressions and audience reactions, such as physical, social, or economic sanctions. These reactions, in turn, serve as feedback which may cause actors to redefine the situation, target a new audience, and/or adjust their performance.

Self-Presentation Through Computer-Mediated Communications

Interaction Opportunity

As noted in the model overview, an opportunity for interaction is a minimum prerequisite for self-presentation. Opportunities for interaction via particular media are constrained by technological and situational factors. For example, face-to-face exchanges can only occur when both parties are located in approximately the same place, at the same time. While geographic and temporal constraints are most pronounced for face-to-face exchanges, they are also operative, albeit to a lesser degree, for other media such as the telephone and postal mail (Culnan & Markus, 1987; Sproull & Kiesler, 1986; Trevino et al., 1987). Geographic distance slows the rate of communication that can be achieved through postal mail. Similarly, differences in time zones limit opportunities to converse by telephone. Moreover, unless receivers have an answering machine, they must be present to receive a message.

By comparison, CMCS are much less constrained by geographical location or temporal availability (Culnan & Markus, 1987; Huber, 1990). "The networked computers might be physically proximate and connected via a local area network, or they might be in different states, countries, or continents, connected via long distance telecommunications" (Sproull & Kiesler, 1986, p. 1493). Like paper mail, the sender is not dependent on the availability of the recipient; messages can be sent at any time and stored at the receiving end until the other party decides to retrieve the information. Unlike paper mail, however, computer-mediated communications can be transmitted in seconds or minutes across these distances. Upon receipt, a message can be read, edited, saved, deleted, moved to another computer file, forwarded to others, or responded to, at the receiver's convenience (Culnan & Markus,1986).

Empirical evidence of the attractiveness of these features to organizational members is provided by Trevino et al. (1987). These authors conducted structured open-ended interviews with 65 managers from a mixture of 11 public and private sector organizations, to determine their reasons for selecting particular media. Their results revealed that 62 percent of the reasons the managers gave for selecting EMS over face-to-face,telephone, or written media involved situational factors such as time anddistance constraints. The six most common reasons, in order, were: 1) it is quick, easy, convenient, and efficient, 2) personal preference, 3) to send simple, routine messages, 4) to provide a permanent record, 5) accessibility, (to reach someone who is busy, frequently absent), and 6) to reach many receivers at one time. Similarly, Steinfield (1986) found that time pressures, need for communication across geographic locations, and the proportion of coworkers outside the work group who also use electronic mail were significant predictors of employees' task-related uses of an EMS.

Despite these advantages of CMCS, many organizational members are unable to use them for self-presentational purposes, simply because this medium is unavailable to them. Furthermore, even when a system is available, access may nevertheless be limited if members are forced to share a computer terminal with other persons, or if its location is not convenient. Indeed, IT research has shown that usage of CMCS is depressed when people are required to share terminals (Edwards, 1977; Johansen, Vallee, & Spangler,1979; Steinfield, 1986). These findings imply:

> *Proposition 1:* The extent to which actors use CMCS for self-presentation is directly related to system accessibility.

> *Proposition 2:* Self-presentations using CMCS are less constrained by geographical location of the audience and/or temporal availability than self-presentations via face-to-face, telephone,postal media.

Social Context Cues

In an effort to explain how social context information effects information exchange, Sproull and Kiesler (1986) make an important distinction between static and dynamic cues. "Static cues emanate from people's appearance and artifacts such as a clock, a private office, a big desk, and a personal secretary. Dynamic cues emanate from peoples' nonverbal behavior which changes over the course of an interaction — for instance, nodding approval and frowning with displeasure" (p. 1495). While we agree with this basic distinction, we contend that dynamic cues are not limited to nonverbal behaviors, since verbal behaviors are also instrumental in establishing the social context. A more detailed examination of the cues for self-presentation follows. We begin by looking at three major sources of static cues: the environmental setting, the actor, and the audience.

Environmental setting. Earlier, it was noted that the environmental setting plays a major role in shaping people's situated identities. Research into the factors which influence people to use CMCS, as well as other media, documents the impact of environmental considerations (Steinfield,1986; Trevino et al., 1987). Pertinent features of the environment include the physical setting, organizational culture, organizational design, and the nature of the work performed.

Important elements of the physical environment include the geographic location and distance between communicators, as well as the"furniture, decor, physical layout, and other background items that supply the scenery and stage props for" the actor's presentation (Goffman, 1959, p. 15). Through cues such as office size and layout, organizational members are made aware of the relative status of their superiors, peers and subordinates.

The culture of an organization includes the norms, values, philosophies, and standards of behavior that are specified for members (Deal &Kennedy, 1982). Organizational stories, myths, strategies, and policies operate to teach and support the attitudes espoused by the organization (Trice & Beyer, 1984). Furthermore, the organizational culture conveys powerful cues for impression management (Wexler, 1983). For instance, the perceived need for ingratiation is greatly impacted by organizational norms and practices regarding objective performance appraisals (Ralston, 1985).

Members situated identities are also shaped by the organization's design. For example, the emphasis that mechanistic organizations place on rules and regulations may cause members to present themselves in less personal ways than members of organic organizations (Daft & Lengel, 1984). In addition, structure has been found to impact on the organization's information processing activities, including members' preferences for communication media (Huber, 1990). Thus design features may encourage some organizations, and hence their members, to be more receptive to CMCS than others.

The self-presentational behaviors of members are influenced by the nature of their work (Gardner & Martinko, 1988b). Politically-based influence attempts, such as ingratiation, are more common when task ambiguity is high, because workers do not see their evaluation as being related to task performance. Similarly, workers who perform specialized versus menial tasks are able to claim higher levels of expertise (Ralston, 1985).

Actor attributes. Both the impression management and the communication literatures stress the role which actor attributes, such as age, race, gender, appearance job title, departmental affiliation, and level in the organizational hierarchy play in the interaction process (Gardner& Martinko, 1988b; Rice & Associates, 1984; Schlenker, 1980, 1985). From a self-presentational viewpoint, these attributes often limit the range of images an actor can successfully claim. For instance, it is very difficult for elderly men to present themselves as young during face-to-face exchanges.

Audience attributes. The situated identities of both the actor and audience are shaped by audience attributes. For example, when a female manager makes a formal presentation to a group of male executives, the gender of both parties typically impacts the situated identities of the interaction participants (Deaux & Major, 1987). Self-presentation research also indicates that certain characteristics of an audience operate to arouse actors' concerns about their image. In the above scenario, the actor may be highly self-conscious since she is the only woman in the group. Furthermore, the fact that she is presenting to a group of superiors will most likely compound her self-consciousness. Indeed, impression management research has unequivocally demonstrated that actors tend to be much more self-aware in the presence of persons who hold high versus low status/power positions (Gardner& Martinko, 1988b; Schlenker, 1980). Clearly, the attributes of both parties have influenced the definition of the situation in this example.

Dynamic cues. In addition to the demographic and environmental cues described above, the preceding and ongoing behavior of the audience may provide cues for actor self-presentations. For instance, many of the self-descriptive statements made by job applicants during an employment interview are offered in response to questions by the interviewer(Fletcher, 1989). Similarly, many of the messages that actors send by CMCS are responses to messages from the target audience (Finholt & Sproull, 1990; Rice,1982).

Communication Media

As Figure 1 indicates, the environmental, actor, and audience cues perceived by the interaction participants must first pass through one or more communication media. Accordingly, it is important to consider the impact which alternative media have on the stimuli that are selected for attention. To do so, we draw on two constructs from the communication and management literatures, respectively: 1) social presence, and 2) information richness.

Social presence. Short et al. (1976) define social presence as "the degree of salience of the person in the interaction and the consequent salience of the interpersonal relationships" (p. 65). A medium's capacity to transmit dynamic cues such as tone of voice, eye contact, facial expression, and body language, and static cues such as age, race, gender, hierarchical level, job title, departmental affiliation, and dress add to its social presence. While all of these cues are apparent during face-to-face exchanges, many are removed by other media. Research by Short et al. (1976)indicates that social presence is highest for face-to-face communications, followed by television, multispeaker audio, telephone audio, and business letters.

Although Short et al. (1976) did not examine the social presence of CMCS, subsequent research suggests that they fall between the telephone and business letters in terms of social presence (Rice & Associates,1984). The speed with which information can be exchanged through CMCS, plus the relative flexibility of this medium, provides it with a higher level of social presence than paper mail. Like paper mail, however, dynamic cues are filtered out by CMCS, and static cues are reduced to a minimum (Culnan & Markus,1987; Finholt & Sproull, 1990; Kiesler, 1986; Kerr & Hiltz, 1982; Sproull &Kiesler, 1986). Indeed, in many CMCS, the only information about actors that is routinely provided is their name and computer address. Consequently, information about senders' meaning and frame of mind, loyalties, symbolic variations, and details embodied in their dress, demeanor, and expressiveness are lost (Kiesler et al., 1984). Moreover, CMCS appear to be largely blind to the status, power, and prestige of the communicators (Kiesler, 1986). Of course, people who use CMCS may still have knowledge of each others' attributes if they have previously met, or learned about one another through a richer medium. Nevertheless, CMCS provide few cues to make this information salient. Furthermore, research indicates that a substantial proportion of persons who interact using CMCS have never communicated with one another using any other medium (Culnan & Markus, 1987; Finholt & Sproull, 1990;Freeman, 1984). Obviously, social cues under these circumstances are especially low, and the interaction participants are largely "invisible" to one another (Finholt & Sproull, 1990; Kiesler, 1986).

Information richness. One concept which appears to be closely related to social presence is information richness. Daft and Lengel(1984) define information richness "as the potential information-carrying capacity of data" (p. 196). When communication of data leads to substantial new understanding, it is considered to be rich in information. Data that produces little additional understanding are low in richness. Daft and Lengel (1984, 1986) argue that media vary with regards to the richness of the information processed. Furthermore, they specify four criteria that determine media richness: 1) feedback capability, 2) communication channels used, 3) source, and 4) language. Table 1 provides a summary of media attributes with regards to these criteria, and their location on a continuum of information richness.

Because face-to-face communication provides immediate feedback, uses visual and audio channels, is personal in nature, and allows for both natural language and body language, it is the richest medium. Each of the other media are lower in richness because they are unable to carry some of this information. As Fulk et al. (1987) note, electronic mail appears to lie between the telephone and written communications in terms of information richness. Although it uses natural language and provides rapid feedback, it filters out cues such as voice pitch, tone, and inflection. We agree with this reasoning, and extrapolate it to apply to other types of CMCS as well.

While much of the information about an audience appears to be filtered out by CMCS, the findings of Finholt and Sproull's (1990) field study of electronic group mail suggest that this medium may provide more information about audience identity than previous researchers realized. These researchers found that organizational members used electronic distributionlists (DL) to form electronic groups at work. Five, high-activity, electronic groups were identified from the following DL names: 1) ComputerScienceGroup (CSG, 35 members), 2) Captain Developers (CD, 25 members), 3)User Feature Forum (UFF, 125

Information Richness	Medium	Feedback	Channel	Source	Language
High	Face-to-Face	Immediate	Visual, Audio	Personal	Body, Natural
	Teleconferencing, Audio & Visual	Immediate	Visual, Audio	Personal	Body, Natural
	Telephone	Fast	Audio	Personal	Natural
	CMCS, Personal, Formal Synchronous	Fast	Limited Visual	Personal or Impersonal	Natural
	CMCS, Personal, Formal Asynchronous	Depends	Limited Visual	Personal or Impersonal	Natural
	Written, Personal or Formal	Very slow	Limited Visual	Personal or Impersonal	Natural
Low	Numeric, Formal	Very slow	Limited Visual	Impersonal	Numeric

Adapted from Daft and Lengel (1984), p. 197.

Table 1. A Matrix of Media Characteristics

members), 4) Cinema (500 members), and 5) Rowdies (98 members). Finholt and Sproull (1990) further classified these groups as required (CSG &CD) versus discretionary (UFF, Cinema, Rowdies), and work (CSG, CD, UFF) versus non-work (Cinema, Rowdies). Importantly, senders in this EMS system mail their messages to the group name, rather than the individual members. Thus, they do not need to know the names and addresses of the DL members, although they can get this information with a simple command. Still, the names of the DLs, coupled with the senders' knowledge of their attributes (e.g.,number of DL members), serve as cues for appropriate self-presentational behavior from the actor. Overall, these findings imply:

> *Proposition 3:* CMCS filter out many, but not all, of the social context cues that are available with face-to-face communications.

Self-Identification by the Actor

The manner in which actors select and interpret social context cues depends upon their internal cognitive, motivational and affective processes (Gardner & Martinko, 1988b). To date, available theory and research on CMCS has devoted relatively little attention to these internal processes. As a consequence, it appears that pertinent theory from social psychology could greatly contribute to our understanding of the role these processes play in electronic communications. Accordingly, the following sections draw on this literature to advance propositions about the internal processes that shape an actor's situated identity, and subsequent self-presentations using CMCS.

To make sense out of the multitude of social context cues that confront an actor in any given situation, people must select relevant information and organize it into a coherent pattern. Stimuli selection is determined by an interaction between the properties of the social context and a schema internal to the perceiver (Lord, 1985). Schema, or schemata, can bedefined as preexisting cognitive structures that focus perceptual activities and are modified by the ongoing perceptual process (Neisser, 1976).

Two kinds of schema are most relevant to social information processing: person schema and event schema. Person schema identify categories of people and the attributes associated with category members. One type of person schema which is critical to our analysis is the self-schema or identity. Schlenker (1985) defines identity as "a theory (or schema) of an individual that describes, interrelates, and explains his or her relevant features, characteristics, and experiences" (p. 68). Event schema, or scripts, involve "a hypothesized cognitive structure that provides a guide to appropriate behavior sequences in a given context" (Gioia & Manz, 1985, p.528).

Schlenker (1985) has developed a theory of identity and self-identification which explains how people use schema to process situational and personal stimuli and develop a situated identity. A basic proposition in this theory is that "an initial assessment and evaluation of self, the situation, and the audience evokes for the actor or prompts the actor to formulate: (a) a goal or set of goals that might satisfy needs and values, (b) a script or plan for goal accomplishment, and (c) a set of desired identity images" (p. 76).

We discuss each of these components of Schlenker's (1985)theory below, while incorporating pertinent social information processing theory. More specifically, we will use the concepts of audience and situation categorization, situated identities, levels of self-awareness, interaction goals, affective states, cognitive scripts, desirable identity images, and self- efficacy expectations to explain the process of actor self-identification. To avoid overcomplicating Figure 1, many of these components are not specified in this figure. The reader should recognize, however, that they are all subsumed under the process labeled actor self-identification in Figure 1. Throughout our discussion of this process, its implications for CMCS are examined.

Audience and situation categorization. As an initial step in the self-identification process, actors must size up the audience and the situation. Recent developments in the literature on cognition suggest that actors process incoming stimuli through either automatic or controlled processes(Feldman, 1981; Lord, 1985). Audience attributes such as their age, race, gender, dress, speech, and height may automatically evoke relevant schema which are then used to encode them in memory. These attributes are probabilistically compared to those of a category prototype, and membership is determined by the degree of attribute overlap. For instance, a young, white,upwardly mobile, male executive may be categorized as a "yuppie" because he possesses many of the attributes that are associated with this category. Situations are likewise categorized; examples of event schema include business meetings, office parties, crises, job interviews, and performance appraisals.

The automatic categorization process may be superseded by a controlled process if the attributes of the person or situation do not satisfactorily resemble available prototypes, or information that is highly discrepant with the original categorization is revealed. Under these circumstances, attributional processes are invoked to categorize the stimulus source (Kelley, 1971; Martinko & Gardner, 1987). Basically, a "naive analysis of variance" is performed to determine if the discrepancy in the incoming information should be categorized using an event schema or a person schema (Feldman,1981; Lord, 1985). It is important to recognize, however, that regardless of whether the controlled or the automatic categorization processes areemployed, both the stimulus person and the situation are ultimately placed in categories using person or event schema. These categories enable the actor to define the situation and their role in it (Feldman, 1981; Schlenker, 1985).

The preceding discussion has important implications with regards to CMCS. Because this medium removes some of the social contextcues that invoke person and event schema, people may define interactions using this medium much differently than would be the case with richer media. For instance, a lack of static and dynamic cues about the audience may increase the likelihood that an event schema or script will be evoked by the actor. Since cues about the audience's physical and social settings are also lacking, however, actors are likely to define the situation from their personal perspective exclusively. Evidence to this effect is provided by research on EMS, which indicates that senders tend to become much more self-absorbed, as opposed to other-oriented, than is the case with other media (Kiesler, 1986; Sproull &Kiesler, 1986).

> *Proposition 4:* People evoke event schemata more often and person schemata less often, when they process social information using CMCS as opposed to richer media.

> *Proposition 5:* People are less sensitive to others' perspectives when they interact via CMCS as opposed to richer media.

It should be noted that a lack of appreciation for the audience's perspective may not be problematic for an actor as long as both parties work under similar conditions. That is, if they both work in the same organization, at the same location, and perform similar work with a common group of colleagues, the scripts they evoke may share many common elements. Still, some discrepancies may arise due to the lack of dynamic cues. If, however, their circumstances differ markedly, dissimilar scripts may be evoked for the same event. While divergent viewpoints would be expected whenever people work in dissimilar settings, regardless of the medium, they may be more problematic with CMCS simply because they are less salient.

> *Proposition 6:* The extent to which an actor's cognitive script for a computer-mediated exchange, converges with that of the audience is directly related to the degree of similarity in their work settings.

A final implication of the social information processing perspective for CMCS is that organizational members will rely most heavily on controlled information processes when they have relatively little experience with this medium. Recall that when people or situations are sufficiently different from existing prototypes, controlled processing is triggered (Lord,1985; Feldman, 1981). Because of their lack of experience, naive users of CMCS are unlikely to have developed cognitive scripts for the events they encounter(Gioia &

Manz, 1985). With experience, however, many of the activities people perform with CMCS are likely to become scripted. Once these scripts areformed, processing will be automatically invoked for familiar interaction episodes (Feldman, 1981). Accordingly, we propose:

> *Proposition 7:* User experience with CMCS is directly related to automatic processing of social information, and inversely related to controlled information processing.

Situated identity. Once an audience and situation are categorized, this information is incorporated into the actor's situated identity. As Schlenker (1985) notes, one's identity encompasses many interrelated images (or schemata) that specify pertinent self-constructs about one's appearance (e.g., height, weight, age, sex, race, physical fitness, style ofdress), roles (e.g., parent, executive), motives (e.g., desire to appear competent or attractive), personal qualities (e.g., intelligence), background (e.g., family tree, education), relatives, friends and associates (one is judged by the company one keeps), aspirations (e.g., goals and dreams), accomplishments (e.g., honors, level of competence, work related achievements), pursuits (e.g., interests and hobbies), surroundings (e.g., home, office), possessions (e.g., automobile, wealth), and standing on construct dimensions relative to others (e.g., more powerful than Tom Smith, less powerful than Sally Jacobs). The unique contents and combinations of this information aboutthe self leads to a distinctive personal identity.

To fully comprehend the notion of the situated identity, it is important to distinguish it from the self-concept. To Schlenker (1985), the self- concept is the theory of self which is invoked when one serves as one's own audience, i.e., when one attempts "to answer a personal question about the self posed by oneself" (p. 67). In contrast, a situated identity is "a theory of self that is wittingly or unwittingly constructed in a particular social situation or relationship" (p. 68). Importantly, the situated identity may differ from one's self-concept because (1) the actor tries to"package" information about the self to make it understandable or agreeable to the audience, (2) the actor attempts to misrepresent information about the self to achieve a desired outcome, and (3) other people attempt to influence the actor to adopt attributes or roles that they consider to be consistent with their personal goals. Thus, an actor's situated identity during an exchange is jointly constructed by the exchange participants. For instance, an executive, who convinces subordinates to use CMCS by sending them messages via this medium and providing them with extensive training, may cause these managers to view themselves as having "expertise" with this technology.

From the above discussion, it is clear that the attitudes and behavior of superiors, peers and colleagues partially determine whether actors see themselves as persons who communicate electronically (Fulk, Steinfield, Schmitz, & Power, 1987). These audiences also influence the types of messages that the actor perceives to be appropriate for CMCS.

> *Proposition 8:* The attitudes and behavior of relevant audiences regarding CMCS influence the extent to which actors view CMCS as an avenue for various forms of self-presentation.

Levels of self-awareness. The extent to which people are conscious of their identity and the impressions others have of them is a function of the situation (Leary & Kowalski, 1990). At one extreme, there are situations that induce acute levels of public self-awareness; under these circumstances, people consciously attend to their image related behaviors. Examples of such situations include job interviews, performance appraisals, and public speeches. At the other extreme are situations characterized by "subjective self-awareness," in which people's attention is drawn away from their identity (e.g. ecstatic joy, deindividuating circumstances). In between these extremes, actors process others' image-related reactions at a "preattentive" or nonconscious level. People operate at this level most of the time. Here, they scan the environment for image-related feedback, without consciously considering others' impressions of them. Self-presentations at this level are often overlearned, habitual, and scripted, (e.g., smiling at an attractive colleague, shaking the hand of a new client).

Because CMCS filter out many personal cues, this medium can lead to deindividuation (Gardner & Schermerhorn, 1988; Kiesler, 1986;Siegel et al., 1986; Sproull & Kiesler, 1986). The above discussion suggests that deindividuation may in turn lead to subjective self-awareness. That is, actors will be less concerned about their public image when interacting via CMCS than they are with richer media. Evidence to this effect is provided by Sproull and Kiesler's (1986) finding that "feelings of anonymity tend to produce self-centered and unregulated behavior. That is, people become unconcerned about making a good appearance" (pp. 1495-1496). Accordingly, we propose:

> *Proposition 9:* Actors are more subjectively self-aware when they use CMCS to communicate, than they are with richer media.

Interaction goals. A basic assumption of Schlenker's (1985) theory of self-identification is that "people are purposive, planning creatures who are always thinking, always acting, and always trying to achieve particular objectives in life" (p. 76). Impression management theoryindicates that motives to appear attractive, competent, credible, trustworthy, morally worthy, dangerous, or even pitiful may enter into one's situated identity (Baumeister, 1982, 1989; Jones & Pittman, 1982; Tedeschi &Norman, 1985).

Research on CMCS reveals that self-presentations via this medium may be used to achieve instrumental or symbolic goals (Trevino et al.,1987). For example, Finholt and Sproull's (1990) finding that employees use EMS to impress their peers and superiors, suggests that these actorswere motivated by a desire to appear competent. In other cases, however, EMS users appeared to be motivated by a desire to be "in-the-know" or to simply have fun (Finholt & Sproull, 1990). Regardless of the nature of the actors' goals, it is apparent that their motives are critical elements of their situated identity, and that they impact upon their self-presentations.

> *Proposition 10:* Actors' motives shape their situated identity, which in turn influences their self-presentations via CMCS.

Affective state. Another factor which influences the situated identity of the actor is his or her affective state (Gardner & Martinko,1988b). Although Schlenker (1985) did not explicitly discuss the impact of affect on identity, it is clear that emotions such as love, joviality, hatred, depression and anger may color one's perceptions of oneself, the audience, or the situation. Of course, the cognitive schemata that are evoked may likewise impact upon the affective state of the actor. Thus, the relationship between the actor's affect, motives, and cognitions is interactive in nature.

In face-to-face interactions, the actor's affective state is often influenced by an audience's emotions, and vice versa. When individuals interact through CMCS, their emotional state is less apparent, since many nonverbal behaviors that reveal emotions are not communicated. Therefore, actor affect is less likely to be influenced by audience emotions when these parties communicate via electronic media, as opposed to richer media.

> *Proposition 11:* The situated identity of CMCS users is influenced by their affective state.

> *Proposition 12:* Actor affect and situated identity are influenced less by the emotional state of the audience when these parties communicate using CMCS, as opposed to richer media.

Cognitive scripts. To achieve their interaction goals, people rely on scripts or plans that are either cognitively constructed oraccessed from memory (Schlenker, 1985). The distinction between scripts andplans arises from their relative specificity (Schank & Abelson, 1977). As our earlier discussion of schema indicated, scripts (i.e, event schemata) define the situation by specifying a relatively detailed set of anticipated actions and events. For instance, many managers invoke scripts when they conduct performance appraisals. Plans, on the other hand, are constructed when pre-established scripts are either unavailable or unacceptable, by interweaving general information about goal directed behavior with specific information about the situation and audience. The ability to create plans enables people to deal with novel situations or to pursue novel goals (Schlenker, 1985).

When actors use CMCS to present themselves to others, they are likely to rely on either scripts or plans to guide their behavior. Because plans are used when incoming cues are unfamiliar, we expect that they will be created most often by CMCS novices, or experienced users who are confronted with novel audiences or messages. Scripts, on the other hand, will be formed as CMCS users gain experience. They will be evoked to guide user self-presentations during routine CMCS exchanges. For instance, actors appear to develop and employ scripts for sending routine messages via EMS (Finholt & Sproull, 1990; Malone, Grant, Turbak, Brobst, & Cohen, 1987; Rice & Case, 1983;Rice & Love, 1987; Steinfield, 1986; Trevino et al., 1987). This reasoning suggests:

> *Proposition 13:* CMCS users evoke cognitive scripts to serve as self-presentational guides for routine messages; plans are developed to guide presentations to novel audiences or to send novel messages.
>
> *Proposition 14:* As actors gain experience with CMCS, the proportion of self-presentational behavior that is scripted increases.

Desirable identity images. To Schlenker (1985, p. 74),"desirable identity images represent what the person would like to be and thinks he or she really can be, at least at his or her 'best.'" Thus, desirable images specify the perceived potential of the individual as opposed to an unattainable ideal. Accordingly, these images must satisfy thedual criteria of being personally beneficial and believable. If the first criterion is violated, actors may unnecessarily limit their accomplishments because they underestimated their abilities. In contrast, violation of the second criteria may cause the actor to pursue tasks that are beyond their capabilities, thereby exposing them to personal failure, disappointment and social ridicule.

It is important to recognize that scripts and plans must be accompanied by a desirable identity image which permits the actor to capitalize on the interaction opportunity. In this way, "desired identity images mediate self-identification on the occasion, acting like subscripts orsubplans that are embedded within the overall script or plan" (Schlenker, 1985, p.76). In CMCS, as in all interpersonal exchanges that have reputational implications, actor performances reflect their salient and desired identity images.

Proposition 15: Actors' self-presentations via CMCS will be influenced by their salient and desired identity images.

Self-efficacy expectations. Self efficacy refers to an individual's belief that his or her efforts will lead to goal accomplishment (Bandura, 1977). Actors develop self-efficacy expectations about their ability to claim particular images (Gardner & Martinko, 1988b). These expectations influence the actor's situated identity and self-presentations. As noted above, actors pursue identity images that are both desirable and feasible. That is, they weigh the positive and negative outcomes that they perceive to be associated with particular images, by their self-efficacy expectations about their ability to claim these images, to determine expected values for alternative images. Subsequent presentations reflect the image with the highest expected value (Gardner & Martinko, 1988b; Schlenker, 1980, 1985).

The above line of reasoning suggests that actors will only use CMCS systems for self-presentational purposes, when they have high self-efficacy expectations regarding their ability to create and maintain desirable identity images with this medium. Accordingly, we advance the following proposition.

Proposition 16: Actors' self-efficacy expectations regarding their ability to claim desired identity images via CMCS are directly related to their usage of this medium for self-presentation.

Actor Personality

The self-identification process is mediated by several personality traits of the actor (Gardner & Martinko, 1988b; Schlenker, 1985). Research indicates that people with high self-monitoring abilities (Snyder, 1987), needs for social approval (Crowne & Marlowe, 1964), social anxiety (Weary & Arkin, 1981), and/or public self-consciousness (Fenigstein,Scheier, & Buss, 1975) tend to be more self-aware than persons with lower levels of these traits. Thus, we can speculate that personality may account for differences in the self-presentational behaviors of CMCS users. Of particular relevance is the construct self-monitoring ability, which involves one's skill at observing and controlling expressive and self-presentational behaviors (Snyder, 1987). High self-monitors may be more attuned to the limited number of social context cues provided by CMCS, and hence more likely and capable of using this medium to strategically present themselves to others. In contrast, research suggests that the predisposition to acute public self-awareness of CMCS users with high levels of approval needs, social anxiety, or public self-consciousness, is more likely to be debilitating (Schlenker,1980).

Proposition 17: Actor self-awareness during computer-mediated exchanges is directly related to self-monitoring ability, needs for approval, social anxiety, and public self-consciousness.

Proposition 18: For high self-monitors, self-awareness is positively related to interpersonal effectiveness during computer-mediated communication; for actors with high levels of approval needs, social anxiety, or public self-consciousness, self-awareness is negatively related to interpersonal effectiveness via CMCS.

The Definition of the Situation

The preceding discussion suggests that objective media properties are not sufficient in explaining people's perceptions and uses of various media. Instead, media behavior is a function of people's situational definitions, including their perceptions of alternative media. Importantly, communication researchers have recently come to the same conclusion (Culnan &Markus, 1987; Fulk et al., 1987; Rice, 1984). Fulk et al. (1987) argue that "media characteristics and attitudes are in part socially constructed"(p. 529). In addition, social presence has been found to be a function of 1)people's "preexisting attitudes, familiarity, and preferences" for various media (Rice, 1984, p. 58), and 2) the context for the task to be performed. Similarly, Finholt and Sproull (1990) concluded from their field study of EMS "that 'richness' is as much a function of the social definition of a medium as it is of any inherent properties of it" (p. 61).

Because media use behavior is driven by people's situational definitions, it is important to consider some of the dimensions of these definitions that shape their presentations. Gardner and Martinko(1988b) have identified four perceived situational dimensions which have been found to impact the self-awareness and self-presentations of organizational members: 1) familiarity, 2) formality, 3) ambiguity, and 4) favorability.

Familiarity. Self-presentation research indicates that there is a complicated relationship between audience familiarity and actor self-presentations. On one hand, actors tend to be acutely self-aware in thepresence of novel audiences because they recognize that first impressions are difficult to overcome (Kleinke, 1975). On the other hand, familiarity may operate to constrain the identity related images the actor can effectively claim. As the exchange participants come to know each other better, however, familiarity can serve as a liberating factor when actors feel accepted by the audience.

The extent to which audiences in CMCS are seen as familiar to actors depends, in part, on the degree to which the exchange participants interact using other media. When people interact via richer media, the influences described above are likely to be operative. For example, Finholt and Sproull (1990) point out that because members of the Computer Science Group(CSG) DL saw each other on a daily basis, social cues were still salient for them. However, if the parties only interact through CMCS, there is a greater likelihood that actors will become deindividuated, and hence subjectively self-aware (Sproull & Kiesler, 1986). Thus, we propose:

> *Proposition 19:* Audience familiarity is inversely related to the public self-awareness of CMCS users, but only when the exchange participants also interact via richer media.

Novel settings and activities likewise tend to increase actors' self-awareness, since they cannot rely on overlearned scripts inpresenting themselves to others (Gardner & Martinko, 1988b). Thus, we anticipate that CMCS novices will be more self-aware when using this medium than experienced users. Support for this proposition is provided by several studies which indicate that the extent to which CMCS are considered appropriate for exchanging messages with socioemotional content, depends upon the experience and familiarity of users with this medium (Hiltz & Turoff, 1981;Rice & Case, 1983). Even experienced users, however, may become self-conscious when confronted with unfamiliar CMCS activities or messages.

> *Proposition 20:* There is an inverse relationship between actor experience with CMCS and actor self-awareness.

> *Proposition 21:* The novelty of CMCS activities and/or messages is directly related to actor self-awareness.

Formality. Self-presentational research indicates that audience characteristics such as status, power, and size (i.e., number of people present) increase actor perceptions of situational formality, which in turn heightens actor self-awareness (Gardner & Martinko, 1988b; Jones & Pittman,1973). Interestingly, studies of CMCS (Kiesler, 1986; Kiesler et al.,1984; Sproull & Kiesler, 1986; Finholt & Sproull, 1990) suggest that because this medium provides weak social context cues, status and power differentials between the interaction participants are equalized. Of course, if the parties interact periodically on a face-to-face basis, this equalization effect is diminished.

> *Proposition 22:* The lack of social context cues provided by CMCS leads to status/power equalization, but only when the exchange participants do not also interact via richer media.

> *Proposition 23:* The extent to which computer-mediated communications equalize status/power differentials among the exchange participants, is inversely related to actor perceptions of

situational formality and public self-awareness.

Earlier, we noted that size is one audience attribute that is at least potentially apparent to CMCS users through cues provided by DLs. Size is a critical audience attribute in electronic exchanges, because the capability of the actor to reach a large audience is expanded by this technology (Culnan & Markus, 1987; Finholt & Sproull, 1986; Huber, 1990; Kerr & Hiltz,1982; Trevino et al., 1987). Indeed, Hiltz & Turoff (1982) found that the number of people organizational members can stay in touch with, on a daily basis, is between two and ten times as great with CMCS as it is with face-to-face, mail, or telephone media. Kiesler (1986) likewise notes that it is as easy to send a message to one person as it is 1000 using CMCS.

Bostrom (1970) has shown that the number of possible relations between members increases dramatically as a group grows larger. This finding suggests that actors may receive more cues for self-presentational behavior as the size of the audience increases. Thus, one might expect actors' self-awareness and self-presentational activity to be directly related to audience size. Both Bostrom (1970) and Gardner and Martinko (1988a), however, found that actors are less active in larger as opposed to smaller groups, apparently because they experience greater competition for "air time."

Finholt and Sproull (1990) have generalized this finding to electronic groups. Specifically, they found that group size was inversely related to the proportion of active members. This relationship held up across work and non-work groups, as well as required and discretionary groups. Perhaps, members of large groups feel that their opportunities to participate are socially — if not technically — constrained, since the ability of other members to process a large number of messages is limited. Thus, CMCS appear to lessen, rather than eliminate, the constraints imposed by group size.

> *Proposition 24:* CMCS reduce, but do not eliminate, the perceived constraints of group size.

After observing that some members of large electronic groups are very uninhibited in their self-presentations, however, Finholt and Sproull (1986) speculate that the audience may seem small to these actors. That is, because users only receive cues about the audience's size upon request, they may be unaware of this attribute. As a result, they may view the audience as small and informal, despite its large size. Still, it is important to note that only a small proportion of EMS users were active. The remaining members may have declined to participate because they defined the group as large and formal, and/or they perceived their opportunities to participate as limited.

Proposition 25: Actor perceptions of electronic audience size are directly related to their perceptions of situational formality, and inversely related to their perceptions of self-presentational opportunities via CMCS.

Proposition 26: To the extent that CMCS filter out social context cues regarding audience size, actor perceptions of situational formality decrease.

Proposition 27: The public self-awareness of CMCS users is directly related to their perceptions of situational formality.

Ambiguity. Gardner and Martinko (1988b) speculate that because ambiguous situations lack clear cut norms for behavior, they require actors to monitor their self-presentations more closely. Given the weak social context cues inherent in electronic communications, one might expect actors to be more self-aware. However, research by Daft, Lengel & Trevino (1990)indicates that actors tend to refrain from using "lean" media, such as CMCS,when the information to be exchanged is ambiguous. Indeed, most of the messages sent via CMCS (e.g., memos, short messages) appear to be routine in nature (Finholt & Sproull, 1986; Rice & Case, 1983; Steinfield, 1986; Trevino etal., 1987). Furthermore, as proposition 14 indicated, we suspect that many of the routine activities users perform with this medium become scripted overtime. For all of these reasons, we speculate that actors perceive most situations that involve electronic exchanges as unequivocal.

Proposition 28: When the messages sent and activities performed via CMCS are unequivocal and scripted, actors define the situation as unambiguous.

Proposition 29: The extent to which situations involving electronic exchanges are defined by actors as unambiguous, is inversely related to their public self-awareness.

Favorability. Self-presentational research indicates that the degree to which actors define situations as threats versus opportunities impacts upon their self-presentations (Gardner & Martinko, 1988b). However, both highly favorable and highly unfavorable situations can be expected to heighten actors' public self-awareness (Schlenker, 1980). Because CMCS filter out many of the social context cues that actors use to define situational favorability, they may have difficulty recognizing image-related threats and opportunities. This analysis suggests the following propositions:

Proposition 30: There is a curvilinear relationship between CMCS users' perceptions of situational favorability and their self-awareness.

Proposition 31: Actors are less attuned to image-related threats and opportunities when they present themselves to others via CMCS, than is the case with richer media.

Actor Self-Presentation

The situated identity of an actor influences the communication medium that is selected and the message sent to the audience. These components of self-presentation, however, are not independent of each other, since the message content can influence the medium chosen, and vice versa. Because we are especially interested in self-presentations by CMCS, we examine how actors come to choose this medium first, and then consider the types of messages that may be sent with it. This order is arbitrary, however, since the message will typically, but not always, precede the choice of media.

Media Choice. As a useful starting point for understanding the media choice decisions of organizational members, we examine a model of managerial information processing developed by Daft and Lengel (1984, 1986). The basic premise of this model is the positive relationship betweensituational and message complexity/ambiguity and media richness described earlier. That is, actors tend to use rich media (e.g., face-to-face meetings) to process equivocal messages (e.g., strategic planning, employee motivation), and lean media (e.g., electronic mail) for unambiguous messages (e.g., routine information exchange). Furthermore, Daft and Lengel (1984) contend that using rich media for simple problems results in overcomplication and aloss of efficiency, while using lean media for complex tasks results in oversimplification. For example, it would be inefficient to hold face-to-face meetings with subordinates to discuss inventory levels. Conversely, efforts to develop strategic plans through written memos would most likely result in oversimplification, since written communications are not well suited to capturing managers' intuitions. Subsequent research has provided empirical support for these predictions (Daft et al., 1987; Trevino et al.,1987).

Together, Daft and Lengel's (1984, 1986) ideas and our preceding discussion of situational dimensions suggest that actors are more likely to select CMCS when they define the situation as familiar, informal, unambiguous, and favorable. To deal with the complexity involved in situations that are perceived as novel, formal, ambiguous and/or unfavorable, however, actors will prefer richer media (Daft & Lengel, 1984, 1986), unless situational constraints (e.g., geographic distance, temporal availability) preclude their use (Culnan & Markus, 1987; Huber, 1990; Trevino et al., 1987).

> *Proposition 32:* Actors' perceptions of situational novelty, formality, ambiguity, and favorability are directly related to the richness of the media they select to present themselves to others.

As an extension of Daft and Lengel's (1984, 1986) model,Trevino et al. (1987) propose that actors sometimes select media for symbolic reasons. For instance, face-to-face communication symbolizes personal concern and interest in the message recipient, while electronic mail symbolizes speed and efficiency. To test these predictions, Trevino et al. (1987)interviewed managers regarding their reasons for selecting particular media, as noted in an earlier section. A content analysis of the managers' reasons provided support for their predictions. Furthermore, the results revealed that face- to-face communications were selected primarily for symbolic and content reasons, while the telephone and electric mail media were chosen as a result of situational constraints. One of the primary situational reasons managers provided for using CMCS was to reach someone who is busy or frequently absent. This finding suggests that actors may select CMCS to reach an audience that is otherwise inaccessible — either physically or as a result of status differentials.

The notion that actors may use CMCS strategically to achieve status equalization is intriguing. It is also at least partially supported by empirical research. Sproull and Kiesler (1986), for example, found that individuals tend to prefer EMS for communicating up the hierarchy as opposed to downward. Similar findings were obtained by Hiltz and Turoff(1978). They speculate that CMCS provide greater opportunities for participation to low status actors, including women and socially anxious persons, who typically defer to others during face-to-face communications. It is even possible that actors would select CMCS to achieve status equalization when interacting with superiors, while opting to interact with secretaries face-to-face to preserve status differentials (Sproull & Kiesler, 1986). Thus we propose:

> *Proposition 33:* The degree to which actors prefer CMCS as a medium for self-presentation is directly related to audiencestatus/ power.

In interpreting their findings, Trevino et al. (1987) speculate that EMS probably have not been available in organizations long enough to acquire the symbolic value of other media. Furthermore, they argue that the extent to which electronic mail possesses symbolic meaning depends upon the organization's culture. Organizational norms dictate which issues should be put in writing, which should be discussed face-to-face, and the kinds of messages that are considered to be appropriate for EMS (Finholt &Sproull, 1990; Fulk et al., 1987; Trevino et al., 1987). Finholt and Sproull (1990), for instance, found that electronic groups possess well-defined norms about the content of messages exchanged among members, and that discretionary groups used EMS to communicate and enforce these norms. Similarly, Rice's (1982) study of CC identified norms of reciprocity which specified that actors who received messages via this medium were expected to use it to respond.

> *Proposition 34:* Organizational norms, including norms of reciprocity, specify the circumstances under which actors should use CMCS.

The preceding review of media choice theory and research is very consistent with the self-presentational model presented in this chapter. Specifically, it suggests that actors' situated identities -which are shaped by social context stimuli such as organizational norms influence their media choice decisions. It also indicates that actors' media choice decisions may be influenced by their self-presentational motives or goals. That is, actors sometimes choose particular media because of a desire to present themselves in a certain fashion. Based on this literature we propose:

> *Proposition 35:* Actors' situated identities, including their interaction goals, influence their media choice decisions.

Message content. Communication research suggests that the content of actors' self-presentations using CMCS can be categorized along the following dimensions: 1) task versus socioemotional content, and 2)purposiveness. We examine pertinent findings with regards to each of these areas below.

At the outset of this chapter, we noted that there is some disagreement among communications researchers regarding the ability of CMCS to transmit socioemotional messages (Kerr & Hiltz, 1982; Hiltz & Turoff,1978; Rice, 1982; Rice & Love, 1987; Short et al., 1976; Steinfield, 1986). While Short et al.'s (1976) social presence theory suggests that this medium is poorly suited for socioemotional messages, communications research indicates that 14 (Hiltz & Turoff, 1978) to 40 percent (Sproull & Kiesler, 1986) of the messages sent via this medium are socioemotional in content. Thus, while it is clear that CMCS are primarily used to send task-related messages, this medium is also used to present actors for social purposes.

Additional insight into the types of task and socioemotional messages actors exchange via computer networks is provided by the results of several field studies (Finholt & Sproull, 1990; Kerr & Hiltz, 1982;Kiesler, 1986; Rice & Love, 1987; Steinfield, 1986). Kiesler (1986) reports that people use CMCS to exchange a wide variety of information including reports,computer programs, ideas, travel plans and gossip. Rice and Love (1987)found that most of the task related messages actors sent using EMS involved giving (57%) or requesting

(6.1%) nonpersonal information. They also found that 28 percent of the socioemotional content could be categorized as positive (e.g., showing solidarity, 18%; giving personal information, 6.1%). Only .4 percent of the socioemotional content was negative in character (e.g., shows antagonism).

The findings of Finholt and Sproull's (1990) field study of electronic groups at work with regards to message content are particularly noteworthy. These researchers discovered that two of the largest groups were formed to exchange non-work related information. One of these groups, called "Cinema," was an extracurricular DL consisting of over 500 members, including professionals, managers, technicians, and secretaries. This DL was used to exchange highly subjective, and often emotionally charged, messages relating to movie reviews. The second extracurricular DL, called "Rowdies," was made up of 98 members, most of whom were male technicians and professionals. Most of the messages sent to this DL pertained to gossip, group activities (e.g., "afternoon beer busts"), and crude humor. Obviously, the official work responsibilities of members did not have a major impact on the messages sent via these DLs. Instead, the primary purpose of these groups was to have fun (Finholt & Sproull, 1990). Clearly, the organizational culture had a profound impact on these groups' perceptions of appropriate content for CMCS messages.

It should be noted that the high level of socioemotional content documented by these field studies of CMCS, runs counter to the predictions of social presence theory (Culnan & Markus, 1987). These results, however, are completely consistent with our self-presentational model, which indicates that the degree to which members use CMCS for task versus socioemotional purposes depends upon their situational definition. Furthermore, the self-presentational model can be used to interpret other findings of CMCS research that are not adequately accounted for by the constructs of social presence and information richness. For instance, Finholt and Sproull (1990) found that people varied the style of their "voice" depending upon the identity of the electronic group their message was directed toward. In one case, the "detached UFF voice discussing 'contention resolution implementation' became a highly personal Rowdy voice" (p. 57). Notice that these differences appear to be attributable to the impact of differential audience characteristics on the actor's situational identity and subsequent self-presentation. That is, when the audience was defined as a formal electronic work group (UFF), the actor's self-presentation was task-oriented. However, when the audience was defined as an informal, non-work group (Rowdies), the content of actor's self-presentation was socioemotional. Apparently, the actor was much more publicly self-aware during the interaction with the formal group, as predicted by proposition 27. Taken as a whole, the preceding discussion suggests:

Proposition 36: The extent to which actors define CMCS as an appropriate medium for task and socioemotional content, depends upon their definition of the situation.

Proposition 37: The content of actors' self-presentations via CMCS is influenced by their perceptions of audience characteristics.

Proposition 38: Perceptions of situational formality during electronic exchanges are directly related to the amount of task-oriented content, and inversely to the amount of socioemotional content.

The second dimension along which self-presentations via CMCS appear to vary is purposiveness. Self-presentational theory suggests that the degree to which actors consciously present themselves to others to achieve specific goals varies across situations (Gardner & Martinko, 1988b; Leary& Kowalski, 1990; Schlenker, 1980). In addition, the extent to which actors knowingly manipulate their self-presentations is directly related to their level of self-awareness. That is, purposiveness increases as they become more acutely self-aware, while it declines under conditions of subjective self-awareness.

This relationship between self-awareness and purposiveness provides the key to reconciling a second set of apparently contradictory findings. As noted at the outset of this chapter, the purposiveness of CMCS users' behavior varies markedly from one situation to the next. An examination of available research, however, reveals that CMCS users tend to exhibit strategic self- presentations most often during exchanges that heighten their public self- awareness, i.e., situations that they define as formal, ambiguous, novel, and/or highly favorable or unfavorable. For instance, Finholt and Sproull's (1990) finding that actors used UFF to demonstrate their knowledge to superiors and peers is consistent with this explanation. Indeed, such self-promotional tactics are most often directed toward formal, high status audiences (e.g., peer and superior members of UFF) (Gardner &Martinko, 1988b; Jones & Pittman, 1982; Schlenker, 1980; Tedeschi & Norman, 1985).

Several other types of strategic self-presentations have been identified by impression management researchers, including ingratiation, exemplification, supplication, and intimidation (Gardner & Martinko, 1988b; Jones& Pittman, 1982). These tactics are used to attain attributions of likability, moral worthiness, helplessness, and danger, respectively. While it is probable that actors occasionally use these tactics while interacting via CMCS, empirical evidence to this effect is not currently available. To the contrary, available theory and research suggests that actors prefer to use richer media for these kinds of complex and ambiguous self-presentations (Daft& Lengel, 1984, 1986; Daft et al., 1987; Trevino et al., 1987).

An important implication of the apparent nonconduciveness of CMCS to purposive behavior is that individuals will find it difficult to exert leadership through this medium (Gardner & Schermerhorn, 1988; Kerr & Hiltz, 1982; Turoff & Hiltz, 1982). Because this effect is compounded by status equalization, heavy reliance on this medium could weaken an organization's leadership. The above discussion suggests the following propositions:

> *Proposition 39:* The purposiveness of actors' self-presentations via CMCS is directly related to their public self-awareness.

> *Proposition 40:* Whenever possible, actors will use richer media than CMCS for strategic self-presentations.

> *Proposition 41:* Because CMCS are not highly conducive to purposive behavior and it encourages status equalization, this medium is not well suited for actor presentations of leadership.

Additional support for proposition 39 is provided by several studies which indicate that, in comparison to persons who interact face-to-face, CMCS users tend to be more self-absorbed (i.e., subjectively self-aware), candid with their opinions, uninhibited, creative, willing to take risks, self-disclosing, amenable to sending bad news, and prone to exhibit unconventional, deviant and irresponsible behaviors, such as "flaming" (Finholt & Sproull, 1990; Hiltz & Turoff, 1978; Kerr & Hiltz, 1982; Siegel et al., 1986; Sproull & Kiesler, 1986). Thus, under some circumstances, the low level of objective social presence which characterizes CMCS causes actors to become subjectively self-aware and perhaps even deindividuated. In these situations, the actors' relative anonymity relaxes their image-related concerns and inhibitions. The consequences of self-absorption may be either positive (e.g., greater candor in decision making), or negative (e.g., anti-organizational flaming). Accordingly, we advance a corollary to proposition 39.

> *Proposition 42:* The subjective self-awareness of the actor is inversely related to purposive self-presentations and directly related to uninhibited behavior.

Audience Self-Identification and Behavior

As Figure 1 indicates, the audience develops a situated identity by selectively perceiving and interpreting static and dynamic social context cues through the self-identification process. Note that this processis basically identical to the self-identification process described above for the actor. As part of their situated identities, audiences possess certain expectations about the roles and behavior of the actor. When the actor exhibits self- presentational behavior, the audience assesses the degree ofcongruence between the expected and actual presentation. The amount of perceived congruence is the key determinant of the audience's impressions of the actor, and the favorableness of its reactions (Gardner & Martinko,1988b). Certain personality characteristics of the audience, however, tend to mediate its interpretation and response to the actor's performance. For instance, research indicates that high self-monitors respond more negatively to ingratiation attempts than low self-monitors (Jones & Baumeister,1976).

The importance of selecting a communication medium that is congruent with the sender's message has been noted by communication and IT theorists and researchers. Trevino et al. (1987), for example, point out that because of its symbolic connotations of efficiency and impersonality, managers tend to view EMS as a poor medium for ingratiatory behaviors signaling personal caring and interest. They note that "although it may be efficient for sending a personal message of congratulations or regret ... the receiver... may be disappointed in the lack of caring shown by the sender" (p. 571). We agree that if the audience defines the situation as being inappropriate for electronic exchanges, the selection of this medium by an actor is likely to lead to unwanted impressions and an unfavorable audience response. Support for this argument is provided by Daft et al.'s (1987) finding that effective managers are more sensitive to the relationship between message ambiguity and media richness than their less effective counterparts.

It is important to understand, however, that the degree of perceived congruence depends on the audience's subjective assessment of media appropriateness, rather than objective media properties. Fulk e tal. (1987) emphasize the importance of "matching" one's choice of media with prevailing organizational norms for appropriate media usage. They argue that these norms produce "a similar pattern of media attitudes and use behavior within groups, even across tasks with different communications requirements" (p.542-543). They also suggest that prevailing norms can produce a"mismatching" between objective media properties and information richness in some cases, e.g., using CMCS to send messages high in socioemotional content. This argument is entirely consistent with the model advanced in this chapter,which suggests that actors will select whichever medium is most congruent with their situational definition when presenting themselves to others.

The audience's assessment of the actor's self-presentational behavior, however, extends beyond the choice of medium. Although an audience may consider a topic to be appropriate for CMCS, they may still form a negative impression if the manner in which the actor is presented deviates from their situational definition. Evidence of the importance of presentational style, as well as content, is provided by the following normative statement sent by a member of the "Cinema" DL described above. This message was sent in response to a colorful movie review that was entered by another member.

> Just because not many people type in Movie reviews is no reason to put stuff on this DL that is better suited to Trash <a catchall DL>. Perhaps all the people that complain that this DL is all but dead should go out to the movies and give us a review or two. — Jim (Finholt & Sproull, 1990, p. 52).

In this example, movie reviews are clearly seen as appropriate content for this medium, but the style of the offending review was considered to be unacceptable. The above discussion suggests the following proposition:

> *Proposition 43:* The congruence between the audience's situational definition and the actor's self-presentational behavior is positively related to the favorability of the audience's impressions and subsequent behavior toward the actor.

Self-Presentation: An Ongoing, Dynamic Process

Once the audience responds to the actor's performance, the actor assesses the degree of congruence between this response and his or her situated identity. The self-presentational process then begins anew. While a congruent response will serve to support the legitimacy of this identity, an incongruent response is likely to lead to adjustments. The actormay redefine the situation by pursuing different identity images, targeting a different audience, selecting an alternative communication medium, or changing the message content (Gardner & Martinko, 1988b). Thus, we propose:

> *Proposition 44:* The perceived congruence between actors' situated identities and the audience's response, is inversely related to the degree to which their situated identity and self-presentational behaviors (including media choice and use) are altered.

The behavior of the actor and audience may lead to changes in the social context as the feedback arrows in Figure 1 indicate. Trevino et al. (1987) point out, for example, that because media have symbolic value, they can impact upon the organization's culture. That is, the preferences of coworkers—especially senior managers — for using particular media to perform certain types of tasks shape the overall culture of the organization. Thus it is clear that self-presentation in organizations is a dynamic, ongoing process.

> *Proposition 45:* Actor self-presentations and audience responses via

Potential Benefits	Potential Liabilities
• Greater accessibility to a wider variety of audiences (Culnan & Markus, 1987).	• Overaccessibility to busy managers as a wide variety of members employ this medium for self-promotion (Finholt & Sproull, 1990).
• Status equalization makes it possible for actors who previously might have been stereotyped to create desired identities (Hiltz & Turoff, 1978; Sproull & Kiesler, 1986).	• Loss of leadership influence due to a perceived incompatibility of this medium for purposive presentations (Gardner & Schermerhorn, 1988).
• Enhanced creativity in decision making due to reduced concern about maintaining face (Hiltz & Turoff, 1978; Kerr & Hiltz, 1982; Kiesler, 1986).	• Reduced flexibility in decision making due to actor reluctance to change their position for fear of losing face, whenever a permanent electronic record is created (Culnan & Markus,1987).
• Enhanced means of socializing since busy workers can make acquaintances and interact without leaving their desk and on their own schedule (Finholt & Sproull, 1990; Kiesler, 1986; Sproull & Kiesler, 1986; Steinfield, 1986).	• Excessive socializing which distracts employs from their work (Finholt & Sproull,1990).
• Greater affiliation and commitment provided by the enhanced opportunities to interact and socialize (Finholt & Sproull, 1990; Kiesler, 1986).	• Reduced affiliation and commitment due to deindividuation and the ease with which disgruntled employees can communicate negative norms and antagonistic attitudes about the organization totheir peers (Finholt & Sproull,1990).
• Increased effectiveness due to heightened levels of commitment (Finholt & Sproull, 1990).	• Decreasedeffectiveness due to excessive socialization and low morale (Finholt & Sproull, 1990).

Table 2. Potential Benefits and Liabilities of Computer-Mediated Communications Systems as a Medium for Self-Presentation

CMCS may lead to changes in the social context.

Issues

A number of issues pertaining to the social and organizational consequences of self-presentation via CMCS are suggested by the preceding analysis. Overall, it appears that these outcomes may be either beneficial or adverse, depending upon the specific technology employed and its interaction with organization's culture. Several of these benefits and liabilities are summarized in Table 2 and discussed below.

Potential Benefits

Because most of the benefits summarized in Table 2 have been alluded to in earlier sections, we will not discuss them further here. The considerable promise of CMCS as an enhanced medium for socializing, however,warrants additional attention. This promise has been observed by several researchers (Finholt & Sproull, 1990; Kiesler, 1986; Rice, 1982; Rice & Love,1987; Sproull & Kiesler, 1986; Steinfield, 1986). Sproull and Kiesler(1986), for example, point out that because EMS are asynchronous, audiences may be enlightened or amused on their own schedule, as opposed to that of the actor. In addition, audiences are not obligated to reciprocate; they can read their message and get back to work. As a consequence, even busy or socially handicapped workers have enhanced opportunities to socialize (Kiesler, 1986). Other researchers have noted that CMCS can serve as a useful means whereby organizational members, especially new employees, can expand their social network by making contacts with persons who share their interests (Finholt & Sproull, 1990; Rice, 1982; Steinfield, 1986). Ric eand Case (1983) found that 43 percent of the managers they surveyed reported exchanging messages with persons via CMCS that they had not previously communicated with through traditional media. Finally, Finholt and Sproull (1990) found that many people used EMS simply to have fun.

If computer-mediated communication does indeed provide members with new avenues for socializing, then it is possible that other benefits such as increased affiliation, commitment, and performance may also accumulate. Unfortunately, these benefits are by no means guaranteed.

Potential Liabilities

As Table 2 indicates, many of the potential benefits ofCMCS could just as easily turn out to be liabilities. For instance, the ready access to new audiences which this medium provides could lead to excessive presentation of information that is useless to the recipient (Kerr & Hiltz, 1982;Kiesler, 1986). This danger is well illustrated by the following example.

One day a message was sent to PHC Today, a DL for company news, announcing that the PHC research division had just been sold to a competitor. It explained how the division would be organized in the new company, who would have management responsibilities and how the transition would proceed. Twelve days later the sender of this April Fool's day message sent another message apologizing for his joke (Finholt & Sproull, 1990, p. 61).

In this incident, that actor's message may have been motivated by a desire to appear humorous. Apparently, however, the sender was subsequently rebuked by the audience, at which time an apology (a type of defensive self-presentational tactic; Snyder, Higgins & Stucky, 1983) was issued to save face. The point is that this rather frivolous incident may havebeen viewed as a waste of time by many recipients, while causing anxiety for others. While this example may be extreme, there is evidence that CMCS users often experience stress and difficulty in providing feedback due to information overload (Hiltz & Turoff, 1985).

There is also a danger that antagonistic messages sent by a few disgruntled employees may undermine the morale and organizational commitment of a large number of their peers. Alternatively, excessive socializing via this medium could distract employees from their work responsibilities, thereby causing a drop in productivity. Thus, it is apparent that organizations are presented with many challenges regarding the management and utilization of CMCS in order to capitalize on their potential.

Effective Management of Computer-Mediated Communciations Systems

The issues discussed above are too important to be left to technicians or programmers by default (Finholt & Sproull, 1990; Kiesler,1986; Sproull & Huber & McDaniel, 1986; Kiesler, 1986). Because of its social and organizational consequences, the design of CMCS involves important policy decisions that are the rightful responsibility of the organization's managers. To fully capitalize on these systems' potential, managers must consider both their technical and social properties. A summary of these considerations follows.

Control Over Access and Participation

A critical set of design decisions revolve around the degree of control the system will provide with regards to user access and participation (Culnan & Markus, 1987). From a technical standpoint, it is possible to incorporate security codes that limit access, as well as filters that enable users to screen incoming electronic mail (Rowe, 1988; Malone et al., 1987; Szewczak & Gardner, 1989). There is a danger, however, that these types of restrictions on access will eliminate many of the benefits of this technology, such as status equalization and efficient socialization (Hiltz & Turoff,1985). Thus, managers face a tricky tradeoff in deciding how much control over access should be built into the system. Communications and IT researchers (Finholt & Sproull, 1990; Kiesler, 1986; McGuire et al., 1987; Sproull &Kiesler, 1986) have inventoried many of the design questions that confront managers. Kiesler (1986) provides a good summary of the key questions.

Will the systems allow group communication? Who will have access to group accounts? Who can create electronic distribution lists or bulletin boards? Who can send computer mail to whom? How closely will messages be monitored? And who passes on or responds to the information contained in these messages? (p.58).

Shaping the Organizational Culture

By carefully weighing the advantages and disadvantages associated with system options, managers have an opportunity to favorably shape the structure, processes, and culture of their organization (Strassman, 1985). While the potential impact of CMCS on the organization's structure and processes is beyond the scope of this chapter, other authors have specified a number of issues that merit consideration (e.g, Child, 1987; Keen, 1987; Huber, 1990).

With regards to the possible impact of CMCS on organizational culture, our analysis suggests several factors for managers to consider. Specifically, it appears that the organizational culture affects members' perceptions and use of CMCS, and vice versa. Therefore, managers need to take into account the impact the culture will have on members' acceptance and utilization of electronic media (Fulk et al., 1987; Trevino et al., 1987), as well as the impact CMCS will in turn have on the organization's culture.

A key question to ask is "How far does management want to go toward creating an electronic community?" (Kiesler, 1986, p. 58). If the answer to this question is "to a substantial degree," managers may want to endorse specific norms and provide incentives to encourage people to use electronic media in a responsible fashion (Finholt & Sproull, 1990). Alternatively, management can take a restrictive approach. For instance, it is technically feasible to implement an EMS that makes it difficult to send group mail, or prohibits the creation of discretionary and extra curricular electronic groups (Finholt & Sproull, 199). Here again, however, there is a danger that the increased control this option provides, may be offset by a loss of benefits (Hiltz & Turoff, 1985).

Encouraging Leadership

In an earlier section, we noted that the objectively low social presence of CMCS does not appear to be very conducive to purposive presentations or the emergence of leadership. For managers who present themselves frequently via CMCS to organizational members who are geographically dispersed and inaccessible for face-to-face meetings, software innovations that provide additional social context cues may have some appeal. For instance, future EMS may augment the messages provided by today's systems with mechanisms for displaying the names of DL members, attaching pictures to names, and informing actors of the identities of the persons who have read their messages (Finholt & Sproull, 1990). It is possible that these actions would enable leaders to highlight their status and exercise greater influence. Unfortunately, because these actions are likely to make interactions appear more formal, the benefits of status equalization and subjective self-awareness would belost.

Conclusion and Future Directions

This chapter has presented a general model of the self-presentation process, and applied it to examine the ways in which organizational members present themselves to others using CMCS. In the process, we have reviewed a wide variety of relevant literature from the fields of communication, management, IT, and social psychology. In applying the self-presentation model, we have attempted to reconcile some apparently contradictory findings regarding the social consequences associated with this medium. For example, the fact that CMCS are used to send messages high insocioemotional content, in some, but not all settings, is consistent with the notion that media choice and use depends on the situational definition. Similarly, the seemingly contradictory set of results that indicate CMCS users exhibit strategic self-presentations in some situations, while displaying counternormative behaviors (e.g., flaming) in others, is easily explained in terms of self-awareness. Specifically, people tend to be most concerned about their imagewhen they are publicly self-aware, and least concerned when they are subjectively self-aware. Through the introduction of these concepts, we have attempted to contribute to the clarity of this literature.

A large number of propositions regarding organizational members' self-presentations through CMCS have been advanced in this chapter. Empirical support for some of these propositions is much stronger than it is for others. Those advanced with regards to the self-identification process are especially speculative and only minimally supported by available research. The tentative nature of these propositions, however, is unavoidable, given the paucity of research into the cognitive, motivational and affective processes that influence the situated identities and subsequent self-presentations of persons who interact via CMCS. While the self-identification propositions are the most speculative, all of the propositions are tentative and require confirmation through additional research. As noted at the outset, the current chapter is intended to provide a micro-level analysis for the purpose of stimulating empirical investigation along these lines. Given the rapid rate at which organizations are implementing wide area and local area networks, the contribution of this stream of research for management theory and practice could be immense. Ultimately, it is hoped that the insights gained from this line of inquiry will enable organizations to reap the benefits of this technology while minimizing its liabilities.

References

Bandura, A. (1977). Social learning theory. Englewood Cliffs,N.J.: Prentice-Hall.

Baumeister, R. F. (1982). A self-presentational view of socialphenomena. *Psychological Bulletin, 91*, 3-26.

Baumeister, R. F. (1989). Motives and costs of self-presentation in organizations. In R. A. Giacalone & P. Rosenfeld (Eds.), *Impression management in the organization* (pp. 57-71). Hillsdale, NJ:Erlbaum.

Bostrom, R. N. (1970). Patterns of communicative interaction insmall groups. *Speech Monographs, 37*, 257-263.

Child, J. (1988). Information technology, organization, and response to strategic challenges. *California Management Review, 30*(1),33-50.

Crowne, D. P., & Marlowe, D. (1964). *The approval motive*. NewYork: Wiley.

Culnan, M. J., & Markus, M. L. (1987). Information technologies. In F. M. Jablin, L. L. Putnam, K. H. Roberts, & L. W. Porter (Eds.), *Handbook of organizational communication: An interdisciplinary perspective* (pp. 420- 443). Newbury Park, CA: Sage Publications.

Daft, R. L., & Lengel, R. H. (1984). Information richness: A new approach to managerial behavior and organization design. In B. M. Staw & L. L. Cummings (Eds.), *Research in organizational behavior* (Vol.6, pp. 191- 233). Greenwich, CT: JAI Press.

Daft, R. L., & Lengel, R. H. (1986). Organizational information requirements, media richness and structural design. *Management Science,32*, 554-571.

Daft, R. L., Lengel, R. H., & Trevino, L. K. (1987). Message equivocality, media selection, and manager performance: Implications for information systems. *MIS Quarterly, 11*, 355-368.

Deaux, K., & Major, B. (1987). Putting gender into context: An interactive model of gender-related behavior. *Psychological Review, 94*,369-389.

Deal, T. E., & Kennedy, A. A. (1982). *Corporate cultures: The rites and rituals of corporate life*. Reading, MA: Addison-Weasley.

DeSanctis, G., & Gallupe, R. B. (1987). A foundation for the study of group decision support systems. *Management Science, 33*, 589-606.

Edwards, G. C. (1978). Organizational impacts of office automation. *Telecommunications Policy, 2*, 128-136.

Estrin, J., & Cheney, K. (1986). Managing local area networks effectively. *Data Communications, 15*(1), 181-189.

Feldman, J. M. (1981). Beyond attribution theory: Cognitive processes in performance appraisal. *Journal of Applied Psychology, 66*, 127-148.

Fenigstein, A., Scheier, M., & Buss, A. H. (1975). Public and private self-consciousness: Assessment and theory. *Journal of Consulting and Clinical Psychology, 43*, 522-527.

Finholt, T., & Sproull, L. S. (1990). Electronic groups at work. *Organization Science, 1*, 41-64.

Fletcher, C. (1989). Impression management in the selection interview. In R. A. Giacalone & P. Rosenfeld (Eds.), *Impression management in the organization* (pp. 269-281). Hillsdale, NJ: Erlbaum.

Foster, L. W., & Flynn, D. M. (1984). Management information technology: Its effects on organizational form and function. *MIS Quarterly, 8*, 229-236.

Freeman, L. C. (1984). The impact of computer based communication on the social structure of an emerging scientific specialty. *Social Networks, 6*, 201-221.

Fulk, J., Steinfield, C. W., Schmitz, J., & Power, J. G. (1987). A social information processing model of media use in organizations. *Communication Research, 14*, 529-552.

Gardner, W. L., & Martinko, M. J. (1988a). Impression management: An observational study linking audience characteristics with verbal self-presentations. *Academy of Management Journal, 31*, 42-65.

Gardner, W. L., & Martinko, M. J. (1988b). Impression management in organizations. *Journal of Management, 14*, 321-338.

Gardner, W. L., & Schermerhorn, J. R., Jr. (1988). Computer networks and the changing nature of managerial work, *Public Productivity Review, 11*(4), 85-99.

Gioia, D. A., & Manz, C. C. (1985). Linking cognition and behavior: A script processing interpretation of vicarious learning. *Academy of Management Review, 10*, 527-539.

Goffman, E. (1959). *The presentation of self in everyday life*. Garden City, NY: Doubleday Anchor.

Hiltz, S. R. (1984). *Online communities: A case study of the office of the future*. Norwood, NJ: Ablex.

Hiltz, S. R., & Turoff, M. (1978). *The network nation: Human communication via computer*. Reading, MA: Addison-Wesley.

Hiltz, S. R., & Turoff, M. (1981). The evolution of user behavior in a computerized conferencing system. *Communications of the ACM, 24*, 739- 759.

Hiltz, S. R., & Turoff, M. (1985). Structuring computer-mediatedcommunication systems to avoid information overload. *Communications ofthe ACM, 1985, 28*, 680-689.

Huber, G. P. (1990). A theory of the effects of advanced information technologies on organizational design, intelligence, and decision making. *Academy of Management Review, 15*, 47-71.

Huber, G., & McDaniel, R. (1986). Exploiting information technology to design more effective organizations. In M. Jarke (Ed.), *Managers, micros, and mainframes* (pp. 221-236). New York: Wiley.

Johansen, R., Vallee, J., & Spangler, K. (1979). *Electronic meetings: Technical alternatives and social choices*. Reading, MA:Addison-Wesley.

Jones, E. E., & Baumeister, R. F. (1976). The self-monitor meets the ingratiator. *Journal of Personality, 44*, 654-674.

Jones, E. E., & Pittman, T. S. (1982). Toward a general theory of strategic self-presentation. In J. Suls (Ed.), *Psychological perspectives on the self* (pp. 231-262). Hillsdale, NJ: Erlbaum.

Keen, P. G. W. (1987). Telecommunications and organizational choice. *Communications Research, 14*, 588-606.

Kelley, H. H. (1971). *Attribution in social interaction*. Morristown, N.J.: General Learning Press.

Kerr, E. B., & Hiltz, S. R. (1982). Computer-mediated communication systems. New York: Academic Press.

Kiesler, S. (1986). The hidden messages in computer networks. *Harvard Business Review, 64*(1): 46-60.

Kiesler, S., Siegel, J., & McGuire, T. W. (1984). Social psychological aspects of computer-mediated communication, *American Psychologist, 39*, 1123- 1134.

Kleinke, C. L. (1975). First impressions: *The psychology of encountering others*. Englewood Cliffs, NJ: Prentice-Hall.

Kraemer, K. L., & King, J. L. (1988). Computer-based systems for cooperative work and group decision making. *ACM Computing Surveys, 20*,115-146.

Leary, M. R., & Kowalski, R. M. (1990). Impression management: A literature review and two-component model. *Psychological Bulletin,107*, 34-47.

Lord, R. G. (1985). An information processing approach to social perceptions, leadership and behavioral measurement in organizations. In L. L. Cummings & B. M. Staw (Eds.), *Research in organizational behavior* (Vol. 7): 87-128. Greenwich, Conn.: JAI Press.

Malone, T. W., Grant, K. R., Turbak, F. A., Brobst, S. A., & Cohen, M. D. (1987). Intelligent information-sharing systems.*Communications of the ACM, 30*, 390-402.

Martinko, M. J., & Gardner, W. L. (1987). The leader/member attribution process. *Academy of Management Review, 1987, 12*, 235-249.

McGuire, T. W., Kiesler, S., & Siegel, J. (1987). Group andcomputer-mediated discussion effects in risk decision making. *Journal of Personality and Social Psychology, 52,* 917-930.

Neisser, U. (1976). *Cognition and reality.* San Francisco: W. H.Freeman.

Nunamaker, J. F., Jr., Applegate, L. M., & Konsynski, B. R.(1987). Facilitating group creativity: Experience with a group decision support system. *Journal of Management Information Systems, 3*(4),5-19.

Ralston, D. A. (1985). Employee ingratiation: The role of management, *Academy of Management Review, 10,* 477-487.

Rice, R. E. (1982). Communication networking in computer-conferencing systems: A longitudinal study of group roles and system structure.In M. Burgoon (Ed.), *Communication yearbook 6* (pp. 925-944). Beverly Hills, CA: Sage.

Rice, R. E. (1984). New media technology: Growth and integration.In R. E. Rice & Associates (Eds.), *The new media: Communication, research, and technology* (pp. 33-54). Beverly Hills, CA: Sage.

Rice, R. E., & Associates. (1984). *The new media: Communication,research, and technology.* Beverly Hills, CA: Sage.

Rice, R. E., & Case, D. (1983). Electronic message systems in the university: A description of use and utility. *Journal of Communication,33,* 131-152.

Rice, R. E., & Love, G. (1987). Electronic emotion:Socioemotional content in a computer-mediated communication network. *Communication Research, 14,* 85-108.

Rowe, S. H. (1988). *Business Telecommunications.* Chicago: Science Research Associates, Inc.

Schank, R., & Abelson, R. P. (1977). *Scripts, plans, goals, and understanding.* Hillsdale, N.J.: Erlbaum.

Schlenker, B. R. (1980). *Impression management: The self-concept, social identity, and interpersonal relations.* Monterey, CA:Brooks/Cole.

Schlenker, B. R. (1985). Identity and self-identification. In B.R. Schlenker (Ed.), *The self and social life* (pp. 65-99). New York:McGraw-Hill.

Schneider, D. J. (1981). Tactical self-presentation: Toward abroader conception. In J. T. Tedeschi (Ed.), *Impression management theory and social psychological research* (pp. 23-40). New York:Academic Press.

Short, J., Williams, E., & Christie, B. (1976). *The social psychology of telecommunications.* New York: Wiley.

Siegel, J., Dubrovsky, V., Kiesler, S. & McGuire, T. W. (1986).Group processes in computer-mediated communication, *Organizational Behavior and Human Decision Processes, 37,* 157-187.

Snyder, C. R., Higgins, R. L., & Stucky, R. J. (1983). *Excuses:Masquerades in search of grace.* New York: Wiley.

Snyder, M. (1987). Public appearances/private realities: *The psychology of self-monitoring*. New York: W. H. Freeman.

Sproull, L., & Kiesler, S. (1986). Reducing social context cues:Electronic mail in organizational communication. *Management Science,3 2*, 1492-1512.

Steinfield, C. W. (1986). Computer-mediated communication in an organizational setting: Explaining task-related and socioemotional uses.In M. McLanghlin (Ed.), *Communication yearbook 9* (pp. 777-804).Newbury Park, CA: Sage.

Steinfield, C. W., & Fulk, J. (1987). On the role of theory in research on information technologies in organizations: An introduction to the special issue. *Communications Research, 14*, 479-490.

Szewczak, E. J., & Gardner, W. L. (1989). Social and organizational impact of local and telecommunications systems — open questions. *Information Resources Management Journal, 2*(1), 14-25.

Tedeschi, J. T. (Ed.). (1981). *Impression management theory and social psychological research.* New York: Academic Press.

Tedeschi, J. T., & Norman, N. (1985). Social power,self-presentation, and the self. In J. T. Tedeschi (Ed.), *Impression management theory and social psychological research* (pp. 293-322). New York:Academic Press.

Thompson, T. (1987). Local area networks. *Byte, 12*(8), 145-198.

Trevino, L. K., Lengel, R. H., & Daft, R. L. (1987). Media symbolism, media richness, and media choice in organizations. *Communication Research, 14*, 553-574.

Trice, H. M., & Beyer, J. M. (1986). Charisma and its routinization in two social movement organizations. *Research in Organizational Behavior, 8*, 113-164.

Turoff, M., & Hiltz, S. R. (1982). Computer support for group versus individual decision. *IEEE Transactions on Communications,30*, 82-91.

Verity, J. W., & Lewis, G. (1987, November 30). Computers: The new look. *Business Week*, 112-123.

Weary, G., & Arkin, R. M. (1981). Attributional self-presentation. In J. H.Harvey, W. J. Ickes, & R. F. Kidd (Eds.), *New directions in attribution research* (Vol. 3, pp. 223-246). New York:Erlbaum.

Wexler, M. (1983). Pragmatism, interactionism, and dramatism: Interpreting the symbol in organizations. In L. R. Pondy, P. J. Frost, G. Morgan, & T.C. Dandridge (Eds.), *Organizational symbolism* (pp. 237-253). Greenwich, CT: JAI Press.

Zuboff, S. (1982). New worlds of computer-mediated work. *Harvard Business Review, 60*(5), 142-152.

Chapter 8

The Use of Expert Systems in Business

Irmtraud S. Seeborg
Ball State University

From time to time, the arrival of a new technology has such an impact on business and society that the entire way of life changes. Over the past century, we have seen many developments which have changed the way we manufacture goods and communicate with each other. Frequently, machines have replaced humans or, at least, made them much more productive.

With the advent of the computer in business, organizations have realized that information is a valuable asset and needs to be managed. Attention has been focused on the office, and today a variety of software is available which makes the office worker more productive (such as word processing, spreadsheets, graphics).

One group has been relatively untouched by all these developments—those individuals whose contribution to the organization's well being comes in the form of expertise or knowledge. In the last 5 years it has been recognized that the management of knowledge is also important and deserves our attention (see, e. g., Brooks, 1989, 37). How can we automate—or at least support—the role of the expert?

This topic has had a certain fascination for researchers in a number of fields since the early days of the computer. In the late 1940s and throughout the 1950s researchers in artificial intelligence developed programs which could play games (such as chess) or solve certain problems (such as small puzzles). However, the problems that were addressed were not relevant to business functions, and after some initial enthusiasm there was little impact of artificial intelligence on management, although the use of robots in manufacturing was actively pursued.

This changed in the 1980s when the development of expert systems moved from universities and research labs into mainstream businesses. During the last 5 years, a large number of expert system applications have been developed in a variety of business settings, and the popular literature is full of enthusiastic endorsements of expert system technology. Some examples:

> "It is clear that 'knowledge management' will be just as important in the 90's as 'information management' has been in the '80s" (Brooks, 1989, 37).
>
> "Whilst computers have assisted companies in making the best possible use of one of their most valuable resources (data or information), expert systems can now help organizations make the best possible use of what is probably its most valuable resource—knowledge or expertise" (Goodall, 1989, 28).
>
> "For many years managers have struggled to improve white- collar performance, with little success. Throughout the world, productivity gains in so-called nontouch work have been near zero. Some companies are now reporting dramatic progress through use of 'expert systems'" (Enslow, 1989, 54).

The Promise of Expert Systems

To understand the purpose of this chapter, it is necessary to have an understanding of the nature and promise of expert systems. A preliminary definition is that an expert system is a computer program which contains the knowledge and reasoning ability of a human expert. This concept will be explored further in the next section.

While the original expectations for artificial intelligence were very high—computers which can "think" and learn—expert systems have more limited goals. Expert systems attempt to recreate the behavior of experts in very limited applications. The functions performed by expert systems can be broadly grouped as interpretation (interpret, diagnose, monitor, predict), generation (plan, design), debug, repair, control (Ege & Sullivan, 1990; Cook & Schleede, 1988).

The use of expert systems is attractive to managers because they can save costs, speed up problem solving, improve quality and consistency of decisions, and preserve and distribute expertise in an organization. Not all expert systems projects fulfill all these expectations, but the potential is there, as will be discussed in more detail later.

Type of Application	No. of Applications	Percent
Business	50	9.4
Computer related	92	17.3
Engineering/technical	132	24.8
Medicine	159	29.8
Other	100	18.6
TOTAL	533	99.9

Based on expert systems listed in Smart & Langeland-Knudsen, 1986; compilation of data by the author.

Table 1. Expert Systems Applications Up To 1986

The Use of Expert Systems in Business

Historically, these benefits have not been seen or been utilized by business. As can be seen in table 1, the vast majority of applications summarized in a 1986 directory of expert systems (Smart and Langeland-Knudsen, 1986) are in medicine, engineering, and computer-related fields. Direct business applications in areas such as accounting, finance, human resource management, or marketing account for less than 10 percent of the total.

This situation is rapidly changing. A survey of expert systems described in the recent literature shows a large number of systems being developed and used in industry. Applications range from banking and insurance to manufacturing and encompass all functional areas of business. Some examples will be discussed in the next section of this chapter.

Issues of Relevance to Management

Why should a manager be concerned with expert systems? Expert systems can touch the role of a manager in a variety of ways: 1. One of the manager's subordinates may be an "expert" in an area for which the manager is responsible, and the manager needs to decide whether to capture his/her expertise and needs to know how to approach the expert. 2. Some of the manager's subordinates may be using an expert system—how does that affect the manager's supervisory and evaluative tasks? 3. The manager may be involved in the creation of an expert system—what choices need to be made, who should do the work? 4. The manager may be able to use expert systems technology to affect the profitability or strategic position of the organization—how does he/she find appropriate projects?

If expert systems will have such a far reaching effect on organizations in the next decade, managers need to understand what they are, how they can be used, and what potential problem areas need to be managed. The purpose of this chapter is to acquaint managers with expert systems concepts on a non-technical level, discuss the potential uses and problems of expert systems as they touch management, and report on some solutions that have been worked out by those already using expert systems extensively in their organizations. Finally, an outlook for the future of expert systems will be given.

SECTION II

Background

In this part of the paper, expert system concepts which are important for the manager will be introduced. We will then discuss a number of examples which illustrate how expert systems can be used in an organization. Following this, we will discuss in general what types of problems are best suited for expert system applications, and what the advantages and disadvantages of expert system usage are. The issues surrounding expert system use will be discussed in more detail in the next section.

Definitions

Publications relating to what we loosely call expert systems can be found under at least three different headings: artificial intelligence, knowledge-based systems or knowledge systems, and expert systems. How are these three related?

Definitions of artificial intelligence can be summarized by saying that it is an attempt to make computers behave like intelligent human beings. This is expressed in terms such as "building computers that exhibit intelligent behavior" (Feigenbaum et al., 1988, 31); "the attempt to reproduce human intelligence on the computer" (Bruce & Adam, 1989, 481); or "the attempt to make machines achieve human-like capabilities such as seeing, hearing, and thinking" (Leonard-Barton & Sviokla, 1988, 93).

Areas of research in artificial intelligence include robotics, natural language processing, and expert systems. This chapter is concerned with expert systems only, which can be seen as a subgroup of artificial intelligence.

The objective of expert systems is more limited than that of artificial intelligence. Researchers soon realized the difficulties inherent in trying to make computers act intelligently which would require them to handle information from the environment and find successful responses to these stimuli in general. Instead, they focus on mimicking the behavior of an expert in a small domain by capturing the factual knowledge of the expert as well as his ways of approaching and solving a problem.

How does an expert system differ from an ordinary computer program, for example a linear programming approach to resource allocation?

The ordinary computer program is based on numerical data. A linear programming package, for example, will not find a proper solution if incomplete or inaccurate data are used, and it cannot easily process symbolic (non-numeric) knowledge. The program cannot explain why it reached a given conclusion, and it will always be executed in a predetermined order.

The expert system, on the other hand, allows the processing of symbolic knowledge. It will work even if some facts are not known, and it can handle unreliable data. The system can express the confidence it has in its solution, thus allowing the user to decide how to proceed. Many expert systems are designed so that they can explain to the user why they ask a certain question or reach a certain conclusion. And the designer of the expert system does not predetermine the order in which the expert system applies its rules—that is decided at the time the user interacts with the expert system.

Thus, interacting with an expert system is more like interacting with a live expert than interacting with a regular computer program is. However, the systems are not truly intelligent.

How does an expert system differ from a knowledge-based system? There is no clear demarkation between these two concepts. Many authors use the two terms as synonymous (for example, Kirrane & Kirrane, 1989; Fordyce, Norden, & Sullivan, 1989). Others draw a distinction by pointing out that a knowledge-based system is a simpler version of an expert system (e.g. Kaiser, 1990).

The distinction, for many people, is one of degree rather than one of substance. The knowledge-based system, in this sense, captures the knowledge which you might find in a manual or textbook; the expert system goes a step beyond by incorporating the expert's way of looking at the knowledge and quickly resolving problems using the knowledge. The expert system is then more complex, more costly, and takes longer to develop.

In this chapter, we will adopt the term expert system to describe all systems which incorporate the knowledge as well as the problem-solving abilities of a human expert and which can, in the majority of the cases, reach the same conclusions the expert would when given the same input.

Parts of an Expert System

An expert system can be visualized as consisting of three essential parts, as shown in figure 1: a knowledge base which contains the expert's knowledge, an inference engine which uses that knowledge plus input from the user to come to a conclusion, and a user interface which allows the user to communicate with the system—providing input, asking questions, receiving answers.

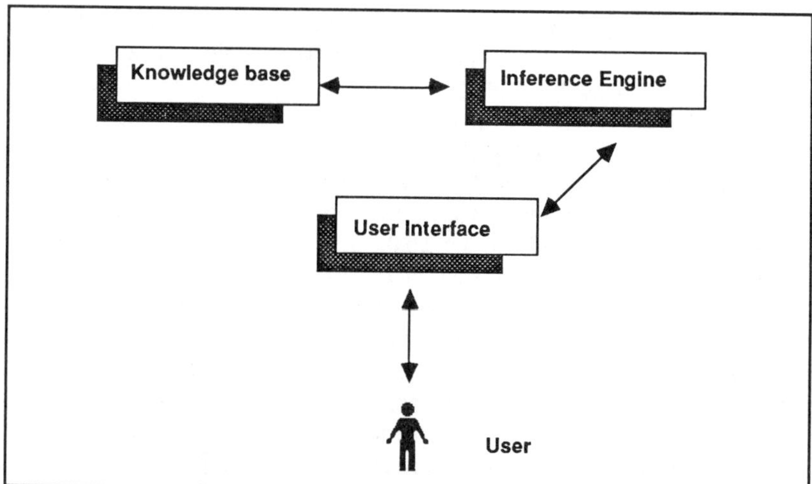

Figure 1: Components of an Expert System

Knowledge

Knowledge can pertain to different objects and come from different sources. Factual knowledge can be found in a book or a procedures manual and is typically available to, and agreed upon, by most people working in the field.

Heuristic knowledge is less rigorous and more experiential. This is where the expert may differ from the novice who has access to the same factual knowledge. Based on experience, the expert may use judgment, shortcuts, or intuition to solve a problem quickly and efficiently, whereas the novice may have to study all the facts and use trial and error to find an efficient approach to solving the problem (Feigenbaum et al., 1988). It is questionable whether a computer can actually "know" anything. Forsyth (1989, 125), for example, states: "Digital computers, therefore, know nothing; they merely store and manipulate information. Thus the very term 'knowledge base' carries a presumptuous philosophical claim." However, it is common practice to talk about the knowledge base as that part of the expert system where both facts and decision rules are stored.

Knowledge representation

The way knowledge is represented is critical in building an expert system. The most commonly used method to formalize and organize an expert's knowledge is a set of production rules. These rules are formulated as IF {condition} THEN {conclusion}. Of those examples in the CRI Directory (Smart & Langeland-Knudsen, 1986) which listed the knowledge representation, 74% used production rules, and a majority of the systems mentioned in the more recent literature are also rule-based systems.

At first glance production rules look like regular IF..THEN statements, and one might wonder how they differ from those used in a regular program. However, in a traditional program the programmer determines a priori at which point in the program flow the statement is executed, and there is a predetermined path leading to that statement. In an expert system, the inference engine determines at execution time and based on user input in which order to execute the rules, and the developers can enter the rules in any order they please (Van Name & Catchings, 1990b). In addition, rules and control statements (where to go next) are interleaved in an ordinary program, whereas they are separate in an expert system. The separation means that a change in knowledge (rules) does not require a change in control; in an ordinary program, a change in a rule may also require a change in program control (Feigenbaum et al., 1988).

Production rules are a natural and intuitive representation for many problems. Since the rules are independent of each other, it is easy to maintain and update the knowledge base. The simple syntax makes some consistency checking possible, and the system can easily provide a record of its "thinking process." This approach works best when there are many rules which are mostly unrelated (Shadbolt, 1989; Van Name & Catchings, 1990b).

There are, however, some drawbacks to production rules. Not all knowledge comes in the IF—THEN format, and the representation of simple facts may be awkward and inefficient in this format. Also, with a large number of rules the computational overhead may be considerable since it is necessary to scan the base to find the next rule to fire.

Another knowledge representation scheme which is particularly suited for linguistic information is called semantic nets. They focus on the relationships between items in a domain. A graphic example of a semantic net is shown in figure 2. Nets are appropriate for special applications but are not efficient for large and complex domains.

A third representation scheme is referred to as frames or scripts. Frames are data structures in whose slots information about an object (for example a car) can be stored. The information may be descriptive, or a slot may contain rules or references to a database. Frames are ordered hierarchically, and some properties of an object can be inherited from the class to which the object belongs and do not need to be stored within each individual frame. For example, every typical car has a steering wheel—that information only needs to be stored once.

The representation of knowledge in frames is indicated when the domain knowledge is naturally hierarchical and a lot of information about objects needs to be stored. The ability to use inheritance is a definite advantage. However, reasoning has to be done in terms of matching and usually requires a more powerful inference engine than a rule-based system (Shadbolt, 1989;

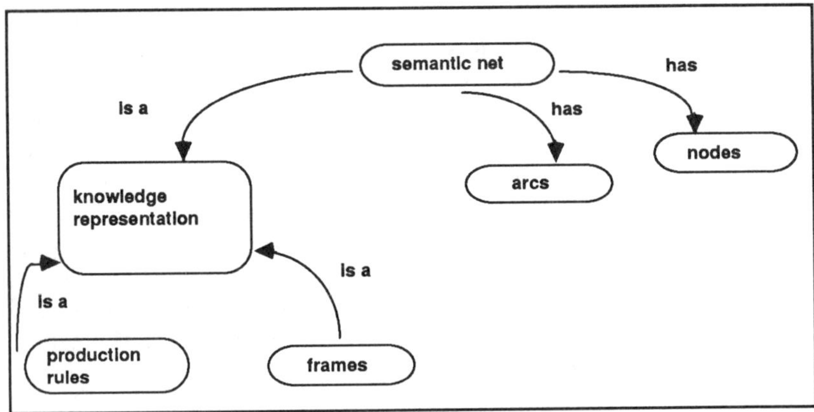

Figure 2. Example of a Semantic Net

Van Name & Catchings, 1990b). Of those applications in the CRI sample which mention the knowledge representation, nine percent used frames.

Other knowledge representation schemes are used in specific applications. The reader is encouraged to consult a book such as Turban, 1990, or Forsyth, 1989, for further information.

The inference engine

The inference engine is the part of the expert system which actually uses the knowledge and makes inferences, i. e., draws conclusions based on the user's data input and the knowledge in the knowledge base. It is separate from the knowledge base, a fact which allows an updating of the knowledge as facts change or more data are accumulated, without having to change the logic of the inference process.

In the case of a rule-based knowledge representation, the inference engine has to decide which rule to apply ("fire") next and, if two or more rules apply at a given time, which one to select. A number of methods are available (see Shadbolt, 1989, for a detailed discussion).

Two reasoning methods the reader should be familiar with are forward chaining and backward chaining. Forward chaining (also called data-directed reasoning) starts with the input provided by the user and fires every rule which this assertion can trigger. If one of the rules leads to a goal (or conclusion) of the system, the search stops; otherwise, the conclusions from all the rules fired are added to the working memory and new rules are fired based on this extended knowledge now available.

Forward chaining is appropriate where data are expensive to collect, but few in quantity (Graham, 1989); where input assertions typically outnumber possible answers (Van Name & Catchings, 1990b); when goals are not defined and options preferred (Smith, 1989). This may require considerable time if questions which allow the system to rule out a large subset of rules are not asked initially.

Backward chaining (also called goal-directed reasoning) starts with a desired goal and works backwards through the rules to see whether this goal can be supported by facts. Thus, a rule is sought out which has the desired goal as a conclusion, and then the conditions are checked to see whether the rule applies. Backward chaining is the preferred approach when there are many possible answers and fewer input assertions (Van Name & Catchings, 1990b); where the quantity of data is potentially very large and some specific characteristic of the system under consideration is of interest (Graham, 1989); where the number of solutions is limited (Smith, 1989). Diagnostic systems are typical of backward chaining applications. The problem here may be to decide in which order to try the available goals and to obtain the necessary evidence.

A good expert system will use a mixture of forward and backward chaining strategies to reach a conclusion quickly by capitalizing on the strengths of both methods.

While knowledge representation and reasoning method are of concern to the developer of the expert system, the user is probably most concerned with the user interface. The user's goal is for the expert system to be easily understood so that he does not have to trade extensive training in the subject matter for extensive training in using the expert system. Thus, questions should be asked clearly, response options should be easy to state or select, and the user should be able to ask why a question is asked or a conclusion reached and receive a reply that is easily understood. Many times, using graphics in the interface makes an expert system more appealing to the user. The solutions section will discuss some ways of creating good user interfaces.

Examples of Existing Expert Systems

The first expert systems were developed by university researchers, and many were non-business applications. One of the examples is the medical system MYCIN which was developed in the 1970s to select appropriate antibiotic therapy for patients with bacterial infections of the blood. This system has been described and discussed extensively (see Smart and Langeland-Knudsen, 1986, 177-185, for an extensive list of references).

Today, there are thousands of expert system applications in industry. We will look at examples from banking, advertising, and manufacturing to get a feeling for the power and potential of expert systems.

American Express

American Express has an expert system called Authorizer's Assistant. There is no fixed credit limit for the American Express credit card. For large purchases, merchants have to call in for authorization. A credit authorizer then has to check in a variety of data files and make a judgment on the spot. In difficult cases this may involve searching up to 13 files and might take up to 30 minutes.

The expert system searches the same databases and makes a recommendation, which the authorizer may or may not take. The decision is still made by a person, but the turnaround timc is cut down to seconds. The system does nothing that the authorizer could not do—the advantage is in the speed with which the system can make a recommendation.

Although no exact data on financial benefits are available, it is expected that the system will raise the authorizers' productivity by as much as 20 % and reduce losses from overextension of credit. It also supports American Express' strategy of product differentiation by offering individualized credit limits (based on Leonard-Barton & Sviokla, 1988).

Power Systems Division

Power Systems Division of Combustion Engineering, Inc. custom-builds steam-generating boilers to fit the needs of their customers—mostly utility companies. These large boilers are manufactured in pieces and then assembled on site. A boiler may be several stories high, contain up to 160 miles of tubing, and cost more than $100 million.

The expert system ICAD contains rules for building boiler components. It takes the customer's specifications and creates an initial design which is then finished with conventional CAD drawings. The system can also assist in the design of some replacement parts; any time savings on replacement parts may lead to large savings for customers by reducing downtime.

It has been estimated that the expert system has reduced the design time for furnace walls (as an example) from 3500 to 2100 man-hours. It has helped to reduce the cost of boiler design from 15% to 12 % of total cost—a significant dollar savings on a $100 million boiler (based on Warner, 1988).

MORE

MORE is an expert system developed for direct marketers. The system supports direct mail decisions by analyzing each individual on a mailing list with regard to last direct-mail purchase date and response to previous mailings (Cook & Schleede, 1988).

Types of Tasks Supported by Expert Systems

Our examples are just the tip of the iceberg as far as expert systems applications are concerned. The CRI listing includes a variety of business-related functions supported by expert systems, such as auditing, evaluation, forecasting, production control, scheduling, selling, tax planning and preparation, training, and underwriting.

Most of the tasks supported by the expert system are in a very narrow problem domain. Often the system will support just one specific type of decision such as: should I grant this loan, should I hire this person. Other systems are help systems—they will support the user in writing programs, using data bases, finding legal precedents. The expert system is not flexible in the same way that a decision support system is flexible; it cannot be applied to a variety of problems. The expert system "takes care of things we understand so we can concentrate on the things we don't understand" (Michael Scott Morton, as quoted in Friedland, 1988, 78).

Benefits of Expert Systems

What are the benefits which the designers/users of expert systems perceive? Obviously, this varies a lot from case to case, but certain patterns do develop as we read the reports about hundreds of systems in use.

Some of these benefits are directly related to the human expert after whom the system is patterned. A motivation for the creation of an expert system is often that there is only one expert in the organization, and he/she cannot meet all the demands on his/her time or is ready to retire (e.g., GE's top locomotive field service engineer, see Bonissone & Johnson, 1984). Capturing that person's expertise in an expert system preserves it even if the expert leaves and also allows access to the expertise throughout the organization. The expert, on the other hand, has the opportunity to devote his/her time to work on difficult problems which might otherwise not be tackled due to lack of time. Thus, both the organization and the expert benefit. Expert systems also benefit the organization by making decisions more consistent. A person may not always consider all the relevant information or may not always consider the same information due to biases or circumstances. The expert system will not forget to ask a question and will not react to personal characteristics of the user, thus reducing bias and increasing consistency. This may be particularly beneficial in human resource applications where decisions may be made based on first impressions or reluctance to deal with sensitive issues (Mitchell, 1990, discusses these issues for applicant screening with the help of an expert system).

Frequently, decisions can be turned over to less experienced employees when an expert system is used. Constant work with the system may help these employees learn much faster than they would have in regular training or in the traditional progression in their job. Thus, the expert system may actually increase the skill level of employees. However, there is no agreement on this—see the "deskilling" issue discussed in the next section.

Expert systems can also be fairly easily maintained once they are fully developed. Since the knowledge and reasoning process are stored separately, a change in the environment may often require simply the exchange of one rule for another. However, in a complex system it may require a full staff to keep the system up-to-date. DEC, for example, has 10 persons working on maintenance of XCON (Bramer, 1989).

As we have seen in some of the examples discussed earlier, expert systems can also have a positive impact on an organization's profits. However, this is not always measurable—unfortunately, the examples which provide figures are usually the spectacular successes, and the majority of published examples do not provide monetary benefit information.

Disadvantages of Expert Systems

With all these positive outcomes, are there any negative side effects of expert systems? Some potential problems are listed below; however, there is no empirical evidence to show that these problems were actually encountered.

Expert systems do not have the ability to notice changes in the environment on their own. Whereas human experts may be attuned to political or legal changes and constantly update their knowledge, the expert system does not recognize when its expertise becomes obsolete. If the expert system has taken the place of the expert and is now used by less experienced employees, this may eventually lead to poor or incorrect decisions—but nobody may realize this until a major problem has occurred (Greenwald, 1988a; Brown & Phillips, 1990).

Consistency of decisions is an advantage in routine cases. However, over-reliance on the expert system may stifle imagination and lead to more uniformity than is desirable. Managers may have so little experience in making independent decisions that they will not have faith in their decision making ability or intuition and will become incapable of making difficult decisions when the need arises (Enslow, 1989).

Although some publications praise expert systems as a cure-all for business problems and as the technology of the 1990s, there are issues which need to be considered before an organization embarks on expert system projects. The next section will examine some of the issues of which the manager needs to be aware.

SECTION III

Issues

Before we engage in an expert systems project, there are a number of issues which need to be considered, ranging from technical through managerial to organizational or even societal concerns. Since the focus of this paper is managerial rather than technical, we will not discuss truly technical issues, but rather focus on issues which are of concern to management and need input from management to be resolved. We will start with the consideration of some very general issues which may affect the organization or, in the long run, even society as a whole.

Organizational Issues

There are four wide-ranging concerns which we will address: how does the use of expert systems affect the skill level of the work force; how does the use of expert systems affect the decisions made in an organization and the long-term strategy of the organization; to what extent is an organization liable for the decisions made with the help of—or automatically by—an expert system; and who owns the knowledge incorporated in the expert system.

Effect of Expert Systems on Skill Level

One of the arguments in favor of expert systems is that the use of the system will automatically train the user and raise the user to the skill level of the expert in a much shorter time than either formal training or ordinary work experience can. The potential for this certainly exists, for the following reasons:

1. The expert system contains not only formal knowledge, but also heuristics. Frequently, the experts formulated the heuristics only after long discussions with the knowledge engineer since they were not even conscious of using these particular decision rules. Thus, novices would have never found out in ordinary conversation how the expert decides, and they would have had to develop these rules through personal experiences over a much longer time period.

2. The user can question the expert system freely and repeatedly. Any time a decision is made or even a question asked by the system, the user has the opportunity to ask "WHY?" A human expert is not likely to encourage that kind of questioning on a routine basis, and the novice will have to pick up subtle cues to arrive at the same information, which may take much longer or may never happen.

3. The human expert is typically accessible to only one or a few other

person(s) at a time. The expert system can be used throughout the organization simultaneously and thus disseminate problem solving information much more effectively and efficiently.

4. More cases requiring decisions can be turned over to a less skilled person in a shorter period of time than in a traditional environment. Thus, the process of gaining experience is speeded up, and learning can occur faster.

5. As opposed to classroom learning which is often more passive, the user of an expert system learns by doing which is a more effective way of learning.

For these reasons, some training professionals enthusiastically endorse the use of expert systems for training purposes (see, for example, Siegel, 1989).

However, arguments are also made on the opposite side of the issue. For example, Carr (1989, 118) argues "Because of the emphasis on replacing human skill with skills embodied in the system, the users end up noticeably less proficient than they would have been without the system." Several factors may contribute to this effect:

1. An expert system which is governed by rules is competent, but not truly an expert—it does not use intuition, it does not truly learn from the results of its decisions, it does not automatically react to changes in the environment. The person learning by using the expert system can learn to emulate the expert to the extent that the program does—but the extra skill which makes a person an expert cannot be learned from a computer program, but only through human interaction. Thus, the overall skill level will be lower in the long run.

2. To learn from the "thought process" of the expert system is not automatic; it requires that the user asks "WHY" and thinks about the answers given by the system. This is by no means automatic. It is entirely possible that junior staff use the system blindly and rely on its decisions without giving the process a thought. This does speed up the decision process and increases their productivity (measured in number of decisions made per day), but it does not help their decision making skills. In those rare cases where a decision has to be made without the system, the staff member may be insecure and actually less effective than before.

3. An expert, to remain one, has to keep up with his or her field. Any changes in domain knowledge will be noted by the expert and built into his/her "internal expert system." Any wrong decisions made will be analyzed and will lead to an increase in skill/experience afterwards.

The expert system is not automatically aware of changes in the knowledge base or environment. It does not find out if a decision was correct and, therefore, cannot learn from its mistakes. A human expert is required to keep the system up to date. If an expert system is built to replace a human expert who then retires or moves to a different job, there is no effective maintenance of the system. Eventually the users will not learn the best approach to solving a given problem, but one that is potentially worse than what they might have done without the system. However, the users do not know this. Thus, the skill level of the users may be considerably less than that of the expert they replace, and their skills will not increase by using the system (Carr, 1989; Bramer, 1989; Juras, 1989).

The arguments on both sides are largely speculation. There is no empirical evidence regarding the skill level in an organization. One reason is that expert systems are so new that many organizations have not used them long enough to assess the impact on their workforce.

We have done some empirical work comparing one group of students who use an expert system to select the appropriate statistical procedure for a problem set with another group who does the same problems using strictly textbook and class notes as an aid. Our evidence so far does not indicate that the use of the expert system raises the skill level of the students; most students do not use the explanation feature and simply copy down the results from the computer. When they later have to make decisions without the computer, they are less able to do so than the students who never used the computer (Seeborg et al., 1990). This tends to support the arguments about the "deskilling" effect of expert systems; however, more research is necessary in this area, especially in the work place.

Effect of Expert Systems on Organizational Decision Making

One of the potential benefits listed earlier for expert systems is that they lead to more consistent decision making in the organization. When we look at decision situations which occur frequently in the organization and for which there is a correct answer if all the relevant information is considered, this is indeed an advantage. Thus, in credit authorization or underwriting decisions, the consistency gained by the use of expert systems (which will always ask all the necessary questions) is a definite benefit of the system.

However, there is also a danger here. By being formalized in the system, the decisions one (or a few) experts would have made are given more weight than they might have before the system was in existence. They have a normative effect which may lead to the exclusion of divergent ideas. Thus, truly innovative persons who may have the chance to become experts by virtue of their divergent ideas may be stifled by the system or may leave (Bruce & Adam, 1989; Feigenbaum et al., 1988).

An important variable in this context may be the level at which decisions are made. Many of the expert systems designed for business use are intended for functions at the operational level where innovativeness may not always be desirable to the organization. Can expert systems be used to support decision making at the strategic level, and how do they affect decision making there?

There are not many reports available on the use of expert systems for strategic decisions. An expert system is for sale (Alacrity Strategy) which assists in the strategic planning process, but it does not make any decisions. The software solicits information from the user about issues ranging from competitive structure of the market to organizational culture, and the result is a lengthy "overview of a business unit's strategic position relative to its market and competitors" (Cook & Sterling, 1989). Expert systems used in this way—pulling together a variety of information in a relatively short time as a basis for decisions made subsequently by human experts—may have a much more beneficial effect on organizational decision making by supplementing rather than replacing human expertise.

Liability

As we mentioned earlier, some expert systems make decisions — possibly to the extent that they control machinery without human intervention—whereas others serve in an advisory capacity. Since the systems are not foolproof, it is entirely possible that they make an incorrect decision. On a pessimistic note, one British study stated that "inappropriate use of expert systems might result in catastrophe, even an accidental nuclear war" (Watts, 1989, 35). Who is liable if there are errors in the system or it makes bad decisions?

Bad decisions may be caused in several ways. First, the experts who supply the expertise may not be infallible—they may be the best the organization has, but their knowledge may not be complete, and the expert system will show a lack of knowledge in the same area. Besides, the experts may have certain prejudices or personal preferences which would also be incorporated in the system and might lead to incorrect decisions.

A second source of error may be the developer of the system —the knowledge engineer who helps extract the knowledge from the expert, or the programmer who actually writes the code. (Of course, in many cases all three are the same person.) Communication problems frequently occur when people from different backgrounds try to 'program' expertise, and the development team needs to be aware of this problem.

Finally, the user may not use the system correctly. If the experts come across a new situation, they may have resources to fall back on, whereas a less experienced user may not realize that the expert system operates outside of its area of expertise. Liability claims could then be directed at the user (who makes a decision affecting a third party), the programmer, or even the expert

(Enslow, 1989; Warner, 1988). There are not enough precedents available yet to know exactly how far liability may reach in these cases, but the issue might become important for an organization which relies heavily on expert systems to make independent decisions.

Ownership

Creation of an expert system is not possible without willing experts who make their expertise available. Several of the descriptions of expert systems contain statements like this: "The experts contributed their knowledge as part of their regular job. No special compensations were made to the contributors" (Turban, 1990, 584).

The question of ownership of knowledge was not very salient in early expert system development projects. The projects were usually conducted by universities, and researchers shared their knowledge freely. However, as we move into the commercial area, an expert's position in an organization is often secure only as long as s/he is the only person with specialized knowledge. If that knowledge is formalized and made available throughout the organization, the expert becomes expendable. Can an organization ask a person under these circumstances to share his/her expertise?

This is not simply an academic question. At least one case has gone to court where an employee provided information for an expert system as part of his job, later lost the job and sued the company for royalties (Warner, 1988; Enslow, 1989).

The issue has further implications. If the expert knowledge is essential for an organization (e.g. because it gives them a competitive advantage), does it hurt the organization's market standing if that knowledge becomes available to competitors? Contractual agreements with the expert may provide that the expert may not work for a competitor for a period of time. However, once the expertise is available inside the organization on the computer, an additional security problem arises—how can we safeguard this program so that it does not reach a competitor? Diskettes are easily removed from the premises. The very benefit of making expertise more widely available may also be a detriment to the organization if this expertise cannot be protected against outsiders (Feigenbaum et al., 1988; Enslow, 1989).

Managerial Issues

The issues we have discussed thus far are very general in nature. The questions raised cannot be answered by one individual and empirical research is required to resolve them. Because expert systems are relatively new, we do not have many empirical results to fall back on to answer these questions (Juras, 1989).

The next set of issues to be addressed is important for managers who will either start an expert systems project or will use an expert system in their department. In this section of the paper, we will simply explain the issues to be considered. Some solutions will be discussed in the next section of the paper. Questions to be asked include: what type of problem should we turn over to an expert system, how do we select the expert, what staff do we need to develop the system, how do we validate the system once it is developed, and what do we need to consider to be successful in the development project?

Problem Type

There are several rules of thumb describing the "ideal" expert system project, and we will discuss these in the next section. However, there is also a fundamental difference of opinion exemplified by the approaches taken by different companies.

DuPont has started out by developing a considerable number of small expert systems. These systems require an average of one person-month in development time, they can be done in-house without AI experts or consultants, and they can typically be done on a PC using fairly simple tools. These systems do pay off, but not in a big way. With this approach many people in the organization become familiar with expert systems (Feigenbaum, 1988).

The opposite approach is taken by DEC, as an example. They have created large expert systems which have taken years to develop, but have also had a large impact on the organization. XCON and XSEL, for example, are used throughout the organization and are reported to save $15 million to $20 million a year (Feigenbaum et al., 1988; Brown & Phillips, 1990).

The developers of big systems argue that you will never find large projects if you start out looking for small ones, and that the small project developers may not realize that the big projects are doable. On the other hand, small projects carry less risk; it is easier to see results quickly which will be encouraging to developer and user alike, and soon it is not necessary to sell the expert system idea—everyone will have seen an application and will be anxious to start one.

The nature of the tasks performed in the organization may have a major impact on the choice of projects. Another question to be considered may be the expertise available inside the organization. DEC started out using outside consultants until there was sufficient expertise available internally; DuPont decided to go with the skills available in the organization which were typically more suited to small projects.

Selection of Experts

Should an expert system reflect the expertise of one or several experts? Some of the earlier expert systems were built to capture the expertise of one person who was getting ready to leave the organization (e.g. GE's Smith, Turban, 1990, 426/427). The expert system then takes on the personality of that person, including that person's idiosyncracies. An advantage of the single expert approach is definitely easier development; the knowledge engineer can tune in to the expert's style of thinking, and there will not be conflicts concerning the contents of the system that may need to be resolved (Lindsay, 1988).

On the other hand, other experts may not totally agree with the system, and we may end up building a system which is only truly useful for one person. In many cases, it is not possible to find a single expert who has all the relevant knowledge, especially as we try to develop systems who go beyond mimicking an existing expert. There may also be problems in communicating with the expert, and if the single expert is not capable of verbalizing his/her decision process, working with one expert may have definite drawbacks.

In the case of GE's locomotive field engineer (Turban, 1990), it was easy to pinpoint the expert. But if a function is currently filled by several people, how do we decide who is the "expert" whose expertise should be the basis for the system? How do we know what is actually valuable expertise and, if there is conflicting advice, who is the "true" expert?

What we would need is a yardstick which allows us to measure the value of expertise, and a person who can use that yardstick. Should that person be the knowledge engineer, or the manager in charge of the project? Can somebody who is not an expert even judge what is valuable expertise?

These questions are not always equally important. In many cases, the expert system is not truly "expert," but is simply a way to automate a task which is being done by a number of people following set procedures with varying degrees of success, and we have organizational measures of success which allow us to identify the best performer. But these are typically the smaller projects. For expert systems which may have a large impact on the organization, the question of the value of expertise is one that needs to be decided before the project is undertaken.

Staffing

Originally, expert systems were mostly developed by university researchers. They were, for the most part, written in special languages such as LISP and PROLOG. Even of the systems described in the CRI directory, some of which are as recent as 1985, over 45 % are written in some variety of LISP or PROLOG (see table 2). Thus, when companies decided to get started with expert systems, they frequently had to hire a programmer who had experience with artificial intelligence, or they had to employ an outside firm.

The situation is different today. There are a number of expert system shells on the market which are comparatively easy to use (a shell is an expert

Language	Number of Systems	Percent
LISP	86	36.3
PROLOG	21	8.9
FORTRAN	16	6.8
PASCAL	14	5.9
BASIC	11	4.6
C	8	3.4
Others	81	34.2
Total	237	100.1

Based on Smart and Langeland-Knudsen (1986); totals differ from table 1 because not all examples give detail about language used.

Table 2. Languages used for Expert System Development as of 1986

system without a knowledge base). Thus, an artificial intelligence specialist is not necessary to develop an expert system; often, the expert can sit down and formulate his/her own system in a relatively short time period. There is still a demand for people who can develop expert systems, but the requirements are less technical today. Bredin (1990) lists the requirements as ability to communicate verbally, personality and analytical ability, some basic technical know-how, inquisitiveness and innovativeness. The technical (programming) skills can be picked up fairly easily on the job. Employees outside of the IS area may have these characteristics, and a manager may be able to identify members of the present departmental staff who are interested and capable of acquiring the technical skills and developing an expert system.

Validation

We have already discussed the possibility that a company or individuals within the company may be liable if the expert system gives incorrect advice. Under these circumstances, it becomes important to validate the system before it is used on a routine basis.

How do we know whether an expert system generates the correct solution? Bramer (1989) states: "In many (probably most) expert system domains the 'correct' solutions are not known ... or are not 'knowable' (e.g. in deciding on a company's long-term financial strategy)."

This issue is important and cannot be settled in general terms. In some cases, there is a 'right' answer, and the system can simply be tested against those cases just like any other information system. However, these may not be truly 'expert' systems, or the domain may not be one requiring a high level of expertise.

The system may be based on a number of past decisions, and one way to validate it is to compare the system's decisions to those actually made. However, there is a possibility that those decisions were not truly optimal in the first place, and the system would just perpetuate a less-than-perfect decision. The same problem arises when the system is validated by comparing it with an expert's solution. The expert may not be infallible, and who makes the decision that the system is better (or worse) than the expert?

The best validation is probably one in which the system is used extensively, and the outcomes are critically evaluated. If the system is right a high percentage of the time (maybe 80 % or more), it is probably a good system. Increasing the proportion of correct answers may require a disproportionately large effort, and no human expert is correct all the time, either. This implies, however, that the decisions of the expert system need to be checked carefully and any deterioration in the system must be noted and corrected quickly.

Measuring Success

The first expert systems were developed as demonstration projects and were not really intended for everyday use. By now the technology is fairly well established, and most organizations expect an expert system to "produce"—either in terms of savings or in terms of better service, better decisions etc.

How do we measure whether an expert systems project has been successful? This is a question which needs to be answered for other information systems as well. One possible measure is usage —either number of users or amount of time used. But this may not always be a valid measure. For example, if there is a training effect, then the system may be used extensively at first, but as users learn how to make correct decisions, they will not need the system (they have become experts) and usage will go down. In that case, decreased usage is actually an indication of success. Having a success measure is important because expectations may be unrealistic and may lead to disenchantment with the system right at the time when support is needed. Using a shell, it is often possible to develop a prototype quickly. Users will work with it and see the possibilities, but the prototype can only deal with a small number of cases. Adding the fine points may take considerably longer, and if the success measure is 100% correct decisions, the system may never be 'successful.' Thus, a user may not pursue expert systems although there may be many useful applications in the organization (Leonard-Barton & Sviokla, 1988).

On the other hand, one may also spend considerable amounts of time and money on expert system demonstration projects which work fine in such a limited domain that they are useless to the organization although exciting to the developers. This is not a good use of organizational resources. Thus, the development of success measures is important to the identification of suitable projects.

Technical Issues

As mentioned earlier, truly technical issues in the development of expert systems are beyond the scope of this chapter. We will discuss two issues related to the development of expert systems which are of interest to the manager, but are more 'technical' in nature than those listed above: acquiring knowledge from experts and user interfaces.

Knowledge acquisition

The crucial part of expert system development is acquiring the expert's knowledge. Difficulties may arise for several reasons:

1. The experts may resist the project more or less openly, for one of the following reasons: they do not see a need for the computerized system; they are afraid that they may eventually lose their job to the system; or they may feel that they may expose the limits of their expertise by working on the systems development project (Lindsay, 1988; Forsyth, 1989). These are problems that may also occur in the development of more traditional systems, and managers need to be prepared to handle them.

2. The experts may be willing to assist in the development, but may not be able to articulate their decision making process. The experts may try to give textbook-type reasoning which does not necessarily correspond to the actual process, or they may not even be aware of the actual reasoning process they apply (Forsyth, 1989; Bruce & Adam, 1989; Leonard-Barton & Sviokla, 1988). Overcoming this barrier requires considerable skill from the knowledge engineer as well as some domain knowledge to get the expert started.

3. The most serious problem may be that experts do not think in the way computers process information. Thus, the knowledge representation method used may not match the expert's decision making or information processing approach, and the resulting system will be at best awkward and inefficient, but it may not be possible to create a working system at all under these circumstances (Forsyth, 1989).

Although it is acknowledged that knowledge acquisition is the major bottleneck in building expert systems, "there have been remarkably few experiments aimed at deciding which criteria ... determine the choice of approach to follow" (Bramer, 1989). "Very little has been written, and we suspect little more than that is known, on how to elicit inference methods" (Graham, 1989). Thus, extracting expertise is an issue which requires future research. Some methods are available to facilitate the process, and these will be discussed in the next section.

User Interfaces

Without adequate methods to extract knowledge from the expert, it is not possible to create a working expert system. On the other hand, without an adequate user interface, an expert system may give perfect answers and yet may not be accepted by the user. For example, Texas Instruments developed the FMS Communications Diagnostic System. It was tested and performed correctly. "However, the operational test was not considered a success because the users did not accept the expert system" (Martin & Oxman, 1988). Only after the user interface was improved (graphics were added) was the system accepted in the operational environment.

Obviously, the user interface should be interactive, transparent, and allow natural language queries. In systems which advise on diagnostics or repair of machinery, the addition of graphics increases the usefulness of the system since even a novice will be able to find the parts if a picture of the part or a clearly marked blueprint is shown.

However, other features of the user interface may be useful. For example, expert systems are able to explain their reasoning, but this is typically done in the form of listing the rules which have led to the current point in the reasoning process. This is not very helpful to the user. If the system could also explain the issues and criteria involved in the rules, the user would be better able to judge the validity of the system's conclusions (Carr, 1989).

Another consideration is the user's level of expertise. A human expert will not explain his decision in the same terms to a colleague who has considerable understanding and to a novice who knows next to nothing. For an expert system to take the place of the expert, this same ability to interact with the user at several levels would be necessary. This would need a more sophisticated approach to creating expert systems than just the creation of a rule base—it needs something like an "expert interface" (Bramer, 1989; Carr, 1989).

Summary

This section has outlined some of the issues which should be familiar to a manager who has to work with an expert system—either its development or its use. In the next section, we will discuss some of the solutions—methods that have been used in successful expert systems projects or have been shown by research to be successful.

SECTION IV

Solutions

This section could also appropriately be entitled "lessons learned" or "success factors." It does not address all the issues mentioned in the previous section; as was discussed, for some of these issues the answers are unknown and more research is needed to provide a solution.

The following topics surface in most discussions of key success factors: proper project definition, the interaction with the expert, the introduction of the expert system into the organization, and the—more technical—issue of the selection of an expert system shell. We will briefly summarize the experiences reported in the literature in the last two years.

Selection of a Project

Selection of a project requires two steps—one, identifying opportunities, and two, assessing the feasibility of the project. As we have seen earlier, many successful expert system projects involve diagnosis, design, or monitoring procedures—similar tasks may be a good start. Leonard-Barton & Sviokla (1988, 94) suggest asking yourself: "Is there a task in my organization that would be improved if ...we had more time? ...the best expert always did the job? ...we had consistent decisions?"

Once a number of potential projects has been identified, several criteria can be used to judge whether expert systems are an appropriate technology for this task. These include:

• complexity—the task should take more than a few minutes, but not more than a few hours to perform (Leonard-Barton & Sviokla, 1988; Goodall, 1989). However, this is not a strict guideline—some of the systems we have mentioned earlier apply to tasks that are much more complex. Also, the task should require the application of a significant amount of knowledge or a complex set of rules (Brooks, 1989).

• availability of an expert. First of all, can we identify how an expert differs from a novice? Is there a person who is clearly an expert by virtue of experience, not just by title or status? Does this person have time to devote to the project? Is this person enthusiastic about working on the project? (Leonard-Barton & Sviokla, 1988; Goodall, 1989).

• impact of project on the organization. The task should occur frequently, or occur infrequently but require rare expertise; the benefits to the organization should be clearly defined through a rigorous benefit analysis (Siegel, 1989;Cook & Schleede, 1988; Vasilash, 1989; Texas Instruments, 1990).

• appropriateness of the technology. If the system could be implemented through table look-ups or a model using mathematical equations, then this is not an expert system project. Likewise, if the system requires a high amount of 'common sense' or the expert needs to see or touch, expert systems may not be appropriate (Goodall, 1989; Leonard-Barton & Sviokla, 1988). When a project has been identified that meets most of these criteria, work on the expert system can begin.

Interaction with the Expert

Next, it is necessary to solicit the cooperation of the expert. This may require education of the experts who have expertise in their domain, but do not necessarily know expert systems. A major challenge may be to deal with the experts' fears of losing their job, having their job changed into one requiring less expertise, or exposing the limits of their expertise.

A system that assists the experts by taking over the more mundane parts of their task and giving them time for the more intricate activities may be an easy one to sell. Also, solid contractual agreements regarding the compensation for the shared knowledge may help overcome an expert's resistance. (Warner, 1988; Enslow, 1989; Goodall, 1989; Lindsay, 1988)

Once the expert's cooperation has been won, the problem of extracting knowledge from the expert needs to be addressed. Several approaches have been suggested, depending on the ability of the expert to express his/her procedures, the availability of test cases and the access to knowledge acquisition software. When a database of past situations and correct decisions exists, it is possible to have the computer generate rules from these examples by induction. This will not lead to a complete expert system, but it will give a good start. Even if experts are not good at verbalizing their decision process, they are typically able to criticize the decisions made by the prototype system and point out how their decision process differs.

An interesting approach was used by Coopers & Lybrand. A team of experts was placed on one side of a curtain and a person with access to all the data on the other side. The experts had to perform the task by asking for all the data they wanted to examine. The process was videotaped and served as a basis for the expert system (Leonard-Barton & Sviokla, 1988).

Another possibility is to train the experts in the building of expert systems using a simple shell and let them build their own system. This eliminates the fear of appearing less expert since no interaction is required. The interaction with the prototype may lead the experts to a deeper understanding of his/her own decision processes. A possible drawback is that an expert who is not aware of his/her own thought processes may not be able to formulate them for a computer program, either, whereas a skillful knowledge engineer may be able to lead the expert to the point where s/he can explain how s/he makes decisions.

Introduction into the Organization

At this point, there may be experts who are willing to share their expertise and a manager who is willing to sponsor the project. However, there are other parts of the organization which need to be sold on the idea, especially if this is one of the first projects. In organizations where expert systems are pervasive, this selling task is no longer essential.

One group that needs to be sold on the idea are the future users of the system. Many times the system will be used by employees other than the expert and the system developer. Since they do not have the expertise, it may be felt that they have nothing to contribute in the early stages of the project. However, end-user orientation is an important key to success, and end-users should be represented on the development team (Vasilash, 1989). The FMS example mentioned earlier illustrates that point; whereas the expert is essential to the development of the knowledge base, the end-user should have input into the user interface from the start.

It is also a good idea to raise general awareness of expert systems technology in the organization. This may be done by education (making short courses available) or by publicity and management support. As more people understand what expert systems can do for the organization, it may be easier to find projects which have a high payoff and contribute significantly to organizational outcomes (Feigenbaum et al., 1988; Goodall, 1989).

Selection of an Expert System Shell

Selection of an expert system shell may seem more like an issue for technical personnel than for management. However, shells make it possible for managers or knowledge workers to create their own expert systems without the help of a technical expert, and the acquisition of the shell may, thus, become an issue for the manager.

The first question to ask is who will use the shell. If the manager or the domain expert will build their own system, the shell needs to be extremely user friendly. The presence of development tools is important, and the intended user should try to get some hands-on experience before purchasing to test out the command language, the handling of errors etc.

If the shell is to be used for a complicated project and by a programmer, other issues become important: can the shell deal with other data formats, e.g. read files from Lotus or dBASE or write into Lotus or dBASE files? is it possible to write programs to communicate with the shell or embed the expert system into a program written in another language? can you fine tune the performance by changing the order in which the rules are processed by the shell?

Also important is the presence of an explanation facility. As we have discussed earlier, this is often a simple listing of rules; the user may need to write a program to provide more detailed explanations.

An important part of the use of an expert system is keeping it up-to-date. Therefore, it is important to check out how easy it is to update the knowledge base using the expert system shell (this discussion based on Van Name & Catchings, 1990a and 1990b). To illustrate the capabilities a shell may have, we will introduce briefly the shell ADEPT developed by SYNERGTECH. Figure 3 shows an overview of the ADEPT architecture from the view of the expert system developer.

As can be seen from Figure 3, this shell allows knowledge to be represented in the form of rules, frames, or semantic nets. The developer has a separate editor available for each type of knowledge representation. The shell also allows a choice of forward chaining, backward chaining, or both, depending on the nature of the problem.

To aid in expert system development, the shell has a rule induction feature which will generate rules directly from a set of data. In addition, as the expert system is used, data will be collected concerning the paths followed in each consultation. For example, in a diagnosis application, statistics are compiled on the frequency with which each problem occurs. The diagnosis

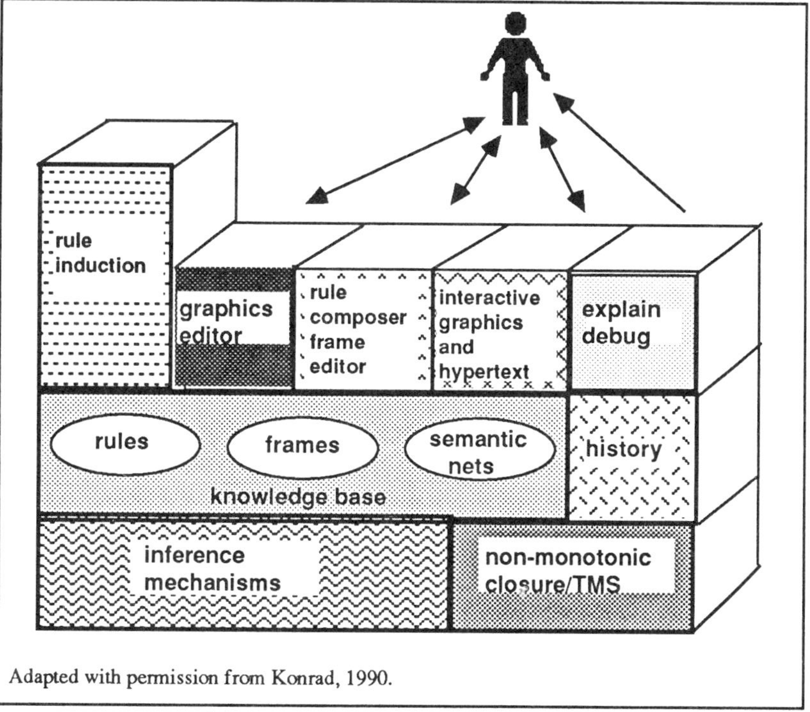

Adapted with permission from Konrad, 1990.

Figure 3. ADEPT Architecture

graph is automatically updated so that future passes through the system can start exploring the most frequently incurred problems first. In this sense, the system can learn from past experiences.

Summary

In the last few years, several articles have appeared which attempt to draw conclusions from actual expert systems implementations and pinpoint the factors which made them successful. The suggestions which were made throughout are reported in this section. Expectations for future developments will be discussed next.

SECTION V

Future Trends and Emerging Technologies

The question of future trends can be considered in three general areas: use of expert systems in business, characteristics of expert systems of the future, and different ways of representing knowledge, especially neural networks.

Use of Expert Systems in Business

The use of expert systems in business has grown exponentially in the last 5 years. While there were few business expert systems listed in 1986, in 1989 "one analyst estimated ... that half of the Fortune 500 companies are now investing in the development and maintenance of expert systems" (Enslow, 1989, 56).

The number of positions in expert system development has doubled or tripled in the last year (Bredin, 1990). "Insurance, banking and finance, and manufacturing are the most prominent industries contributing to the acceptance and growth of expert system applications" (Smith, 1989, 90). The growth in the insurance field has been particularly impressive: whereas at a convention in 1988 only two out of 50 companies represented had even heard of expert systems, estimates are that by 1990 half of all companies are using expert systems in some manner (Greenwald, 1988). But expert systems applications in human resource management, particularly in training, are also on the rise. In the coming years expert systems will be used in all functional areas of business and throughout all sectors of the economy. Not only will expert systems be used more frequently, they will also be more visible. Predictions are for expert systems to move into front offices and to be used in interactions between sales personnel and customers. Already, companies are developing expert systems to be used with their product to simplify maintenance or training for the customer (Feigenbaum et al., 1988).

The work place may be quite a bit different in a few years. People with little education may be able to do tasks which are now reserved for "knowledge workers." Large knowledge bases will give experts access to information from many fields. Executives will be able to draw on the expertise of many experts—captured in expert systems—to support them in their decision making. Computer and user will work together like colleagues, each performing the task at which they are best, and intelligence will be situated in the man-machine system rather than in the person alone. This will probably not be a revolution, but a gradual change (Feigenbaum et al., 1988; Siegel, 1989).

Characteristics of Future Expert Systems

To make expert systems truly function like colleagues, changes will have to come in the user interface. There is a need for user and system to interact in speech rather than writing, and for systems to accept sensory inputs in addition to texts and numbers (see, hear, touch). Also, as we discussed earlier, the interface needs to be intelligent enough to realize at what level of detail and complexity to interact with the user.

Expert systems will also have to learn to reason more like human beings. Today most reasoning methods are based on logic and probabilities. People often reason by analogy, and more systems will embody this form of reasoning. Also, many business situations are not clear-cut, and the use of fuzzy sets may be more appropriate in expert systems of the future (Feigenbaum et al., 1988).

Knowledge Representation

To date, most expert systems built and most expert system shells use a rule-based knowledge representation. However, this representation scheme limits the applications suitable for expert systems. There is a need for different ways to represent knowledge since not all facts come in IF...THEN form and not all reasoning processes can be formulated in this way, either. Besides, the dream of expert systems is to let them learn on their own, to benefit from past experiences and constantly update themselves. A truly intelligent system would be able to develop its own knowledge base from a given set of examples, without requiring an expert.

A possible solution to some of these problems may be in the use of neural networks. Although a thorough discussion of neural networks is outside the scope of this chapter, a few comments are in order at this point.

Neural nets can be thought of as a series of inputs and outputs which are linked by nodes or "neurons." Weights are attached to the connections of the nodes, and when an input exceeds a threshold value, a neuron "fires," i.e., sends out a message. The neuron summarizes the information it is receiving at a given point in time by adding up the weighted sum of all its inputs (Vasilash,

1989; Forsyth, 1989).

Neural nets are not expert systems. Vasilash (1989, 81) distinguishes the two as follows: neural nets are example based and domain free; they find rules, need little programming and are easy to maintain; they are fault tolerant, adaptive, dynamic, run in real time, and keep problems from occurring. On the other hand, expert systems are rule based and domain specific; they need rules, much programming, and are difficult to maintain; they are not fault tolerant, need reprogramming, are static, run in batch mode, and can diagnose problems after they occur.

Expert systems are designed to explain to the user how they reached a decision. The user can follow the reasoning process step by step and assess the validity of the conclusions.

On the other hand, neural nets are more like a black box: they cannot explain why they come to a certain conclusion, and the user will not be able to follow the reasoning process and evaluate the outcome. This is a danger in neural nets—if they "learned" from a non-representative set of data, the results will not be very useful in the real world (Aronson, 1988).

Whereas neural nets will not replace expert systems in the next few years, they are certainly an alternative technology that deserves to be watched and considered in some applications. Projections for market penetration are optimistic: while total US sales in 1987 were less than $5 million, estimates for 1990 are $100 million and for 2000 total US sales of $500 to $1000 million are expected (Domke, 1990).

SECTION VI

Conclusion

In this chapter we have discussed the use of expert systems in business. We have shown that expert system use is growing in organizations worldwide and that applications are not limited to certain industries, functional areas, or even levels in the organization. Until now, the majority of applications have been in areas of diagnosis, design, or training (helping, tutoring), but expert systems can assist with decision making or make and implement decisions on their own.

Although expert systems have been developed for some twenty years, the technology has not been used widely for every day applications until the 1980s. Therefore, there has not been time for empirical research exploring the longrun effects of expert systems on organizations. Fears have been expressed by some that a proliferation of expert systems will lead to a loss of skills in the labor force as we rely more on the computer and less on our own thinking ability and intuition. On the other hand, the argument can be made that expert systems will help to spread expertise throughout the organization and will thus allow decisions to be made at lower levels in the organization—and will also help train employees much more quickly, thus raising the level of skill. Research is needed to resolve this issue. Another issue which will need attention as the use of expert systems spreads is that of liability. Are expert systems a product—which would imply strict liability for the producer—or a service where liability is based on negligence? The safe procedure may be to use expert systems in an advisory capacity only and let the ultimate responsibility rest with a human expert.

Whereas the first expert systems were developed by artificial intelligence specialists, today almost anybody can develop a system quickly by using an expert system shell. This enables managers to develop systems in their own departments, using their own staff. This development makes it necessary for each manager to be familiar with expert system concepts, to understand the issues and the potential benefits of expert systems. Organizations need to decide whether they need to develop standards for expert systems and who decides what resources to allocate to their development and maintenance.

Expert systems carry the potential for large monetary benefits for an organizations, as we have seen in examples such as DEC's XCON and XSEL or American Express' Authorizer'Assistant. Not all systems have such an impact on the organization, but it is worthwhile for many managers to study the potential of this technology and see whether it could make a contribution in their area.

References

Aeh, R.K. (1989). Quality and Productivity. *Journal of Systems Management*. August, 23.

Aronson, D.R. (1988, December). What Neurocomputers Can Do for Wall Street. *Institutional Investor*, 22-23.

Ashmore, G.M. (1989). Applying Expert Systems to Business Strategy. *The Journal of Business Strategy*, Sept./October, 46-49.

Besser, J., & Frank, G.B. (1989). Artificial Intelligence and the Flexible Benefits Decision: Can Expert Systems Help Employees Make Rational Choices? *SAM Advanced Management Journal, Spring, 54* (2), 4-9,13.

Bijker, W.E., Hughes, T.P. & Pinch, T. (Eds.). (1989). *The Social Construction of Technological Systems*. MIT Press.

Bonissone, P.P. & Johnson Jr., H.E. (1984). Expert System for Diesel Electric Locomotive Repair. *Human Systems Management, Vol. 4*, p. 255.

Bramer, M. (1989). Expert systems: where are we and where are we going? In R. Forsyth (Ed.), *Expert Systems: Principles and case studies* (2nd ed.). London, New York: Chapman and Hall Computing.

Bredin, A. (1989, February 19). AI: Techies need not apply. *Computerworld*, p. 123.

Brooks, N. (1989, August). A. L. Expert Systems—So that's what they can do! Bank Administration, pp. 36-37.

Brown, C.E., & Phillips, M.E. (1990, January). Expert Systems for Management Accountants. *Management Accounting*, pp. 18-20, 22-23.

Bruce, M. & Adam, A. (1989, October). *Expert Systems and Women's Lives: A Technology Assessment.* Futures, pp. 480-497.

Carr, C. Expert Support Environments (1989) . *Personnel Journal*, pp. 117-126.

Chen, Y. S. (1989). Organizational strategies for decision support and expert systems. *Journal of Information Science, 15*, 27-34.

Connolly, J. (1989, July 17). Expert Systems Promising. *National Underwriter—Life & Health/ Financial Services*. pp. 6; 11.

Cook, D.A. & Sterling, J.W. (1989, November/December). Alacrity: Software That Asks Shrewd Questions. *Planning Review*, pp. 22-27.

Cook, R.L. & Schleede, J. (1988). Application of expert systems to advertising. *Journal of Advertising Research*, pp. 47-56.

Domke, M. (1990, February). Werden Neuronale Netze langsam praxisreif? *Online*, pp. 40-43.

Ege, G. & Sullivan, W.G. (1990). Expert Systems Update. *Management Accounting*, p. 21.

Enslow, B. (1989, January/February). The Payoff from Expert Systems. *Across the Board*, pp. 54-58.

Feigenbaum, E., Nii, H.P., & McCorduck, P. (1988). *The Rise of the Expert Company*. New York: Times Books.

Ferranti, M. (1990, April 16). Expert systems find niche as tools for sales, support. *PC Week*, pp. 27-28.

Fordyce, K., Norden, P. & Sullivan, G. (1989). Artificial Intelligence and the Management Science Practitioner: One definition of Knowledge-Based Expert Systems. *Interfaces*, 19:5, pp. 66-70.

Forsyth, R. (Ed.). (1989). *Expert Systems: Principles and Case Studies* (2nd ed.). London, New York: Chapman and Hall Computing.

Forsyth, R. (1989). The expert system phenomenon. In R. Forsyth (Ed.), *Expert Systems: Principles and case studies* (2nd ed.). London, New York: Chapman and Hall Computing.

Friedland, J. (1988, July). The expert-systems revolution. *Institutional Investor*, pp. 76-87.

Goodall, A. (1989). An introduction to expert systems. In R. Forsyth (Ed.), *Expert Systems: Principles and case studies* (2nd ed.). London, New York: Chapman and Hall Computing.

Graham, I. (1989). Inside the inference engine. In R. Forsyth (Ed.), *Expert Systems: Principles and case studies* (2nd ed.). London, New York: Chapman and Hall Computing.

Greenwald, J. (1988a, October 31). Artificial intelligence aids reinsurance underwriters. *Business Insurance*, pp. 34, 36-38.

Greenwald, J. (1988b, October 31). Expert systems won't replace human underwriters: Observers. *Business Insurance*, p. 38.

Hochron, G. (1990). Capture that information on an expert system. *Journal of Business Strategy*, pp. 11-15.

Iida, J. (1990, February 8). Expert System to Control Network. *American Banker*, p. 3.

Jones, D.C. (1989, June 26). Outlook for Expert Systems 'Unlimited.' *National Underwriter—Property & Casualty/Risk and Benefits Management*. p. 38.

Jones, D.C. (1989, July 31). Outlook for Expert Systems 'Unlimited.' *National Underwriter—Life & Health/Financial Services*. p. 14.

Jones, P. (1989). Uncertainty management in expert systems. In R. Forsyth (Ed.), *Expert Systems: Principles and case studies* (2nd ed.). London, New York: Chapman and Hall Computing.

Juras, P.E. (1989). The Next Generation of Decision Support. *The CPA Journal*, pp. 72-74.

Kaiser, R.W. (1990). Knowledge-Based Systems. *Journal of Accountancy*, pp. 111-113.

Kirrane, P.R. & Kirrane, D.E. (1989). What artificial intelligence is doing for training. *Training, 26*, 37-41, 43.

Konrad, W. (1990). *SYNERGTECH/ADEPT: Werkzeug zur Entwicklung von Expertensystemen.* Muenchen: SYNERGTECH.

Leonard-Barton, D. & Deschamps, I. (1988). Managerial Influence in the implementation of new technology. *Management Science, 34* (10), pp. 1252-1265.

Leonard-Barton, D. & Sviokla, J.J. (1988). Putting Expert Systems to Work. *Harvard Business Review*, pp. 91-98.

Lindsay, S. (1988). *Practical Applications of Expert Systems*. Wellesley, MA: QED Information Sciences.

Martin, J. & Oxman, S. (1988) *Building Expert Systems*. Englewood Cliffs: Prentice Hall.

Mitchell, B. (1990). Interviewing Face-to-Interface. *Personnel*, pp. 23-25.

Naylor, C. (1989). How to build an inferencing engine. In R. Forsyth (Ed.), *Expert Systems: Principles and case studies* (2nd ed.). London, New York: Chapman and Hall Computing.

Seeborg, I.S., Green, J., & Lim, S.D. (1990). *The Use of Expert Systems in Teaching: Supporting Hypothesis Testing*. Working Paper.

Shadbolt, N. (1989). Knowledge representation in man and machine. In R. Forsyth (Ed.), *Expert Systems: Principles and case studies* (2nd ed.). London, New York: Chapman and Hall Computing.

Siegel, P. (1989). What expert systems can do. *Training & Development Journal*, pp. 71-73.

Smart, G. & Langeland-Knudsen, J. (1986). *The CRI Directory of Expert Systems*. Oxford: Learned Information (Europe) Ltd.

Smith, J.J. (1989). *The Future Belongs to ES. Best's Review—Property/Casualty Insurance Edition 90* (6), pp. 88-92.

Texas Instruments. (1990, January). Growing role of expert systems in process control. *Supervision*, pp. 24-25.

Turban, E. *Decision Support and Expert Systems*, 2nd ed.. New York: Macmillan Publishing Company, 1990.

Van Name, M.L. & Catchings, B. (1990a, March 5). Expert-system shell users advise forethought. *PC Week*, p. 78.

Van Name, M.L. & Catchings, B. (1990b, March 5). Choices abound in expert-system shells. *PC Week*, pp. 77, 79.

Vasilash, G.S. (1989, September). Expert Hackers, Brain Connections. *Production*, pp. 79-81, 84-89.

Warner, E. (1988, October). Expert Systems and the Law. *High Technology Business*, pp. 32-35.

Watts, S. (1989, May 20). Expert systems can seriously damage your health. *New Scientist*, p. 35.

Chapter 9

The Impact of Information Technology on Work Group Innovation and Control

Soonchul Lee
Boston University

P.J. Guinan
Boston University

A large pharmaceutical manufacturer recently installed a lap top computer system for its national sales force, but it has not had the productivity impact that was expected. The system was introduced in the hope of decreasing routine tasks and subsequently giving the sales force more time and energy for the more creative and innovative parts of their jobs. Unfortunately, many sales representatives percieved that the system allowed management to more closely monitor and control their activities. Therefore, the sales force resisted using the system. Viewed as a "corporate eye" for management, IT was clearly unable to improve productivity much less facilitate creative and innovative behavior.

Within this same company, a few districts were happily adopting and using the system. These sales representatives believe the electronic mail features have substantially improved coordination within their districts. Further, they felt that the information contained within the system has made them less dependent upon and less controlled by corporate information. Specifically, one representative felt that the system indentified opportunities and facilitated the retrieval of relevant information for innovative marketing ideas, giving him more time and energy to pursue new sales initiatives.

How can we sort out these apparently contradictory results? After all, we are talking about a single type of computer support system installed within a single organization. In part, the answers may lie in understanding the complex relationship between innovation and control. Although we do not often think of the two in conjunction with each other, the relationship between innovation and control may prove important to our understanding of complex issues such as contradictory reactions to the same information system.

The focus of this chapter, therefore, is to examine the ways in which IT can both facilitate innovation processes and control work processes in a single work group. By detailing the rationale method and results of an empirical study, we will first examine how IT can, by acting as a control mechanism, help the work group to manage tasks and achieve organizational goals, followed by an examination of the perceived effect of IT usage on a work groups' ability to innovate. Finally, we will attempt to compare the effects of tighter control with the ability to innovate, providing insight into the complex relationship between innovation and control.

Although the relationship between control and innovation is not clear, there is extensive work in both areas to guide the rationale for this chapter. Hence, the next section begins by defining control and innovation, followed by a brief review of related literature. Susequent hypotheses, methods and analyses are followed by concluding remarks and a discussion of the results.

Definition of Control and Innovation
Control - Defined

Control emphasizes the decision - making process taken to ensure that organizational members will internalize and act on organizational goals. Often used synonomously with power and influence, control systems have been studied by a number of researchers from various disciplines (Flamholtz, 1985; Weber, 1947; Ouchi ,1979; Tannenbaum, 1968; Smith and Brown, 1954; House and Dessler, 1974; Jermier and Benkes, 1979; Schniesheim et. al., 1976; Thompson, 1967; Reeves and Woodward, 1970). According to the control perspective, the hierarchical nature of the organization is believed crucial to effective coordination of action (Smith and Brown, 1964). Further, the patterns of influence existing in the organizations' culture can be manipulated to promote the best outcomes for the organization as a whole. Researchers in the area have suggested a number of methods that can be used to achieve organizational goals. Typically referred to as Control Systems, we argue that any operationalization of Control Systems should include both self-control as well as managerial control.

Managerial Control refers to the manager's attempts to influence organizational members to behave in accordance with organizational directives. For example, in the IS design context the project manager influences team members to design computer based systems which meet organizational standards, goals and directives.

The extent to which he/she is successful is often referred to as the manager's instrumental behavior or the amount of time a manager initiates activity in the work settings and organizes and defines the way work is to be done. Empirical research validates the power of managerial control as a mechanism for an increased performance of employees (Lee,1989).

Self-Control

More recent research suggests that managerial control accounts for a small portion of the criterion variance, such as perfromance and job satisfaction (Howell and Dorfman, 1981; Kerr, 1977; Kerr and Jermier, 1978). One reason for this may be the fact that a team member's self-control may be acting in such a way as to prevent or neutralize manager influence (Lee, 1989). While there is little empirical research specifically addressing the relative importance of self-control to organizational effectiveness, the literature concludes that as tasks become increasingly complex, a change from managerial control to self-regulated control is desirable (Manz and Sims, 1980: Mills, 1983). In addition managerial control and team member control operate simultaneously (Lee, 1989).

Innovation - Defined

Change, invention, creative behavior and adaptation, are often times operationalized to capture the construct, referred to as innovation. A brief review of the literature illustrates the range of definitions and diverging conceptualizations of innovation (see Pierce and Delbecq [1977] for more detail).

Mohr (1969) differentiated innovation from invention: invention is the creative act of development of a physical item that is relatively new to organizations and innovation is the successful introduction of invention into an applied situation of means or ends that are new to that situation. Most researchers have defined innovation in terms relative to Mohr's. For example, Gerwin (1981) claimed that innovation is a process with two major steps: 1) initiation, in which the decision to adopt is made, and 2) implementation, in which the new item is brought to commercial readiness. Becker and Whisler (1967) focused on innovation as an organizational or social process. They saw innovation as a process that follows invention in time. Invention is the creative act, while innovation is the first or early employment of an idea by one organization or a set of organizations with similar goals. Damanpour and Evan (1984) claimed that while technical innovations are ideas for new products, new services, or the introduction of new elements in an organization's production process or service operation, administrative innovations are those that occur in the social system of an organization. Therefore, innovation should include rules, procedures and structures that are related to the functioning of organizations or individuals.

From this limited review, it is obvious that innovation is not a single construct but one that is conceptualized by many definitions including: 1) the relative newness of ideas; 2) the development or introduction of new ideas on products; and 3) the process of implementing the output of innovation. This multidimensional view of the properties of an innovation corresponds to Bigoness and Perrault's (1981) operationalization of innovation. According to the authors, innovation has three dimensions: content domain, reference domain, and innovativeness domain. Content domain refers to the specific of innovation (i.e., a single new product or sets of innovation). The reference domain refers to the relative newness of the innovation to the organization or individual. The innovativeness domain refers to the time frame for implementing the new product.

For our purposes, innovation is defined as the conception of ideas new to the organization or individual. We include new procedures as well as new products. Further, incorporating Bigorness and Perrault's (1981) notion of a reference domain, innovation includes other people's conceptions of ideas or procedures as long as those ideas are relatively new to the work group. Since the implementation of new ideas and procedures is most often beyond the information system's domain, but depends upon the social system and the strategy of management, we exclude the implementation stage of innovation from our discussion. Clearly, the number of factors that can affect innovation is so great that innovation cannot be explained by any simple model. We will, therefore, limit our discussion to the percieved effect of IT usage on a work group's ability to innovate. Congruent with our understanding of the current design philosophy, IT supports human information processing activities such as the ability to be innovative. What we are interested in, therefore, is not the absolute measure of innovation, but how IT can help the work group innovate.

Review of the Literature
Control Strategies

As was previously stated, our notion of control systems includes dimensions of both self-control and managerial-control. We are using leadership theories to explore these two types of control strategies. The first body of leadership theory suggests that control can be accomplished through a leader's structuring/initiating behavior (leadership behavior theory). This strategy emphasizes that an effective leader provides some type of guidance and/or positive feelings for subordinates as they carry out their tasks(Howell and

Dorfman 1981). The theory focuses on managerial-control systems. The other body of theory arrives at the somewhat unexpected conclusion that control can be achieved by minimizing the roles of leaders (leadership substitute theory). A leader's role can be substituted for under certain circumstances. For example, several variables that characterize professionals such as expertise, independence, and motivation have been suggested as potential substitutes for hierarchical leadership (Kerr, 1977). The focus is on managerial-control systems.

Conceptualizing information technology as a control mechanism, we will attempt to use leadership behavior theory and leadership substitute theory to build a framework of IT's impact on control. Specifically, one of the goals of this chapter is to examine how IT impacts the work group's ability to manage tasks and achieve organizational goals.

Leadership Behavior Theory and IT's Impact on Control

Leadership behavior theory surfaced in the 1950s and generated considerable interest in training leaders to exhibit desirable behavior. Leadership behavior theory seeks to identify the process side of leadership attributes instead of focusing on the psychometric attributes of leaders.

Leadership behavior theory identifies three dimensions of instrumental behavior which clarify what is expected of subordinates in their work roles (House and Dessler 1974; House and Mitchell 1974; Schriesheim 1978); (1) task clarification—clarifying management expectations of subordinates in their work, (2) work assignment—assigning subordinates to specific tasks, and (3) procedure specification—enforcing rules, procedures, and work methods. IT can augment and support these leadership behaviors in controlling and coordinating a work group. Therefore, we hypothesize that task clarification, work assignment, and procedures specification are factors through which IT can affect control.

Task Clarification: Task clarification refers to the information provided by a leader on roles and tasks associated with a given position to help the subordinate understand a given job. IT can provide the work group with a comprehensive view of the business, opportunities for detailed analysis, and consideration of alternate courses of action.

Work Assignment: IT can help assign group members to specific tasks through flexible time scheduling and work distribution. Typical examples are project management tools which focus on improving organizational efficiency by automating the scheduling process. These tools reduce the costs of communication and coordination by carrying out activities in partnership with the users. Also, electronic mail can make coordination easier, especially for geographically dispersed groups.

Procedure Specification: If procedures are explicitly defined and readily measured, effective control can be accomplished by standardizing operating procedures. Although standardization often leads to inflexibility in the work group, it can streamline work procedures, resulting in improved efficiency. In addition, the increased efficiency frees time for the work group to explore new ideas. Saunders (1981) claimed that because IT automates and routinizes simple tasks, the work group can devote more time to a variety of complex tasks through enhanced processing capability.

Leadership Substitution Theory and IT's Impact on Control

House and Mitchell (1974) argued that under circumstances when both goals and paths to goals are clear, attempts by the leader to clarify paths and goals will be both redundant and seen by subordinates as imposing unnecessarily close control. Several variables that characterize professionals working in organizations have been suggested as potential substitutes for hierarchical leadership because professionals can often perform their jobs effectively without a leader's supportive functions (Kerr 1977). Kerr and Jermier (1978), Jermier and Berkes (1979), Miles and Petty (1977) added task and organizational characteristics to the list of possible leadership substitutes. Building on the substitution concept, Howell and Dorfman (1981) empirically tested hypotheses that match specific substitutes with a given leader behavior. They reported that although their operationalizations of leadership substitutes cannot be a strong substitute for leadership, these operationalizations may be viewed as substituting for the potentially important explanatory power of instrumental leader behaviors in the absence of any leadership.

These researchers claimed that many individual, task, and organizational characteristics have the capacity to act as substitutes for hierarchical leadership in that they often serve to neutralize or substitute for a formal leader's ability to influence satisfaction and performance for either better or worse (Kerr 1977). We maintain that information technology can play the role of these substitutes for hierarchical leadership by clarifying the purpose of goals and by suggesting paths to goals. For example, standard operating procedures introduced with an information system may be a substitute for a leader's structuring function. Also, IT can help to reduce the level of turbulence and uncertainty in organizational goals for lower level subordinates and thus serves as a substitute for a formal leader's instrumental and participative function (Moynihan 1982).

Therefore, leadership substitute theory suggests other factors through which IT can affect control. These variables include the ability and knowledge of subordinates, task induced motivation, and task provided feed back, methodological invariance, and cohesive interaction (Kerr, 1977).

Ability: Zuboff (1983) examined IT's impact on managerial jobs. The results suggested that monitoring and decision-making based on fairly routine information was added to the jobs that had the greatest operation proximity to the relevant decision. Because of IT there was increased information processing capability which increased the level of knowledge of lower level subordinates. IT, therefore, enhances the information processing capability of an individual and thus enables her to better understand the impact of a decision. Furthermore, information systems may enable a group member to learn specific skills outside of the jobs to which she is assigned. Gerrity (1971) studied the impact of portfolio management information systems on bank managers and concluded that the managers developed extended knowledge due, in part, to working with Information Systems.

Information Support: The need for information support in carrying out complicated job tasks has been noted in many studies of the limitations on human cognition (Tversky and Kahneman 1974; Slovic 1976). These studies noted that computerized analytical tools can augment the capabilities of the human mind. Many empirical findings reported that interactive computer support facilitated group decision making effectiveness by providing analytical decision support tools and by increasing the number of sources of information (Adelman 1984; Kull, 1982).

Intrinsic Motivation: Curry et al. (1986) reported that people who were satisfied with their jobs were more likely to remain in the organization as well as expend considerable effort for organizational performance. Thompson (1965) argued that internal commitment and a sense of intrinsic rewards are important to carrying out tasks effectively. IT can affect motivation by creating challenging, meaningful tasks (Malone 1985). Malone (1982) suggests a number of ways to include motivational factors such as challenge, fantasy, and curiosity in a computer system to augment child learning. Early office automation projects reported that bank clerks had their jobs restructured so that they no longer performed isolated clerical steps in a process they did not understand (Matteis 1979). Since they could improve their perception of the "bigger picture" with IT usage, bank clerks were motivated to handle all of the steps required in dealing with customers on their own, resulting in overall better performance.

Task Feedback: Lawrence and Lorsh (1967) noted that the relatively greater uncertainty facing research development results, in part, from the longer time span needed for definitive feedback. IT can reduce the span of task-related feedback by making goals explicit and then providing feedback on performance in relation to these goals (Moynihan, 1982).

Unambiguous Procedure: Van de Ven, et al. (1976) claimed that under conditions of low task uncertainty and low task interdependence the most frequent and effective coordination strategy was programming through impersonal modes. In addition to making the procedures explicit, IT can make targets and achievements explicit and visible, resulting in unambiguous procedures.

Collegial Interaction: The need for better interaction is strongly supported by Allen's (1970) empirical study of a communication network in R&D laboratories. According to his study, high performers reported significantly more time in their discussions with colleagues. Furthermore, they relied on more people both within their own technical specialty and in other specialties. Information systems, especially those supported by telecommunications capabilities, can facilitate communication network build-up among experts. Recent studies of computer-based message systems or electronic mail systems (Montgomery and Benbasat 1983; Crawford 1982; Hiltz and Turoff 1981) all reported that these systems improved communications by reducing the dysfunctional features of meetings, conversations, written mail, and telephone calls.

Control Related Hypotheses

In the previous section, leadership behavior theory and leadership substitute theory were described and used to establish two frameworks to help explain IT's impact on control. We will empirically examine the two frameworks in order to determine the extent to which the frameworks explain criterion variance and identify the important variables.

We focus on two dependent variables, which relate to the use of IT as a control mechanism. First, we examine the notion of self-control. An information system can enable the work group to assume more responsibility or autonomy, because the system can provide more rapid and more comprehensive feedback on organizational and managerial performance. Thus, control by evaluation of outcome can be obtained on a more timely basis and more decision-making authority can be delegated (Pfeffer and Leblebici 1977). Second, we focus on managerial control and, specifically, the aspects of planning and control. Our focus is on the short term aspects of operational planning and control.

Since IT is conceptualized as a mechanism for control, IT's impact on the control variables derived from each theory should contribute positively to self-control and to planning and control. Therefore, the first hypothesis states:

> H1: The greater the IT's impact on a control variable (i.e., task clarification, work assignment, procedure specification, ability, information support, intrinsic motivation, task feedback, unambiguous procedure, and collegial interaction), the greater the effect of IT usage on self-control and on planning and control.

In addition, since the leadership substitute theory largely refers to conditions in which leadership behavior is unimportant, we hypothesize that the leadership substitute theory version of IT impact on control framework better explains the criterion variance of self-control. Therefore, the second hypothesis states:

> H2: The leadership substitute theory more clearly explains IT's impact on self-control than does the leadership behavior theory.

Innovation Strategies

Rogers' (1962) diffusion theory stated that the innovation process is influenced by the nature of four interrelated constructs: 1) the innovation; 2) the relevant social system; 3) time; and 4) communications about the innovation. Since the publication of the theory, a broad literature has developed on the process of innovation. Several studies have identified factors affecting or conducive to innovation.

At a micro level, individual characteristics or individual differences have been studied in parallel with the availability of individuals capable of producing new ideas([Steiner 1965), the attitude of individuals toward innovation (Hage and Aiken 1970, Cyert and March 1963), and resource availability (Mohr 1969). In additon to individual capabilities, the structure of the work group or organization has been studied. For example, Guetzkow (1965), and Hage and Dewar (1973) focused on the relationships between organization variables and innovation. Thus, these studies and other theoretical and empirical studies identify two determinants of innovation: 1) individual/group capabilities and 2) group interaction.

We will briefly review past research which studies these determinants of innovation and their relationships to IT usage.

Individual and Group Capabilities

Individual and group capabilities have been the primary focus of several research articles. Carter and Williams (1957) noted that if a firm is to make rapid and effective use of scientific ideas, it must include people capable of evaluating or interpreting these ideas in accordance with the needs of the firm. Von Hippel (1983) claimed that for new product development, user segmentation is critical. The segments are: 1) those who have real-world experience with the novel product concepts and attributes of potential interest, and 2) those who would enable clearer understanding of selected product attributes outside of the normally experienced range and clearer understandings of future profitable markets.

The innovation literature emphasizes the importance of individual or group characteristics in creating and implementing innovative ideas in the context of a firm. An organization may augment individual/group capabilities in several ways: 1) by increasing information about customer needs, environmental changes, and other relevant data from various sources (information support); 2) by enforcing the motivation of the group members (motivation support); and 3) by providing resources the work group can use in an innovation process (resource support).

Information Support: As was previously stated, the need for information support is strongly supported by Allen's [1970] empirical study of a communications network in R&D laboratories. Furthermore, Pelz and Andrews (1966) reported that the variety of contacts and their frequency contributed independently to performance. Colleague contacts both within the immediate work group and with other groups in the organization were positively related to worker's performance. Souder (1977) argued that interdepartmental cooperation is needed at various stages of the product innovation process. Information systems, especially those supported by telecommunications capabilities, can facilitate the growth of communications networks among experts. Recent studies of computer-based message systems or electronic mail systems (Montgomery and Benbasat ,1983; Crawford, 1982; Hiltz and Turoff ,1981) all reported that these systems improved communications by reducing dysfunctional features of meetings, conversations, written mail, and telephone calls.

Firms that possess open communication networks relative to their environments are more likely to innovate than firms that maintain closed communication systems [Rogers and Schoemaker 1971, Zaltman et al. 1973]. This stems from the belief that an innovation process consists largely of responses to customer needs [Utterback 1974, Gilpin 1975]. Even though IT cannot change market demand for technological innovation, IT can be helpful in collecting the customer needs in a more timely and accurate fashion.

Motivation Support: Hage and Aiken (1970) noted that people who were satisfied with their jobs were more likely to be committed to the organization and their jobs. As a result, they were more likely to search for ways to improve conditions as well as be more receptive to new ideas. In contrast, March and Simon (1958) claimed that innovative behavior is derived from a motivational concept. They claimed that dissatisfaction initiates search behavior which, in turn, may lead to innovative behavior. While there are conflicts in the literature regarding the role of satisfaction for innovation, motivational reinforcement is positively related to innovation. Thompson (1965) argued that an internal commitment and a sense of intrinsic rewards reinforce motivation of innovative ideas.

As was previously stated, there is some evidence that IT can affect motivation by creating challenging and meaningful tasks (Malone 1985; Matheis, 1979).

Resource Support: The need for resource support in the innovation process has been discussed in a number of studies on the limitations of human cognition (Tversky and Khaneman 1974, Slovic 1976). According to the research, the human mind is often limited in its ability to appraise the effectiveness of a new idea through a rational or analytical process. Slovic (1976) suggests that computerized analytical tools can help to expand the human mind. The design philosophy of decision support systems reflects this idea. Many empirical works suggest that interactive computer support facilitates such as group decision making improves effectiveness by providing analytical decision procedures (Adelman 1984; Kull 1982).

Additional evidence for the importance of resource support can be explained by the concept of "slack" introduced by Cyert and March [1963]. They argued that the subunit which has extra resources can devote more effort and energy to critical work. For instance, users have increased time and energy for tasks that they are uniquely qualified to perform due to computerization of routine tasks, and the increased availability of time and energy facilitates innovation (Foster and Flynn 1984).

Group Interaction

As discussed in the previous section, increased use of organizational colleagues for information is strongly related to scientific and technological innovation (Allen, 1970). There are some strategies for properly structuring the flow of technical information to augment the innovation process. One of the strategies is building informal relations through the formation of project teams and intergroup transfers (Allen, 1970). However, these strategies require that the organization radically change the existing structures and that management intervene. The change in structure cannot be done directly by an information system (Robey, 1983). Therefore, integral group interaction processes which can be changed with IT usage and may prove relevant to innovation are discussed. Structural variables which require management intervention are not included in this discussion.

Structuring of group interaction embodies a variety of concepts which are typically divided into component parts, referred to as dimensions of structure. Dimensions of structure have been extensively studied in organization science. Campbell et al. (1974) differentiated structuring from structural dimensions. The structural qualities of an organization are its physical characteristics, such as size, span of control, and level of hierarchy, which are apparent by studying its organizational chart. In contrast, structuring refers to interactions and activities occurring within the organization that prescribe or restrict the behavior of organization members. Based on Dalton et al.'s (1980) structuring classification, we will examine the following three dimensions: 1) specialization, 2) decentralization, and 3) standardization.

Specialization: Specialization forces the work group to organize around disciplines or technologies, and enables the individuals to stay in touch easily with new developments in disciplines or technologies. As was previously stated, Gerrity's (1971) study suggests that the Information System helps to extend job knowledge. We maintain that part of the new knowledge helps the work group to innovate. It is, therefore, suggested that IT impact on specialization facilitates the innovation process of the work group by making the work group members more knowledgeable, and inducing them to think more broadly, more creatively.

Decentralization: Hage and Aiken (1967) found a positive relationship between innovation and decentralization, and a negative relationship between innovation and close hierarchical supervision. Pfeffer and Leblebici (1977) reported that IT enables the apparent delegation of decision making authority because it provides the management with comprehensive information on performance. As was previously mentioned, Zuboff's (1983) field survey suggests that the lower level subordinates assumed more power and increased responsibility because they had greater operational proximity to relevant decisions and were provided with accurate data and procedures. We maintain that decentralization via IT can lead to innovative behavior. Specifically, subordinates may have a greater number of sources of information, generate a greater number of ideas, and commit more easily to implementation of changes.

Standardization: Although standardization often leads to inflexibilities in the work group, standardization can streamline the work procedures, resulting in efficiency. This increased efficiency frees time for the work group to explore new ideas. Saunders (1981) claimed that because IT automates and routinizes simple tasks, the work group can devote more time to a variety of complex tasks. Therefore, we suggest that standardization is positively related to organizational innovation.

Innovation Related Hypotheses

In the previous section, the literature on individual and group capabilites and group interaction was described and used to establish a framework to help explain IT's impact on innovation. Since IT is conceptualized as a mechamism for innovation, IT's impact on the innovation variables derived from each theory should contribute positively to innovation. Therefore, the third hypothesis states:

> H3: The greater the IT's impact on an innovation variable (i.e., information support, motivation support, resource support, specialization, decentralization and standardization), the greater the effect of IT usage on innovation.

The impact of IT on individual and group capabilities may directly impact innovation. On the other hand, the impact of IT on structuring variables leads to increased group interaction behavior that are conducive for innovation. The impact of IT on innovation through structuring variables is therefore secondary in nature. Thus, an additional hypothesis states:

> H4:The perceived effect of IT usage on ability to innovate through enhancement in individual and group capabilities is more profound than the effect through changes in group structuring variables.

The Relationship Between Innovation and Control

One of the goals of this chapter is to examine the ways in which IT can both facilitate innovative processes and control work processes in a single work group. In order to look at the relationship between the two, we attempted to compare the effects of tighter control with the ability to innovate.

As was previously stated, innovation is defined here as the conception of ideas new to the organization or individual (Lee and Treacy 1987; Damanpour and Evan 1984). Research in the innovation area suggets that if increased control is employed by an organization, the individuals or work groups will innovate less. The explanation is that tightly imposed control will restrict openness in the system and will strictly enforce work procedures, resulting in greater prediction of prescribed behaviors and less innovation (Pierce and Delbecq 1977; Hage and Aiken 1967). However, since IT is a supporting tool for human information processing activities, we do not believe that the IT supported work group will be less innovative. Rather, we hypothesize that the greater the IT's impact on control, the greater will be the ability to innovate. Note that IT's impact on control was conceptualized as a control device for more effective self management. Specifically, the theoretical rationale from leadership substitute theory suggests that if control variables are supported by the information system, the work group can control their activities effectively without the presence of formal intervention from upper management. Therefore, the following hypothesis is stated:

> H5:The greater the IT impact on control, the greater will be the ability to innovate.

Methodology

Sample

The sample in this study consisted of information system users at seven case sites. These sites were various departments including legal support, sales support, corporate planning, legal service, engineering, purchasing, and computer support, in several large manufacturing firms. In selecting the sample, only sites with extensive information systems usage were considered.

Procedure

The principal instrument was a questionnaire completed by the information systems users. The users could answer questions with a Likert response format with seven response alternatives ranging from strongly disagree to strongly agree. Each user was to act as a key informant to report the perceived impact of IT on control and innovation in his or her small work group. The responses from each site was subjected to ANOVA tests and no significant differences were found to exist across case sites when we controlled for environmental complexity.

Questionnaires with attached cover letters and stamped return envelopes were mailed to 180 users of information systems at the seven case sites. Of the 180 questionnaires sent, 136 were completed and returned, representing a response rate of 75.6%. The sample size was later reduced to 126 by deleting respondents who had left an excessive number of items unanswered, and those who had responded consistently in a specific scale over a successive number of items. Pair-wise elimination was used in the treatment of individual missing data.

Measures

We constructed a new set of questions because we could not find existing ones which were appropriate for our purposes. We wanted to measure the change in ability to control and to innovate due to IT usage. Perceptual measures were used because key informants were in the best position to judge the effects of IT on their work group's ability to control and to innovate. Objective outcome measures, such as number of innovations, are influenced by many contextual variables outside of the control of this study. Therefore, perceptive measures rather than hard measures were appropriate.

IT's Impact on Control Variables from Leadership Behavior Theory

There were three factors measured: task clarification (TC), work assignment (WA), and procedure standardization (PS). The Cronbach alpha reliability coefficients for TC, WA, and PS were 0.82, 0.73, and 0.78, respectively. A multi-trait, multi-method correlation matrix (Campbell and Fiske, 1959) was used to examine convergent and discriminant validity. Convergent validity was tested by examining whether the correlations between measures of the same theoretical concept were different from zero and sufficiently large. The smallest within-variable correlation was 0.55, which was significantly different from zero ($p < 0.01$). Discriminant validity was tested by counting for each theoretical construct the number of times that an item measuring the same theoretical construct correlates more highly with an item of another theoretical construct than with the response on other items of the same theoretical construct. Less than 10% (four out of 42) comparisons violated the discriminant validity test. Therefore, the measures for all three factors displayed good convergent and discriminant validity.

IT's Impact on Control Variables from Leadership Substitute Theory

There were six factors measured: ability (AB), information support (IS), intrinsic motivation (IM), task feedback, (TF), unambiguous procedures (UP), and collegial interaction (CI). The Cronbach alpha reliability coefficients for AB, IS, IM, TF, UP, and CI were 0.73, 0.80, 0.86, 0.80, 0.79, and 0.81, respectively. The convergent validity was significant (the smallest within variable correlation was 0.46). Less than 10% (nine out of 270) of comparisons violated the discriminant validity test.

IT's Impact on Innovation Variables

Six factors of innovation which can be affected by IT usage include: information support (IS), motivation support (MS), resource support (RS), specialization, decentralization (DE), and standardization (ST). The Cronbach alpha reliability coefficients for IS, MS, RS, SP, DE, ST, and INNOV were 0.80, 0.86, 0.82, 0.77, 0.57, and 0.79, respectively. The smallest within-variable correlation was 0.38 between decentralization measures. For a sample size of 126, this was significantly different from zero ($p< 0.01$), and thus convergent validity was significant. Less than 10% (15 of the 268) comparisons violated the discriminant validity test. In other words, each of the correlations between different measures of the same construct was greater than the correlations with measures of different constructs. Therefore, the measures for all six factors displayed good convergent and discriminant validity.

Dependent Variables

Three sets of questions were asked to measure self-control, planning and control, and innovation with IT usage. The Cronbach alpha reliability coefficients were 0.79, 0.88 and 0.81, respectively.

Analysis and Results

The analysis will focus on the development of the relationships between the effect of IT usage on control and the sets of control variables identified from the two leadership theories, and between the effect of IT usage on innovation and the set of innovation variables. The normalized average scores of the item measures were used as single measures for all variables in this study.

Determinants of Control

H1 states that the greater the IT impact on a control variable, the greater the effect of IT usage on self control, and on planning and control. The results are as shown in Tables 1 and 2. All of the correlations between control variables were positive and statistically significant ($p < 0.01$). Also, there were generally stronger associations between control variables and planning than there were between control variables and self control.

Multiple stepwise regression analyses were used to assess the independent contribution of each control variable to the dependent variables. The regression results are incorporated with the correlation results in Tables 1 and 2. Note that each set of control variables derived from each theory had similar predictive power (similar R-Squares) for both self control and planning and control. However, planning and control were better explained than self control by the control variables from both models. The tables show a remarkable difference in the relative importance of individual control variables in predicting self control and planning and control, and hence each dependent variable will be discussed separately.

Self Control: The regression results in Tables 1 and 2 show that task clarification were significant in the leadership behavior theory model, while ability and unambiguous procedure were significant in the leadership substitute theory model. The two model's R-Squares were 0.392 and 0.421

Variables from Leadership Behavior Model	Self-Control		Planning and Control	
	Correlation Coefficient	Regression Beta Coefficient	Correlation Coefficient	Regression Beta Coefficient
Task Clarification (TC)	.587**	.469**	.811**	.621**
Work Assignment (WA)	.410**		.695**	.294**
Procedure Standardization (PS)	.471**	.248**	.448**	
(R-Square)		(.392)		(.708)
(Adj. R-Square)		(.379)		(.702)

* p<0.05
** P<0.01

Table 1. Correlation and Multiple Stepwise Regression Coefficients for Control Variables from Leadership Behavior Theory

Variables from Leadership Substitute Model	Self-Control		Planning and Control	
	Correlation Coefficient	Regression Beta Coefficient	Correlation Coefficient	Regression Beta Coefficient
Ability (AB)	.465	.205*	.529**	
Information Support (IS)	.302		.547**	
Intrinsic Motivation (IM)	.368		.623**	.278**
Task Feedback (TF)	.395		.763**	.385**
Unambiguous Procedure (UP)	.624	.522**	.683**	.267**
Collegial Interaction (CI)	.342		.583**	.153*
(R-Square)		(.421)		(.738)
(Adj. R-Square)		(.408)		(.726)

* $p<0.05$
** $p<0.01$

Table 2. Correlation and Multiple Stepwise Regression Coefficients for Control Variables from Leadership Substitute Theory

respectively. Since the stepwise regression analyses identified two statistically significant variables for both models, the predictive power can be easily compared.

$$1/2*\ln[(1+r)/(1-r)] \qquad (1)$$

For large samples ($n > 25$), the statistic in (1) is approximately normal with mean $1/2*\ln[(1 + ro)/(1-ro)]$ and variance $1/(n-3)$, where n is the size of the samples, r is the sample correlation coefficient, and ro is the population correlation coefficient (Myers 1986). The two correlations were compared using the standardized statistic in (1) and there was significant difference between each model for predictive power ($p < 0.01$). Therefore, we conclude that H2, the leadership substitute theory can better explain IT impacts on self control, and cannot be rejected in our study, even though the difference in coefficients of determination was small.

Planning and Control: Both models could better explain the dependent variable, planning and control, than the dependent variable, self control. R-squares for planning and control were 0.708 and 0.738 for the leadership behavior model and the leadership substitute model, respectively. Task clarification was the most significant, followed by work assignment in the leadership substitute theory model. For the leadership substitute theory model, task feedback, intrinsic motivation, unambiguous procedure, and collegial interaction were significant. There was no basic difference in predictive power in the two models, suggesting that both can be used for examining the IT's impact on control.

Relationship between the two leadership models: If both models have the same predictive power, the two sets of control variables should correlate highly. Table 3 shows that they were indeed correlated significantly. Task clarification was highly correlated with all the control variables derived from leadership substitute theory. This is not surprising if we consider the effects of IT as clarifying the tasks the work group performs. Task clarification can be enhanced if the work group members become more knowledgeable, obtain more information and task-related feedback, and get more expert advice, which were all reflected in the control variables from leadership substitute theory.

Determinants of Innovation

H3 states that the greater the effect of IT impact on an innovation determinant, the greater will be the perceived effect of IT usage on innovative ability, where the innovation determinants are information support (IS), motivation support (MS), resource support (RS), specialization (SP), decentralization (DE), and standardization (ST), respectively.

To test the hypothesis, we performed correlation analysis of the innovation determinant measures with the perceived innovation measure as shown in Table 4. The correlation table provides general support for the hypothesis, H3. IT impact on motivation support was the most strongly correlated with the innovation measure INNOV, followed by IT impact on resource and information support. IT impacts on specialization (SP), standardization (ST), and decentralization (DE) were less significantly correlated to INNOV. For a sample size of 126, however, all the six correlation coefficients were significant at the 1% level.

The multiple regression analysis was used to assess the effect of each innovation determinant on perceived effects of IT on ability to innovate. Table 4 also shows the result of the multiple regression analysis. Three factors, motivation support (MS), information support (IS), and resource support (RS), were significant ($p<0.01$) in the regression model. The three factors explain 80.0% of the total variance. Motivation support is the most important factor with a coefficient of 0.60. Information support and resource support are almost equally important with coefficients of 0.17 and 0.24 respectively. In past literature, the diversity, frequency, and centrality of information through communication are the most cited facilitators of innovation (Allen 1970, Pelz and Andrews 1966, Ebadi and Utterback 1984). Our results show that motivation support is the most important in the context of IT. In other words, the major impact of IT on innovation is not through the channel most frequently cited in the literature. Instead as Thompson (1965) claimed, an internal commitment and a sense of intrinsic reward contribute more to the initiation of ideas. IT can enhance motivation and commitment by making work more exciting. Further, IT builds experimentation into the work process, thus generating new ideas. IT facilitates experimentation through fast feedback and by instilling confidence in experimentation.

Our results showed that information support and resource support contribute almost equally to innovation. The importance of information support to innovation had been widely cited in the literature (Allen 1970, Pelz and Andrews 1966, Ebadi and Utterback 1984). In Allen's study of communication networks in R&D laboratories, he reported that high performers generally seek more information from their colleagues both within and outside their own technical specialty. Pelz and Andrews (1966) found that a person's performance was positively related to both the frequency and variety of contacts with people in the organization. IT enhances information support by facilitating information access through electronic mail and database access. As well as the ease of access, IT has the added attribute of anonymity. Allen (1970) claimed that it may be very costly for a project member to admit to a colleague that he needs help. Sharing IT resources can thus promote interaction between

	Ability (AB)	Information Support (IS)	Intrinsic Motivation (IM)	Task Feedback (TF)	Unambiguous Procedure (UP)	Collegial Interaction (CI)
Task Clarification (TC)	.591**	.664**	.662**	.657**	.655**	.521**
Work Assignment (WA)	.304**	.517**	.454**	.704**	.551**	.592**
Procedure Standardization (PS)	.498**	.329**	.442**	.362**	.658**	.187**

* $p < 0.05$
** $p < 0.01$

Table 3. Correlation Coefficients between Control Variables from Leadership Behavior Theory and Control Variables from Leadership Substitute Theory

colleagues and does so in a less personal manner if desired.

The importance of resource support was evident in the early uses of IT. Early information systems were primarily efficiency-oriented. The improvement in efficiency constitutes a slack resource for use in other purposes (Cyert and March 1963). In addition, IT provides analytical tools that stretch the limited ability of human minds to appraise the effectiveness of a new idea through a rational and analytical process (Slovic 1976). Specialization (SP), decentralization (DE), and standardization (ST) were not significant in the multiple regression model. Nevertheless, we can make some observations from the correlation matrix among the factors and the general question (INNOV) shown in Table 4. While they were all significant at p=0.01, the correlation coefficients of SP, ST, and DE were much smaller than those recorded by information support, motivation support and resource support. Furthermore, when we controlled for the variables, information support (IS), motivation support (MS), and resource support (RS), the correlation coefficients of specialization, centralization, and standardization with INNOV were no longer significant (p>0.20).

Our results, therefore, suggest that standardization, decentralization, and specialization are positively related to innovation. Nevertheless, the results are insufficient to indicate that they may be causal variables to innovation. The impact of IT on innovation through group interaction, therefore, appears secondary in nature. This is not surprising if we consider the measures of standardization, decentralization, and specialization. These measures do not reflect general structuring or interaction constructs, but changed structuring or interaction due to IT usage. Furthermore, the dependence of innovation on IT impact on interaction was partially reflected in other constructs. For instance, information support through electronic mail and database access captures certain aspects of decentralization. The effect of standardization can be reflected by resource support through efficiency gain. In other words, changes in group interaction due to IT are of secondary importance to innovation. Therefore, our results provide support to H4: the perceived effect of IT usage on ability to innovate through enhancement in individual/group capabilities is more profound than the effect through changes in group structuring variables.

In summary, theoretical concepts and our empirical results provide support that IT enhances innovation through motivation support, information support, and resource support. Our results also showed that IT impact on motivation support is the most important factor for ability to innovate.

Control and Innovation

Self control and planning and control measures were positively correlated to innovative ability perceived with IT usage. The correlation

Variables from Innovation Theory	Correlation Coefficient	Regression Beta Coefficient
Information Support (IS)	.611	.173••
Motivation Support (MS)	.856	.599••
Resource Support (RS)	.746	.236••
Specialization (SP)	.448	
Decentralization (DE)	.287	
Standardization (ST)	.399	
(R-Square)		(.800)
(Adj. R-Square)		(.786)

• p<0.05
•• p<0.01

Table 4. Correlation and Multiple Stepwise Regression Coefficients for Control Variables from Innovation Theory

analyses showed that planning and control was more highly correlated to innovation (r=0.683) than self control (r=0.383). In addition, we regressed the ability to innovate on control variables derived from leadership substitute theory, since these control variables may be relevant to the innovative ability of the work group by providing information, resources, and motivation (Lee and Treacy 1988). In the stepwise regression analysis, intrinsic motivation, information support and task feedback were statistically significant ($p<0.01$), in that order, and R-Square was 0.72. The result provided support for H5: the greater the IT impact on control, the greater will be the ability to innovate. It also confirmed our belief that IT's impact on the control variables derived from leadership theories is positively related to a work group's ability to innovate.

Discussions and Conclusions

The focus of this study was to examine the ways in which IT can both facilitate productive processes and control work processes in a single work group. The first set of hypotheses were designed to look at the ways in which IT acts as a control mechanism for the organization. The second set of hypotheses examined the perceived effect of IT usage on a work group's ability to innovate.

Implications for Control Strategies

Based on leadership theory, two frameworks for IT's impact on control were developed. One emphasizes leader's substitution, the other leadership structuring and initiation. Contrary to our belief, the data did not allow us to reject either of the frameworks. The results were particularly strong with regard to explained variances for both self control and planning control.

Perhaps because IT is conceptualized as a mechanism for control, not as a formal leader who dictates group members to follow machine-induced instructions, both the leadership behaviour framework and the leadership substitute framework are equally powerful in explaining criterion variances. There were no significant differences between the explanatory or predictive power of either framework (the R-squares are almost identical). However, because leadership substitute theory largely deals with the conditions in which leadership behavior is less important, the leadership substitute framework explains the criterion variance of self control slightly better. Furthermore, the set of control factors from one framework is significantly correlated to the set of control factors from the other. Therefore, it is reasonable to use either framework when studying the impact of IT on control.

The framework from leadership behavior theory identified task clarification as the most important factor of IT impact on control. In the empirical literature on leadership, there were consistent findings that role clarification was the most important variable in explaining criterion variance (Schriesheim,

House and Kerr 1976; Schriesheim and von Glinow 1977). Jermier and Berkes (1979) studied police organizations and reported that role clarification was the only leadership behavior that explained organizational commitment when task characteristics and leadership substitute variables were controlled for. It may be reasonable to assume that IT can provide a broad range of information on how the work group should function and make goals explicit.

As well as task clarification, procedure standardization was significant for self control and work assignment was significant for planning and control. If the organization implements standardized operating procedures, rules, and specialized roles, it can delegate much of the decision-making authority to the work group since decision making can be monitored by higher level management through enhanced MIS data. Perhaps this occurs because higher level management can be more confident in allowing subordinates to take charge when decision rules are prescribed (Carter 1984; Pfeffer and Leblebici, 1977). IT can also help the work with distribution tools, such as project management tools, and more flexible communication media, such as computer conferencing and electronic mail.

Leadership substitute theory suggests that a number of individual, task, and organizational characteristics can act as a control mechanism for coordinating the work group's activities, reinforcing the control mechanism in existence. Ability and unambiguous procedure were the most important factors of IT impact on self control. These two factors were similar to task clarification and procedure standardization in the leadership behavior framework (the correlation coefficients between task clarification and ability and between procedure standardization and unambiguous procedure were above 0.55). Task feedback was the most important factor followed by intrinsic motivation, unambiguous procedure, and collegial interaction for IT's impact on planning and control. Salancik (1977) suggested that if the goal is made explicit and feedback on the achievement of the goal is provided, an employee is induced to make a public commitment and this commitment creates implications for future behavior. Increased motivation resulting from the usage of IT can also make a work group commit its efforts to operational planning and control. This task-induced motivation was one of the factors affecting an individual's organizational commitment (Curry et al. 1986).

Planning and control measures were positively correlated to the impact of IT on innovative ability. The result contradicts the innovation literature which suggests that if more control is employed by an organization, then the work group will innovate less (Pierce and Delbecq, 1977; Hage and Aiken, 1967). This may be true because IT is conceptualized as a supporting tool for control (not as a hierarchical control device) which can enhance the work group's ability to innovate. Thus, IT can increase the control ability and innovation ability of a work group by augmenting human information processing ability.

Implications for Innovation Strategies

In general, the empirical data did not allow us to reject the general framework that the impact of IT perceived ability to innovate can be explained through its impact on individual and group capabilities and group interaction. Correlation analysis of the data as shown provided general support for the framework. However, multiple regression analysis revealed that the effect of IT usage on perceived innovative ability through individual and group capabilities is more profound than the effect through group structuring variables. All three individual and group capabilities factors were significant in the regression equation while the three group interaction variables were not. IT impact on motivation support was shown to be the most important contributor to perceived ability to innovate, and IT impacts on resource and information support were next in order of importance.

Our results, therefore, show that IT impacts on individual and group capabilities are a better set of variables to predict innovation than that of group structuring variables. The results are not surprising since our group structuring variables measured changed interaction due to IT usage, not the innate group structures in place. Even though group structuring differences may have significant impacts on innovation as predicted in the literature, the changes in group interaction brought about by IT usage are not sufficient to carry over to affect innovative ability. Furthermore, the IT impacts on group interaction were partially reflected in other variables. For instance, information support through telecommunications capabilities captures certain aspects of decentralization.

The item pools provide a useful instrument for further studies of the impact of IT on innovation and control. The instrument appears to be robust, although minor modifications could be made based on the empirical analysis that we performed.

Concluding Remarks

This chapter provides a useful framework from which to study IT's ability to both faciliate innovation processes and control work processes in a single work group. First, the results point to the ways in which IT can help individuals channel their efforts so that the results are more consistent with the work group organization. Second, the study suggests the perceived impact of IT usage on a work group's ability to innovate. Third, we suggest that IT can both increase the control of, and innovate behavior of, a work group by augmenting human information processing ability.

The study should, however, be regarded with the following limitations in mind. The basis of our analysis was the perceptions of the users. In regard to control dimensions, it is necessary to go beyond users' perception of IT's impact on control by relating perceptions to actual measures of control. Likewise, users' perception of innovation should be related to behavioral anchors of innovation. Further, more rigorous research design is warranted. A research design incorporating both pre- and post-test measurements over time would give us a better understanding of innovation and control.

There are several potentially important task and organizational characteristics that were not directly dealt with in this investigation. For example, Jermier and Berkes (1979) and Cheng (1983) examined the link between interdependence and coordination in the work group. Their results indicated that interdependence relates to coordination. Eisenhardt (1985) and Ouchi (1979) found that the effectiveness of a control strategy depended on task and organizational characteristics. Therefore, task characteristics need to be studied to allow greater generalization of this study's results. IT can change task characteristics and environmental impacts. For example, if a task is uncertain in nature, IT can reduce uncertainty by providing forecasting data, resulting in a work group perception of low uncertainty. Continued work in IT support of innovation should likewise be broadened. For example, studies of innovation may be a major part of studying effective problem solving. The link between innovative behavior and effective problem solving is unclear.

Implications for managers can be drawn. The data suggests that IT can both control and help innovate work group behavior. Hence, if properly introduced, the perception of "corporate eye" could promote experimentation with tools and freedom to "play" with technology. Second, managers should emphasize ways in which IT might improve task clarification, task feedback and intrinsic motivation. For example, recent project management tools emphasize task clarification and feedback. Increased motivation may be improved by acquiring the right tool for the right person. The relationship between control and innovation is complex. Broader studies are required to ensure effective deployment of IT for increased performance.

References

Adelman, L. Real-Time Support for Decision Analysis in a Group Setting: Another Class of Decision Support Systems. *Interfaces,* 1984, 14(2), 75-83.

Aiken, M., & Alford, R. Community Structure and Innovation: The Case of Urban Renewal. *American Sociological Review,* 1970, 35(4), 650-665.

Allen, T.J. Communication Networks in R&D Laboratories. *R&D Management,* 1970, 1(1), 14-21.

Attewell, P., & Rule, J. Computing and Organizations: What We Know and Don't Know. *Communication of the ACM,* 1984, 27(12), 1184-1192.

Barnett, H. *Innovation*. New York: McGraw-Hill, 1953.

Becker, S.W., & Whisler, T.L. The Innovative Organization: A Selective View of Current Theory and Research. *The Journal of Business*, 1967, 40(4), 511-518.

Bigoness, W.J., & Perreault, W.D., Jr. A Conceptual Paradigm and Approach for the Study of Innovators. *Academy of Management Journal*, 1981, 24(1), 68-82.

Campbell, D.T., & Fiske, D.W. Convergent and Discriminant Validation by the Multitrait-Multimethod Matrix. *Psychological Bulletin*, 1959, 56(2), 81-105.

Campbell, J.P., Bownas, D.A., Peterson, N.G., & Dunnette, M.D. The Measurement of Organizational Effectiveness: A Review of the Relevant Research and Opinion. Report Tr-71-1, *Navy Personnel Research and Development Center*, San Diego, CA., 1974.

Carroll, J. A Note on Departmental Autonomy and Innovation in Medical Schools. *The Journal of Business*, 1967, 40(4), 531-534.

Carter, C.F., & Williams, B.R. *Industry and Technical Progress*. Oxford: Oxford University Press, 1957.

Carter, N.M. Computerization As a Predominant Technology: Its Influence on the Structure of Newspaper Organizations. *Academy of Management Journal*, 1984, 27(2), 247-270.

Cheng, J.L.C. Interdependence and Coordination in Organizations: A Role-System Analysis. *Academy of Management Journal*, 1983, 26(1), 156-162.

Clegg, S. Organization and Control. *Administrative Science Quarterly*, 1981, 26(4), 545-562.

Cooper, R.G. The New Product Process: An Empirically-Based Classification Scheme. *R&D Management*, 1983, 13(1), 1-13.

Crawford, A.B., Jr. Corporate Electronic Mail - A Communication Intensive Application of Information Technology. *MIS Quarterly*, 1982, 6(3), 1-13.

Curry, J.P., Wakefield, D.S., Price, J.L., & Mueller, C.W. On the Causal Ordering of Job Satisfaction and Organizational Commitment. *Academy of Management Journal*, 1986, 29(4), 847-858.

Cyert, R., & March, J. *A Behavioral Theory of the Firm*. Englewood Cliffs: Prentice-Hall, 1963.

Dalton, D.R., Todor, W.D., Spendolint, M.J., Fielding, G.J., & Porter, L.W. Organization Structure and Performance: A Critical Review. *Academy of Management Review*, 1980, 5(1), 49-64.

Damanpour, F., & Evan, W.M. Organizational Innovation and Preference: The Problem and Organizational Lag. *Administrative Science Quarterly*, 1984, 29(3), 392-409.

Ebadi, Y.E., & Utterback, J.M. The Effects of Communication on Technological Innovation. *Management Science*, 1984, 30(5), 572-585.

Eisenhardt, K.M. Control: Organizational Economic Approaches. *Management Science, 1985,*

31(2), 134-149.

Fayol, H. *General and Industrial Management.* London: Sir Isaac Pitman & Sons, 1949.

Foster, L.W., & Flynn, D.M. Management Information Technology: Its Effects on Organizational Form and Function. *MIS Quarterly*, 1984, 8(4), 229-236.

Gerrity, T.P., Jr. The Design of Man-Machine Decision Systems: An Application to Portfolio Management. *Sloan Management Review*, 1971, 12(2), 59-75.

Gerwin, D. Control and Evaluation in the Innovation Process: The Case of Flexible Manufacturing Systems. *IEEE Transactions on Engineering Management*, 1981, EM-28(3), 62-70.

Gilpin, R. Technology, *Economic Growth, and International Competitiveness. Study* Prepared for the Subcommittee on Economic Growth of the Congressional Joint Economic Committee, U.S. Government Printing Office, Washington, D.C., 1975.

Guetzkow, H. The Creative Person in Organizations. In G. A. Steiner (Ed.), *The Creative Organizations*, Chicago: The University of Chicago Press, 1965.

Hage, J., & Aiken, M. *Social Change in Complex Organizations.* New York: Random House, 1970.

Hage, J., & Aiken, M. Program Change and Organizational Properties: A Comparative Analysis. *American Journal of Sociology*, 1967, 72(5), 503-519.

Hage, J., & Dewar, R. Elite Values Versus Organizational Structure in Predicting Innovation. *Administrative Science Quarterly*, 1973, 18(3), 279-290.

Hiltz, S.R., & Turoff, M. The Evolution of User Behavior in a Computerized Conferencing System. *Communication of the ACM*, 1981, 24(11), 739-751.

House, R.J., & Dessler, G. The Path-Goal Theory of Leadership: Some post hoc and a priori Tests. In J.G. Hunt & L.L. Larson (Eds.), *Contingency Approaches to Leadership*, Southern Illinois University, Carbondale: Southern Illinois University Press, 1974, 29-55.

House, R.J., & Mitchell, T.R. Path-Goal Theory of Leadership. *Journal of Contemporary Business*, 1974, 3(4), 81-97.

Howell, J.P., & Dorfman, P.W. Leadership and Substitutes for Leadership Among Professional and Nonprofessional Workers. *Journal of Applied Behavioral Science*, 1986, 22(1), 29-46.

Howell, J.P., & Dorfman, P.W. Substitutes for Leadership: Test of a Construct. *Academy of Management Journal*, 1981, 24(4), 714-728.

Jermier, J.M., & Berkes, L.J. Leader Behavior in a Police Command Bureaucracy: A Closer Look at the Quasi-Military Model. *Administrative Science Quarterly*, 1979, 24(1), 1-23.

Kerr, S. Substitutes for Leadership: Some Implications for Organizational Design. *Organization and Administrative Sciences*, 1977, 8(1), 135-146.

Kerr, S., & Jermier, J.M. Substitutes for Leadership: Their Meaning and Measurement. Organizational Behavior and Human Performance, 1978, 22(3), 375-403.

Kull, D.J. Group Decisions: Can Computers Help? Computer Decisions, 1982, 14(5), 70-82.

Lawrence, P.R., & Lorsch, J.W. Organization and Environment: Managing Differentiation and Integration. Homewood: Richard D. Irwin, 1967.

Lee, S., & Treacy, M.E. The Impact of Information Technology on Control: A Leadership Theory Perspective. *Proceedings of the Eighth International Conference on Information Systems,* Pittsburgh, December 1987, 442-453.

Lee, S., & Treacy, M.E. Information Technology Impacts on Innovation. *R & D Management,* 1988, 18(3), 257-271.

Malone, T.W. *Computer Support for Organizations: Toward an Organizational Science. Management in the 1990s* Working Paper, #1711-85, Sloan School of Management, M.I.T., Cambridge, MA., September 1985.

Malone, T.W. Heuristics for Designing Enjoyable User Interfaces: Lessons from Computer Games. *Proceedings of the CHI Conference on Human Factors in Computer Systems,* Gaithersburg, March 15-17, 1982.

March, J., & Simon, H.A. *Organizations.* New York: Wiley, 1958.

Matteis, R.J. The New Back Office Focuses on Customer Service. *Harvard Business Review,* 1979, 57(2), 146-159.

Miles, R.H., & Petty, M.M. Leader Effectiveness in Small Bureaucracies. *Academy of Management Journal,* 1977, 20(2), 238-250.

Mohr, L.B. Determinants of Innovation in Organizations. *The American Political Science Review,* 1969, 63(1), 111-126.

Montgomery, D.C., & Peck, E.A. *Introduction to Linear Regression Analysis.* New York: Wiley, 1982.

Montgomery, I., & Benbasat, I. Cost/Benefit Analysis of Computer Based Message Systems. *MIS Quarterly,* 1983, 7(1), 1-14.

Mowery, D., & Rosenberg, N. The Influence of Market Demand upon Innovation: A Critical Review of Some Recent Empirical Studies. *Research Policy,* 1979, 8(2), 102-153.

Moynihan, T. Information Systems As Aids to Achieving Organizational Integration. *Information and Management,* 1982, 5, 225-229.

Myers, R.H. *Classical and Modern Regression with Applications.* Boston: Duxbury, 1986.

Ouchi, W.G. A Conceptual Framework for the Design of Organization Control Mechanisms. *Management Science,* 1979, 25(9), 833-848.

Palumbo, D.J. Power and Role Specificity in Organization Theory. *Public Affairs Review,* 1969, 29, 237-248.

Pelz, D.C., & Andrews, F.M. *Scientists in Organizations.* New York: Wiley, 1966.

Perrow, C. A Framework for the Comparative Analysis of Organizations. *American Sociological Review,* 1967, 32(2), 194-208.

Pfeffer, J., & Leblebici, H. Information Technology and Organizational Structure. *Pacific Sociological Review,* 1977, 20(2), 241-261.

Pierce, J.L., & Delbecq, A.L. Organization Structure, Individual Attitudes, and Innovation. *Academy of Management Review,* 1977, 2(1), 27-37.

Reeves, T.K., & Woodward, J. The study of Managerial Control. In J. Woodward (Ed.), *Industrial Organizations: Behavior and Control,* London:Oxford University Press, 1970, 37-56.

Reif, W.E. Computer Technology and Management organization. *Bureau of Business and Economic Research.* Iowa: University of Iowa, 1968.

Robey, D. Information Systems and Organizational Change: A Comparative Case Study. *Systems, Objectives, and Solutions,* 1983, 3, 143-154.

Rogers, E. Diffusion of Innovation. New York: Free Press, 1962.

Rogers, E., & Schoemaker, F. *Communication of Innovations, A Cross-Cultural Approach.* New York: Free Press, 1971.

Salancik, G.R. Commitment and the Control of Organizational Behavior and Belief. In B.M. Staw & G.R. Salancik (Eds.), *New Directions in Organizational Behavior,* Chicago: St. Clair, 1977, 1-54.

Saunders, C.S. Management Information Systems, Communications, and Departmental Power: An Integrative Model. *Academy of Management Review,* 1981, 6(3), 431-442.

Schriesheim, C.A. *Development, Validation, and Application of New Leadership Behavior and Expectancy Research Instruments.* Doctoral Dissertation, College of Administrative Science, Ohio State University, Columbus, OH., 1978.

Schriesheim, C.A., House, R.J., & Kerr, S. Leadership Initiating Structure: A Reconciliation of Discrepant Research Results and Some Empirical Tests. *Organizational Behavior and Performance,* 1976, 15(2), 297-321.

Schriesheim, C.A., & von Glinow, M.A. The Path-Goal Theory of Leadership: A Theoretical and Empirical Analysis. *Academy of Management Journal,* 1977, 20(3), 398-405.

Slovic, P. *Toward Understanding and Improving Decisions. Science, Technology, and the Modern Navy: Thirtieth Anniversary,* 1946-1976. Office of Naval Research, Arlington, VA., 1976.

Souder, W.F. Effectiveness of Nominal and Interacting Group Decision Processes for Interacting R&D and Marketing. *Management Science,* 1977, 23(6), 595-605.

Steiner, G.A. *The Creative Organizations.* Chicago: University of Chicago, 1965.

Taylor, F.W. *The Principles of Scientific Management.* New York: Harper & Row, 1911.

Thompson, J.D. *Organizations in Action.* New York: McGraw-Hill, 1967.

Thompson, V.A. Bureaucracy and Innovation. *Administrative Science Quarterly,* 1965, 10, 1-20.

Tversky, A., & Kahneman, D. Judgment under Uncertainty: Heuristics and Biases. *Science,* 1974, 185, 1124-1131.

Utterback, J.M. Innovation in Industry and the Diffusion of Technology. *Science,* 1974, 183, 620-626.

Van de Ven, A.H., Delbecq, A.L., & Koenig, R. Determinants of Coordination Modes within Organizations. *American Sociological Review,* 1976, 41(2), 322-338.

Von Hippel, E. *Novel Product Concepts from Lead Users: Segmenting Users by Experience.* Sloan School of Management Working Paper #1476-83, December 1983.

Weber, M. *The Theory of Social and Economic Organization.* Glencoe: Free, 1947.

Zaltman, G.R., Duncan, R., & Holbek, J. *Innovation and Organizations.* New York: Wiley, 1973.

Zuboff, S. *Some Implications of Information Systems Power for the Role of the Middle Manager.* HBS 84-29, Harvard Business School, Boston, MA., May 1983.

Chapter 10

Information Technology and Group Effectiveness:

Managerial and Research Implications of Introducing GDSS in New Situations

Holly R. Rudolph
Southern Illinois University

John R. Schermerhorn, Jr.
Ohio University

The computer's effect on the decision making process of organizations has been well documented empirically in organizational literature (Joyner & Tunstall, 1970; Benbasat & Schroeder, 1977; King & Rodriguez, 1978; Chakravarti, Mitchell, & Staelin, 1979; Benbasat & Dexter, 1982; McIntyre, 1982; Dickmeyer, 1983; Eckel, 1983; Aldag & Power, 1986; Goslar, Green, & Hughes, 1986; Cats-Baril & Huber, 1987; Sharda, Barr, & McDonnell, 1988). With its capabilities in the area of organizing, synthesizing, analyzing, and retrieving information and data, many managers now believe that both efficiency and effectiveness of the decision making process in organizations can be enhanced through computer mediation. Researchers, however, have been careful to point out that computer mediation may also interject new problems such as a de-emphasis in social exchange and an increase in overt expressive behavior by participants during decision making(Siegel, Dubrovsky, Kiesler, & McGuire, 1986).

As advocated by Bonczek and his colleagues (1979), the primary impetus for interest in computerized decision assistance has come from the data-base field and not from any further understanding of the decision making process *per se*. This impetus applies to both individual and group decision making situations. Given the growing importance of group-oriented management approaches, it is the purpose of this paper to investigate specifically the behavioral and social effects of Group Decision Support Systems (GDSS) as they relate to the group decision making process and group effectiveness in organizations. In particular, we are interested in the research and managerial implications of introducing GDSS in new situations.

Background and Issues

Gerrity's (1971) original concept of decision support systems (DSS) refers to an interaction of human intelligence, information technology, and software to solve complex organizational problems. Since 1971, researchers have offered their own various definitions of DSS but there seems to be general consensus that DSS involves an interactive computer-based system which facilitates solution of unstructured and non-routinized problems (Vazsonyi, 1978; Bonczek, Holsapple, & Whinston, 1979; Neuman & Hadass, 1980; Sprague, 1980).

Building on this definition of DSS, DeSanctis and Gallupe (1985; 1987) have posited a foundation for applying DSS to groups. According to these researchers, a group decision support system (GDSS) is "an interactive computer-based system which facilitates solution of unstructured problems by a set of decision makers working together as a group" (DeSanctis & Gallupe, 1985, p.3). The foundation work by DeSanctis and Gallupe (1987) recognizes four distinct GDSS environments — each of which is an attempt to provide a group with the correct configuration of computerized decision support. These computer-assisted group environments are the decision room, the legislative session, the local area decision network, and the computer-mediated conference.

The decision room or electronic boardroom. This configuration of decision support is generally used with small groups to aid and facilitate the decision making process (DeSanctis & Gallupe, 1985; Kull, 1982). Interaction among group members can occur verbally or through computer mediation. A large screen can be used to display group decisions, ideas, comments, voting results, etc. A group facilitator is used in most situations to coordinate group efforts. Such an environment could have practical appeal for an organization board of director's meeting or a task force meeting.

The legislative session. This configuration of decision support is used with large groups to coordinate the decision making process during a common session (DeSanctis & Gallupe, 1987). Due to the large number of group members, group interaction via the computer is somewhat limited and a group facilitator is necessary in this configuration to coordinate group efforts and display data on a large screen. Rules and group norms may also need to be established concerning the chain of communication so the decision process does not become ineffective. Computer intervention can be used to determine an orderly progression of speakers and even to activate microphones of speakers (Stodolsky, 1981). Stockholders meetings could conceivably be conducted in this type of group environment.

Local area decision network. This configuration can be used to link physically or geographically dispersed group members together in order to facilitate decision making (DeSanctis & Gallupe, 1987). Such a network can be used to link group members from various offices and dispersed locations or even to link small groups together who are geographically dispersed. Therefore, networks can be used to facilitate communication and decision making among individuals and small groups who find it inconvenient or impossible to meet face-to-face. A company's sales force could possibly be linked by this type of decision network.

Computer mediated conference. This type of configuration can be used to link large, geographically dispersed groups together via long distance telecommunications (DeSanctis & Gallupe, 1987). As mentioned in conjunction with the Legislative Session, large groups present certain problems to the computer mediated decision process. In order to prevent chaos in such a session, certain decision rules and norms must be established. Each group needs a facilitator to coordinate communication and decision making within his group, and the overall conference needs a facilitator to keep the conference moving and effective. Even though it is conceivable to conduct large conferences in this manner, it seems unlikely that the Democratic or Republican National Convention is going to be held by teleconference in the very near future.

Moving beyond the prior sketch of basic applications, Siegel and his colleagues (1986) have categorized research on group processes in computer-mediated communication into four categories: (1) technology assessment studies, (2) organizational studies, (3) technical capabilities studies, and (4) social psychological studies. The framework used in the present analysis incorporates aspects of organizational studies (group 2 on Siegel's list) and social psychological studies (group 4) to develop an open systems model of the group decision making process as it relates specifically to GDSS.

The model in Figure 1 depicts the work group as an open system where group processes interact with various inputs or stimuli to produce group effectiveness. This model will be used to explain the major tenants of group decision support systems (GDSS); to review the relevant decision support literature; and to tie together what is known about GDSS in proposition form.

As suggested by the figure, any group can be viewed as an open system that transforms a variety of inputs into outputs. An effective group is one that achieves high performance (that is, accomplishes its task objectives) and creates an environment in which members experience personal satisfaction (that is, maintains its social fabric for long-term performance potential).

With a specific focus on the group throughput of decision making, the figure can be applied to the situation of introducing a GDSS with the objective of improving group decision making capabilities. The GDSS in this instance can be used to support the overall decision making process by changing the way group members work together with various input factors and/or changing the way they process inputs into outputs.

In terms of input factors, the GDSS implications apply to such issues as:

- the nature of the decision
- the types and demands of the decision task
- group size
- group member proximity

In terms of the throughput factors, the GDSS implications apply to such issues as:

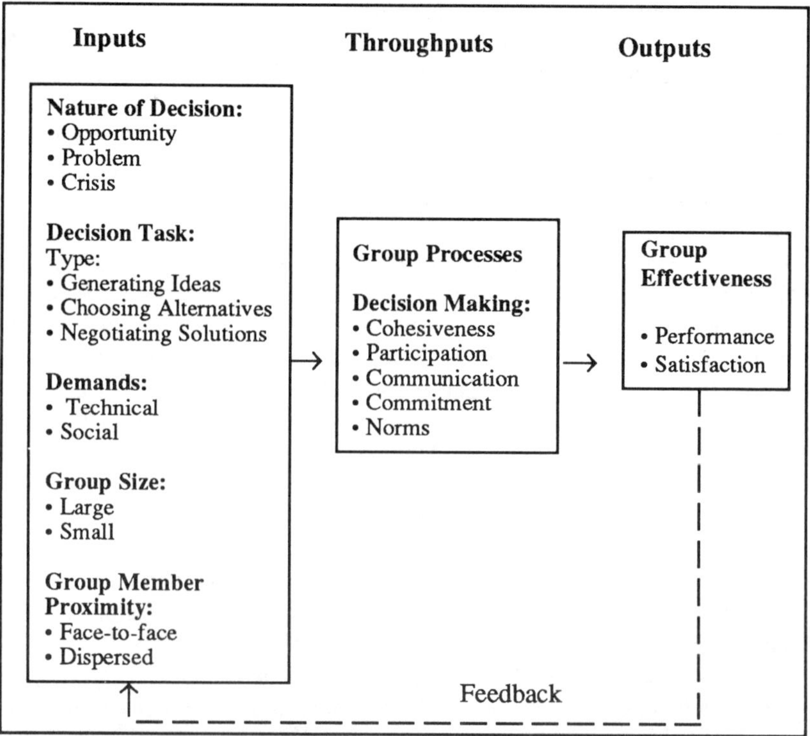

Figure 1. An Open System's Approach to Group Decision Support Systems

- group cohesiveness
- participation within the group
- communication channels
- decision commitment
- group norms

Of specific interest to this paper is how the GDSS "fits" when introduced into a new group situation. Does it fit positively and in such a way that these input and throughput factors can have an enhanced impact on group effectiviness? Or, is it a "forced" fit that leads to input and/or throughput problems that may actually reduce group effectiveness? Even more directly stated, the managerial question of most significance is: "What must and/or can be done so that the introduction of a new GDSS enhances the positive influence of important input and throughput factors on group effectiveness?

GDSS Utilization Suggestions

Maximizing Group Inputs

Here, we begin to apply the open systems model of groups to develop ideas on best utilizing GDSS in organizations. These ideas are first presented relative to maximizing the positive impact of group input factors on group effectiveness.

GDSS and Type of Decision

Many organizational decisions are routine as well as frequent in nature. Examples include hiring decisions, advertising decisions, sales-mix decisions, or dividend decisions. Due to the repetitive nature of these types of decisions, solutions are fairly structured and learning from past mistakes aids the decision maker or decision-making group in reaching an optimal solution. In such decisions, GDSS avails the decision process very little (Neumann & Hadass, 1980; Bonczek, et. al, 1979).

The greatest benefits of GDSS accrue from applying various group decision support aids to decisions that are made infrequently and where their use alters normal un-supported group decision making processes. It follows, therefore, that computer mediation and group decision support aids enhance group effectiveness most when the decision is unstructured and nonroutine (Siegel, et. al, 1986; DeSanctis & Gallupe, 1987; United Communications Group, 1987). These types of decisions fall on the continuum depicted in Figure 2 and include: (1) opportunity decisions, (2) crisis decisions, or (3) problem decisions (Mintzberg, Raisinghani, & Theoret, 1976; Shangraw, 1986). The following propositions are offered concerning the use of GDSS by groups

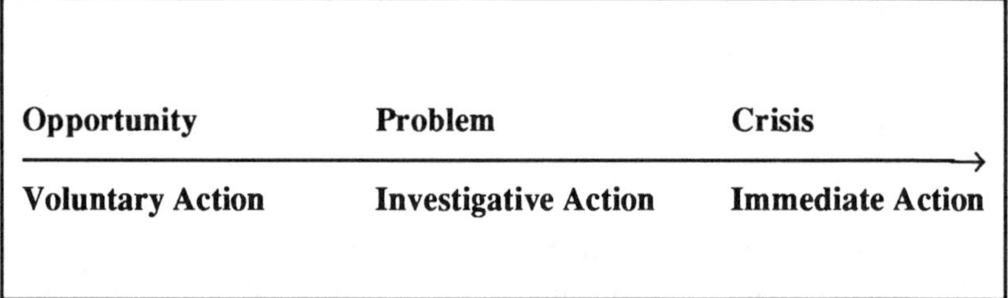

Figure 2 Nature of Decision and Required Group Action

facing these three different types of problems.

> P1: The impact of a group decision support system on group effectiveness is greater for unstructured and/or non-routine decisions than for structured and/or routine decisions.

> P2: The type of decision faced by a group —opportunity, problem, or crisis decision — can influence both the design and effectiveness of a group decision support system.

Opportunity decisions call for voluntary action by the manager or group since they are initiated only to improve an already satisfactory situation (Mintzberg, et al., 1976). Generally the result of a single stimulus, opportunity decisions include adding a new product line, expanding into new markets, acquisitions, and certain investment decisions. Often, these types of decisions involve discrete information calling for a high degree of information security. The parties involved in such decision processes may very well be dispersed geographically rather than in face-to-face communication with each other.

Due to the critical nature of such decisions, timeliness of forthcoming solutions, and lack of proximity of group members, GDSS can play an important role in facilitating the group decision making process. Success often depends on computer support through evaluation and consensus gathering techniques. The modeling capabilities of new "groupware" facilitates the evaluation process by providing the group with pro-forma data on various scenarios (Richman, 1987). Computer support also facilitates the decision process by providing both intergroup and intragroup communication channels, as well as procedures that can assure anonymity of group members' positions and security of sensitive information.

Crisis decisions call for immediate action by the group since they are initiated as a result of a severe situation (Mintzberg, et al., 1976). The result of a single stimulus, crisis decisions include strikes by labor unions, tragedies involving manufacturing facilities, terrorist activities, hostile takeovers, expropriation by foreign governments, or financial crises such as the recent savings and loan failures. Again, these types of decisions may involve discrete information calling for a high degree of information security. The situation is further complicated by the fact that parties involved in crisis decisions may be dispersed geographically and the critical nature of the problem generally calls for prompt action by organizations, and therefore, very timely decision making techniques.

Organizations may utilize GDSS in crisis situations to assist crisis decision groups or teams (Huber, 1984). Experts can be used to build and test crisis scenarios and teams can be trained in evaluating proposed solutions to crisis situations. A danger in any crisis is the possible tendency of "groupthink" to occur and limit the group to examining too few alternatives (Janis, 1977). GDSS can be utilized to improve an organization's chances of arriving at effective solutions to critical situations by providing a means to simulate responses to such events and broaden the review of alternatives.

One of the advantages of GDSS's for organizations rests in their ability to link physically dispersed decision makers together when a group consensus is called for. However, some managers may feel that a GDSS approach takes too much valuable time in cases where extreme urgency is called for in the decision making process. This may not be a valid criticism of GDSS, rather, this may be a question of whether an individual or a group decision is best in a crisis situation. A GDSS still appears to be an efficient means of connecting dispersed group members together, providing appropriate support aides, and assuring security of information when a group consensus is called for and time is sufficient. Due to the proliferation of crisis decisions in the public sector, considerable research has focused on the use of computerized support systems in public administration (Kraemer & King, 1986; Shangraw, 1986; Rubin, 1986; Hurley & Wallace, 1986; McGowan & Lombardo, 1986).

Problem decisions arise due to multiple stimuli resulting from pressures from many sources and involve investigative group action. Examples include problems with suppliers and customers, labor management disputes that haven't reached the crisis stage, or cost-overruns or production problems that haven't reached a critical point. Groups are often brought together to settle such problems before they reach the crisis stage.

GDSS mediation of problem decisions can aid by linking various dispersed groups and by enhancing the decision process with various decision aids such as modeling techniques, spreadsheet analysis, or graphical representations. By providing faster, more efficient, information retrieval, computer support can save groups both financial and human resources when evaluating potential problems as well as when resolving problems of a more immediate nature.

GDSS and the Decision Task

Another variable shown as a stimulus to the group decision process in Figure 1 is the type of decision task. Hackman and Morris (1975) explain that tasks have been used as moderators of group processes and performance, or as stimuli in most empirical studies. Poole, Seibold, and McPhee (1985) report the importance of task type since it often accounts for as much as 50% of the variance in group performance. Numerous task typologies have been offered by researchers over the last several years and are summarized in Table 1.

DeSanctis and Gallupe (1987) used McGrath's (1984) typology in their recent foundation work on GDSS. According to McGrath (1984), tasks can be defined as what must be accomplished during a group meeting: (1) generating ideas and actions; (2) choosing among alternatives; or (3) negotiating solutions. A fourth dimension, execution, was apparently omitted from the model because it deals with execution of manual tasks. Perhaps a stronger task concept for purposes of GDSS is a framework incorporating both McGrath's (1984) typology and Herold's (1978) typology.

Herold (1978) categorizes tasks according to technical demands (availability of solutions or data needed to respond adequately to task demands) and social demands (amount of social interaction among group members). Herold's (1978) typology is diagnostic in nature in that the objective is to pinpoint the type of task demand (technical or social) required and apply an appropriate intervention strategy that will increase performance. When inter-

Task Typology/Year	Task Dimensions
Carter, Haythorn, & Howell (1950)	Clerical, Discussion, Intellectual Construction, Mechanical Assembly, Motor Coordination, and Reasoning
Bass, Pryer, Gaier, & Flint (1958)	Easy/Difficult
Thibaut & Kelley (1959)	Steady/Variable, Conjunctive/Disjunctive, Correspondence/Noncorrespondence
Hackman (1968)	Production/Discussion/Problem-solving
Cohen (1968)	Sensor/Control/Effector
Steiner (1972)	Unitary/Divisible, Maximizing/Optimizing, Conjunctive/Disjunctive/Additive
Shaw (1976)	Simple/Complex, Intellective/Manipulative, Solution Multiplicity/Specificity, Intrinsic Interest, Population, Familiarity; and Cooperation Requirements
Davis, Laughlin & Komorita (1976) **Laughlin** (1980)	Cooperative/Competitive/ Mixed Motive
Herold (1978)	Technical Complexity/Social Complexity
Tushman (1979)	Routine/Environmental Uncertainty/Interdependence
McGrath (1984)	Generating/Choosing/Negotiating/Executing

Table 1. Task Typologies

vention strategies are correctly matched with task demands, overall group effectiveness is improved.

By using Harold's task typology, the two most apparent consequences of GDSS—generation of data and intervention into the social processes of the group—can be evaluated. When combining Herolds' and McGraths' typologies, as in Figure 3, each of the three types of group decision tasks (generating ideas, choosing alternatives, or negotiating solutions) can be evaluated according to its technical and social demands. The intervention strategy to be applied by the organization consists of the appropriate GDSS environment and group decision aids consistent with task demands. As will become apparent in later sections of this paper, GDSS often provide technical capability at the expense of group social interaction. It is therefore important for management to assess both technical and social demands of a particular decision prior to recommending computer intervention. This review concerning group task requirements leads us to offer the following propositions:

> P3: Type of decision task — generating ideas, choosing alternatives, and negotiating solutions — impacts both the design and effectiveness of a group decision support system.
>
> P4: Task demands — technical and/or social — interact with type of decision task to determine the appropriate level of group decision support.
>
> P5: There is a direct relationship between technical task demands and level of group decision support and an inverse relationship between social task demands and level of group decision support.

GDSS and Group Size

As suggested in our open systems model of groups, group size also interacts with the nature of the decision and the decision task to affect group behavior. Most researchers agree that the effects of group size on performance depends on the type of group task (Steiner, 1972; Shaw, 1976). Larger groups seem to be better when any one member can solve the problem or make the decision (disjunctive tasks) and smaller groups are superior when each group member must solve the problem in order for the task to be completed (conjunctive tasks) (Steiner, 1972). Intuitively, larger groups appear to offer organizations some real advantages—increased pool of talent, knowledge, creativity, and diversity of viewpoints. These advantages should enhance group decision making. However, larger groups are not without their problems—as group size increases, participation of group members often decreases and true group consensus may become harder to achieve.

Undoubtedly, group size should be considered when designing a group decision support system. Since the features of computerized decision

Task Types	Task Demands
Generating Ideas & Actions	Technical / Social
Choosing Alternatives	Technical / Social
Negotiating Solutions	Technical / Social

Figure 3. Types and Demands of Decision Task

support are varied, organizations should choose the appropriate decision aids and GDSS environment consistent with the size of their group. GDSS may even have a greater effect on larger groups than on smaller groups due to the many communication problems inherent in larger groups. However, small groups may feel more comfortable making decisions in an atmosphere of anonymity which can be easily created through the use of GDSS.

Various GDSS environments have been posited by DeSanctis and Gallupe (1987) including the decision room, the legislative session, decision networks, and teleconferencing. These group configurations will be discussed later in this paper. Based on both group dynamic literature and the available GDSS literature, the following proposition concerning group size is offered:

> P6: Group size is an important determinant of the GDSS environment configuration and the appropriate decision aids to be used by the group.

GDSS and Group Member Proximity

Proximity of group members is also an important stimulus to group decision making in organizations. Though many definitions of proximity have been offered, Monge and his colleagues defined it as "the extent to which people in an organization share the same physical locations at the same time providing an opportunity or psychological obligation to engage in face-to-face communication" (Monge, Rothman, Eisenberg, Miller, & Kirste, 1985, p. 1133). Within an organizational setting, face-to-face group involvement is often desired but not always practical while on occasion dispersed group participation may be superior to face-to-face group participation.

Group decision support system features have been developed to enhance both of these types of group interactions. For example, computer intervention can provide dispersed group members an efficient means of communicating with other group members; can quickly analyze and feedback results to group members; and can even provide video capability allowing group members to see each other . Computer intervention can also support face-to-face meetings by allowing group members a means of orderly and efficient communication, by providing modeling and graphical techniques, and by expediting the tallying of votes. However, organizations should also consider the effect of computerized decision support on face-to-face groups on "perceived" proximity. Computers often have a distancing or isolating effect on face-to-face group members and the opposite effect on dispersed group members (DeSanctis & Gallupe, 1987). This effect can serve either to depersonalize or personalize the group environment (Kull, 1982). The following propositions are suggested concerning group member proximity:

> P7: Group member proximity greatly impacts the type of GDSS environment configuration used by the group.

> P8: There is a "perceived" distancing effect between group members when computerized decision support is used in face-to-face group meetings and a "perceived" converging effect when computerized decision support is used in dispersed group meetings.

GDSS Utilization Suggestions

Maximizing Group Throughputs

At the second level of analysis, the open systems model of groups can be used to develop ideas on maximizing the positive impact of GDSS on group throughput factors. In particular we can develop insights into how the introduction of GDSS to the group process can facilitate the transformation of inputs into outputs. Since decision making involves the interaction of stimuli on various group processes as originally depicted in Figure 1, this section of the paper discusses the effect of computer intervention on cohesion, participation and communication, commitment, and norms.

GDSS and Group Cohesiveness

Defined by Lott and Lott (1961), cohesiveness is "that group property which is inferred from number and strength of multiple positive attitudes among the members of a group" (p. 408). As mentioned earlier, computer intervention into the decision making process affects "perceived" proximity of group members which results in increasing or decreasing positive attitudes among group members. Since highly cohesive groups in organizations are often tagged "dysfunctional" and susceptible to "groupthink" (Janis, 1977), the effect of cohesiveness on the design and impact of group decision support systems need to be considered. According to DeSanctis & Gallupe (1987), perceived proximity and a sense of cohesiveness can either be increased or decreased through computer intervention. Physically dispersed groups can be brought together electronically or face-to-face groups can be somewhat distanced through computer intervention. Conceivably, managers could enhance the decision making capabilities of various work groups by adjusting the level and type of decision support depending on the amount of group cohesiveness desired.

Richman (1987) suggests that computer intervention can actually bring cohesiveness to meetings. By using a decision room atmosphere, group members can use their keyboards to generate ideas about a project which automatically appear on other participants' screens. Ideas can then be structured into related categories by group members. Finally, the group can work together to evaluate it's proposals. There is, however, some research that indicates that group members are more likely to attack ideas of other group members when communicated electronically than when exchanged in face-to-face conversation causing conflict among group members (Siegel, et al., 1986). The following proposition relates to group cohesion:

> P9: Computer intervention through a GDSS can be used to influence perceived proximity among group members which, in turn, will affect group cohesion.

GDSS and Group Member Participation

Participation refers to the distribution of communication within a group (Siegel, et al., 1986). Generally speaking, higher status group members often maintain higher group participation rates than do lower status group members (DeSanctis & Gallupe, 1987; Siegel, et al., 1986). This results in unequal participation rates among group members. But what happens when status-bearing cues are removed? The electronic element in GDSS can remove impressive titles, one's position in the organization chart, one's usual seat at the head of the board table, and even eye contact during discussions. By removing status symbols such as these, participation rates should become more equal among group members. In a more formal manner, the following proposition is posited for group participation:

> P10: When using computer-mediated communication to make decisions, group members will participate more equally in group discussions than will group members engaged in unsupported face-to-face group discussions.

Siegel and his colleagues (1986) used three groups (face-to-face, computer used anonymously, and computer used nonanonymously) to empirically test the above proposition on a task requiring group consensus. Group participation equality was measured by each group's standard deviation of individual remarks. As predicted, face-face groups showed unequal participation (larger standard deviations) while computer-mediated groups showed more equalized participation rates (lower standard deviations).

Furthermore, one could contend that if participation is not dominated by one particular individual or group faction but is shared equally by all group members, true group consensus is more likely to emerge. One of the major advantages of group decision making over individual decision making is the diversity of viewpoints available in a group setting. Computer intervention in the decision making process allows more of these viewpoints to be communicated during the group meeting. This leads to the following proposition:

> P11: The effects of GDSS on group member participation has a moderating effect on the group decision making process.

GDSS and Group Communication

In addition to equalization of participation rates among group members, computer mediation of the group decision process can also alter the communication among group members in other ways. Face-to-face groups have been found to be more argumentive than computer mediated groups suggesting that computer mediated communication limits the amount of group verbal exchange and exchange of social cues (McGuire, Kiesler, & Siegel, 1987). Research has revealed both advantages and disadvantages of computer mediated communication. For example, increased speed and flexibility of computer communication and an emphasis on only written words lead to "task" centered communication and decision emphasis (Siegel, et. al, 1986). However, some research indicates that a lack of nonverbal information such as an exchange of social cues can interfere with the exchange of feelings (Carnevale, Pruitt, & Seilheimer, 1981).

The following propositions are offered concerning computer-mediated communication and the decision making process of groups:

> P12: By equalizing participation and encouraging different viewpoints, groups using computer mediated communication to make decisions may require more time to reach consensus than groups involved in only face-to-face unsupported communication.

> P13: In computer mediated groups, a higher proportion of the

discussion will be task centered than social centered.

P14: Negative social communication will dominate positive social exchanges in computer mediated groups.

Siegel and his colleagues (1986) found that communication efficiency (based on number of communication exchanges and length of time to reach consensus) was lower in computer-mediated groups than in face-to-face groups. However, communication efficiency based on the number of actual decision proposals delivered per communication exchange was greater in computer mediated groups than in face to face unsupported groups. The researchers also suggested that group members tend to be more critical of each other's ideas using computer mediated communication than face-to-face communication which lends support to P14.

GDSS and Group Commitment to Decision

Over the last several years, both empirical and conceptual research has led us to believe the computer is more precise, more credible, more relevant, and more timely than conventional printed information (Shangraw, 1986). It follows that users of computer mediated information will be very confident of their decisions due to the "halo" effect overshadowing computer technology. It has been suggested that groups using computer mediated information for decision making may be more committed to their decisions than groups using face-to-face unsupported decision making processes (Shangraw, 1986; McGuire, et. al, 1987). Little support, however, has been found for this proposition. Shangraw (1986) found only limited support when comparing hard copy information and computer generated information to decision commitment. McGuire et.al (1987) found that groups using computer intervention compared to face-to-face interaction were equally convinced and confident of their decision choices. The following proposition relates to group decision commitment:

P15: Group members are equally committed to their decisions whether the decision process involves computer mediation of information or unsupported face-to-face communication processes.

GDSS and Group Norms

Group norms have been found to possess the following major characteristics (Hackman, 1976):

(1) Norms are structural characteristics of the group which summarize and simplify group influence processes.

(2) Norms apply only to behavior—not to private thoughts and feelings.

(3) Norms generally are developed only for behaviors which are viewed as important by most group members.

(4) Norms usually develop gradually, but the process can be shortened if members want.

(5) Not all norms apply to everyone.

Face-to-face decision making groups generally adhere to a set of group norms in order to make the decision making process more efficient. However, established norms for computer mediated decision groups have been slow to take shape. This has led to problems such as aversive behavior on the part of some group members (Siegel, et.al, 1986).

Most group decision support systems include an element which facilitates the orderly flow of group meetings. However, the anonymity feature of GDSS may encourage rather than discourage affective behavior. The following proposition states the effect of established norms on groups using computerized decision support systems.

P16: Due to the anonymity feature of most GDSS systems, groups using computerized support are less influenced by norms than are face-to-face unsupported groups.

GDSS Technology Utilization and Group Effectiveness

As has been noted in the previous section of this paper, the stimuli to the group decision making process—nature of the decision, type of decision task, group size, and group member proximity—all interact with various group processes to influence group effectiveness. Intervention of the appropriate decision support aids into the group decision environment is contingent on the level of each of these stimuli. Since the impact of GDSS on group effectiveness depends in part on the ability of the computerized support to change normal unsupported behavior, management should periodically assess the needs of each decision making group. For example, providing computerized vote tallying capability to a small group will probably have no effect on their behavior, but providing computerized support to a large group session that has the capability of facilitating communication channels and minimizing group conflict will greatly affect group behavior and effectiveness. As has been noted, a group working with a decision task that requires a high degree of social interaction among group members rather than technical support will probably respond negatively to computer support, while a group working with a very difficult task that requires a high degree of technical support will most likely respond favorably to computerized support. Figure 4 summarizes the four input or stimuli factors of the decision making process discussed in this paper in a contingency model.

GDSS and Group Performance

Since 1971, empirical studies investigating the effectiveness of decision support have primarily dealt with individual decision making. From this reference point, the effects of DSS on effectiveness are equivocal at best. Results have ranged from increasing organization profits (Eckel, 1983; McIntyre, 1982; Benbasat & Dexter, 1982); to decreasing organization profits (Chakravarti et al., 1979); to increasing decision time (Benbasat & Dexter, 1982; Benbasat & Schroeder, 1977). Cats-Baril and Huber (1987) even found improved decision quality on a career planning task with the interaction of a decision aid, whether or not the decision aid was computerized.

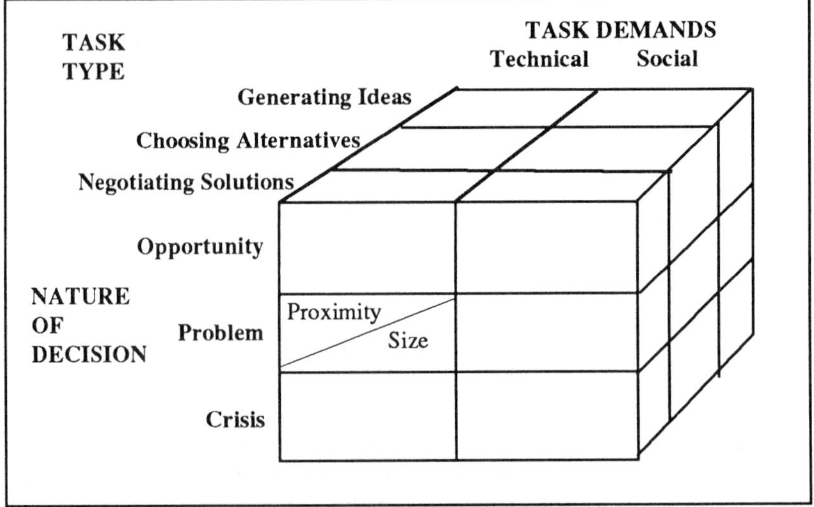

Figure 4. A Contingency Model for Evaluating GDSS

Few empirical studies have used the group as their unit of analysis when researching the effect of decision support on performance (Joyner & Tunstall, 1970; Sharda, et al., 1988). Sharda and his colleagues (1988) conducted an empirical study to assess the effects of decision support on group effectiveness. The task encompassed multiple decisions involving the purchase or sale of securities, investments in plant and equipment, and a series of decisions involving different product lines. In a laboratory setting, student participants performed the task in small groups of three. The researchers found significant improvement in performance (measured by net profit) of the computer assisted groups over the unassisted groups. In addition, the researchers also assessed performance in terms of volatility or variance of performance (net earnings). This measure of performance was included since the researchers hypothesized that the intervention of decision support into the group process should decrease uncertainty thus decreasing variability of performance among group decision makers. Again, significant improvement in variability of performance was noted in the computer supported group over the unsupported group. No overall differences in average time spent in decision making, the number of alternatives examined, or in confidence in the decision were noted between the computer supported and the unsupported group.

From an intuitive viewpoint, it is understandable why organizations might feel that computer intervention in the decision process would enhance group performance. After all, computers have massive data generation and analysis capability, as well as the capability to organize and synthesize group ideas. Modeling techniques can be used to enhance the group decision making process in several ways: (1) by keeping the group focused on what is actually part of the decision problem and what is not which facilitates group discussion; (2) by imposing discipline on the analytical process; and (3) by improving skills of decision makers (Kraemer & King, 1986). As discussed earlier, computer intervention facilitates communication channels among dispersed group members; equalizes participation rates among group members; and can even be used to manipulate group cohesion. By accentuating the positive group effects, computer intervention should enhance the group decision making process and therefore, increase both group and organizational effectiveness.

Maier (1967) has addressed group effectiveness from a "group asset—group liability" approach. One way to consider the effect of GDSS on group effectiveness is to consider their impact on each of these issues. Simply stated, by increasing group assets and/or decreasing group liabilities, overall group effectiveness should be enhanced by the GDSS. When it comes to the relationship between GDSS and the assets of group decision making, the following considerations apply.

1. **GDSS can contribute to a greater sum total of knowledge and**

information. Most managers would agree that there is more knowledge available in a group than there is in any one individual group member. The problem arises in tapping this great wealth of information. Computer intervention can aid in forcing group participants to focus on the issues and the decision task at hand (Kull, 1982). Computer technology can also expand the group knowledge base by providing group members access to various data bases and analytical tools that they might not otherwise use. It is conceivable that this exposure will sharpen their own analytical and decision making abilities.

2. **GDSS can provide greater number of approaches to a problem.** As pointed out by Maier (1967), individuals often get into ruts in their thinking. This problem is also applicable to groups, but at least the group provides a mechanism for considering different alternatives to a problem. The computer can enhance this group advantage by providing group members an environment in which to explore their own individual approaches to problem solving. One of the major advantages researchers feel a decision support system provides the decision process, is an increase in decision quality by allowing more alternatives to be considered (Alter, 1980; Cats-Baril & Huber, 1987; Sharda, et al., 1988).

3. **GDSS can stimulate participation in problem solving and increase acceptance.** By providing a means for individuals to personally participate in the decision making process, GDSS can enhance the likelihood of acceptance of the solution by individual group members. This, in turn, may result in group members accepting more responsibility for making the solutions work. As has already been established, computer intervention into the decision process tends to equalize participation rates among group members by removing status symbols. A group decision support system can, therefore, be used as a vehicle for management to stimulate participation among group members who might be very creative but who contribute very little in the normal group environment due to social pressures.

4. **GDSS can provide for better comprehension of the decision.** Needless to say, decisions are oftentimes better comprehended by those who make them than by those who implement them. Group decision making processes can be used to help solve this problem. The electronic component in GDSS can be used to assemble various dispersed group members and thus, facilitate a group decision that might otherwise be an individual decision. The support element of GDSS can be utilized to allow group members to witness first-hand the actual decision making process, resulting in a greater appreciation for both thought and time that goes into a managerial decision. When it comes to the relationship between GDSS and the assets of group decision making, the following considerations apply.

GDSS may buffer social pressure to conform. Certainly one disadvantage of group decision making is the tendency for group members to favor consensus. By giving in to social pressure, true group consensus is rarely reflected in the decision. Computer intervention through a GDSS reduces this group pressure by removing many social cues from the group environment. Therefore, one of the major barriers to achieving true group effectiveness is eliminated.

GDSS may reduce limits to the valence of possible solutions. Maier (1967) used an index to assess the maximum number of discussion points allowed by group members prior to calling for group consensus. Solutions introduced after this point, receive little to no attention. Since a GDSS allows group members to store in computer memory relevant information about each alternative, this inhibiting group factor may be lessened.

GDSS may reduce individual domination. As has been mentioned earlier, one of the major advantages of a GDSS is its ability to equalize participation rates among group members. Dominant individual influence is, therefore, curtailed as a wide range of group members participate in the decision making process.

GDSS may assist in meeting time requirements. One of the major disadvantages cited for a decision support is that it often takes longer to reach consensus than in normal unsupported face-to-face group settings. It is conceivable, however, that there is a "learning" effect to decision support and that the process can become more efficient as groups become more comfortable and knowledgeable with this type of decision environment. It is also important to point out that decision support systems can save valuable time when group members are geographically dispersed and a group consensus is called for.

GDSS and Group Member Satisfaction

Another major consequence of GDSS can be a decrease in the satisfaction of group members. One reason for this result, is the depersonalization promoted with the intervention of computers into the decision making process. Group participants may feel that the success or failure of a meeting only depends on the state of computer technology rather than on personal contributions of group participants.

Another factor promoting group dissatisfaction could be an over-reliance on written communication (Siegel, et al., 1986). By removing social cues such as eye or voice contact, a particular seating arrangement, or even laughing and joking from the group environment, the group may no longer satisfy one's social needs and thus lead to enhanced dissatisfaction with the whole group process. On occasion, computer intervention could also prevent early closure and thus, maintain tension longer (Cats-Baril, et al., 1987). It is conceivable that as more and more ideas and alternatives are generated with the aid of the computer, the group may become frustrated with the decision making process and have real difficulty trying to decide on the best course of action.

There is some research that indicates that computer intervention into the group decision process can lead to overt, expressive behavior (Siegel, et al., 1986). Norms have been very slow to develop concerning computer mediated communication. The anonymity feature available to groups using electronic communication can even be viewed as promoting affective group behavior. Group participants using computer intervention have been shown to be very blunt and uninhibited in their language during group discussions, as compared to group participants meeting face to face and not linked electronically (Siegel, et al. 1986). Another possible danger is an over-reliance by groups on computer technology rather than on their own individual abilities and skills as decision makers. Some organizations feel that a GDSS can even have detrimental effects since it promotes task-centered communication at the expense of social interaction.

Perhaps the best way to deal with potential dissatisfaction among group participants is for the manager to stress the "support" function and "facilitator" role of computer intervention. Participants should be made aware that the purpose of decision support systems is to support, not replace, judgment (Kull, 1982; Wedley & Field, 1984). The idea is to combine both electronic and human components to enhance the decision beyond what would be possible when either element is used individually.

The following propositions concerning group effectiveness are offered:

> P17: Initially, decision making groups using GDSS may be both less efficient (time) and less effective (performance) than unsupported groups, but after allowing for a learning period should become more effective and efficient.

> P18: Decision making groups using GDSS may have lower satisfaction rates than unsupported groups.

Conclusion

The concept of Group Decision Support Systems (GDSS) is still a fairly new and somewhat untested decision tool. This paper has explored the major tenants of group decision support, has reviewed relevant decision support literature, and has proposed a framework with which to consider both the determinants and consequences of applying decision support in group environments. It was the intent of this paper to discuss these issues in relation to their impact on organization behavior and ultimately on organizational effectiveness. Insights have been made drawn from a diverse literature, including that dealing with individual decision support literature, decision theory, and organizational behavior.

The major impact of GDSS concerns their ability to actually change the way group members work together. The model presented in this paper conceptualized this as the way inputs (stimuli) are transformed into outputs (group effectiveness) through an open systems approach. Even though GDSS cannot and should not be applied in all group-based environments, they appear to be an effective means of supporting the organizational decision process. Since this concept has enormous practical appeal and potential to today's group-based organizations, further conceptual and empirical research on GDSS is needed and encouraged.

References

Aldag, R.J., & Power, J. (1986). An empirical assessment of computer-assisted decision analysis. *Decision Science, 17*, 572-588.

Alter, S.L. (1980). *Decision Support Systems: Current Practice and Continuing Challenges*. Reading MA: Addison-Wesley.

Bass, G.M., Pryer, J.W., Gaier, E. L., & Flint, A.W. (1958). Interacting effects of control, motivation, group practice, and problem difficulty on attempted leadership. *Journal of Abnormal and Social Psychology, 56*, 352-358.

Benbasat, I., & Dexter, A.S. (1982). Individual differences in the use of decision support aids. *Journal of accounting Research, 20*, 1-11.

Benbasat, I., & Schroeder, R.G. (1977). An experimental investigation of some MIS design variables. *MIS Quarterly, 1*, 37-50.

Bonczek, R.H., Holsapple, C.W., & Whinston, A.B. (1979). Computer-based support of organizational decision making. *Decision Sciences, 10*, 268-291.

Carnevale, P.J.E., Pruitt, D.G., & Seilheimer, S.D. (1981). Looking and competing: Accountability and visual access in integrative bargaining. *Journal of Personality and Social Psychology, 40*, 111-120.

Carter, L., Haythorn, W., & Howell, M. (1950). A further investigation of the criteria of leadership. *Journal of Abnormal and Social Psychology, 45*, 350-358.

Cats-Baril, W.L., & Huber, G.P. (1987). Decision support systems for ill-structured problems: An empirical study. *Decision Sciences, 18*, 350-372.

Chakravarti, D., Mitchell, A.A., & Staelin, R. (1979). Judgement-based marketing decision models: an experimental investigation of the decision calculus approach. *Management Science, 25*, 251-262.

Cohen, G.B. (1968). Communication network and distribution of "weight" of group members as determinants of group effectiveness. *Journal of Experimental Social Psychology,* 302-314.

Davis, J.H., Laughlin, P.R.., & Komorita, S.S. (1976). The social psychology of small groups: Cooperative and mixed-motive interaction. *Annual Review of Psychology, 27,* 501-541.

DeSanctis, G., & Gallupe, B. (1985). Group decision support systems: A new frontier. *Data Base, 16,* 3-10.

DeSanctis, G., & Gallupe, B. (1987). A foundation for the study of group decision support systems. *Management Science, 33,* 589-609.

Dickmeyer, N. (1983). Measuring the effects of a university planning decision aid. *Management Science, 29,* 673-685.

Eckel, N.L. (1983). The impact of probabilistic information on decision behavior and performance in an experimental game. *Decision Science, 14,* 483-502.

Gerrity, T.P. (1971). Design of man-machine decision systems: An application to portfolio management. *Sloan Management Review,* 59.

Goslar, M.D., Green, G.I., & Hughes, T.H. (1986). Decision support systems: An empirical assessment for decision making. *Decision Science, 17,* 79-91.

Hackman, J.R. (1968). Effects of task characteristics on group products. *Journal of Experimental Social Psychology, 4,* 162-187.

Hackman, J.R. (1976). Group influences on individuals. In M. Dunnett (Ed), *Handbook of Industrial & Organizational Psychology.* Chicago: Rand McNally.

Hackman, J.R. & Morris, C.G. (1975). Group tasks, group interaction process, and group performance effectiveness: A review and proposed integration. In L. Berkowitz, ed. *Advances in Experimental Social Psychology, 8,* 45-49.

Herold, D.M. (1978). Improving the performance effectiveness of groups through a task-contingent selection of intervention strategies. *Academy of Management Review, 3,* 315-325.

Huber, G.P. (1984). The nature and design of post-industrial organizations. *Management Science, 30,* 928-951.

Hurley, M.W., & Wallace, W.A. (1986). Expert systems as decision aids for public managers: An assessment of the technology and prototyping as a design strategy. *Public Administration Review, 46,* 563-571.

Janis, I.J. (1977). *Decision Making.* New York: Free Press.

Joyner, R., & Tunstall, K. (1970). Computer augmented organizational problem solving. *Management Science, 17,* 212-225.

King, W.R., & Rodriguez, J.I. (1978). Evaluating management information systems. *MIS Quarterly, 2,* 43-51.

Kraemer, K.L., & King, J.L. (1986). Computing and public organizations. *Public Administration*

Review, 46, 488-496.

Kull, D.J. (1982). Group decisions: Can computers help? *Computer Decisions, 14*, 70-84+.

Laughlin, P.R. (1980). Social combination processes of cooperative, problem-solving groups as verbal intellective tasks. In M. Fishbein (Ed), *Progress in Social Psychology*. Hillsdale, NJ: Erlbaum.

Lott, A.J., & Lott, B.E. (1961). Group cohesiveness, communication level, and conformity. *Journal of Abnormal and Social Psychology, 62*, 408-412.

Maier, N.R.F. (1967). Assets and liabilities in group problem solving: The need for an integrative function. *The Psychological Review, 4*, 239-249.

McGrath, J.E. (1984). *Groups: Interaction and Performance*. Englewood Cliffs, NJ: Prentice Hall.

McGowan, R.P., & Lombardo, G.A. (1986). Decision support systems in state government: Promises and pitfalls. *Public Administration Review, 46*, 579-583.

McGuire, T.W., Kiesler, S., & Siegel, J. (1987). Group and computer-mediated discussion effects in risk decision making. *Journal of Personality and Social Psychology, 52*, 917-930.

McIntyre, S. (1982). An experimental study of the impact of judgement based marketing models. *Management Science, 28*, 17-23.

Mintzberg, H., Raisinghani, D., & Theoret, A. (1976). The structure of "unstructured" decision processes. *Administrative Science Quarterly, 21*, 246-275.

Monge, P.R., Rothman, L.W., Eisenberg, E.M., Miller, K.I., & Kirste, K.K. (1985). The dynamics of organizational proximity. *Management Science, 32*, 1129-1141.

Neumann, S., & Hadass, M. (1980). DSS and strategic decisions. *California Management Review, 22*, 77-84.

Richman, L.S. (1987). Software catches the team spirit. *Fortune, 125*, 125+.

Rubin, B.M. (1986). Information systems for public management: Design and implementation. *Public Administration Review, 46*, 540-552.

Shangraw, R.F. (1986). How public managers use information: An experiment examining choices of computer and printed information. *Public Administration Review, 46*, 506-515.

Sharda, R., Barr, S.H., & McDonnell, J.C. (1988). Decision support system effectiveness: A review and an empirical test. *Management Science, 34*, 139-159.

Shaw, M.E. (1976). *Group Dynamics: The Psychology of Small Group Behavior* 2nd edition. New York: McGraw-Hill Book Company.

Siegel, J., Dubrovsky, V., Kiesler, S., & McGuire, T.W. (1986). Group processes in computer-mediated communication. *Organizational behavior and human decision processes, 37*, 157-187.

Sprague, R.H. (1980). A framework for the development of decision support systems. *MIS Quarterly, 4*, 1-26.

Steiner, I.D. (1972). *Group Process and Prodcutivity*. New York: Academic Press.

Stodolsky, D. (1981). Automatic mediation in group problem solving. *Behavior Research Methods and Instrumentation, 13*, 235-242.

Thibaut, J.W., & Kelley, H.H. (1959). *The Social Psychology of Groups*. New York: Wiley.

Tushman, M.L. (1979). Impacts of perceived environmental variability on patterns of work related communication. *Academy of Management Journal, 22*, 482-500.

United Communications Group. (1987). Group decision support systems. *EDP Analyzer, 25*, 1-16.

Vazsonyi, A. (1978). Information systems in management science. *Interfaces*, 9, 72-77.

Wedley, W.C., & Field, R.H.G. (1984). A predecision support system. *Academy of Management Review, 9*, 696-703.

Part IV

Information Technology Impacts on Japanese and American Organizations

Chapter 11

The Impact of Information Technology on Organizations in Japanese Companies*

Tatsumi Shimada
Yokohama College of Commerce

About forty years have passed since the computer was first used in business. Over 30 years ago the concept of information technology involving computers, and the effect it might have on business organizations, was first introduced when Leavitt & Whisler wrote in 1958 that information technology would progress remarkably (Leavitt & Whisler, 1958). In this chapter, we discuss empirically how information technology affects business organizations, which hereafter we will refer to as simply organizations, mainly from the point of view of Japanese companies. We wish to make clear the process by which information technology has an impact on those organizations which have implemented it. In addition, we will explain a little about the effect information technology will most likely have on organizations in the near future. We will cover the following two topics in detail.

First, we will discuss the mutual relationship between information development and the process of organizing by giving a model of the relationship between information technology and organizations. Then, while we are reviewing the growth stage of information systems up to now, we will make clear that there are four types of mutual relationships between information systems development and organizing, which involve "passive organizing", in

*Based on, and translated by permission of Soshiki Kagaku(Organizational Science), Jyoho Gijyutu ga Keiei Soshiki ni ataeru Eikyo[The Impact of Information on Organizations], by Tatsumi Shimada(April 1990).

which organizing follows information technology, and "active organizing", in which information technology follows organizing. Second, we will explain organizations which are being affected by information technology, divided into the two sub-groups of End-User Organizations and a computer-based management information system (MIS) department. In the case of End-User Organizations, we will deal with issues which have been discussed, some unresolved issues, and some ongoing issues. These include the hierarchical structure, the unification and elimination of jobs, the allocation of decision making, the communication (vertical/horizontal), the dynamics of organizations, the number and the role of middle managers, and the ways in which they are interrelated. On the other hand, for the MIS departments, we will deal with the function, the relationship with End-User Organizations, and the organizational forms.

The work consists primarily of two parts. We will explain the theory and concepts which serve as the framework we developed for analyzing the data from our field survey. Then the results of the survey will be presented. This data is extremely valuable to our research, since very little work has been done in Japan in this field to date. In other words, we will present the data of cases in which information technology was introduced into the organizations, which were then changed through "the Impact Process."

Other factors in addition to information technology also exist which have contributed to change in the organizations. Because the impact of information technology is indirect, it is not easy to isolate and extract its effect on the organizations. Therefore, we decided to analyze which types of information technology affect which facets of organizations.

In order to do this, we sent preliminary questionnaires to 45 Japanese companies from many of Japan's primary industries, and received 31 replies. Based on the results of these preliminary questionnaires, we did further follow-up interviewing in person and by telephone. This survey was conducted between April and July of 1989. The respondents and interviewees who participated in this survey were CIOs and middle managers in MIS departments. The 31 companies were large, with 2,067 to 76,447 employees, and featured advanced MIS. For example, the two indicators of the ratio of annual revenues to annual IS (Information System) expenditures and the ratio of employees to PCs/terminals show the advanced MIS of these companies in Japan. They showed an average ratio of annual revenue to annual IS expenditures of 0.99% (the valid respondents being the 23 companies excluding financial institutions) which was higher than the 0.5% average shown in a survey of 654 of Japan's large companies (excluding financial institutions) (JIPDEC, 1989). Then, the ratio of employees to PCs/terminals is higher at 17.1% (the valid respondents being 29 companies) than 5.4% (an average 105 of Japanese large companies) (Shimada, 1989a).

Furthermore, although the term "information technology" is currently used to express a variety of meanings, I fundamentally follow Leavitt & Whisler's definition. They used the term information technology, as early as 1958, to include computer methods for handling vast amounts of information quickly, applying statistical methods to decision making problems, and stimulating via computer higher-order thinking (Leavitt & Whisler, 1958). Whisler (1967) later rearranged his definition to have three components; the computer, telecommunications, and management science techniques. And then, he further defined information technology to be the technology of sensing, coding, transmitting, and transforming information (Whisler, 1970).

In this chapter, we define information technology with a focus on the process, as follows: Information technology is the technology to process/calculate, record/store, and to exchange/transmit information. More specifically, computer systems have the capability to process/calculate information, database has the capability to record/store information, and telecommunications has the capability to exchange/transmit information. In addition, the processing/calculating of information involves technologies for decision making such as OR (Operations Research), simulation methods, statistical methods, and so on.

For the purposes of this work, the term "MIS department" means a division which develops, implements, uses, and controls MIS from the standpoint of the company as a whole. On the other hand, the term "End-User organization" means a division which enters input data into computers and uses the output from computers. In general, this means a division which consumes or produces information by using computers installed in the MIS department or in the users' work places.

Background

In the early stages of computer use, the article "Management in the 1980's" by Leavitt & Whisler became very well-known. They said in the article that the advent of information technology would cause planning functions to shift from the middle managers to operation researchers and system analysts, middle manager ranks to shrink and sink, top management to take on more of a creative role, and large organizations to centralize. And then Anshen (1960) expressed the following ideas. Changes will not progress as fast or as far as some enthusiasts have been professing. The professional experts in the new approaches will hold a position of growing significance in the organization, and will have an increasing power on the whole system of information processing and decision making. At the same time, the role of middle managers will change and become more important. Decision making in large organizations will differentiate to the extreme into centralization or decentralization.

Now that the decade of the 1980's has come to a close, it is not difficult to judge which of the two predictions turned out to be better suited to the actual situation. Needless to say, Anshen's predictions were more suitable, on the whole, although his predictions did include some parts which proved to be incorrect. At least his predictions were more relevant to the situation in Japan. During the 1980's, however, since Leavitt & Whisler's predictions were more progressive, even though they met with strong criticism, their article was talked about and often quoted (Attewell & Rule, 1984; Huff, 1986; Applegate & Cash, Jr., 1988). On the contrary, Anshen's predictions, which took a more conservative view, were not widely talked about at the time. However, they have proven to be well suited to describe conditions even today. As a whole, in these kinds of predictions it does not matter whether the outcome is entirely correct or not. Generally speaking, it seems that a progressive view tends to attract more attention and create more of a sensation than a conservative view would. Perhaps researchers are subconsciously aware of this fact, and tend to choose a more progressive hypothesis as a result, but this often leads to inaccurate conclusions.

Now that thirty years have elapsed since these articles were published, problems regarding information technology and organizations are being discussed again. There are big differences between the immature information technology of thirty years ago and the more mature information technology of today, but some views overestimate the impact of information technology on organizations. These views predict a decrease in the number of hierarchial levels, flat structure, decline in the number of middle managers to date, or rapid flattening and decrease in the number of levels or in the number of middle managers in the near future (Anonymous, 1983; Anonymous, 1984; Nakatani, 1984; Yaeger & Targowsky, 1989). Furthermore, there is another view that the network organization is created as a form of the dynamics of organizations (Wave, 1987). Some examples of these cases which we feel overestimate the impact of information technology follow.

First, in the United States there is a report based on research done in 60 United States companies which use information technology. It showed that companies which had invested heavily in information technology were able to reduce the number of levels to 4 and showed a higher profitability than the companies which had invested less, had an average of 8 levels, and showed a lower profitability (Goleman, 1988). As another example, in the United Kingdom the number of levels decreased from 13 to 4, while in the United States the number of levels decreased by 6 or 7 levels in the same period of time (Keen, 1988; Rockart & De Long, 1988). Last, a large oil company was able to reduce middle managers to 40% by a comprehensive restructuring, and a large manufacturing company recently thinned middle managers by 30%, using sophisticated telecommunications (Applegate & Cash, Jr.,1988).

We disagree, because we question the validity of these assertions that information technology was the sole contributing factor in the impact on these organizations, because information technology still has many problems in terms of user-friendliness and in the inadequacy of existing computer and communications tools. However, the predictions regarding the flattening of levels and the use of the network organizations do seem to be correct in terms of organizational directions. Of course, information technology today has advanced remarkably, since the above-cited reports and it does have a strong impact, but other factors also exert important influences on the overall change in the organizations. At any rate, we do not believe these views and predictions could come to pass immediately. Even if it were somehow true that the major factor in change was information technology, the above examples are exceptional cases, and do not reveal any general trends.

Very few empirical studies have been done regarding the impact of information technology on organizations in Japanese companies. Here we will present one of the few recent studies, the Study Group on Microelectronics Technology Innovation (Okubayashi, 1990). In this research, questionnaires and interviews were carried out with the manufacturing companies listed on the Japanese stock market. The Study Group's findings regarding changes in organizational structure occurring between 1980 and 1985 in about 130 companies are as follows.

1. The majority of cases (78 companies: 60.0%) showed no change in the number of levels, which shows a relative stability of hierarchial structure.
2. In the 52 companies (40.0%) which showed some change in the number of levels, the majority showed a decrease (33 showed decrease, 19 showed increase). This indicates a flattening of levels.
3. Regarding decision making, the number of companies which shifted decision making to the lower levels(58) was higher than the number of companies(21) which shifted it to the upper levels, and than the number of companies(51) which showed no change. Therefore, there is a tendency for companies to change organizations to delegate authority to the lower levels.
4. However, the tendencies toward centralization/decentralization differed according to the content of decision making. In the accounting, planning, and personnel management departments, the highest numbers(60) showed no change, while the number of companies(48) which concentrated on headquarters was much higher than the number of companies(22) which delegated to branch offices.
5. Regarding mutual relations among the sectional divisions (Ka) within the branch offices, the section (Ka) which tended to be on good terms with the other sections (Ka), resulting in unclear boundaries, was more prevalent(52) than the sections(Ka) which tend to be more

independent of each other and to utilize more clear, definite boundaries(27). Yet, many companies answered no change(50).

On the whole, these conclusions agree with our own findings, with the exception of 2. For example, the result seen in 1, that generally the number of levels show no change, is in accord with the results of our field survey, which we will explain later. The results of 2, however, seem to have two methodological weaknesses.

The first weakness is whether or not it is reliable that the Study Group's survey showed a higher percentage of companies which decreased in the number of levels than we found in our survey. We suspect that the Study Group's higher percentages were caused by the fact that the term "microelectronics technology" has a slightly broader definition than does "information technology", but of course this discrepancy is inevitable. We also suspect that other factors besides just information technology were responsible for the change in the number of levels, and hence, their higher percentages. In order to properly assess these multi-faceted ambiguous factors, a more complete questionnaire would be needed whereby respondents could differentiate between the influences caused by information technology and the influences caused by other factors. Furthermore, a follow-up interview could have helped to ensure that the respondents clearly understood and distinguished among the many complex facets of this issue, making the findings more accurate.

The second weakness found in point 2 makes us wonder if companies really use microelectronics technology for an increase in the number of levels and, if so, exactly what was the nature of the microelectronics technology they used. There were 19 companies which increased in the number of levels, according to this research, but since this study does not cite any definite cases, their results are not conclusive.

So far, we have discussed the accuracy of the views represented by Leavitt & Whisler (1958) and Anshen (1960), as well as recent trends showing the impact of information technology on organizations. Recent empirical studies in this field are still lacking in both quality and quantity, and this field is still open to much further research before more complete evaluations can be made, especially in regard to research on Japanese companies. We need to ask whether organizations were really changed by the information technology developed in microelectronics innovation from the second half of 1970's. Moreover, in order to measure the change in organizations, a framework for understanding the mutual relationship between information technology and organizing is needed as a premise. Next, this framework will be discussed.

Information Technology and The Impact Process

Information Technology and Organizing

Leavitt(1965) explained the mutual relationship between technology and organizations by using four variables, which are technology, people, structure, and task. These four variables are highly interdependent and if one of the four changes, the other three variables also change. For example, using technological tools, such as computers, can cause changes in structure (in the communication system or decision map), in people (their numbers, skills or activities), and in performance. Sometimes the introduction of technology may even cause changes in definition of task, with some tasks becoming feasible to accomplish for the first time, and others becoming unnecessary. This model is also valid for the impact of information technology on the organization. In fact, since the four variables are interdependent, this model does not assume a beginning or an end, and does not focus on a stage of change but on equilibrium (Kraemer, King, Duncle & Lane, 1989). Since this model is simple and too abstract, a model is needed which would deal with the specific technology of information technology, rather than with general technology, and which is broken-down into more detail in order to explain the more complicated impact process.

Scott-Morton (1986) gave a framework for the organizational and social impacts of technology, based on the model by Leavitt. He used four variables, similar to the four used by Leavitt, which were technology, people, structure, and task, but Scott-Morton used "individuals and roles" instead of "people" and "organizational structure and the corporate culture" instead of "structure". Then, in addition to these four interdependent variables, other variables were set up as intervening variables, such as management process, planning, budgeting, and reward. Yet, even so, this model had basically the same framework as Leavitt's and is still abstract.

Laudon & Westin (1988) presented a model of systems development and of the impact process. In this model, organizations adopt information technology because of environmental factors (uncertainties and opportunities) or internal institutional factors (values, norms, and interests). When the system is developed, information technology has an influence on organizational variables (organizational politics, organizational structure, organizational culture, work) and management decision making through the implementation process of technology. Then, the impact feeds back to the environment and to the organization by creating forces of change and resistance.

However, although systems development influences organizational variables, organizational variables influence systems development only indi-

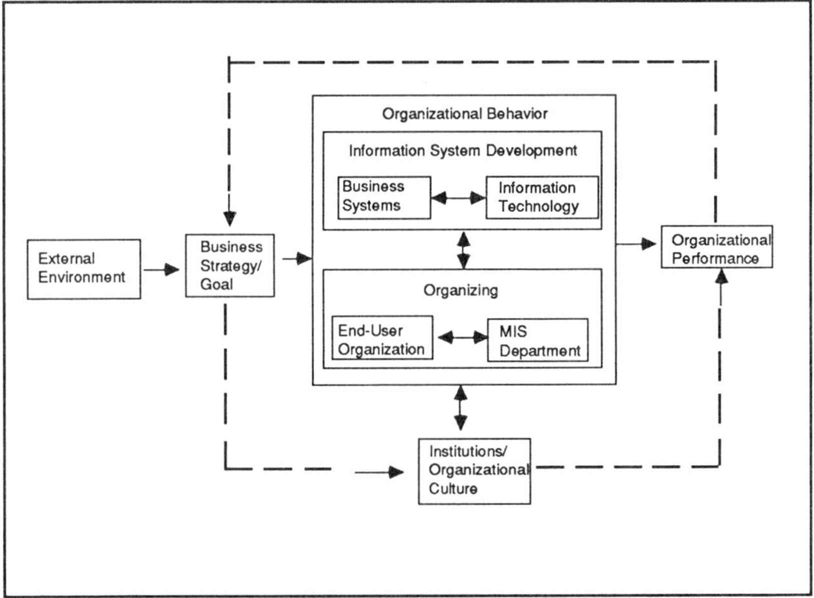

Figure 1. The Impact Process of Information Technology on Organizations

rectly. The relationship between these two should be interactive. We have reviewed the typical models of the relationship between information technology and organizations. Next, we will show our model, based on these models.

In general, since changes in organizations are caused by many factors, it is not easy to determine which of the factors caused the changes. We can not say that the implementation of information technology in organizations immediately effects changes in the organizations. Our framework for describing the impact process of information technology on organizations is shown in Figure 1.

The organization recognizes the external environment by scanning, and plans the business strategy/goal in order to adapt itself to the recognized environment. As a result of the organization recognizing the environment, information technology itself becomes one of those factors which the organization absorbs. Then, information technology affects not only business strategy/goal, but also organizational behavior and institutions/organizational culture. In order to realize the business strategy/goal, many kinds of organizational behaviors are developed. Of these, we will focus primarily on information system development and organizing in this chapter. "Information system development" is defined to be the integration of business systems and information technology, and "organizing" means the formation of an MIS department or End-User organizations taking place before or after information system development. When the organization adopts information technology, organizational performance differs according to the kinds of technology used, the kind of attitude taken — active/leading, or passive/following — and the methods of integration of business systems and information systems attempted. In other words, organizational behavior causes organizational performance. We have explained above the direct mutual interrelatedness of all the factors in the management process. In addition, some institutions such as personnel management and organizational culture, are indirectly related to business strategy/goal, organizational behavior, and organizational performance.

The Growth Stages of Information Systems

In order to grasp the impact process of information technology on organizations, we would like to explain information system development and organizing, the two subgroups of organizational behavior seen in Figure 1. In this case, it is necessary to outline the growth stages of information systems as a premise, since we will be discussing the relation between information development and organizing.

Concerning the stages of the assimilation of information technology by organizations, there exist the model of 4 growth stages by Gibson & Nolan (1974) and the model of 6 growth stages which was modified by Nolan (1979) according to changes in the environment. This model of 6 growth stages is composed of Initiation, Contagion, Control, Integration, Data Administration, and Maturity. Gibson & Nolan's model was affected by the model of 4 growth stages (Investment, Technology Learning and Adoption, Rationalization, and Maturity) of information technology assimilation by Shein (1961). Nolan says that the period from the third stage (Control) to the fourth stage (Integration) is the time of technological discontinuity, and is also a turning point from the time of narrow-ranging DP (Data Processing) to the time of wide-ranging, advanced information technology.

A summary of his opinions is as follows (Nolan, 1984). Office Automation (e.g. a word processor, a personal computer), CAD/CAM, and Robotics, which are the results of microelectronics beginning from the latter half of the 1970's, were introduced and used as stand-alone devices for the first time. However, the result of requiring more efficient usage from these devices was the integration of DP and new technology. In order to actualize this integration, database as a resource was indispensable, unlike the poor database infrastructure during the period of DP. Information technology used at the turning point was said to have been built by data resource administration. Then, this model of growth stages was adapted to the period from the later half of the third stage (Control) to the earlier half of the fourth stage (Integration) in U.S companies (Nolan, 1987). It became difficult to adequately explain the revolutionary growth of information systems with Nolan's 4 stage model. Therefore, he switched to the model of 6 growth stages, which showed when the big turning point came to information systems.

Although information systems in Japanese companies basically follow the model by Nolan, their development lags slightly behind that of information systems in the U.S. In other words, almost all Japanese companies are now in the third stage, with just some advanced companies in the fourth stage. As we review the growth process of Japanese information systems, the process is divided into two stages; information systems beginning from the latter half of the 1950's, which were mainly general purpose computers, and the integration of computers whose core was microelectronics technology and telecommunications, beginning from the latter half of the 1970's. In this chapter, the former stage is labelled the "first information system", and the latter stage is labelled the "second information system"(Shimada, 1988). The difference between the former and the latter is shown in Table 1.

As you see in Table 1, during the second information system a lot of technology expressed by English acronyms appeared one after another. Users

	First	**Second**
Period	From the latter half 1950's	From the latter half 1970's
Purpose	Efficiency	+Effectiveness
Type of Information	Coded Information	+Text, Image, Voice Information
Type of Decision Making	Structured	+Unstructured
Type of Processing	Centralized	+Decentralized
Scope	Internal Organization	+Inter Organization
Tools	Computer / System Engineering	+Telecommunications/ Management Science
Main Leading Concept	ADP(Automatic Data Processing), IDP (Integrated Data processing), MIS (Management Information System)	DSS(Decision Support System), OA(Office Automation), ES(Expert System), SIS(Strategic Information System), CIM(Computer Integrated Manufacturing)
Main Technologies	General Purpose Computer, Advanced Language, Digital Transmission, DB/DC POS, OCR	PC/WS, WP, 4GL, RDB LAN, VAN, MML, EPBX DD/DS, ISDN, CAD/CAM Teleconference,E-mail Electronic Filing

Table 1. The First and the Second Information System

got the opportunity to select the appropriate one from a variety of kinds of information technology. OA in particular has spread to organizations and promoted end-user computing by decentralized systems. However, the somewhat chaotic spread of OA caused its widespread use to be reconsidered in terms of efficiency due to the fact that more integration between information resources and information technology by sharing was needed. Technology which meets this need includes RDB(Relational Data Base), DD/DS (Data Dictionary/Directory System), MML (Micro Mainframe Link), LAN (Local Area Network), VAN (Value Added Network), EPBX (Electronic Private Branch Exchange), and so on. In addition, by the improvement of the database and telecommunications as an infrastructure, companies which use information technology for discriminating against the competition and for the effectiveness of organizations, such as decision support for managers, have appeared.

The Impact Process of Information Technology on Organizations

Next, we discuss the impact of information technology on organizations in the terms of the mutual relation between information system development and the end-user organization aspect of organizing, from Figure 1.

The impact of information technology on organizations is not direct. In order to carry out a business strategy/goal of adjusting to the external environment, information system development and organizing are carried out. In other words, the direct aim is to realize the business strategy/goal, such as shortening of R & D time, production time, lead time for products/services, differentiation of products/services, cost reduction, and/or labor saving. Thus, the impact on organizations is indirect. Information technology is a tool for supporting the acts of organizing and operation. Furthermore, the information system itself has no power to change organizations. Companies use information technology as a tool to accomplish their subjective intentions for transforming organizations. Based on the above, there are 4 types of mutual relations between information system development and organizing (Figure 2).

Type I has information system development, but does not have organizing. Thus, information technology does not affect End-User organizations. Although information technology does not affect End-User organizations, it does affect the organizing of the MIS department. This type is the case in which information system development is used, but organizing is not, for example, in the improvement of customer services, cost reduction, or small-scale applications of information system development. Moreover, this type, in many cases, is applicable to the initiation stage or contagion stage of Nolan's stage model. In other words, the MIS departments develop each application system whenever End-User organizations require, and also store data accord-

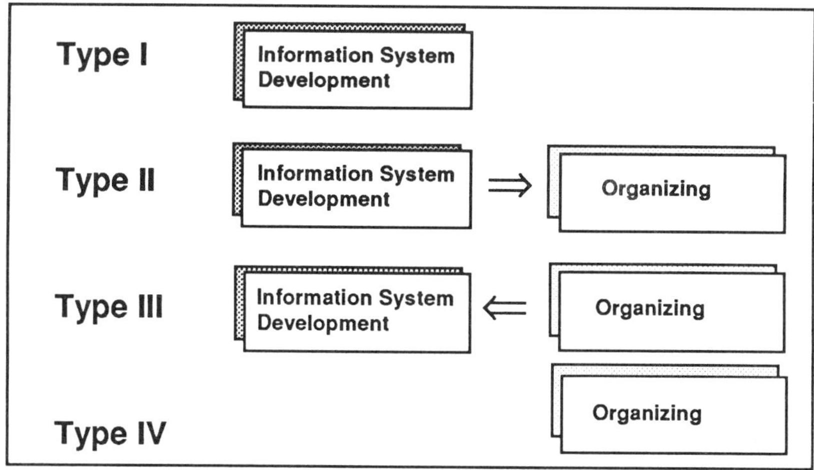

Figure 2. The Types of Impact

ing to each application.

In type II, organizing follows information technology because the information system development causes organizing to occur. This type includes cases in which information system development is developed for the specific purpose of reduction of labor, speedy communication, decision support, etc. It does not directly affect organizational transformation, but rather, the effect on organizational transformation comes about later as a secondary result. In this case, after a certain period during which the information system is introduced, the above aims are accomplished, which then causes organizing to occur, such as in the unification and elimination of jobs, change in the role of middle managers, and the dynamics of organizations. We can call this case "passive organizing", which means a change in organizations which occurs after companies have been using information technology. Most of this type is seen in the control stage of Nolan's stage model. In other words, although data processing dealing with coded information is still the most common usage, the scope of information systems is spreading from small and sub-applications to main fundamental applications. Moreover, information systems can make use of common accumulated data with plural applications for supporting decision making.

Next, type III is the case in which organizing causes information system development. In type III, information technology follows organizing, because organizing which takes in information technology plays an active role in various organizational transformations. In this case, in order to accomplish type III, the following is a premise: The data is accumulated as a resource, telecommunications are arranged, and the information technology environment in which end-users access a shared database through time and space, is prepared. Since the data is not affected so much by business systems or by organizational change and is stable, the premise for organizing is to have a database and a communications infrastructure. Therefore, this type is applicable to the integration stage of Nolan's stage model, and beyond, into what is called the second information system. Although at present this type is seen in only a few cases, as information systems grow the numbers will be increasing. If so, organizing and information system development will have deeper interdependence.

Finally, type IV is the organizing which does not use information technology as a tool. Since there are many factors in organizational transformation which are not related to information technology, this type was used in the majority of cases and for a while in the future will continue to be the most prevalent. However, as companies advance from the first information system to the second information system, information technology will be deeply related to organizing. Then, type III and type II will gradually increase.

In addition, Markus & Robey (1988) examine the causal structures found in theories about the relationship between information technology and organizational change. Their work evaluates theory and research in the field by focusing on three dimensions of causal structure: 1. causal agency, 2. logical structure, and 3. level of analysis. Then, they refer to Pfeffer (1982), and identify three conceptions of causal agency in the literature on information technology and organizational change, and label these: 1. the technological imperative, 2. the organizational imperative, and 3. the emergent perspective. In the technological imperative, information technology is viewed as a cause of organizational change. In the organizational imperative, the motives and actions of the designers of information technologies are a cause of organizational change. In the emergent perspective, organizational change emerges from an unpredictable interaction between information technology and its human and organizational users.

If the three conceptions by Markus & Robey are applied to our classification of types of mutual relations between information system development and organizing, the technological imperative is similar to type II, and the organizational imperative is similar to type III. But, there is no parallel to the emergent perspective in our classifications. The emergent perspective can be useful as a concept, but in practice it is difficult to distinguish from type II or type III. This is because in type II organizing follows information technology, and in type III information technology follows organizing, and thus each type is relative. Therefore, both type II and type III involve unexpected interactions between information technology and humans/organizations or between users and other users, thus making the emergent perspective's existence as an independent type meaningless.

When we use the term "the impact of information technology on organization" in the following section, it involves type II and type III.

The Impact of Information Technology

End-user Organization and the MIS Department

Information Technology affects both End-User organizations and the MIS department of the company. We would like to discuss how information technology affects End-User organizations, mainly in regard to the following topics; the hierarchial structure, the unification and reduction of jobs, the allocation of decision-making, communication(vertical/horizontal), the dynamics of organizations, the number and the role of middle managers, and the relationships between each of these topics. On the other hand, since the MIS department is an end-user of information technology, it is also affected by information technology, while at the same time the MIS department acts on End-User organizations. As a result, the MIS department is uniquely affected by information technology through the staff which introduces information technology. In the next section, we will explain how the role of the MIS department changed in regard to End-User organizations.

Before we explain about that change, however, some caution is in order. As we explained earlier, there are other factors besides information technology which affect organizations. Then, we also explained that we should exclude these other factors as much as possible, and that we should develop a method for understanding the impact on organizations made by the factor of information technology alone. So, when we designed the form of our preliminary questionnaires, we devised them so that the respondents could answer by thinking in relation to information technology (Appendix A). But, we confirmed at the survey interview that some respondents had answered involving the impact on organizations caused by factors other than information technology, and we corrected some figures of the questionnaires results according to the interview.

Moreover, we should pay attention to the following point. "Information system development" means the integration of business systems and information technology. A typical process of information system development begins with business systems analysis, and goes through requirements specification, logical design, physical design, coding, system test, and maintenance. In that case, how the range of business systems analysis is set up determines the development period and organizational performance. There are two cases. One case is that in which information technology is introduced after approval of the existing business systems. The other case is that in which elimination, simplification, or/and reconstruction of existing business systems is done when information technology is introduced. In the former case, it is easy to grasp the impact of information technology on organizations, while in the latter case, the improvement of business systems and the factor of information technology are compounded, and therefore it is difficult to distinguish between them.

In Japan, there are many companies which improve the business systems before the introduction or renewal of information technology. These companies try for a synergistic effect between the improvement of the business systems and information technology. Since the improvement of the business systems is coupled with information technology in that case, both of them should be treated as a factor of information technology.

Main Change in End-User Organization

The Hierarchical Structure. Generally it is said that information technology causes a decrease in the number of levels and causes a flat structure. However, so far, information technology seldom affects the hierarchical structure in Japanese companies. The Japanese companies which decreased in the number of levels were just 3 of 31 companies (Kao, Honda Motor, Yamaha) (Table 2).

According to the results of the survey concerning the hierarchical structure, affirmative answers were shown concerning the increase in the average span of control, increase in autonomy at the clerical level, and strengthening vertical communication (Table 3).

If we consider these affirmative results, the question of why the number of levels did not decrease will be raised. We will consider some reasons as follows: the first reason is the increase of information load in each hierarchical level because of the companies' continuous growth, the second reason is that although decision making when it concerns critical problems is done through the hierarchical structure, a bypass communication is used for all other problems. Finally, the third reason is that Japanese companies use human resources effectively in their companies or in the group companies. For instance, in instances in which Japanese companies have surplus personnel, they do not lay off, but instead implement job rotation. For these reasons, the number of levels does not decrease. Next, we will look at those few companies which did decrease in the number of levels through information technology.

First of all, Kao decreased in the number of levels in the factories. Kao has 280 plants in Japan, the majority of which are located in the Wakayama factory complex, where chemical products are made. Each plant is connected by telecommunications and products are controlled centrally from the operation center in the factory complex. Furthermore, the factories and the Kao sales companies (Hansya) are connected by telecommunications, and this integration of manufacture and marketing is an efficient system. Up to now,

Question	Answer	Number of companies	Percent of total
Decrease in the number of levels	1)Yes	3	9.7%
	2)No	26	83.9%
	3)Don't know	2	6.5%
	4)No answer	0	0.0%
	Total	**31**	**100.0%**

Table 2. Decrease in the Number of Levels

Question	**Answer**	**Number of companies**	**Percent of total**
Increase in the average span of of control	1)Yes	15	48.4%
	2)No	11	35.5%
	3)Don't Know	5	16.1%
	4)No answer	0	0.0%
	Total	**31**	**100.0%**
Increase in autonomy at the clerical level	1)Yes	18	58.1%
	2)No	8	25.8%
	3)Don't Know	4	12.9%
	4)No answer	1	3.2%
	Total	**31**	**100.0%**
Strengthening vertical communication	1)Yes	13	41.9%
	2)No	7	22.6%
	3)Don't Know	10	32.3%
	4)No answer	1	3.2%
	Total	**31**	**100.0%**
Strengthening horizontal communication	1)Yes	22	71.0%
	2)No	3	9.7%
	3)Don't Know	5	16.1%
	4)No answer	1	3.2%
	Total	**31**	**100.0%**

Table 3. Increase in the Span of Control, Increase in Autonomy, and Strengthening Communication

supervisors in the factories did the product planning, but with the introduction of telecommunications the product planning is now done by operators. The operators need to have good judgement, thus they are called "operation engineer". On the contrary, the supervisors are said to mainly work to solve their subordinate's problems, and act as counselor for problems both in the work place and in the workers' families. Then, as a result, in the Wakayama factory, the rank of middle manager (Bucho), lower manager (Kacho), and supervisor (Kakaricho) changed into plant manager and supervisor, which is only a two-level system.

Kao sales companies (Hansya) installed facsimile ("fax") in the homes of sales workers and a wireless terminal in their cars. In this way, sales workers are able to work directly from their homes and go directly to their customers and back home without the necessity of going to the office every day. Although the managers go with the sales workers for direction once or twice a week, the sales workers are required to do self management because Kao wants to increase their autonomy. Moreover, Kao sales companies (Hansya) are planning to achieve a system of only two levels, the president and the sales workers, in the near future.

Next, Honda Motor abolished all 6 branch offices in the marketing division, so branch function was taken over by the regional sales department. The majority of the 400 branch office members moved to the sales department and the parts department. Honda Motor first planned a draft of the organizational transformation, then developed an information system of direct order entry between the headquarters and the dealers, and soon abolished the branch offices.

Finally, Yamaha reduced from 3 levels to 2 levels in its divisions (Jigyobu). The 3 levels had consisted of division manager (Jigyo Bucho), departmental manager (Bucho), and sectional manager (Kacho), and Yamaha changed to 2 levels consisting of departmental manager (Bucho) and specialist.

The Unification and Elimination of Jobs

According to our findings, there are many cases of unification and elimination of jobs at the individual level, but not so many at the departmental(Bu) level or sectional(Ka) levels. Moreover, increase in autonomy at the clerical level was also prevalent (Table 4).

The unification and elimination of jobs at the individual level began from the period of the first information system when DP was at the forefront, and has advanced rapidly with the progress of information technology. There are two methods of introducing information technology. One is the replacement method, in which repetitive manual task workers are replaced by machines. Another is the support method which supports decision making. In the first information age, the replacement method was mainly used, but later the

Question	Answer	Number of companies	Percent of total
Unification and elimination of jobs at the individual level	1)Yes	21	67.7%
	2)No	6	19.4%
	3)Don't know	4	12.9%
	4)No answer	0	0.0%
	Total	**31**	**100.0%**
Unification and elimination of jobs at the departmental (Bu) and the sectional(Ka) level	1)Yes	12	38.7%
	2)No	15	48.4%
	3)Don't Know	4	12.9%
	4)No answer	0	0.0%
	Total	**31**	**100.0%**

Table 4. Unification and Elimination of Jobs

support method gradually gained popularity. At any rate, in both methods individual jobs change a great deal and the unification and elimination of jobs occurs.

For instance, in banking many tellers' jobs have been replaced by such machines as the CD (Cash Dispenser) or ATM (Automatic Teller Machine). At the same time, the tellers' role has changed from that of subdivided-unit specific-job workers to versatile multi-job workers, doubling as operators, resulting in the unification of jobs. Unification and reduction of jobs is also advanced in the manufacturing industry. In the electronics manufacturing industry which makes control equipment, for example, the design job changed very much because of the introduction of CAD. Up to now, the design processes were composed of basic design, detailed (circuit) design, structure design, and interface design, and each design process had its own section(Ka). However, through CAD, detailed (circuit) design, structure design, and interface design were done simultaneously, and these three sections(Ka) were combined into one section(Ka) called the production design section(Ka). Then, the basic designer was required to have higher knowledge and technical skill, while the production designer was not.

The reason that the unification and reduction of jobs at the departmental or sectional level is not as advanced as at the individual level, is this. While information technology directly affects jobs at the individual level in most cases, it only indirectly affects them at the departmental and sectional level. In other words, the purpose of reforming existing departments(Bu) and sections(Ka) is not necessarily in agreement with the purpose of using information technology. Moreover, in the case of small applications, information technology does not also make departments(Bu) and sections(Ka) reform.

Increase in autonomy at the clerical level is accepted in many companies. As the structured routine jobs are replaced by the machine, unstructured thinking jobs are considered to be more important among white-collar jobs. This is the support type of information system that accesses internal and external databases, retrieves necessary information, develops alternative solutions to the problem, and selects the best alternative by using workstations and/or PC. This type helps in developing the ability of decision making in quantity and quality and makes autonomy at the clerical level increase. This type also has an important influences on self control at the clerical level, and makes the span of control spread.

Furthermore, end-user computing also contributes to the increase in autonomy at the clerical level.

Allocation of Decision Making

According to our survey, the companies were more likely to answer that decentralization of decision making was not caused by information technology than that it was (Table 5).

Many researchers have often discussed whether information technology has brought about centralization or decentralization of decisions up till now. In the period of the first information system when data processing was of the centralized type, there was once an argument that information technology brought about centralization of decisions by relating it to the centralization of data processing function. In the period of the second information system, in addition to the centralized type of data processing system, the decentralized system appeared and became popular. Therefore, we could conclude that centralization or decentralization of data processing does not necessarily mean centralization or decentralization of decision making. The nature of information technology is neutral and today there are many kinds of information

Question	**Answer**	**Number of companies**	**Percent of total**
Decentralization of decision making	1)Yes	11	35.5%
	2)No	15	48.4%
	3)Don't know	4	12.9%
	4)No answer	1	3.2%
	Total	**31**	**100.0%**

Table 5. Decentralization of Decision Making

technology available. Therefore, although allocation of decision making may look like centralization or decentralization in part, and temporarily, it depends upon the factors other than information technology on the whole and in the long run.

Next, we will introduce some examples in which information technology is used for both centralization and decentralization of decision making.

Nippon Seiko recently developed a sophisticated CIM system that integrated the production system, marketing/distribution system, and engineering system. In this company, the boundaries between the production department and marketing department, such as in making/adjusting a product plan, changed very much, because both departments shared a consolidated database. For many years this company had been assigning product planning to a planning section(Ka) in each plant, but recently they abolished the planning section(Ka), organized a general administration department(Bu) in the headquarters, and centralized product planning in this department. This department plans and changes the production schedule for every assembly line in each plant, based on on-line real time data.

Since the marketing department(Bu) and the production department(Bu) have conflicting or differing goals, they had at times been known to submit false data in an attempt to sabotage the work of the other department, resulting in a large loss by the company. However, the loss was reduced when the data which was accumulated during each day's work was stored in a common database, from which workers could retrieve factual information and make accurate judgments. Now, the general administration department(Bu), using consolidated management, can dynamically cope with the conflicting demands of push/pull between the marketing department(Bu) and the production department(Bu). Therefore, the administrative jobs in each plant mainly evolved into management by exception. Through centralization, this product planning staff of nearly 100 was amazingly able to be reduced to only 10.

This phenomenon of product planning centralization can also be observed at the Kao Corporation's Wakayama factory. These examples show that highly interdependent organization between each department, such as logistics organization, have a tendency to be centralized by replacing interactive processes with machines.

On the contrary, another information system at Nippon Seiko was used to promote decentralization in decision-making. This company produces more than one hundred thousand kinds of bearings. In the past when customers inquired of the sales department which product would be suitable for their needs, the answer took about a month because of the necessity of dealing with inquiries between so many different departments. Today, however, it takes only a few minutes to retrieve the right bearing for the job, by using an Expert

System. Moreover, a facsimile is used instead of pen plotter for printing a drawing, which is soon shown to the customer. Making an approved drawing traditionally came under the authority of the design department manager, not only in Nippon Seiko but in most Japanese companies. However, through information technology, sales workers are now able to offer the customer an approved drawing themselves.

Next, banks have been shifting structured decision making out to the sales branches by making information simultaneously available to both the sales branch offices and the headquarters. In addition, they have been shifting the responsibility for verification from the managers to the clerical workers ever since the period of the first information system. After the second information system, banks began to decentralize decision making primarily to money market departments, rather than to traditional departments such as loan departments, because in the money market departments, immediate decision making is imperative. Information technology and the innovation of financial technologies such as forward operation, swap transaction, and option forwards, make it possible. As another example, Yamato Transport, an express service company, calls their drivers "sales drivers" and enlarges the scope of their jobs. In the past, a sales worker estimated the price for important customers, but now the driver does that job. In both banks and transport companies operational decision-making tends to be carried out at the location of the information source.

The Dynamics of Organizations

There are three methods by which the dynamics of organizations adjust according to the environment (Morimoto, 1987). One of them is "simplification of organization," which means simplifying the vertical/horizontal limitations of task assignment. The elimination of the section(Ka) system is included in this category. Another is "flexibility of organization," which includes flexible task assignment and personnel allocation. Project organization belongs in this category. The other is "multi-dimensional organization", a system based around a vertical central axis, which is acted upon by multi-dimensional coordination. Matrix organization belongs in this category. Network organization, which recently is being discussed, also belongs in this category.

According to our survey, many companies do not believe that information technology affects the simplification and multiplication of organizations, while some companies believe that information technology affects the flexibility of organizations and others do not (Table 6). We also asked whether vertical/horizontal communication, which seems to promote the dynamics of organizations, was strengthened by information technology. From the results, we can conclude that many companies feel that communication was strengthened (Table 3). In particular, many companies pointed out strengthening of horizontal communication. This is because of telecommunications technology, such as electronic mail, LAN, and teleconference, and a consolidated database that many people can share.

For example, in the period of batch processing during the first information system, each department, such as the sales department, production department, and accounting department, had different data and communicated with the other departments through meetings. The data often conflicted concerning timing, range, and mistakes, so they argued on different planes, missing the point because of the inconsistent data. Moreover, some departments sometimes deliberately made "mistakes" in order to gain a departmental advantage. However, the consolidation from sales/production/inventory files into a common database and real time processing solved these problems.

Budget planning is another example. Nowadays in many companies, end-users can plan a budget using spreadsheet language. Mutual understanding between departments has been enhanced since the introduction of this system, since files can be transferred and inquiries made easily by electronic mail when the original budget plan is made or later modified.

However, in spite of this strengthening in communication, the simplification of organization is not being done. This seems to be occurring for the same reason that the number of levels is not decreasing, as discussed previously. The companies which believed information technology affected the simplification of their organization were Dai-ichi Kangyo Bank, Sumitomo

Question	Answer	Number of companies	Percent of total
Simplification of organization(e.g., elimination of section system(ka))	1)Yes	8	25.8%
	2)No	20	64.5%
	3)Don't Know	3	9.7%
	4)No answer	0	0.0%
	Total	**31**	**100.0%**
Flexibility of organization(e.g., project organization)	1)Yes	13	41.9%
	2)No	13	41.9%
	3)Don't Know	5	16.1%
	4)No answer	0	0.0%
	Total	**31**	**100.0%**
Multiplication of organization(e.g., matrix or network organization)	1)Yes	4	12.9%
	2)No	16	51.6%
	3)Don't Know	10	32.3%
	4)No answer	1	3.2%
	Total	**31**	**100.0%**

Table 6. Dynamics of Organizations

Bank, Dai-ichi Mutual Life Insurance, Eisai, Yamaha, Nippon Seiko, Kao, and Marui. The relation between information technology and the flexibility of organizations has the characteristics (that the development of information technology itself created) of many multi-functional products made by computers and telecommunications. For example, these include such products as LAN and EPBX in the computer and telecommunication industries, electrical and mechanical engineering products in the electronics industry, the complex products used by travel agencies, and cross-business products at banks and nonbank financial institutions. Research and development, production, and marketing of these products make cooperation between each department indispensable and promote the flexibility of organizations like project organizations. At the same time, information sharing through a consolidated corporate database contributes to efficient project organizations.

Then, many companies in the survey did not feel that information technology affected the multiplication of organization. One of the reasons for this could be the fact that so few companies in the survey had practiced multiplication of organization. Only 4 companies believed that information technology affected the multiplication of organization (NEC, Sony, Sumitomo Bank, and Yamaha). Most matrix organizations employ two dimensions; product organization/market organization and functional organization. The relation between information technology and multiplication of organization seems to be the same as flexibility of organizations that I explained above.

The Number and The Role of Middle Managers

There were few companies which felt that the number of middle managers had decreased because of information technology, but many companies felt that the role of middle managers had changed (Table 7). Only 6 companies felt that the number of middle managers had decreased (Honda Motor, Kao, Nippon Seiko, Sumitomo Metal Industries, Suntory, and Dai-ichi Kangyo Bank).

For example, at the Dai-ichi Kangyo Bank, during the ten year period from 1975 to 1984 the number of transactions increased by 75%, while the work force was reduced by 35%. At the Sumitomo Bank during the same 10 year period the number of transactions doubled, while the work force was reduced by 50%. Both companies have not laid off workers, of course, but have compensated instead by reducing the ratio of the number of new recruits to the number of natural retired employees each year. Most of the routine jobs were replaced by machines, and the number of clerical workers decreased substantially, mainly in the branch offices. To offset the reduction in clerical workers, the number of sales workers at the branch offices and specialists at the headquarters increased accordingly.

Furthermore, every function, such as deposit, loan, exchange, or

Question	Answer	Number of companies	Percent of total
Decline in the number of middle managers	1)Yes	6	19.4%
	2)No	22	71.0%
	3)Don't know	3	9.7%
	4)No answer	0	0.0%
	Total	**31**	**100.0%**
Change in the role of middle managers	1)Yes	15	48.4%
	2)No	13	41.9%
	3)Don't Know	2	6.5%
	4)No answer	1	3.2%
	Total	**31**	**100.0%**

Table 7. The Number and The Role of Middle Managers

foreign exchange was represented by its own section(Ka) at every branch office in those days. Then, multi-job workers increased by using multi-functional terminals, resulting in the unification and reduction of jobs. On the contrary, at the headquarters the number of middle managers did not decrease, but the section(Ka) system was abolished and a team system was adopted for managing change with flexibility.

Kao and Honda Motor reduced the number of middle managers by reducing or abolishing the function of branch offices. For example, Kao was able to reduce the number of middle managers from 80 to 26, a two thirds reduction, because of introducing CIM at the Wakayama factory.

A summary of the reasons why information technology caused a decline in the number of middle managers is as follows.
1. By integrating the computer and the telecommunications, direct communication can be achieved between a sender and a receiver of information, or between a worker proposing problems and a final decision maker. 2. As consumer needs become more diversified, information load increases at every level. Then, the level as a vertical point of information-exchange becomes a hindrance for coping with increased information load, and bypass communication is done.

However, those companies in which information technology affected a decline in the number of middle managers were in the minority. The reason for that seems to be similar to the reason information technology did not cause a decrease in the number of levels. In addition, another reason is that the role of middle managers has changed. In the traditional management style, a large part of the middle manager's time was spent managing subordinates. By replacing routine jobs with information technology, clerical workers became more specialized and more self-motivated. Moreover, middle managers were required to develop a working environment which enhanced self-direction on the part of the subordinates, and middle managers themselves were requested to specialize.

In other words, the role of middle managers changed from a vertical point of information-exchange to one of contributing as a point of information-node in unit organization. Then, as sharing information advances, unstructured decision making will become more important. Therefore, the following roles of middle managers will gradually decline; the routine gathering of source data, mechanical reporting to top management, and acting as a vertical point of information-exchange to transmit the directions of top management to subordinates.

Main Change in the MIS Department

The Function of the MIS Department

The MIS department has various functions; data processing service, information retrieval support, maintenance and update of applications, consulting, end-user training, improvement of business systems, strategic planning support, coordination, and systems development. During the first information system, a great deal of importance was placed on data processing. However, during the second information system, systems development, strategy planning support, and information retrieval support became more important (Shimada, 1989b). The reason why systems development gained in importance was that it is essential to End-User organizations, and also that backlogs increase so that this function must solve the problem of backlogs. The function of strategic planning support is one which uses information technology, SIS (Strategic Information System) and CIM (Computer Integrated Manufacturing), in order to win in the competitive market. This function is necessary for companies which are already successfully using functions such as data processing services and maintenance/update of applications in order to reform information systems.

The Relations with End-User Organizations

There are 3 types of coordination between the MIS department and End-User organizations (Table 8). Type A is one in which the MIS department leads End-User organizations in information systems development. This type applies to centralized processing and process-oriented design in large scale applications. Next, Type B is one in which End-User organizations themselves develop systems. This type applies to decentralized processing, and process-oriented design in small-scale applications, and at the same time is data oriented, in terms of retrieving corporate database. Finally, Type C is one in which both organizations cooperate. This type applies to the integration of cen-

Type	**A**	**B**	**C**
Coordination Means	MIS department Leads	End-User Organization Leads	Both Organizations
Processing	Centralized	Decentralized	Corporate Integration of Centralized & Decentralized
Medium (Process/Data)	Process Oriented (a large scale application)	Process Oriented (a small scale application) Partly Data Oriented (retrieval)	Data Oriented

Table 8. Types of Coordination between MIS Department and End-User Organizations

tralized processing and decentralized processing, and to data oriented design.

During the first information system, centralized processing systems, such as ADP, IDP, and MIS were mainly developed by an MIS department, because a specialized technique was required for system development, and because hardware was expensive. However, although the staffs of the MIS departments were rich in knowledge of and experience with information technology, they were lacking in the area of business systems. Then, the systems development led by the MIS department often caused conflicts between the MIS department and End-User organizations because of the communication gap between them and the lack of autonomy in the End-User organizations.

Then, during the second information system, ME (microelectronics) technology innovation produced user-friendly and economical machines for decentralized processing. Thus, systems development by End-User organizations rapidly became popular. As a result, many local applications were developed, and the autonomy of End-User organizations increased. However, too much decentralization caused the situation to appear to be out of control, and resulted in wasting information technology resources. Advanced types of OA, such as MML and LAN, were produced to solve these problems, and these systems were aimed at the integration of centralized and decentralized processing. In order to accomplish this, it is required that the MIS department and End-User organizations coordinate with each other. In addition, the more multi-functional the system becomes, such as SIS and CIM, the more coordination becomes necessary. Therefore, the main type of coordination for information systems development has changed from Type A to Type B, and is changing to Type C.

Today most companies have an End-User support division in the MIS department in order to promote end-user computing. Furthermore, when the MIS department develops large or global applications, they employ Type C. The MIS department does not proceed sequentially from the requirements-specification of End-User organizations to the final process, but instead proceeds by building a prototype and repeating changes of the specifications.

The Position of the Mis Department in the Overall Organization

In the days when companies first introduced information technology, information system development belonged to the department(Bu) or section(Ka) such as accounting, general affairs, and planning, and the position was mainly administrated by the sectional manager (Kacho). After that, however, the position gradually was taken over by the departmental manager (Bucho), because as the role of the MIS department increased, the MIS department was organized as an independent organization in the company. Moreover, the names were changed from computer section(Ka)/department(Bu), system section(Ka)/department(Bu), calculation section(Ka)/department(Bu), EDP section(Ka)/department(Bu), office management section(Ka)/department(Bu) to information system department(Bu) or MIS department(Bu) in many companies. In addition, the instances are gradually increasing in which the departmental manager (Bucho) becomes promoted to a director or doubles as a director.

In the past, information technology had less relation to business strategy, but as companies accumulated information resources, the development of SIS or CIM came to be necessary. Moreover, a new position called CIO (Chief Information Officer) has recently appeared. Subordinate to the CEO (Chief Executive Officer), this executive builds and performs a strategy which effectively uses information resources from the standpoint of the company as a whole. So far in Japan there are still few companies which use a CIO, but their numbers are definitely increasing.

In the 31 companies we surveyed, all the MIS divisions are independent divisions, managed by departmental managers (Bucho). It is remarkable that some financial institutions, in which information systems are quite advanced, have even instituted multiple departments, such as the systems planning department(Bu), systems development department(Bu), and operations department(Bu) (Dai-ich Kangyo Bank, Sumitomo Bank, Dai-ichi Mutual Life Insurance, Sumitomo Marine and Fire Insurance, Yasuda Trust and Banking).

Furthermore, as computers and telecommunications become more inter-connected, the trend is for many companies to begin absorbing communications jobs which were scattered in other divisions (e.g., general affairs division) into the MIS department, or into the role of organizing a project team which carries out the role of communications in the MIS department (Marubeni, Marui, Meidensya Electric, Nippon Seiko, Sony, Suntory, Dai-ichi Kangyo Bank, Yasuda Trust and Banking, Tonami Transportation, Topy Industries).

In addition, the companies which have installed a CIO, or which practically have CIO functions, are as follows; Kao, Nippon Seiko, Dai-ichi Kangyo Bank, Tonami Transportation, and Yamato Transportation.

Organizational Form

As information technology is assimilated into companies and information resources are accumulated, the MIS department aims at a profit center instead of a cost center. One of the ways this is done is to set up the MIS department as an independent company, the examples of which have been increasing recently among Japanese companies. Even when the MIS department is not separated as an independent company, the profit center system could be employed. However, many companies conclude that in order to perform this profit center system more effectively, separating the MIS department as an independent company would be the best alternative. The purpose of separating the MIS department is based on diversified business, facilitating personnel management, and promoting cost consciousness. Companies which separate the MIS department range from those which separate all of the MIS department to those which separate only a part. However, many companies leave the job of planning and logical design processes to the parent company.

According to our field survey, although there was no company which separated the entire MIS department, the companies which did separate the MIS department with the exception of planning and logical design processes, were as follows; Isetan, Marui, Sumitomo Metal Industries, Yamato Transport. The companies which have software houses as a dependent company were 14; Dentsu, Japan Travel Bureau, Kao, Marubeni, Meidensya Electric, NEC, Nihon IBM, Sony, Chugoku Electric Power, Dai-ichi Kangyo Bank, Dai-ichi Mutual Life Insurance, Sumitomo Bank, Tokyo Electric Power, Yamaha).

Conclusion

We discussed the impact of information technology on organizations, based on our field survey and divided into End-User organizations and the MIS department. Our findings were that on the whole information technology does not significantly affect the End-User except, as we said, in some advanced companies. On the other hand, information technology does affect the MIS department very much, and the status and the role of the MIS department increase remarkably.

We pointed out the following aspects of End-User organizations; although the span of control is extending and autonomy at the clerical level is increasing, the number of levels is not decreasing. Although unification and elimination of jobs is occurring at the individual level, it is not necessarily occurring at the departmental(Bu) and sectional(Ka) levels. Although the dynamics of organizations does not affect simplification and multiplication a great deal, it does affect flexibility to some extent.

When companies were asked about their predictions on the impact of information technology between now and the year 2001, on the whole the companies agreed that the impact would increase. Many affirmative predictions were made toward decentralization, unification and reduction of jobs at the departmental(Bu) and sectional(Ka) levels, dynamics in organizations, and decrease in the number of middle managers.

Furthermore, the predictions that the organizational structure would change from a hierarchical structure to a flat type, followed by a network type was made largely by those companies in which the impact of information technology on the organizations had been great (Honda Motor, Kao, Marui, Nippon Seiko). For example, Maruta (1989), chairman of Kao commented as follows:

> A new direction in corporate management is renovation of organizations. The vertical and hierarchical structure is no longer effective and a multi-interrelated network is being sought so that many different staff members can better communicate with one another.

Kao places great emphasis on the functional organization management as a living organism apart from the conventional vertically-structured system. Kao's organization is designed to form a highly intelligent group of the entire working force. People can hardly get any job done quickly if they have to go through the long hierarchical passage from foreman to section and then to department head.

We agree with the respondents to the questionnaire that these directions regarding organizational structure will prove to be correct, but disagree with their predictions as to the speed with which the change will occur. We feel it will be the turn of the century before we see these changes actually taking place in the majority of Japanese companies, with the exception of a few progressive companies. In addition, we must realize that the flat type and the network type do not replace the hierarchical structure, but instead are gradually formed to make up for the disadvantages of the hierarchical structure.

Indeed, many respondents believe that companies will not decrease in the number of levels even in the future but advance in the multiplication of organization (e.g., matrix or network organization). In fact, in the companies with sophisticated database and telecommunications, people communicate with each other beyond the hierarchical structure by using new media, such as electronic mail, and formally/informally, vertically/horizontally, and inside/outside the country. However, a hierarchical structure definitely exists in terms of critical reports and decision making such as budget planning and personnel affairs.

In order to realize network organization, a worker will need to have 3 terminals as a base of operations; one in the office, one in the home, and one portable one to be used in the car or train. However, even though information technology develops rapidly, organizational change will advance slowly because of the tendency of human beings to cling to familiar ways, or because of conflict between constituent members in the organization. In other words, network organization will come about later due to the fact that human organization does not change easily.

References

Anonymous. (1983, April 25). A new era for management. *Business Week*, 50-80.

Anonymous. (1984, February 6). The recovery skips middle managers. *Fortune*, 112-120.

Anshen, M. (1960, November-December). The manager and the black box. *Harvard Business Review,38*(6), 85-92.

Applegate, L.M., Cash,J.I, Jr, & Mills, D.Q. (1988, November-December). Information technology and tomorrow's Manager. *Harvard Business Review, 64*(6), 128-136.

Attewell, P., & Rule, J. (1984, December). Computing and organizations: what we know and what we don't know. *Communications of the ACM, 27*(12), 1184-1192.

Gibson, C.F., & Nolan, R.N. (1974, January-February). Managing the four stages of EDP growth. *Harvard Business Review, 52*(1), 76-88.

Goleman, D. (1988, February 7). Why managers resist machines. *The New York Times*, p.8.

Huff, S.L. (1986, November). Computer impacts on organizations -unanswered questions. *Business Quarterly, 51*(3), 16-17.

JIPDEC (1989). *Information White Paper.* Tokyo: Japan Information Processing Development Center.

Keen, P.G. (1988). *Computing in Time.* New York: Harper & Row.

Kraemer, K.L., King, J.L., Dunkle, D.E., & Lane, J.P. (1989). *Managing information systems: Change and control in organizational computing*. California:Jossey-Bass.

Laudon, K.C., & Laudon, J.P. (1988). *Management information systems: A contemporary perspective*. New York: Macmillan.

Leavitt, H.J. (1965). Applied organizational change in industry: Structural, technological and humanistic approaches. In J.G. March (Ed.), *Handbook of organizations* (pp. 1144-170). Califolnia: Irwin.

Leavitt, H.J., & Whisler, T.L. (1958, November-December). Management in the 1980's. *Harvard Business Review, 36*(6), 41-48.

Markus, M.L., & Robey, D. (1988, May). Information technology and organizational change: causal structure in theory and research. *Management Science, 34*(5), 583-598.

Maruta, Y (1989, Spring). Business management in response to the advanced information age. *Management Japan, 22*(1), 3-9.

Morimoto, M (1987). *Keiei soshiki ron* [Management organization theory]. Tokyo: Nihon Hoso Syuppan Kyokai.

Nakatani, I (1984, January 24). *2001 nen no Kigyo Syakai* [Company society in 2001]. Nihon Keizai Shinbun.

Nolan, R.N. (1979, March-April). Managing the crises in data processing. *Harvard Business Review, 57*(2), 115-126.

Nolan, R.N. (1984). Managing the advanced stages of computer technology. In F.W. McFarlan (Ed.), *The information research challenge* (pp.196-216). Massachusetts: Harvard Business School Press.

Nolan, R.N. (1987). *Soshiki henkaku ni motomerareru toppu no yakuwari* [Organization transformations: Executive leadership]. In Nolan, Norton & Co.(Ed.), 21 seiki no kigyo wo mezashite: *Information Technology* niyoru kigyo henkaku wo mezashite [Building the 21st century corporation: Redesigning Business with information technology(pp.15-47).

Okubayashi, K (1990, April). ME Gijyutu Kakushin to Kigyo no Soshiki Kouzoo [Microelectronic technology and organizational structure of business enterprises]. *Soshiki Kagaku, 23*(4), 58-65.

Pfeffer, J. (1982). *Organizations and Organization Theory.* Massachusetts: Marshfield.

Rockart, J.F., & Long, W.D. (1988). *Executive Support System: Top Management Computer Use.* California: Irwin.

Scott-Morton, M. (1986, April). *DSS revisited for the 1990s.* Paper presented at the conference of DSS 1986, Washington,D.C.

Shein, E. (1961). Management development as a process of influence. *Industrial Management Review, 2*, 59-77.

Shimada, T. (1988). *Nihon-teki OA no koso to tenkai* [The perspective and development of Japanese style of Office Automation]. Tokyo: Hakuto Syobo.

Shimada, T. (1989a). *Information Systems on Local Government.* Tokyo: Hakuto Syobo.

Shimada, T. (1989b). SIS tojo no haikei[Background on the advent of SIS]. In Shimada, T., & Ebisawa, E. (Eds.), *Senryaku teki joho sisutemu: Kotiku to tenkai* [Strategic information systems: an approach and development] (pp.63-86). Tokyo: Nikkagiren Syuppansya.

Wave, J.P. (1987). Information technology no shinten to atarashii soshiki kozo [Information technology and organizations: new structures and management practices]. In Nolan, Norton & Co. (Ed.), *21 seiki no kigyo wo mezashite: Information Technology niyoru kigyo henkaku wo mezashite* [Building the 21st century corporation: Redesigning Business with information technology (pp. 15-47).

Whisler, T.L. (1967). The Impact of Information Technology on Organizational Control. In C.A. Myers (Ed.), *The impact of computers on management* (pp. 18-19). Massachusetts: The MIT press.

Whisler, T.L.(1970). *Information Technology and Organizational Change.* California: Wadsworth.

Yaeger, J., & Targowsky, A.S. (1989, Spring). The flattening of middle management in the computer-based environment. *The Journal of Information Systems Management, 2*(2), 47-50.

Chapter 12

Information Technology in Japanese Organizations:
A Report From the Field

Allan Bird
New York University

William G. Lehrman
Virginia Polytechnic Institute and State University

> "We were never given a choice of whether or not to adopt information technologies. It simply became a requirement of existence in our industry."
>
> "I have 9,000 profit centers within my department."
>
> "What we're doing is creating a seamless link between raw materials suppliers at one end of the production chain and end-users at the other."
>
> "The further down you go (in the organization) the faster the decision making gets."
>
> "Increase in information volume? At one point we we had so much we couldn't handle it all. We started throwing it out."

Vast changes are taking place in the way Japanese firms are carrying out their business operations. In companies where ten years previously we found personnel records kept in cardboard boxes next to a manager's desk, we now discovered intelligent terminals linked to headquarters-based mainframes which provided up-to-the minute data on employees anywhere in the world. In companies where orders previously were placed through personal sales calls, we now discovered global value-added-networks (VANs) which placed orders electronically from anywhere in the world and processed them automatically.

In this chapter we will explore the effects of information technologies (ITs) on the structuring of Japanese organizations. Our objective is twofold: to report on recent findings and preliminary conclusions from a field study of the effects of IT introductions conducted among Japanese firms, and to reflect on the implications of those findings for future theorizing about IT and organizations in general.

The chapter is divided into three parts. In the first section we will define IT, discuss reasons for the recent upsurge in interest in IT, and explore the rationale for studying its effects in Japanese organizations. In the second section we will outline our research program and present our preliminary findings. The final section explores theoretical implications, tying the conclusions of our research effort to the existing body of theory.

Looking for the Field

The quotes above provide evidence of dramatic changes taking place within organizations. Whether we are on the forefront of a societal information revolution, or witnessing the transformation of organizations in response to an increasingly turbulent business environment, clearly IT is a part of that phenomenon.

The research effort at hand, however, is not about the big picture of societal or industrial change. Our attention is focussed more toward the ground than the sky. Quite simply, we wanted to learn about what happens when organizations introduce IT.

Previous speculation has led to the drawing of battle lines. On the one hand, modern-day Luddites decry the introduction of IT as deskilling, demeaning, and potentially life threatening. On the other, technological utopians paint a picture of organizational bliss, efficiency wedded to effectiveness, where employees are freed from the drudgery of routine work to pursue the good life of value-adding, creative activities. Both sides offer persuasive arguments, occasionally buttressed with empirical evidence, but the reasoning and data on either side are far from conclusive. We decided to go take a look for ourselves.

What We Looked for?

Companies are constantly introducing new ITs. Whether it's a new PBX or the latest graphic imaging technology, it is safe to conclude that its appearance in the workplace has an effect on how work is carried out. Nass and Mason (1990) suggest that different types of IT have different types of effects on organizations and that researchers need to account for variation in IT if their findings are to be useful.

While recognizing the utility of their suggestion, we opted for a different tack. Rather than focussing on variations across technologies, we chose to look at wholesale adoption of IT in different firms and different industries. Similarly, we chose to use a definition that would include more, not less.

Information technology has been broadly defined as "the technologies and applications which combine the data-processing and storage powers of computers with the distance-transmission capabilities of telecommunications (Child, 1987: 43)." Using this definition, one could conclude that IT has been widely used within organizations since as far back as the 1960's. At that time many firms introduced large mainframe computers to handle the electronic processing of data related to bookkeeping and accounting activities. Data was gathered and key-punched then transmitted to the mainframe through dumb terminals, some of which were located a great distance from the mainframe. However, the cost, bulk, and general usability of IT in its early stages discouraged its application to strategic purposes.

Subsequent technological improvements along five dimensions have enhanced the economies of IT and increased the range of contexts in which it can be applied. *Speed* of transmission and processing has surged, with firms moving beyond batch processing approaches to real time processing. A plant manager, for example, can now watch as an order is placed, raw materials and parts ordered, and a production run scheduled. *Accuracy* of data coding and processing has improved as intermediate data handlers and steps have been removed. For instance, point-of-sale (POS) systems adjust inventory figures at the moment an item is sold, eliminating the need for shelf counts and the possibility of errors and delays. The *power* of data processing has reached the point where small personal computers (PCs) are able to process and analyze larger volumes of data than did the mainframe computers of just ten years ago. One firm we visited had installed over 3,000 PCs in the past year, each with the capacity to independently gain access to the headquarters mainframe, extract data from it, and analyze extracted data. *Capacity* for data storage and transmission has also experienced major advances with the appearance of digitized fiber optic networks and new media for data storage and retrieval. This enhanced storage and transmission capacity is making it possible for companies to create huge databases linking customer profiles and purchasing habits with inventory controls and production schedules. Finally, *access* to information has grown as the cost of installing PCs, smart terminals, and a variety of telecommunication systems have plummeted. In a number of companies we found that personnel from top executives down to ordinary employees are able to access all information in storage except for sensitive data relating to personnel files and client financial data.

Each of the above dimensions has implications for organizational structuring. A few examples will suffice. Increased speed allows managers to more closely monitor their own performance, that of their subordinates, and possibly that of their colleagues and superiors. Heightened accuracy in data processing can lead to the transformation or elimination of various jobs related to information monitoring and controlling functions within organizations. Enhanced computing power may encourage decentralization of decision making, as managers at lower levels in organizations are able to do more with less direction from above. Enlarged capacity for storage can result in the creation of new sections charged with finding uses for the accumulated data. Enlarged transmission capacity may encourage more frequent and different types of communication both within and between organizations. Finally, expanded access to IT suggests potential shifts in power relationships as personnel are more aware of and have greater knowledge about the performance and behavior of other organizational factors.

Why Japan?

The question remains, however, of why study information technology effects in Japanese organizations. Historically, the majority of innovations in IT have appeared and first been implemented in the U.S. From mainframe computers with geographically dispersed terminals to local-area-networks (LANs) to POS systems, adoption and diffusion has occurred first in the U.S. Being early adopters, the experience of U.S. firms is both longer and more extensive, thus the the U.S. may provide a more comprehensive perspective of IT effects than Japan.

Nevertheless, a study of IT effects in Japanese organizations has value for several reasons. From a practical perspective, Japanese firms are major players in the world business community and we need to know more about them and how they are influenced by IT. Japanese firms are the pacesetters in a broad range of industries (Abegglen & Stalk, 1986). In 1990, eight of the fifty largest industrial corporations in the world were Japanese (*Fortune*, 1990). In electronics, where they have established themselves as world leaders on the basis of sales, Japanese companies are poised to takeover technological leadership (Riddell, 1990). The ten largest commercial banks in the world in terms of funds deposited are Japanese, as are seventeen of the twenty-five largest (*American Banker*, 1990).

Japanese firms are perhaps the most prominent developers and producers of IT products, and thus are intimately aware of the organizational effects of IT through both their own experiences and those of their clients. Companies such as NEC, Fujitsu, Hitachi, and Toshiba are strong forces in the global computer and telecommunications markets. Omron is a major player in control systems and equipments. Murata Machinery and Okuma are pacesetters in production-oriented IT such as flexible manufacturing systems (FMS), computer integrated manufacturing (CIM) and just-in-time (JIT) inventory control systems.

Focussing on Japanese firms may have even more significance from a theoretical perspective. Prior theoretical (Aoki, 1988; Ouchi, 1981) and empirical (Kagono, Nonaka, Sakakibara, & Okumura, 1985) work strongly suggests that Japanese organizations differ from their U.S. counterparts in distinctive and important ways. Pascale and Athos (1981) suggest that by looking at these differences, through comparison and contrast, we can learn more about U.S. firms in particular and about all organizations in general. This may be particularly so with regards to the effect of IT on organizations.

Much of organizational theory is predicated on a view of organizations as information processors. March and Simon (1956) direct attention to the bounded rationality of decision making, noting that information acquisition and usage are shaped by organizational goals and human capacities. Cyert and March (1963) elaborate on this theme, stressing the selective and truncated nature of information searches and the political context of competing coalitions within which information is interpreted. Weick (1978) develops more fully the interpretive perspective, contending for a conception of information as an input and outcome of socially constructed views of reality. The introduction of IT has had demonstrable effects on organizational information processing and decision-making activities by enhancing and supplementing human capacities, by allowing more extensive information searches, and by altering the nature and volume of informational inputs.

Both computers and organizations provide means of information processing, decision and control. It has been suggested that, to a large extent, computers and organizations are homologous, i.e., both are information processors. Beniger states, "... formal organizations might be seen as the best known means to make a computer before the development of electronics, or as a prototypic computer that uses multiple brains as components, or as the computer that results from the coupling of various brains by means of formal rules and channels of communication" (Beniger, 1990:32).

However, just as there are differences between types of computers in the ways in which the above-mentioned functions are performed, so there are differences between organizations. And some of the most important differences cleave along national lines. In contrasting Japanese and U.S. organizations, Ouchi (1981) establishes a distinction between control predicated on hierarchy and control predicated on clan. He suggests that informational requirements will differ between hierarchical and clan organizations.

Empirical evidence supporting fundamental differences between Japanese and U.S. organizations is provided by Kagono, et al. (1985). In a comparative study of 227 U.S. companies and 291 Japanese companies they found statistically significant differences between the two countries' firms on a variety of strategic, structural, and process factors related to organizational adaptation. In terms of information-related behaviors they found several major differences that are pertinent to our research.

Japanese firms tend to rely on extensive exchanges of information and values between individuals and between relatively autonomous groups to achieve effective organization. Frequent interactions which ignore hierarchical and functional boundaries facilitate informational exchange. By contrast, U.S. firms lean toward a reliance on hierarchy, emphasizing authority relationships, specialization of roles, and standardized policies and operating procedures to achieve organizational effectiveness.

The strategic orientation of Japanese firms is characterized by an internal focus. Their concern is with efficient internal operations and they place strong emphasis on production. In contrast, U.S. firms adopt a product orientation and are characterized by an external, marketing focus. Their concern is with effective products and responsiveness to market forces. The result is that while Japanese firms seek information relevant to their day-to-day operational activities, much of which is acquired through direct contacts with customers and clients, U.S. firms seek information relevant to strategic activities, much of which is quantitative and can be acquired independent of customer or client interactions.

Many IT systems are designed to handle routine activities such as those found in a firm's operational activities. Japanese firms, with their abundance of information sharing activities and their focus on operational information would seem fertile ground for IT introduction and proliferation.

Kagono, et al. (1985) also found important differences in the structuring of Japanese and U.S. firms' external relationships. Japanese firms exhibited a far higher frequency of reliance on networks, establishing networks between buyers and suppliers, networks within channels of distribution, and networks of spin-off subsidiaries. Japanese firms' extensive experience with interorganizational networks suggests that they may be well-positioned to take advantage of telecommunications networks in their external relationships. To continue Beniger's (1990) line of analysis, Japanese interorganizational networks may represent an interpersonal or institutional counterpart to LANs. If that is indeed the case, then introducing telecommunications and computer links among firms would be a straightforward proposition, essentially consisting of putting into cable and computer chip what already exists in personal contacts and formal contracts. IT presents the prospect of increased efficiency in communication activities and expanded volume of communication within established or emergent networks. The establishment of physical connections of cable and terminal may, by their concrete nature, solidify network relationships.

Because of these differences, some argue that Japanese firms are better positioned to take advantage of IT and realize its advantages, while avoiding some of its pitfalls (Barras, 1990). It is not our intention in this chapter to explore the comparative advantages of Japanese or U.S. firms in exploiting IT. Nevertheless, the possibility that Japanese organizations may be better able to obtain competitive advantages through IT offers a final rationale for looking at Japanese firms.

A Trip to the Field

We undertook our research effort with a single question in mind: What effect does IT have on Japanese organizations? To make pursuit of an answer to this question manageable, we broke it down into two 'smaller,' but broadly stated questions: 1) What effect does IT have on the internal structure of Japanese organizations? 2) What effect does IT have on the external relations of Japanese organizations? By "effect on internal structure" we mean changes in the nature, location, and speed of decision-making, realignment of internal communication and reporting structures, changes in departmentalization (i.e., creations, mergings, downsizings, or cancellations), and changes in job roles. "Effect on external relations" refers broadly to changes in a company's network of inter-firm relationships with buyers, suppliers, distributors, subcontractors, competitors, subsidiaries, and affiliates.

Our objective was to learn as much as we could about the recent experiences of major Japanese firms in introducing IT. We focused on environmental stimuli encouraging the adoption of IT, organizational expectations regarding the benefits and costs of IT adoption, actual changes in organizational structures and processes contemporaneous with IT adoption, and subsequent outcomes. Our intention was to look at organizations as whole entities, rather than focus on a single managerial level or a particular divisional or functional subunit. At times our attention was directed to experiences in a specific section or job category. Even then, however, our interest was in the larger issue of what these experiences meant for the organization as a whole.

Planning the Trip

Deciding What to See. We began by identifying the ten largest firms in each of four major industries listed on the first rank of the Tokyo Stock Exchange (hereafter, TSE). Public domain data is particularly important in doing research in Japan, where acquiring information is viewed as a virtue, imparting information a waste of time (Halberstam, 1986). The availability of public domain information was insured by selecting firms from the TSE. This also negated the need to rely exclusively on company sources for data. The forty firms in the survey represent approximately three and a half percent of the firms listed on the TSE. The firms were not randomly selected, but were chosen through a concerted effort to capitalize on what was already known about the population (Kidder, 1981).

The selection of the four industries—food processing, electronics, large retail, and banking—required special consideration. Barras (1990) postulates a relationship between industry and the degree to which firms are able and willing to implement IT. The selection of firms was an important dimension of the research design. Failure to consider the mix of industries could result in misleading conclusions. Careful selection of industries increases the likelihood that subsequent findings will be more applicable across industries. The challenge was to determine the appropriate number of industries and, within industries, the appropriate number of firms to study.

There are 1,129 firms listed on the first rank of the TSE. Of these, 700 are manufacturing and 400 are service firms. To keep the total number of firms manageable but broadly representative, a split of two industries in each sector was chosen. To include fewer than two industries from a sector would greatly increase the risk of non-representativeness. The two industries in each sector, food processing and electronics in manufacturing and large retail and banking in service, were selected because they were evaluated as differing substantially from each other along dimensions of firm age, size, maturity of the sector, and environmental turbulence. A brief description of each industry follows.

Food Processing. The food processing industry can be characterized as a mature industry, domestically oriented, and populated by relatively small firms. Firms range in age from 30 to 80 years old, with a rough division between those founded near the turn of the century and those founded in the late 1940's. Industry development has been leveling off, with growth steady but slow . The industry experienced solid expansion through the 1960's and first half of the 1970's as consumer incomes rose. The first oil shock in 1973 raised factor costs, particularly energy costs. Because many of the raw materials in the industry are imported, in the latter half of the seventies and into the eighties factor costs continued to rise slowly as the value of the yen appreciated. As a result of cost increases and a maturing of the industry, competitive pressures have moved in two directions. To move up the value curve and offset higher factor costs, firms have been introducing new, higher quality and higher priced product lines. They have also maintained their low-end product lines and competed on the basis of price. The result has been steady, but slow, growth in profits. The food processing industry was included as one of the four industries because it reflects that portion of the manufacturing sector comprised of small, mature, and relatively stable firms.

Electronics. The electronics industry represents a manufacturing industry in sharp contrast with food processing. Here firms are relatively large (Matsushita Electric is the largest electronics manufacturer in the world), heavily involved in international markets and foreign direct investment, and somewhat older. Many of the firms were founded in the 1930's during Japan's first economic miracle. The period from the 1970's to 1980's represented a time of solid growth, despite two oil shocks and the appreciation of the yen. This is reflected in an average annual growth in profits of close to twenty percent. Though growth has been strong, with the exception of a few iconoclastic firms such as Sony and TDK, most of the firms in this industry were found to have defender strategies (Bird, 1989). Here the marketing strategy was to first retain market share, then explore possibilities for expansion. Firms in this industry can be divided along geographical lines between the Osaka and Tokyo areas. The electronics industry was chosen because it typified that portion of the manufacturing sector comprised of large, growing firms operating in an environment of more turbulent change.

Large Retail. The large retail industry, one of our two service industries, is actually two sub-industries in the process of merging. One sub-industry consists of the old, large, well-established department store chains typically located in central downtown districts in large cities such as Tokyo, Osaka, and Nagoya. The second sub-industry consists of young, aggressive superstore chains (combination department/grocery stores) usually found in the suburbs and near larger railway stations. They are spread out over large areas of the country, less geographically concentrated than the department stores. In the mid-1970's, as city-dwellers were moving to the suburbs and the incomes of families, both urban and rural, were rising, superstores and department stores found themselves on a collision course. In order to stay close to consumers, department stores found it necessary to move out to the suburbs. At the same time, superstores found it necessary to upgrade the quality and price of their department store goods as incomes and living standards rose.

The result of merging competition is a volatile large retail industry. Of the ten firms included in the sample, four are department stores and six are superstores. The distinction between them is now sufficiently blurred that even executives from either type of store view the collection as one industry. Because of the merging of the sub-industries, growth has been low and performance varies widely across firms, with some firms experiencing declines. The general strategy of both types has been to imitate the other. Because Japanese firms operate as internal labor markets, however, neither type possesses the managerial talent in-house to engage in imitative strategies. The result has been extensive headhunting among each others' managerial ranks. The superstores have generally been more aggressive in both market strategy and recruiting practices. Their actions have effectively forced department stores to follow suit. The large retail industry was included in the sample because it typified that portion of the service sector comprised of progressive firms operating in a turbulent environment.

Banking. The banking industry can be characterized as a conservative, mature industry comprised of older firms experiencing excellent growth in a relatively stable environment. Because of the success of the Japanese economy, the banking industry experienced high growth between 1973 and 1989. Because the industry is tightly regulated by the government, firms are unable to engage in highly competitive market strategies. Industry analysts depict firms in the industry as classic bureaucracies with defender or analyzer strategies (Bird, 1989). The one area where banks do compete is in profit performance. As a result of liberalization by the Ministry of Finance over the past five years, competition within the industry has been heating up. Foreign banks have pushed for a greater presence in Japan at the same time that Japanese and foreign security firms have begun to encroach on areas previously regarded to be the sole province of banks. Continuing liberalization will ensure a further increase in competitive behavior. The banking industry was included in the sample because it typified that portion of the service sector comprised of large firms operating in a stable, munificent environment.

In Pursuit of Sights to See

After their initial selection, we contacted each of the 40 firms, indicating our interest in their IT-related activities and requesting annual reports and other informational materials related to their current activities. We also solicited interviews with key managerial personnel familiar with IT activities. We received responses and materials from 25 firms, 16 of which consented to be interviewed. Of the remaining nine firms, one asked for permission to respond to interview questions in writing rather than in person, three declined an interview, and five indicated that they were either not active in IT implementation or were about to undertake major IT installations. A number of firms expressed a desire to speak with us after their new IT systems were in place.

We used the Nikkei NEEDS database to search fourteen Japanese business periodicals and newspapers over the previous 18 month period to supplement the information we received directly from the companies in our sample. With this database we hoped to determine the number of times firms had been mentioned in conjunction with IT-related activities.

It is common in Japan for business periodicals to closely monitor individual firm activities. For example, when a department store installs POS terminals in a store this is frequently noted in an industry journal such as *Nikkei Ryutsu Shimbun* (Nikkei Distribution News). Thus, it was possible to estimate indirectly the general amount of IT-related activity of a given firm simply by noting the number of citations in this database. Not surprisingly, a few electronics firms individually garnered in excess of 1,000 citations for the 18 month period. By contrast, in the food processing industry there were only two active firms, each of which was cited more than fifty times.

Of the 16 firms that participated in the field study, six were in electronics, five in large retail, three in banking, and two in food processing. The two food processing firms, perhaps not surprisingly, were the same two that were most cited in the NEEDS database. A similar pattern appeared to hold across the other three industries: active firms were more willing to participate than less active ones.

The Adventure Gets Under Way

Company visits were carried out over a three week period in July of 1990. On average, the companies we visited provided three managers for the interview, although on occasion we spoke with as few as one and as many as eight. Usually we spoke to the manager of the firm's information systems department and one or more lower ranking managers. Interviews averaged one-and-a-half hours in length and followed a structured list of questions. Questions were structured but open-ended, and interviewees were encouraged to elaborate on their responses.

Our questions focussed on the most recent major IT adoption. Nearly all of the firms we interviewed indicated that sometime during the 1980's they had implemented a company-wide IT system. These implementations did not represent first-time experiences, rather they were attempts to upgrade, standardize, and unify disparate systems operating in different divisions, and at different levels.

The IT experiences of the companies we studied followed a certain pattern. Their first experiences with IT had come during the mid-1960's when they adopted computerized accounting and bookkeeping systems. In the early 1970's systems were brought on-line, replacing the batch approach previously employed. During this same period new information systems were introduced for inventory and manufacturing activities. In the late 1970's and early 1980's problems of communication between information systems in different functional or divisional areas or in geographically distant locations spurred firms to explore ways to unify systems. Coinciding with this development was the wave of technological improvements in IT mentioned earlier. Consequently, though all firms were continually engaged in upgrading and redesigning their information systems, system unification represented a watershed event.

In some cases this implementation entailed massive redesign and restructuring of existing systems. In others it constituted an integration and consolidation of current systems. In all instances, implementations were preceded by a three to four year preparatory period during which project teams evaluated the requirements of the implementation, developed action plans, and supervised preliminary installation activities. In short, IT system unification was a carefully considered process implemented over a period of years and built on the base of earlier information systems and experiences.

The results discussed in the following section refer specifically to changes associated with major IT implementations.

Back Home with Stories of What We Saw

We begin our survey of the effects of IT on Japanese organizations by considering the forces or pressures influencing firms to engage in major IT implementation efforts. Then we consider the expectations companies had about IT, that is, how did organizations think IT was going to affect them. Next, we consider the nature of the technology introduced. What were its general characteristics and its immediate informational effects. Having laid this foundation we then proceed to explore IT effects in areas of decision-making, departmental change, role change, and interorganizational relations. Finally, we examine the companies' perceptions of outcomes: were their expectations met, what were the unanticipated consequences.

Pressures to Implement IT

There are numerous pressures in the Japanese business environment that encourage a high degree of IT implementation (Barras, 1990). Managers in the firms we interviewed spoke of pressures to implement, expand, or differentiate information systems that came from many quarters, but these can be categorized as the work itself, the organization, and the environment.

In terms of the work and jobs, companies felt the pinch of a tightening labor market, a more demanding work force, and rising labor costs, all of which became more acute in the mid-1980s. In response, companies sought to automate administrative jobs, rationalize dirty jobs, reduce personnel, improve recruitment, and increase the productivity, efficiency, and creativity of their employees. One retailer reported that because of increases in product variety, its cashiers needed to know the price of over 500 items. However, such people were hard to find, and part-timers difficult to train. The firm's decision to introduce POS systems was motivated more by the desire to reduce training costs and check-out errors than the need for more detailed data on products and customers.

Pressure to adopt IT also originated within the firm, for instance, security concerns, shorter product life cycles, the need to increase daily production, and, especially, growth. A manager's comment in a large electronics firm represented many: 'Growth made computers necessary for control,' especially of sales. Companies could not keep up with demand without faster information processing. Another electronics manufacturer stated: "the bigger we get, the more information we have to share, and the more important that we share it quickly." They felt that Japanese firms did not face the same demands to justify automation as did American firms, that perhaps in Japan the drive to automate and rationalize is an end in itself.

Pressures to implement, expand, and integrate IT could also come from the the environment and in many different guises: globalization of markets, easing of monetary restrictions, rising value of the yen, and the changing needs and demands of consumers in a computerized society.

The recent introduction of a broad consumption tax forced retailers to differentiate between taxed and non-taxed items, a task for which point-of-sale technology is ideally suited. IT was widely viewed as a strategic tool in responding to competition. One food processing company said that it competes with other firms on the basis of computer power, terming its IT a "requirement of competition."

Finally, and more diffusely, the technology seems to have imperatives of its own, and to implement it is to become subject to its own logic. The increasing pace of technological innovation and pressure to use information effectively, coupled with the fact that several of the electronics firms created and manufactured IT, added further impetus. One firm, again speaking for many, came to the realization that materials embody information, and that it now needed more information about information.

In general, the environment of the companies we investigated is complex, turbulent, and munificent. Encouragement came from outside sources to implement IT systems, especially for large retailing firms. For them, the product bar-code system is virtually universal, and point of purchase information is available from a central source, the Distribution Systems Research Center.

The labor market in Japan is becoming tighter and more mobile. These conditions prompted firms to either adopt IT, or modify it to meet changing labor conditions. IT holds the promise both of attracting and retaining knowledgeable workers, and making them less necessary for the operation of stores, factories, and offices.

Lastly, the environment of Japanese business is notable for its high degree of organization and inter-connectedness, conditions that may facilitate the sharing of information among firms belonging to the same group, as well as impede such sharing across group boundaries.

Expectations

What did companies expect from their investment in IT? Responses to this question were as varied in their details as the industries and companies we visited, but two strands or directions are discernible: the conservative and the innovative.

By and large, expectations about IT were conservative. That is, most firms expected their investment in and implementation of IT to assist them in doing what they were already doing, only quicker, more precisely, more efficiently, and on a larger scale than ever before. Only a few firms envisioned the radically transformative potential of IT and took actions to accommodate their structures to this potential. More often, structural change was an afterthought, designed not in advance but hastily drawn in reaction to the unforeseen exigencies of a new technology that had been acquired to meet limited ends within the unmodified confines of the pre-existing framework. A food processing firm put it this way: "Japanese organizations don't often set out to change an organization's structure. It happens as a result of taking action. IT gets introduced, then things change."

Indicative of this conservative, non-reflective approach to IT is the experience of another food processing company. They told us that they don't pay much attention to organization structure or decision making. "Our skills lie in coordinating distribution and production. Information helps us do that." They introduced computers not to change people, but to leave things the same but make them more convenient.

Expectations of IT often centered on efficiency, internal and external. One area in which companies sought improvement was the speed and accuracy of their ordering systems. Here, point-of-sale and automated ordering systems assist retailers in getting the "Right item, right time, and right price." Companies also expected to obtain more complete information on sales activities, more than just totals, that would aid in grasping a better understanding of details. Another expectation was the reduction of customers' waiting time. Capturing an ambition shared by others, one electronics firm told us that their expectation was nothing less than to "perfectly integrate sales and production."

External efficiency was another widely-shared concern. One food processing company's expectations included strengthening its inter-firm information network, improving the sharing of information along it, and developing new businesses, such as direct marketing. A bank expected IT to help build the trust and reliance of their customers through its quick responsiveness and provision of information. An electronics manufacturer expected faster reaction to global market changes, reduced lead time, reduced total inventories, and reduced losses from unsold or unfinished goods. But these companies had not lost sight of the bottom line. Several told us quite directly that the real promise of IT was greater profitability through all of the previously mentioned devices.

While the majority of companies we visited seemed to have appended new IT to their previous structure, some evinced more thoroughgoing alterations in anticipation of more thoroughgoing effects. Rather than viewing IT as simply a new tool to help them do the old job better, a few "visionary" companies saw it as putting them into a new ball game, or "dohyo," the sumo wrestlers'' ring. These firms realized that it was vital to switch from their initial relatively passive posture to a proactive, strategic use of information systems. Doing so called for both a change in organization infra-structure and a change in their way of thinking about information systems. A bank manager sums up well the essence of this position: "You can't mix old and new easily, people will prefer the old. You have to remove the old first, then introduce the new." One bank viewed its new information system as the foundation for future improvements. Before installing, it studied its administrative activities and put into place a new organizational arrangement that would enhance the capabilities of information technology. In the bank's view, "It's never just an importation of IT, it's a change in organization that IT helps facilitate." For them, as for other visionary companies, IT was a tool for change, but the organization had to be changed and prepared to accommodate and exploit the potential of IT. An electronics firm told us that information was the focus. They expected their information systems to increase the effectiveness of information usage. Another ambitious firm, a food processor, is re-making itself into an "information company," that is, building a strategy based on information and its strategic uses.

Technology

By the time of our investigation in the summer of 1990 almost all of the firms had fully integrated their information systems. Typically, a firm had already gone through several phases of acquiring, integrating, expanding and updating their information systems. The date of information systems integration ranged from 1980 to 1990, with 1986 being the median year.

Regardless of the uses to which they put information systems, all companies in our sample shared two two things in common: they had invested heavily in information technology, and they generated or absorbed massive increases in information. The 16 firms were unanimous in terming their investment in IT as huge, which is, of course, why they were in our sample. Several complained that their ever-escalating costs did not reduce as the scale of the investment grew, and one electronics firm likened the cost of information systems to "gold eating bugs that strike continually."

In terms of increased amounts of information, managers' estimates ranged from 25%, to two times, four times, ten times, to being "immeasurable." The new information came in much faster than before. For example, one electronics manufacturer told us that before introducing IT data was entered into the system monthly. Now, yesterday's data is entered by the following morning.

The new information was often far too much to handle. "Bakeries may generate 200 new items a month. No one can keep track of all the numbers," complained one retailer, a member of an industry whose near-universal system of product bar codes had generated a smothering avalanche of data. Like many others, they had more information than personnel with the skill to reduce it. Another retailer said that their point-of-sale technology had resulted in ten times more information, "far more than anyone would ever want." One retailer confessed that information now got in the way. They had to figure out what to do with it, and a lot of it was simply thrown out. For another, having data available on-line meant that errors could be corrected in real time, but an increase in information meant an increase in error-checking, and thus an increase in work, particularly in newly computerized sections.

While IT may raise the floodgates of data, the incapacities this causes may only be temporary. A retailer whose information system had been in place longer than most said the quantity of information had actually decreased since the system was introduced. This firm learned just what type of information was necessary to collect, then was able to selectively gain access to it in their data base.

The newly acquired information was different in ways other than sheer volume. Seven firms noted that the new information was more detailed, accurate, and of higher quality. Four mentioned that it had grown in variety. Previously, an electronics manufacturer had only simple, "yes/no" information. Currently, they generate trend data on each part they use, and key performance indicators are now widely available across functional areas. Before implementing IT one retailer had only generic information on the items it sold. Afterwards, its headquarters had information on totals, brands, and even each item.

Information had changed in another way: as it had become more rapidly and widely available, more discrete, and more precise, it had lost some of its meaning. But the system had also became to some a "black box," and this created ambivalence about information.

There are several dimensions to the new ambivalence about information systems. The first derives from an increasing dependence on those who can maintain and understand the new systems. Crozier (1964) discovered in a French factory, an entire organization can become dependent on personnel who personally possess knowledge of how to install, operate, and maintain technology. While the new technologies could vastly increase the rate at which work was performed, they could also immobilize or marginalize and perhaps alienate those who did not understand it. The technology could also stymie new employees trained only on the new system because they lacked the skills and knowledge of how things were done before. IT seems to have created more dependency within the firm. If the system stopped, no one could do his or her job. A retailer explained that by doing the records manually, clerks had a good idea of how things worked. "Now it is a black box. Clerks don't know where the problem is if something goes wrong. Information system people have to work it out."

The second dimension of ambivalence about IT stemmed from something in the nature of the new information. It was not the same. It was remote, shallow, and sterile. "New data doesn't mean as much. If you do it (the job) yourself, you know it better," complained one retailer. Another retailer explained that while all items were coded, now they knew about the code, not about the products it represented. "We're learning more, getting better information but also losing information and learning less." The context of the information and the empathetic understanding of those who have to use it had been stripped away in the process of automation. Information was becoming harder to use as it became easier to acquire.

Some companies found that they now had a greater appreciation of the value of information, and more confidence in what can be done with it. But even these firms faced the problem of how use the new information, especially how to interpret it.

A third vexing aspect of the new IT stemmed from its potential to distribute information quickly and indiscriminately. In the past, information was winnowed and condensed as it rose through the organizational hierarchy, what Beniger (1990) refers to as "preprocessing." Now, with access as close at hand as a computer terminal, information was as thick atop the company as it was where it entered. The essential triage, condensation, and distribution processes could be circumvented through open channels. Some companies maintained they permitted open access to all information except personnel data. Others, citing security concerns, had made information available only on a need-to-know basis.

But open access to information did not necessarily cause problems. At one food processing company any employee can access the database. Access to information is as open as before, but the number of people with access has grown, as has their reach. In the past, headquarters personnel had to know where information they needed was kept, and then go and get it, often physically. Now, their grasp is both wider and quicker.

External access to information tended to be rather limited among the companies we studied. When asked, only five firms called their internal access to information "open." Within companies there was often greater access to information than before, though typically this was constrained by a member's functional area, position in the hierarchy, or both , a situation we have labelled "hierarchical" access. For instance, in one bank all information is classified and only the top management has access to all of it. Below the top level there is tight control, with access limited by division.

IT systems tended to be centralized at the top of the firm or in its headquarters; only three companies had decentralized information systems, and two others described their level of centralization as "moderate." Not surprisingly, the focus of the information system was usually internal. Managers sought first to better understand what was happening, or could happen, within their company. One retailer mentioned that the new information was mostly by mid-level managers (*kakaricho*). Though centralized, we suspect that the new information had less relevance at the top of company than elsewhere. The new information lacks the substance that is especially valued in the higher echelons of Japanese firms. In Japan, top management is said to rely heavily on the "feelings," impressions, and personal obligations, what Matsuda calls "tacit" information, when making decisions. These factors have not yet been entered into formal and discrete information systems, and perhaps never will.

Though the information systems tended to be centralized, there was a good deal of information sharing within companies across departmental, sectional, and geographical boundaries. New information resulted in much more communication and closer ties between headquarters and divisions, strategic business units, and branches; only four firms termed their degree of boundary spanning "limited" or "self-contained." Yet boundary spanning that went beyond the organization to connect it to other entities was rare, called 'extensive' by only one firm, and 'moderate' by another.

Decision-Making

IT brings with it the potential to alter both the decision making process and its location in an organization. Many authors have pointed to the promise of IT in allowing more autonomy and more control to those located lower in the hierarchy, or further from the organization's hub. While the combination of high-speed computers and pervasive communication networks which characterize IT may have an inherent technical capacity to dislodge decision making downwards and outwards from the top of the organization, our field work revealed that decentralization was only one route, and not necessarily the most popular. Five of 16 firms said that decision making had become more decentralized since the introduction or integration of IT, but seven reported no change in location, and two said that decision making had become more centralized (two others reported shifts of an unspecified nature). One bank said that there had been only minor changes in decision making location and that these were comparatively unimportant. A retailer said that, while the amount of information had changed, decision making location hadn't. With information more widely accessible within the company, "People have become less important because more people have detailed information." Seven companies told us that while formal process of decision making had not changed since IT, the actual practice had.

At one electronics firm decisions about routine matters such a reserving rooms and scheduling meetings had been pushed way the hierarchy. Now, "the office ladies do it all." A bank said that there was decentralization at least for data checking, but that tasks were moving from managers (*kacho*) to computers. An electronics manufacturer used IT to shift decision making and responsibility from the ordering to the manufacturing division, and from the factories to centralized steering committees.

We did not find clear-cut evidence of singular trends in the autonomy of decision making. Six firms reported no change, formal or informal, in the autonomy of decision makers subsequent to IT, and two others said that autonomy actually moved higher in the hierarchy. However, eight firms reported that decision makers had more autonomy after the implementation of information systems, though several pointed out that IT was not the sole or primary cause of this shift. One bank told us that its managers were now taking actions more quickly, aggressively, and self-confidently because of better information. A retailer said that their information system allowed them to push the power to change policy and adjust to local conditions down to local stores. Decision making autonomy was given to salesmen and store clerks to help them become more creative. But only product decisions were decentralized. The technology that allowed greater decentralization and autonomy also provided headquarters with a quicker, clearer view of what decisions were being made at lower levels. Another retailer said that while their stores' managers always made decisions, now they were making decisions about more and more detailed parts of the process.

While supervisors can keep closer track of what is happening below them, they can lose their way in the flood of information now pouring in. An electronics firm found that an increase of information and its availability resulted in a need for more help and assistance from others to locate necessary information. While the formal decision making process did not change, the actual behavior of decision makers did in order to cope with the superabundant, speedily delivered data.

For several companies, the influence of IT on decision making autonomy was difficult to identify. One retailer created a new department to use point-of-sale data, which has changed the way other departments interact and make decisions. Managers can now get information whenever they want, which makes it easier to make decisions. Perhaps too easy. A general manager at the headquarters of this company created for himself the new and needless task of changing the price of peanuts twenty times a month.

Information systems can also dissolve autonomy. In one retail chain, the headquarters took from local store managers the ordering of "regular items," which amounted to about 70% of all items. With the electronic information systems and network it has constructed, headquarters has relieved store and department managers of making these "unnecessary decisions."

When it comes to the effect of IT on the speed of decision making, the evidence from Japanese organizations is far less equivocal: it was much faster. Twelve companies said that decision making was much faster, three others reported that, while the acquisition of information was faster, the speed of decision making had not changed. Each type of response tells us something about the nature of IT.

In a fairly typical response, a food processor said that in the past a month was needed for sales figures to come in. Now they come in the next day. The information they need is available in real time, and decision making speed has increased in magnitude. An electronics manufacturer reports that factory lead time was cut from six to two weeks, and weekly plans can be changed on a daily basis. Another electronics firm found that the real pay-off of their IT investment is the speed of information from consumers, orders, planning, production, and distribution, and the ease with which such information can be combined and applied across the entire organization.

Yet faster information does not necessarily lead to faster decision making, as several firms pointed out in their own experience. For instance, a food processor reported that while their sales-orders-production-distribution cycle had been shortened by a third, decision making had not become noticeably faster.

A retailer offered a perceptive insight to this matter: 'The lower you go, the faster decision making has become." That is, decision making speed varies with level of management. At the level of store managers and department heads, decision making speed has been greatly increased because of instant access to information pertinent to local concerns: sales, orders, inventories, and the like. High above in the hierarchy, top level managers have access to the same information, but here the information has less value. Top level management decision making is concerned with issues other than day-to-day operations. Here more than elsewhere data begs interpretation and integration. Here we find more importance attached to the ritual of decision making, which is the prerogative, if not obligation, of the office holder. And perhaps here more than elsewhere, managers bring with them, or are given greater license to indulge, personal biases, contextual understandings, and sentiments.

Despite the benefits IT can bring to decision making in organizations, the speed, the flexibility, the inclusiveness and imperviousness to boundaries, other purposes may be served by maintaining at least the outward appearance of formal and traditional modes of executive action. Matsuda (1990) points out that the qualitative nature of information varies with hierarchical level in the organization. Factual, quantifiable information that is suitable for computer processing and manipulation is more commonplace at lower organizational levels. In the upper reaches of the organization more context-dependent "sentimental" information is used. This information, dependent on tacit understandings, attachments, and obligations, is less susceptible to computerization, thus the decision making processes here are less likely to be hastened by IT.

A dilemma ensues. When decision making speed increases at the level of operations but is no faster at the level of company strategy, the organization soon finds itself strained and stressed. If top level decision making is more form than substance, essentially the ratification of ongoing processes, then the differing rates of decision making speed are probably of no great consequence. However, great strain could be generated within the hierarchy should strategic decision making be charged with guiding operational decisions. As top level decision making slowed relative to that at lower levels, operating decisions would be bereft of strategic guidance.

Departmental and Role Changes

Information technology has had an effect on organization structure, and this was perhaps most obvious in the new departments and jobs that companies had created to house and staff their expanded information activities. Nine firms reported that they had created an information systems department, while several others merged different units to enlarge their information department. Several companies had created temporary project teams to guide the development and implementation of the IT system. One firm told us that they had created "wisemen sections" to figure out how to use new information. The visibility of these departments attested to their increasing centrality in these companies. Several managers we spoke with mentioned that the information systems department had gained power and importance in their company as other departments became increasingly dependent on the collection, processing, dissemination, and interpretation of information.

IT seems to have had an impact on the centralization or decentralization of functions within the firm and the autonomy of departments, yet no single, clear pattern emerged. A food company executive's reply to our inquiry about this matter was a familiar one: "It's hard to say, things are very fuzzy in the organization. Probably there has been change, but it is hard to see at an overall level." And it was hard to predict whether decentralization would occur. One food company told us that real-time networks allowed them to both centralize routine work and decentralize other functions. The clerical functions of branch stores were concentrated in the company's headquarters, and all billing is now handled at a central location.

Simultaneous with this centralization of comparatively routine tasks, some business and sales activities have been decentralized to the stores where, presumably, local managers have a better understanding of local conditions. Sharing and circulation of information increased subsequent to IT implementation. Monthly meetings of branch managers are now conducted by electronic mail; physical meetings now occur only quarterly.

A retailer told us of 'massive' decentralization of merchandising triggered by an information network that connected all divisions and allowed information to flow freely. The firm became much more focused on the market, the product, and the store than ever before. A bank reported that their information system allowed them to centralize forecasting functions that were formerly handled by various sections. As is often the case in complex organizations, an adjustment in the structure of one department required adjustments in others with which it frequently interacted.

Roles and jobs in the companies we visited were affected by IT in several ways: new roles were created, the type of work people did changed, and the number of employees could change. Where we found increased demand for information analyzers, systems analysts, and other information-related jobs. New roles, such as coordinators, master schedulers, chief information officers, and check-up personnel were created to collect, process, analyze, disseminate, and oversee information.

While the importance and stature of information analyzing jobs grew, other information jobs were subcontracted. Such jobs included data entry, software design, programming, and other specialized, but separable, information processing tasks. A retailer found that they had to spend less time training part-time workers who were hired during the peak sales periods.

Loss of certain employees and roles tended to be balanced by additions in other areas. One electronics firms that had just installed a new IT system told us that "Jobs will disappear, but not people." Instead of laying off personnel, they expected an increase in creativity and output. One bank told us that the focus of their information systems department had narrowed and more attention was given to new information products. At the same time, software design and management tasks were subcontracted. In another bank, the information system freed up the full-time work force to develop sales promotions for important clients and scout for new sales opportunities. At the same time, a part-time, temporary labor force was being used, along with computers, to handle the more routine tasks. Though their sales have climbed steadily, a food processing firm found that overall employment remained much as before: an increase in market research and data analysis personnel was coupled with a decrease in clerical workers. After IT integration there was more data to analyze, but collecting it took fewer people because of electronic point-of-sale devices. An integrated information system allowed one retailer to branch into nontraditional areas of merchandising. This resulted in an expanded headquarters staff, but the number of employees in their stores was reduced at a greater rate because of store-based IT. Overall, company personnel was reduced by about 10%, though they did not attribute all of the reduction to IT alone.

Though IT can change the character of work in an organization by changing the profile of the work force from clerical to analytical, Japanese firms are constrained not to change the size of their work force too quickly. An electronics manufacturer found that while IT resulted in a reduction in blue collar jobs, it created a far greater demand for systems engineers and software programmers. Employees increased from about 2800 in 1986 to about 4100 in 1990, however sales were up 70% over the same period. Despite rapid increase in sales, they have held back on hiring. As one manager explained to us: "Japanese companies are limited in their ability to absorb a lot of new employees. So, despite growth, the desire is not to bring on too many, but to train people from within." However, a large retailer cut their accounting personnel and sales clerks by half, reducing their total work force from 10,000 to 8,000 (though not all of that was due to IT).

Accountability and Responsibility

Roles in the companies we studied changed in ways other than relative numbers, type, and location. They changed in character. Specifically, many roles that were modified by IT had become easier to oversee. An electronics firm told us that the issue of accountability was seldom talked about but had probably changed. "You have to clarify jobs for automation," and this increases individual accountability. The new technology brought with it the capability to oversee individual action and to hold individuals accountable. No fewer than ten firms told us that individual accountability had increased as a result of IT or the forces that led to its adoption. More supervision, quicker oversight, new types of monitoring, and stricter controls on behavior were mentioned, though responsibility for an action or decision could only be established after the fact.

In some companies the new oversight seemed rather benign. One electronics firm emphasized that its headquarters strives for standardized reporting arrangements because, with so many having direct links to information, someone must play traffic cop." Another firm said simply that responsibility followed after decision making, and when decision making became more centralized, accountability also went up to a higher level.

But another firm put the issue bluntly: "After making a decision or taking an action people look for responsibility, and IT makes it easier to find them." An electronics firm's comment reiterated the new clarification and standardization that accompanies IT: "There will be more support and advice from other people in the organization. But there will be less possibility of making individual adjustments. Things will become much stricter. People (top executives) will look at daily figures and notice problems sooner and act sooner." But the long reach, sharp focus, and undimmed memory inherent in many types of information systems may clash with the group-based work orientation prevalent in Japan.

Interorganizational Relations

The effect of IT on interorganizational relations could be profound. One electronics firm told us that their information network and its interconnections with other firms was transforming the way they did business. With it, they are "building fences around affiliates, suppliers, buyers." Business has changed: "Before it was contracts, now it is partnerships. Before there were trade-offs, now there are mutual benefits."

IT has the capacity to link individuals, departments, organizations, and networks of organizations. Thus it is not surprising to find, as we did, that one clear result of IT implementation was stronger, tighter, and more numerous interconnections. Ten firms reported closer relations with external buyers, suppliers, and distributors. Other firms spoke of tighter links to clients, quicker response to competitors, and increased numbers of suppliers. Two firms told us that their information systems had the intended effect of intimidating competitors. Others said that their IT investment was now a necessary cost of doing business. Some said that IT made cooperation with other firms easier, more desirable, and sometimes essential, as when the enormous cost of IT systems were prohibitively expensive for individual firms. In these instances, banking in particular, IT resulted in both increased competition and greater inter-firm cooperation.

The task environment of organizations consists of other organizations that can directly affect a firms goal setting and goal attainment. Typically these include customers, suppliers, and competitors (Thompson, 1968). We will consider each sector in this order.

IT has had an impact on the way goods are sold, and on the organizations that buy them. Food processing companies have developed closer ties with the buyers of their goods. "IT has become a basic requirement of stores and wholesalers." An electronics firm told us that the quality of buyers and what they want has changed. "Before they were persuasive. Now they rely on numbers and a rational approach." A firm's IT could have direct benefits for its customers. As one banker explained, "When banks increase their efficiency they increase the efficiency of their clients. And it makes us closer to them."

IT also affected an organization's suppliers. Connections were quicker and denser, and sometimes mandatory. An electronics manufacturer explained their new arrangements with suppliers this way: "We are using a lot of outside parts. When sub-contractors finish parts they want to bring them here. We don't want them to bring it if we don't need it, and we don't want them to make it if we don't need it. This means that we have to work closely with suppliers to see that parts ordering operates smoothly." The introduction of a pervasive and integrated information system could also force suppliers to adopt the same system. For example, another electronics firm put pressure on their suppliers to computerize, which resulted in faster information exchanges and better relationships.

Information systems could cause a vast increase in the number of suppliers a firm employed. Before installing an information system in 1984 one retailer used only about 100 suppliers and placed all orders by telephone. Now, with their own online system linking them directly to over 1,100 suppliers, they receive priority treatment and can order more promptly—before their competitors. Another larger retailer now has over 3,000 companies online for ordering , and 1,000 online for billing purposes. Their ordering cycle is much shorter, but frequency of placing orders has accelerated from twice a week to once a day, twice a day for prepared foods.

Finally, integrated information systems affect relations with competitors. On one hand, IT could be used as a tool of competition, or even intimidation. A bank's information system allowed them to respond very quickly to any moves by competitors and vice versa. On the other hand, IT could spur more cooperative efforts through the sharing of data and the ease of inter-connection. A food company hoped that their information system would intimidate competitors, but they also have to cooperate with other firms in the development of industry-wide VAN standards. "We have an open system, so competitors know what we are doing.

The theme of competition and cooperation was reiterated by a bank which now faces increased competition in information services. They must offer a broader menu of products and services at a lower cost. However, there is increased cooperation among banks. The information systems that interconnect them and allow them to offer new products and services also create the channel for a problem in one bank's system to ramify in others. In addition, banks are working together to avoid duplication of information systems because of the huge investments such require.

IT systems have the capacity to create links both within and beyond organizational boundaries. Given the tradition of interfirm linkages in Japan, the benefits of cooperation when investing in expensive new systems or developing new and distant markets, and the rising cost of IT, we should not be surprised to find tighter interorganizational relations as a wide-spread outcome.

Outcomes: Positive & Negative; Anticipated & Unanticipated

A final topic of interest to us was managers' assessments of the nature and utility of the outcomes associated with IT. We were interested in both positive and negative outcomes, and whether these were expected or unanticipated. In effect, our brief review of IT outcomes describes a simple 2 X 2 table: positive or negative; anticipated or unanticipated. We will begin our brief review with positive, anticipated outcomes.

The managers we interviewed sprinkled their remarks with many and varied examples of how newly implemented or integrated IT created the positive outcomes they had anticipated. But perhaps this was to be expected given the huge investment in time, energy, and money to implement or integrate information systems. Firms mentioned that IT had brought lower labor costs, greater employee productivity, lessened workloads for employees, and more aggressive employees. We were told by various firms that customers were more satisfied, profitability had increased, strategic thinking was more prevalent, operational capabilities had expanded, and new services and products were being offered. Companies benefitted from a heightened awareness and appreciation of the importance of information, as well as more accurate data and quicker acquisition and entry.

A food company executive remarked that IT led to a greater appreciation of the value of information across the organization: "Before there were two groups: inputters and users. Now they are the same." In one bank greater understanding of the importance of information took place throughout, all the way to the top, as employees became more aware of the uses and limitations of computers. For one food company, more information about individual behavior allowed for more detailed and fine grained management.

One retailer found that an integrated information system allowed them to proffer new, nontraditional goods and services such as entertainment, ticket sales, and vacation planning. Another retailer found that a fortified customer data base permitted them to expand their specialty shops. A bank told us their information systems freed full-time personnel to devote more time for sales promotion activities with preferred customers, and that: "All of this has lead to a more strategic focus and to a more active stance. We are more aggressive."

New innovations can spark enthusiasm in a firm. One firm said that an unanticipated positive outcome was the feeling of excitement and anticipation and higher levels of activity that accompanied IT implementation. However, IT was also the source of over-built expectations when the firm's information system could not do all that was promised.

The variety of positive outcomes firms mentioned was impressive, but it is possible that some may be making necessity into a virtue. As one retailer said, "We needed this IT system to compete. We have to be able to do things to keep up, if not get ahead." Another retailer perhaps summed up this view best, saying that discussions of the positive impact of IT on his company were pointless because now no one could do without information systems.

Information technology also had less positive and even deleterious effects on the companies that implemented it, some of which were anticipated, some not. Many new difficulties were essentially logistical in nature and were solved without great difficulty, such as the problem a retailer faced in breaking up bundles of food items that had a bar-coded price as one unit. Other problems had to do with customer resistance to new products that were made possible by the application of IT, such as a banking instrument that did away with vouchers but was rejected by consumers.

Difficulties of a different nature also erupted and their solution has proved more elusive. However, these "problems" might be more accurately termed "conditions" of the new information-dependent era, and companies may simply have to have to co-exist with them. The chief such complaint about IT, and one that was anticipated, though perhaps underestimated in degree, was its huge cost. This concern was voiced across the broad range of companies we visited. One manger said simply, "It burns up money!" Another manager complained that the costs never go down for a new system. One retailing executive noted wryly that the makers of IT systems were doing very well. "They always say, 'Well, you didn't make a lot of money on the last unit, but you'll make it with our new system.'" He wondered aloud whether the "tail was wagging the dog."

The amount, character, and accessibility of data produced by new information systems also created difficulties, some of which might have been expected, others perhaps not. The most apparent and predictable problem was the sheer volume of data. Four firms complained that they now had too much data, and that handling it caused new problems. Perhaps less foreseeable were problems that stemmed from the sterile and limited nature of the data generated by new information systems. Two firms said that while their new information system was quite adept at collecting factual and numerical data, it ignored other types of vitally important information.

With varying degrees of systematization, companies have attempted to capture information that could not be readily computerized: the hunches, intuitions, empathetic understanding and sentiments of managers often gleaned from interaction with customers, long experience, or knowledge of personal obligations. A retailer must respond to sensitive customer needs and pay attention to details. In the past they had direct communication with customers, but no records. Now they must allow for machine intervention which is far quicker but more distant so they have implemented a supplemental system to respond to customers' needs. To one retailer, "A machine is a machine."

Another unanticipated, though familiar, outcome was the dependency the new technology created. More and more functions, jobs, and people became dependent on the individuals or departments that installed and operated the information system. A food processing company told us: "MIS people knew everything. People using the system didn't know its workings. It was a black box."

The pervasive new technology could cause a bifurcation of a firm's work force between users and non-users. Not all employees were comfortable with the new technology, and newer employees sometimes had very little understanding of how the previous system worked. A food company told us that their information system had created "a clear division between users and non-users. Users got a lot out of it. Non-users became isolated." Furthermore, "the allergy was stronger than we thought." People who could not understand the system could no longer do their jobs, and one company found that it had more and more of them. The promise of wider availability and easier access to information, two of the more heralded virtues of IT, could also cause problems. Not only does the greater volume of data make information analysis and interpretation more time consuming, but wide accessibility may make consensus more difficult to achieve. An electronics firm told us that, "Opinion leaders shape employee attitudes. Before, only leaders had information and no none could argue with them." But since their information systems have been integrated, "Everyone has information and everyone wants to decide."

Given the emphasis placed upon hierarchical relationships and consensus decision making (at least in some matters) in Japanese organizations, the democratic access implicit in IT may create new problems as it solves others.

Thinking About Where To Go Next

Reports from the field are valuable not only for telling us about the current state of affairs, but also for the insights they can provide about where to go next. However, before going anywhere, it would be wise to place this trip in the context of other trips, reflecting on how this experience can inform both past and future wanderings. The findings led us to a number of conclusions about IT-organization structure effects that are relevant to a context broader than that of just Japanese firms. Painted in broad brush strokes the conclusions are of two varieties: reflections on the nature of organizations and their approach to IT, and reflections on some paradoxical effects of IT.

What Are We Looking At?

One recurring theme that we found in our study was that of organizational choice. Simply put: organizations make choices about the type and amount of IT they will adopt and about how they will use it. This is not a novel idea. Lucas (1986) notes that one of the most obvious effects of IT adoption, centralization or decentralization of decision making, is a direct function of what the organization or, more appropriately, top management prefer.

Through our interactions with the firms in our sample we were able to identify two fairly stable and recurrent orientations toward IT: visionaries and bureaucracies. Visionaries and bureaucracies differed along several dimensions, as can be seen in Table 1. There did not, however, appear to be any important industry differences, and we found both visionaries and bureaucracies in each of the four industries.

The clearest indication of organizational choice differentiating visionary and bureaucratic firms is found in their purposes in adopting IT. Bureaucratic firms invariably used terms such as "increase in efficiency," "rationalization of administrative efforts," and "reduction in clerical staff." Their principal concern was efficiency: how to work faster, more accurately, and with fewer people. Their primary goals seem to have been getting more of the same kind of work out of the existing workforce and preexisting organizational structures.

While visionary firms were not uninterested in the efficiency dividend of IT, they seemed far more concerned with using IT as a means for other organizational ends. Indeed, we labeled them "visionaries" precisely because managers in these firms had a picture of where they wanted the organization to be and viewed IT as the vehicle to take them there. One manager put it this way:

> "Our belief has always been that if you hire and train good people and give them the right information, they will make good decisions. The problem was that, in the past, they had to know where the information was and then be able to get it. Our new IT system has removed those two obstacles."

Visionary and bureaucratic firms viewed their technologies differently. Bureaucratic firms took a narrow view of the uses of IT. They either ignored broader application possibilities or viewed them as something worth investigating, but in the future. Three of the bureaucratic firms spoke of Strategic Information Systems (SIS), currently a buzzword in Japanese business circles, as something to consider in the next two or three years. Visionary firms had focussed on how to implement SIS yesterday. Our conversations with such firms took place in an atmosphere of excitement: responses were punctuated with animated gestures, blackboards were overrun with diagrams and pictures, managers stepped on each other's comments with clarifications and elaborations. Visionary firms opt for a broad focus on IT applications, searching for unique ways of applying it or new areas in which to apply it.

Besides this difference in orientation, the most visible difference between visionary and bureaucratic organizations is their size. There was a vast disparity in size between bureaucratic and visionary firms, whether measured in sales or in number of employees. In three of the four industries, bureaucratic firms were larger than visionary ones. The lone exception was the large retail industry, where the older and more traditional department stores were bureaucratic but smaller than the younger, more visionary-like superstores.

We speculate that size is important because of what it allows a firm to do. Small firms were in a much better position to try dramatic shifts in IT usage because they had fewer people and physical resources to re-direct. One electronics firms put it succinctly. "Either we made the move now or we didn't make it at all. Another year or two and we would have been too big to even try."

Another distinction between the two approaches can be found in the underlying values of their respective corporate cultures. Visionary firms possessed cultures which were progressive, open to new ideas, and willing to take risks. By contrast, bureaucratic firms tended to be conservative, closed to new ideas and risk averse. Though both engaged in experimentation and employed trial-and-error approaches, bureaucratic firms were more likely to place limits on the levels of experimentation and the degree to which new approaches could differ from existing ones.

Using the Miles and Snow (1978) typology of strategic orientations, bureaucratic firms are, comparatively speaking, defined as defenders and visionaries as prospectors. However, across industries there are important differences which caution against viewing all visionaries as equally prospector-like. For example, what is defined as prospector-like behavior among banks may seem more timid and less venturesome to any firm in the large retail industry. The same is true for defenders. Nevertheless, both types demonstrated behaviors generally consistent with the Miles and Snow (1978) descriptions of these two strategic archetypes.

Another important dimension along which the two differed was the amount of previous investment in IT. Bureaucratic firms had invested huge sums of money in prior IT adoptions and as a consequence felt tied to those technologies. Having spent so much money, they were reluctant to abandon older technology when there was a possibility of exploiting it further through the addition of peripheral equipment or integrating softwares. Visionary firms had spent less money previously and, hence, felt themselves to be in a position to capitalize on the most recent technological developments.

The issue of sunk costs raises a "catch-22" dilemma for firms. As a manager we quoted above noted, IT has become a requirement of continued existence. At the same time, technological improvements are occurring at such a rapid rate that to adopt technology now is to lock into a system that may be inferior to one available six months to a year down the road. The catch is that to wait six months or a year is to cede a competitive edge to those who adopt state-of-the-art systems now.

To avoid slogging into the dangerous inertia caused by sunk costs, visionary firms have tried to pursue a policy of designing flexible IT infrastructures, configurations that can be easily altered over time as a company's needs change. One firm noted that the most challenging dimension of the IT adoption effort, and the one which they believe will have the greatest long-term payoff, was the architecture of IT interfaces they had developed to merge the various IT subsystems. They referred to IT subsystems as "components" which could be rearranged or unplugged and replaced without unduly disturbing the IT system as a whole.

There was nothing inherently different about the types of IT that visionary and bureaucratic firms employed. The differences were in the choices that these organizations made about how to think about them and what to do with them. Choices were shaped by organizational histories, as reflected in culture and strategic orientation, and organizational constraints, revealed in size and sunk costs.

The practical implications of visionary and bureaucratic organizations remain unclear. However, two observations are warranted. First, if society is on the verge of an "information revolution" similar in magnitude and effect to that of the industrial revolution, then visionary firms may have a distinct competitive advantage. The industrial revolution had far-reaching and unanticipated effects on society and the organization of work. Visionary firms, with their wider range of applications, willingness to take risks, prospector strategic orientation, and openness, would seem better poised to take advantage of emerging opportunities and dodge unanticipated threats. Indeed, their adoption of IT appears to be directed toward that end. Second, conflicting findings among previous research efforts into IT effects may be due to a failure to recognize how differences in organizational choice influence effects. Distinguishing between visionary and bureaucratic firms offers one approach to resolving such conflicts.

Going in Two Directions at Once

Notwithstanding our recommendation to consider the type of firm that is implementing IT, we can empathize with researchers who have uncovered or hypothesized contradictory effects of IT. Our search for IT effects frequently led us to opposing conclusions. By the end of three weeks of interviewing we concluded that opposing effects were part and parcel of IT

	Visionary	**Bureaucratic**
Organizational Size	Relatively small	Relatively large
Organizational Culture	Progressive, open, risk-taking	Conservative, closed, risk-aversive
Strategic Orientation	Prospector	Defender
Previous IT Sunk Cost	Minimal	Substantial
Purpose of IT Adoption	Achieve organizational goals	Efficiency
Focus of Applications	Broad	Narrow

Table 1. A Comparison of Visionary and Bureaucratic Organizations

adoption. We will discuss five of these paradoxical effects.

1. Decentralization of Decision-Making with Tighter Supervision and Control

The nature of current IT is that it tends to increase access to information within the organization. Even in conservative, closed firms it was generally the case that personnel at lower levels had access to more information than before. Another way of looking at it is that managers were able to give more information to subordinates, thereby allowing them to make more of the routine decisions that bubble up from lower in the hierarchy. In the case of one large retail firm, purchasing decisions at the departmental level could be turned over to sales clerks because it was possible to provide them, through the firm's POS data, with information necessary to making purchasing decisions.

Simultaneously, because much of the decision-making involved operational activities—purchasing, production scheduling, inventorying, distributing—which were being tracked by the new IT, superiors were better able to monitor the behavior of subordinates, intervening when they felt it necessary. The same POS data that sales clerks used to make purchasing decisions for their department was also being used to monitor their performance.

A particularly striking effect of IT is that by using it the behavior of individuals can be monitored in real time. Where previously there had been a lag between performance and transmission of reports on that performance, monitoring can now be contemporaneous with actual job performance To return to the example of our sales clerk, prior to the installation of POS terminals the departmental manager would submit monthly sales reports to headquarters. The effect of purchasing decisions would not be known there until those reports arrived. If the departmental manager made a bad purchasing decision, there might still be time to modify the outcomes. Conversely, after a good decision, a time lag might allow the manager to heighten the effects. With the introduction of POS, decision-making is pushed down to the sales clerk, and the decisions he or she makes may be reported to and evaluated by higher management almost instantly. This leaves far less room for error and may discourage innovation or risk-taking.

It is unclear whether the primary effect of IT is to increase individual autonomy (sales clerks now have more control in their work), or lessen it (sales clerk performance is now monitored on a daily basis, and possibly a minute-by-minute basis).

2. Speeding up decision-making while slowing it down.

One of the most obvious effects of IT has been to speed up operational decision-making. For example, purchase orders once placed on a monthly or weekly basis are now placed daily or twice daily. In some instances, decision-making has been computerized, with distribution decision rules kicking in automatically when sales hit a certain number, or, in the case of a manufacturer, inventory controls being activated when orders achieve a set level. These effects occur low in an organization's hierarchy.

However, there has been no accompanying increase in the speed of strategic decision-making at the top of the organization. We found no evidence of a quickening of the decision process at the top of the hierarchy. The net effect of this speeding up of operational decision-making while maintaining the existing pace of strategic decision-making is the creation of a gap between the two. This gap widens as operational decisions are made more quickly while strategic decisions are being made at the previous pace.

What is the meaning of strategic decision-making in such a context? Potentially, operational decision making could take on a life of its own, moving the organization ever further down an unplanned path while executives debate what direction the organization ought to be heading. The possibility gives new meaning to the term "emergent strategy."

3. Learning more, while learning less

In most firms, the introduction of IT brought about an increase in information on the order of ten times more than before. As one manager noted:

> "In the past, we could tell you whether or not we had an item in stock. Now we can tell you in real time whether or not we have it and, if we have it, where it is, when it was made, who ordered it, and where its going next."

Companies are now able to acquire mountains of fine-grained information about their own products, competitors' products, and the behavior and characteristics of buyers, suppliers, and distributors. The amount of information they are compiling is truly staggering. Moreover, data storage and computing power have made it possible for them to create huge databases and manipulate the data within them in new and complex ways.

At the same time, firms have less information, and they are losing more each day as older employees retire. A manager in one retail operation complained:

> "These damn bar codes are worthless! We bar code everything now. Thanks to bar codes we can tell you when an item was ordered, when it was made, shipped, delivered, and sold. But I doubt if anyone looking at that data can tell you what it was we sold. They can bring up a description of the item on a monitor. But they don't know what the item looks like, what it feels like. They have a code number for the store where it was sold, but they don't know what the store looks like."

While acquiring sterile data on an item's specifications and economic characteristics, firms are losing contextual data rich in the type of information necessary for the development of products and marketing strategies consistent with the era of "mass customization."

A similar phenomenon is occurring with regard to employees' knowledge of their jobs and the firms they work in. In many ways it is clear that IT-generated information is useful in helping them do their jobs. When bookkeeping activities are automated, bookkeepers are available to work on other tasks, such as monitoring the performance of the computers. However, because computers handle an ever increasing share of the work, employees understand less and less about the automated procedures. This new division of labor can be especially risky when the computers fail, or when new procedures must be added or amended.

4. Reducing ambiguity while increasing ambiguity.

Organizations confront ambiguity along three dimensions (Okumura, 1982): ambiguity with regard to their relationship to their external environment—external ambiguity, ambiguity with regard to the nature of relationships and operations within a complex organization—internal ambiguity, and ambiguity with regard to the dynamic interaction of organizations with their environments—dynamic ambiguity. The acquisition of detailed information about buyers, suppliers, distributors, and competitors made possible by IT systems such as centralized POS databases or VANs reduces external ambiguity by providing data previously available only at great cost in money and timeliness. The use of CIM, FMS, and JIT systems provides information on the internal workings of firms and helps to reduce ambiguity regarding internal relationships. Indeed, one systems analyst noted that until firms clearly identified relationships and operations it was impossible to computerize them or set up telecommunication systems linking them. Databases reduce dynamic ambiguity by retaining data on firm-environment interactions over time, thus permitting the inspection of past trends and the extrapolation of future ones.

But the introduction of IT also creates ambiguity. People are overloaded with information. As previously mentioned, in many firms volume has grown by a factor of ten. Overload creates ambiguity when managers are unsure about what to do with the information, which information is important and which is not, or how to interpret it.

Overload also occurs through information inflow, or overflow. One effect of IT is to greatly increase the number of gateways through which information can enter the organization. If sales reports are submitted on a monthly basis, then gateways through which information can enter the firm are few, stable, and predictable. However, if salespersons submit daily reports via on-line terminals or portable modems, the number of gateways increases greatly and the information entered may be more unpredictable and perhaps less preprocessed.

A third form of overload may occur when there are frequent inputs. Prior to online systems, information was entered by hand in a batch-like process. Batch processing encouraged accumulation of sufficient amounts of data to achieve economies of scale when inputting. Batch processing creates discrete and predictable intervals during which decisions can be made or actions taken. However, the use of online technologies resulted in a shift from batch to continuous processing of data. Discrete intervals have disappeared and it is increasingly unclear what an appropriate length of time is for different types of decision-making. For instance, should prices be set on a monthly, weekly, daily, hourly, or minute-by-minute basis. POS data allows a manager to know immediate results of various pricing strategies. But how long should a price stay in effect before the manager determines that it should be adjusted? One marketing strategist jokingly commented:

> "We can change prices as fast as we want. Wherever it needs to be to make a sell, we can set it in an instant. In fact, if we get real fast, we can sell everything in the store before they have chance to re-stock."

IT reduces ambiguity in terms of what organizations up until now haven't known because it costs too much to obtain, i.e., ambiguity that stems from not having access to information that is theoretically knowable, but practically unobtainable. But it does so by increasing ambiguity about the data it now reaps in abundance.

All information, almost by definition, has strategic value in the right context. And some information has strategic value in a variety of contexts. Given the limited capacity of humans to attend to and process information, the questions the organizations we studied now face are: What should be done with all of the information? How should it be interpreted? How should it be used? To what strategic end?

When information is limited, the range of possibilities is limited and the choice is simpler. As information and informational sources increase, possibilities proliferate and choices become more complex.

5. Competing while cooperating.

The final paradox associated with IT is a product of the increasing cost and necessity of information in a turbulent and resource scarce environment. Many firms seek competitive advantage through superior information acquisition and utilization especially when they experience high levels of uncertainty due to frequent and unpredictable changes in their environments. While resource scarcity leads to increased competition in both factor and product markets, better information regarding either can bestow crucial advantages over competitors. However, the cost of installing IT systems to monitor product and factor markets may be beyond the capabilities of a single firm. Needed information can be obtained only if firms cooperate.

The underlying incentive to cooperate with rivals is the realization that it is cheaper to cooperate on IT than to compete on IT. Scarcity encourages movement across industry boundaries in search of more resources. In the face of encroachment by outsiders, one way that firms can continue to compete against current rivals (and to exist) is to cooperate with them. Japanese banks have experienced this phenomenon over the past decade. Incursions into their domestic markets by foreign banks and into their traditional domestic and international business domains by Japanese security firms has inspired this cooperation.

The relationship between cooperating competitors is ambivalent and potentially unstable. In the case of Japanese city banks, a common ATM system is shared that allows for a variety of transactions including electronic funds transfers. Firms with advanced technology or unique software are reluctant to share them. At the same time, networks are only as strong as their weakest member. One bank official noted that when a rival bank's ATM system went down it hurt his bank and inconvenienced its clients. Yet to provide technical expertise to rivals runs the risk of giving up strategic knowledge and technology or, at the very least, improving a rival's competitive posture.

As IT systems increase in sophistication, complexity, and cost, the inability of individual firms to both pay for and exploit them will further establish the paradox of competing while cooperating.

Conclusion

Japan provides a fertile field for the study of IT effects on organizations. The rapidity with which firms are embracing new IT systems and the various adjustments which they are making suggest a rich environment for further theoretical and empirical forays. Our experience, as demonstrated by the findings presented here, indicate that, as has been the case in recent past, Japanese firms may yet have much to teach us.

References

Abegglen, J. & Stalk, G. (1986). *Kaisha.* New York: Basic Books.

Barras, R. (1990). Interactive innovation in financial and business services: The vanguard of the service revolution. *Research Policy, 19,* 215-237.

Beniger, J.R. (1990). Conceptualizing information technology as aorganization, and vice versa. In C.W. Steinfeld & J. Fulk (Eds.), *Organizations and communication technology* (pp. 29-45). Newbury Park, CA: SAGE.

Bird, A. (1989). *Career advancement in Japanese top management teams: Who gets to the top?* (Working paper No. 89074). New York: New York University, Stern School of Business.

Child, J. (1987). Information technology, organization, and the response to strategic challenges. *California Management Review, 12,* 211-223.

Crozier, M. (1964). *The bureaucratic phenomenon.* Chicago: University of Chicago Press.

Halbertsam, D. (1986). *The reckoning.* Tokyo: Yohan.

Kagono, T., Nonaka, I., Sakakibara, K., & Okumura, A. (1985). *Strategic vs. evolutionary management: A U.S.-Japan comparison of strategy and organization.* Amsterdam: Elsevier.

KJidder, L. (1981). *Research methods in social relations* (4th ed.). New York: Holt, Rinehart, and Winston.

Lucas, H.C. (1986). Managing the revolution in information technology. *Sloan Management Review,* Fall.

Matsuda, T. (1989). *Advanced organizational intelligence for integrated quality management.* SANNO College Bulletin, 10.

Miles, R. E., & Snow, C.C. (1978). Organizational strategy, structure, and process. New York: McGraw-Hill.

Nass, C. & Mason, L. (1990). On the study of technology and task: A variable-based approach. In C.W. Steinfeld & J. Fulk (Eds.), *Organizations and communication technology* (pp. 46-68). New bury Park, CA: SAGE.

Okumura, A. (1982). *Nihon no toppu manejimento* [Japanese top management]. Tokyo: Daiyamondosha.

Ouchi, W. (1980). Markets, bureaucracies, and clans. *Administrative Science Quarterly, 25,* 129-141.

Riddell, P. (1990.June 1). Japan set to take lead in electronics, says US study. *Financial Times.* p. 4.

Staff. (1988a, August). The international 500. *Fortune,* pp. D7-D25.

The top 500 banks in the world. (1990, July 27) *American Banker,* 18a-31a.

Thompson, J. (1968). *Organizations in actions.* New York: Free Press.

Chapter 13

Health Care Information Systems: A Management Perspective

Binshan Lin
Louisiana State University

The health care industry needs an effective information system injection. The medical climate is becoming more receptive, as the numbers of computers increases in physicians' offices and homes. The success of health care organizations will depend much more on how well they understand and manage the information technology. Most health care information systems (HCIS) are employed to monitor health care processes: the collection of patient charges, lab orders without results, or dollars due specific vendors (Fera & Finnegan, 1986). Kim and Michelman (1990) has illustrated four examples of HCIS.

Organizations can be viewed as highly interrelated systems consisting of tasks, people, technology, and structure (Leavitt, 1965). When one of these components is changed directly through the introduction of information technology, the other components will also be affected, as the organization adjusts to the impact of the technological intervention. Galbraith (1973) offers an information processing view of the organization. He argues that firms are organized around the need to process information, a view that fits nicely with our interest in HCIS.

Much progress in information technology has been realized by the application of computers to organizational and analytical problems, which lies outside the field of specific clinical health care technologies (Banta, 1990). As health care organizations become more information-intensive, managers find themselves in a fascinating but confounding world of sophisticated information technology. For HCIS there is a new way of thinking about the health care environment. The adoption of HCIS will change the way a hospital performs health careservices, the way in which managers must consider their services, and the way in which responsibilities are allocated within the hospital. HCIS can also affect the strategy of the hospital, its revenues and expenses, the structure of the hospital, and the individuals working in the hospital.

Although this is an important contribution to understanding the information technology in the service industry, Miles (1987) points out two limitations. First, it implies that services are relatively passive recipients of information technology compared to the manufacturing industry. Second, the distinction between product and process in service is not as clear as it is in manufacturing.

A major concern in such efforts is the recognition by management of the critical issues to be raised and resolved throughout the HCIS system design and implementation processes.

The HCIS Solution

HCIS are increasingly intrinsic to the organization's proper function, serving a critical mission in hospitals. HCIS become providing company-wide computing—which is to say that HCIS lie at the heart of the health care organization—including patient test results, physician management reports to billing, and the major patient care areas, such as radiology, laboratory, pharmacy, medical records, and nursing.

The computer has brought changes in the health care industry in areas ranging from diagnostic testing, such as computer-aided temography, to organ donation networks, which can match an organ donor with the prospective recipient during the life of donated organ (Swasey, 1990). Computers have also emerged at the patient's bedside, so that the diagnosis and treatment can be tapped directly into a computer for access by health care providers during subsequent visits. The rapid growth in computer technology is related to three trends in health care: the growing appreciation of the wide range of possible uses of this technology, the tremendous growth in information, and the rapidly evolving technologies themselves (Kulikowski, 1983).

Osborn, et al. (1989-1990) suggest seven core services be included in HCIS: electronic mail, referrals service, labstatus, pharmacy service, pre-admission and operating room scheduling, consultation request service, and library application service. In general, the structure of HCIS can be based on the use of two sets of computer-based tools: generic and function-specific. The generic tools are divided into four elements: communication (e.g., E-mail, voice mail), data handling (e.g., capable of handing analytical or graphical data), process management (e.g., providing help in building groups or resolving tasks), and public data management (e.g., allowing for displaying, modification and maintenance of public data).

The function-specific solution that the HCIS department came up with could be a program that automatically puts a patient through the health care data base. Most hospital managers believe that strengthening loyalties between physicians and the hospital isan issue of strategic importance. HCIS can perform a lot of strategic functions in the health care environment. For example, HCIS can be used to facilitate clerical visits. A patient used to fill out a form that would go in triplicate to three departments in the hospital. Then the appropriate member of the clergy would receive notice of the patient's messages.

HCIS are conceptualized as consisting of a public and a private environment. A wealth of HCIS exists in electronic format. Two major types of computer files are bibliographic database and factual data banks. The former contains references for public use, while the latter is more of an online handbook, containing actual data or facts. Individual users would have access to the system for their individual or health care needs. Then the need for flexibility of HCIS becomes crucial. Flexibility means software compatibility across a range of hardware platforms, whether it is in the arenas of mainframes workstations, or personal computers.

Many advances in information technology could become potent forces leading to changes in the future (Van Bemmel, 1985): (1) networks for communication within the hospital and with the outside world; (2) huge mass memories for the storage of medical records, data, and literature data bases; and (3) powerful microcomputers for automation and personal computing. In general, four major types of information technology can be applied to general HCIS objectives. Table 1 lists these types, each of which is discussed in a subsection below.

1. Communication Technology: to foster team, health careorganization, home, and interorganization communication. All parts of HCIS, including the patient, can be linked into one network. Networks of computers between institutions, health care workers, and private individuals will permit a continuous flow of information with feedback (Spyker, et al., 1984). Several

1. Communication Technology
2. Coordination Technology
3. Decision Making Technology
4. Monitoring Technology

Table 1: Information Technology Types for HCIS

communication networks will develop between the various health care providers, including national or regional systems, not only for epidemiological studies, planning, management, and evaluation, but also for consultation (Mohan & Caley, 1983).

2. Coordination Technology: to coordinate resources, facilities, and projects in the hospital. An example is the drug delivery system, a term that has come to refer to any of several current or potential information technologies for administering drugs, whether implanted or external, continuous or intermittent (Check, 1984). Delivery systems, such as pumps and implantable reservoirs, are already in use for coordination of delivery operations in hospitals.

3. Decision Making Technology: to improve the effectiveness and efficiency of individual and group decision making. For example, the HELP system (Pryor, et al., 1983) has been operational for the past 16 years at the LDS Hospital, a 520-bed teaching facility for the University of Utah School of Medicine. In the HELP system, physiological data, including systemic arterial blood pressure, pulmonary artery pressure, cardiac output, and core temperature, are all recorded automatically (Gardner, et al., 1982). Nurses do their charting on the HELP system soon after the data are collected (Johnson, et al., 1984). This system, along with many other tools, permitted implementation of the individual and group decision making for patients.

4. Monitoring Technology: to monitor the status and quality of health care operations and/or health care industry trends. In hospitals, monitoring and evaluation can be improved through the collection of routine operational data within HCIS. For example, hospitals can use HCIS to track nurses' performance (Helmer & Suver, 1988). Many managers are concerned that trends in computer networking with HCIS may encourage their superiors to snoop.

HCIS Impact on the Organization

Duncan (1981) defines an organization as a collection of interacting and independent individuals who work towards common goals and whose relationships are determined according to a certain structure. Ackoff and Emery (1972) define a system as a set of interrelated elements or process. At the core of this approach to organizational analysis is the fundamental postulate that organizations are open systems that need careful management to satisfy and balance internal needs and to adapt to environmental circumstances. Three key conflicting organizational forces were found to have significant impact on the overall corporation: autonomy, evolution, and integration (Madnick & Wang, 1988). The changing nature of today's health care delivery is forcing the managers to redefine their roles and relationship, and to rethink tradeoffs among these factors within the hospital.

Three issues of HCIS impact on the organization are listed in Table 2. They are addressed as follows.

1. Uncertainty and Coordination
2. Centralization or Decentralization
3. Distribution of Power

Table 2: HCIS Impact on the Organization

1. Uncertainty and Coordination. Management theory is concerned with two basic elements that influence how an organization is structured. The notion of uncertainty is extremely important in understanding organizational design (March & Simon, 1958). Design strategies of HCIS to cope with different levels of uncertainty in the health care environment will be suggested. The second important element organizational variable is coordination. As health care tasks become more specialized the HCIS has to be concerned with coordination.

All elements of health care organizations are being broken down into smaller and interactive units. HCIS, then, must respond with smaller and interactive information components. HCIS should be able to incorporate a wide range of interfaces and communication protocols, enabling it to integrate with other systems in other departments, and with systems installed in the future.

2. Centralization or Decentralization. Leavitt and Whisler (1958) suggest that firms would recentralize as a result of new computer technology; the availability of more information than was previously possible would allow management to centralize. Lucas (1990) suggests centralization and decentralization as variables in the information systems design process.

Decentralization seems to be the movement of the present and the future for HCIS (Van Beekum & Banta, 1989). Decentralization has obvious desirable aspects in making health care services more accessible and giving people the feeling that they can influence the nature of health care services (Banta, 1990). For the health care managers, the focus is that management should specify the goals of the hospital and the degree of decentralization desired in HCIS.

3. Distribution of Power. Information is control (Pettigrew, 1972). Those who control information can influence the balance of power within their organization. As control and power are directly related, conflicts will arise over who exercises control of information systems in the organization. Power is a key factor in the process of attitude change. This is the power which resides in the top executive or the customer to influence attitude change. Lee and Treacy (1987) have documented how information technology impacts the distribution of power and influence in organizations. Hickson et al. (1971) suggest a model with four major determinants of power in an organization. Lucas (1984) conducts a study of the power of information services.

In every major health care organization there are certain individuals who control the destiny of the corporation. HCIS cause a major shakeup in the flow of health care information. Information now flows from the analyst or database directly into the system for distribution to all top executives in the hospital. Since middle managers no longer have control over the content or distribution of the health care data, it becomes much harder to hide a problem in the hospital. In a hospital, HCIS should be viewed as the natural evolution of information technology in the health care environment. Hospital management should develop new organizational paradigms to guide the new development approach.

In summary, managers, with their knowledge of managerial skills, have power and authority over the lower employees. Usually, managerial skills shall include arriving at decisions from information, and then implementing them in order to achieve the organization's goals. HCIS can be used to identify problem areas and then to suggest feasible solutions for them. This increased ability to evaluate information may give top management more direct control of their organizations' activities and reduce its reliance on middle management. The use of HCIS will cause a trend towards more centralized as opposed to decentralized power distribution in a health care organization.

HCIS Impact on Individuals

Alderfer (1972) classifies individual needs into three groups: existence, relatedness, and growth. Existence needs are the physiological and material needs. Relatedness needs are those for interpersonal relationships, love, and belongingness. Growth needs are desires for self-esteem and self-fulfillment. The physician's need for hospital services encompasses house staff, physician assistants, marketing of staff physicians, practice-management services (e.g., billing, recordkeeping, collections, monitoring), and formation of an integrated system (Goldberg & Martin, 1990). To achieve growth need satisfaction, managers should strive to create an environment that challenges and encourages HCIS personnel to extend themselves within their roles, allowing them to make use of HCIS, learn new information technology, and develop new skills.

Physicians are faced with increasing competition, greater consumer price awareness, insurance cost sensitivity, a movement toward alternative methods of service delivery, physician extenders, and an increasing emphasis on quality. Computers have reduced clerical employment in some service industries and reduced the need to expand the clerical staff in others (Kraut, 1987). Hospitals invest in some types of information systems for the express purpose of reducing labor costs in nursing or holding them constant. One of the important changes in the U.S. health care system is an increasing interest in information on the costs and quality of medical care services provided by individual physicians (Glanz & Rudd, 1988). The ways in which HCIS are used to affect the health care process of people who worked in the hospital and to affect service delivery and quality are as much an organizational decision as a technical one.

Keen (1981) claims that systems designers must operate in a political and cultural environment since information systems which they create redistribute data, disrupt patterns of communication and reallocate authority. As a result, it requires the direct participation of the management and department heads in the design and development processes, to make sure the development conforms with the information needs of the health care organization.

HCIS design effort should be created toward a system that is flexible, friendly, and that provides a variety of options. The more industry standard the HCIS' interfaces are, the easier and more cost effective it will be dealing with systems integration issues. It can be predicted that information professionals will be far more successful, and probably much more satisfied, as consultants and advisors to managers and users in the health care organization. However, HCIS that serve the needs of all individuals remain controversial with regard to the proper balance of control over decision making between physicians and organizations (Alexander & Morrisey, 1988).

HCIS management will also fulfill more of a coordinating and less of a directing role. Information professionals will exert influence on members of the hospital instead of control over information technology. It seems clear that if health care organizations intend to assure their long-term survival, proactive management approaches with HCIS are essential. Managers can keep an alert eye out for opportunities to build in health care performance as marginal additions to basic health care systems.

HCIS Management Issues

The challenge of management is not only to learn to live with new age systems, but also to facilitate the development of HCIS. Dale (1986) points to the need for goodness of fit and coherence between elements making up the organizational system. In this view, HCIS will only perform well if other parts of the information system are appropriately designed to be consistent with them. Changes in technology will require changes in people (e.g., skill requirements), and in systems and procedures (e.g., for recruitment, training). New organization structures may be required and more subtle cultural adaptation may also follow (Clegg, 1988). The major benefit for information technology lies in its catalytic effect, promoting or indeed forcing changes throughout the rest of the organization (Clegg & Corbett, 1987).

The field of socio-technical systems design (Mumford, Land, & Hawgood, 1978) stresses that both technical and human aspects of the design need careful considerations. This gives an organizational view of HCIS which is seen as supporting the activities of an organization consisting of people. Such a view of the organization has to cope with the impact of an information system on a psychological, sociological, and political environment. When considering HCIS there are five management issues which hospitals need to address. Table 3 lists the five issues.

• Cooperative Decision Making
• Middle Management
• Long-Term Planning
• Assessment Analysis
• Control of Information

Table 3: HCIS Management Issues

1. Cooperative Decision Making. There is an increasing need to change the cooperative decision making in health care organizations. Decisions must be based on the total benefits of the whole organization not just for a department. The sum of local optima is not an optimum for the whole. Functional department relationships must become more flexible if HCIS are to be successful. Horizontal communications, the internal information flows that occur both within and between departments of an organization (Daft & Steers, 1985), must be enhanced. Responsiveness to physician desires for involvement and support can greatly enhance a hospital's ability to meet the demand of today's health care environment by developing HCIS.

This issue addresses the need to allow explicitly for the coexistence and usage of a variety of health care components (e.g., different types of health care database systems, models, and applications). It is crucial to recognize that HCIS increasingly involve the integration of separate systems that have been developed and administrated independently. HCIS provide vehicles through which interested physicians may have input to hospital decision making and encourage their involvement. Steering committees which include senior line managers are required (Keen, 1981). These committees will delegate to technical with staff responsibility for HCIS that do not only have significant organizational impact but also be involved with ones that are part of politics of HCIS.

2. Middle Management. In the health care environment, middle management refers to the department heads. The performance of middle management is crucial for the success of a health care firm; it is crucial for the way in which health care service is provided. Mid-level managers are a critical link to the HCIS in hospitals. These front-line managers must understand the relationship between their regular operational decisions and the cost of hospital services. To achieve this, middle management must be reorganized to respond to and to take advantage of the new way in which HCIS dictates that health care service be performed.

The concept of the flattened organization is so pervasive in the health care industry that managers should understand that career movement may be more horizontal than vertical. With organizations flattening, there will necessarily be new perspective in HCIS management. Health care staff relationships will change. HCIS should create an infrastructure that allows people to work in teams to identify patients requirements. The whole health care effort will be completely integrated. Hybrid skills must be developed in middle management; they cannot dismiss organizational and political issues as irrelevant or not their responsibility, but must be able to operate in the manager's world and build credibility across the organization (Keen & Scott Morton, 1978). HCIS are designed to preserve the management integrity of their physician partners in providing health care services.

3. Long-Term Planning. HCIS reflect the organization's long-term strategy, which must be accounted for in order to select the most effective technology solution for the health care organization. Hospital planning is information planning. In order to achieve integration, HCIS require the adoption of a long-term view. Although chief executives have the power to bring about change in their organizations, they often lack the vision to do so. HCIS planning requires management to be oriented to the long term, determined to reach organizational goals, and to have an evolutionary plan of migration using the information system as its basis.

HCIS planning should be focused on meeting customer needs using the principles of total quality management (Feigenbaum, 1983), which encapsulates four elements (i.e., setting standards, appraising conformance to these standards, acting to ensure standards are met, and planning for improvement in standards). Total quality management involves a series of steps that ultimately leads to an alignment of management and customers. HCIS provides an integrated system for managing all functions within the hospital.

The database managed and controlled by IS specialists of various subunits of an organization constitutes the information utility (Madnick, 1977).

The end-users, through their desktop terminals, access the information utility for health care service what is either directly usable or indirectly usable. HCIS will become part of the health care organization's infrastructure to facilitate long-term planning.

4. Assessment Analysis. In general, the purposes of assessment are: (1) it helps put the information system developing process in the mainstream of management strategy formulation; (2) it helps managers better understand the problems and opportunities that may face the health care organization in the future; and (3) it focuses on trends and developments in the health care organization's environments that should be monitored over time so that HCIS can be maintained intelligently and improved. A multiplicity of attributes constitute the assessment of HCIS, and they can be operationalized contextually. The assessment analysis is required to become an integral and explicit part of the HCIS design process.

Three criteria are employed to reflect an important aspect of the impact of HCIS assessment in hospitals. These are (a) quality of care, (b) economic importance, and (c) expectation effects:

a. Quality of Care

Berwick (1987) developed five indicators for assessing the quality of care: coordination of care, patient and staff satisfaction, ambiance, interpersonal relationships. Management should not only use quantitative methods to monitor carefully information to identify and remove any special causes of low quality, but also should develop HCIS that allow employees to minimize waste, delays, and mistakes.

b. Economic Importance

Technologically, innovative hospitals need an integrated framework that enhances the assessment function and control systems that balance the HCIS' creative activity with the company's financial resource. If such a factor is indicated as the cost of the health care problem to individuals and to the organization, there will be a greater effect of the HCIS on cost.

c. Expectation Effects

In many cases, the expectation effects will play a crucial role in assessment of HCIS. Such effects include the uncertainty about use of the HCIS, the existence of controversy, the receptivity of practitioners to new health care information, and the timeliness of the assessment.

Some health care organizations providing patient-centered health care services have remained traditionally organized - by the physician's specialities, not the patients' needs. The assessment of HCIS makes sure that the systems consist of a structure that divides the organization into the units responsible for carrying out the health care services, and an accounting and quality system that continuously evaluates their performance.

5. Control of Information. As physicians attempt to understand the changing health care environment, many of them view HCIS as "control systems" whose purpose is to monitor and report on their behavior (McFarlan & McKenney, 1983). While one contribution of HCIS is to strengthen control systems in hospitals, HCIS can create tremendous control problems and management challenges. With thousands of transactions being processed in a short period of time, an error can spread through an immense number of transactions in minutes. The reliance on HCIS and rapidly changing technology forces health care organizations to implement comprehensive information security and contingency plans to protect their information assets. Since HCIS force a hospital to store all its vital data and information as a single entity on a computer, HCIS make the control of information systems a critical issue.

HCIS reflect the health care organization, and if the system breaks down or malfunctions, the hospital can be fundamentally impaired. The question is not really IF a hospital will have some types of disaster in HCIS, but WHEN will the disaster occur and HOW it will impact the hospital. Qquestions including how does a hospital prevent information corruption, disaster, theft and sabotage must be addressed.

A systematic approach to information control planning may save the hospital time, money, and personnel - it may even save the very life of the hospital. A total organizational contingency planning is suggested for hospitals. Driving the need is the implementation of new systems and the greatly expanded use of microcomputers and minicomputers in almost every departments, where it is apparent that HCIS must address the issues for corporate-wide information security policy. HCIS will become an integral part of the ongoing management process and will provide controls and policies throughout the health care organization. This new outlook will lead to more effective solutions in the areas of end-user authentication, encryption, contingency planning, security standards, and corporate-wide security policies.

The Development and Implementation of HCIS

The health care industry is a particularly difficult arena for successful system implementation. Large health care firms are decentralized institutions comprised of separate departments with distinct disciplines and functions. The organizational contradiction reflects hospital managers with information technology and cultural challenges.

The HCIS system development process can be characterized as an organizational change process. The description of the change process begins with a system of work patterns that are in place. Lewin (1952) and Schein (1964) describe the process as moving through three stages: unfreezing, changing, and refreezing. The current performance of the system—at the individual and organizational levels—is a result of the actual fit among the tasks to be done, the individuals responsible for completion of the tasks, and the information technology used to support the individuals (Chakravarthy, 1987; Verkatraman, 1989). Based on the concept that individuals and health care organizations seek stability in their work patterns, successful change involves activities that disturb the current patterns and replace them with new ones.

Ottaway (1979) suggests a change agents theory on the implementation of change. The change agents can be divided into three categories: change generators (unfreezers), change implementers (changers), and change adopters (refreezers). Change generators have the initial task of motivating change. Once the organization realizes that it needs to change, the unfreezing phase is complete and a new type of change agent is required for the second phase: the change implementers. Once the final phase phase is reached, lasting change is dependent on the commitment and ability of the change adopters to adopt and maintain the changed behavior. Ottaway's concept of change agents is all-embracing in that the adopters or recipients of change are included as well as the implementers.

Rockart and Scott Morton (1984) developed a model for information technology and organizational change based on Chandler's "strategy drives structure" (1962). Their model suggests that information technology is a key enabler of strategic direction and that an important problem is to find the relationship between strategic ideas and the application of information technology.

There are two major approaches to the study of implementation: the factors approach and the process approach. The factors approach attempts to identify static forces which lead to successful HCIS implementation. The process approach focuses on the dynamic of implementation, examining the behavior of stakeholders over time. Implementation process characteristics of HCIS include the strategies, policies, and procedures employed in the development of the HCIS. In addition, organizational change and participative design are concern issues in HCIS.

The organizational change perspective is particularly applicable to HCIS implementation. The creation of HCIS is a significant organizational intervention; the designer is trying change the power structure by changing the way critical decisions affecting core activities of the organization are made. Linder (1985) suggests that if management can identify the cultural and organizational characteristics that it wants to encourage, then the selection and implementation of information technology can be powerful tools in taking the initiative in the control of change. Markus and Robey (1988) discuss three approaches of investigating the relationship between systems implementation and organizational change. First, the technology imperative assumes information technology is a cause of organizational change. Second, the organizational imperative views the motivation of the designers of information technologies as a cause of organizational change. Last, the emergent perspective infers that organizational change emerges from an interaction between information and people. The first two perspectives imply that a direction of causality does not exist. However, the third perspective suggests the causal link.

Investigating organizational change and HCIS implementation as a unified process leads to several issues. First, we shall study what organizational conditions must exist to ensure successful HCIS implementation. Second, this study will give valuable insight into the way the HCIS designer can manage the change process. Third, researchers have widely disseminated the idea that information technology is a driving force for enabling competitive strategy in the health care environment.

Health care administrators are faced with a dilemma: How can we encourage involvement and still control processing for the overall benefit of the organization? There are two strategies for a hospital to provide this kind of control in HCIS. These are: (1) through support and (2) to make the reasons for coordination clear so that users do not see control as arbitrary (Locas, 1990). Whatever the strategy, the bottom line is managers must be well versed in information technology in general and in the technological infrastructure of their hospitals.

Within hospitals it is crucial to ask if there is too much reliance on the information technology experts for designing HCIS. Would a more effective system result if its design is appropriated by its users? A range of motives exists for participation in developing HCIS, from a democratic and moral commitment, to a desire for health care people to be involved, through to pragmatic goals aimed at incorporating as broad a range of valid expertise in health care as possible. Practical problems regarding participation should be considered seriously. Implementing HCIS requires understanding in its environmental and organizational contexts. Health care organizations have a heavy information processing load. However, health care organizations are also characterized by high levels of differentiation manifested in functional, structural, and behavioural differences across departments. The major impediment to participation in this context is the process of health care. This problem is compounded when expanding organizations seek to increase market share through networking and community outreach programs.

For these reasons, the very first step toward successful implementation of HCIS is incorporating management's strategic plan into the HCIS plan. The strategic plan should explicitly address strategic threats and opportunities, multiple stakeholders, and consider future environmental changes in health care. This reflects the growing trend toward tighter and broader coupling of HCIS with the organization, its processes, cost containment efforts, and long-term objectives.

Future Trends

With the coming environment of the 1990s, health care organizations place demands on many HCIS to be designed in ways that enable them to be modified easily and remain effective in conditions of dynamic environmental change. We are in an era of end-user computing (EUC) combined with new demands from health care management. Users of HCIS may program, enter data, operate, and have responsibility for entire information systems (Rockart & Flannery, 1983). Greater distribution of responsibility for HCIS, including design, development, control, and implementation, creates more challenges in the health care organization. User participation in HCIS development is a necessary condition for increasing the acceptance of change and decreasing resistance to change.

To date, the public access to HCIS is likely to increase. The HCIS offers a rich opportunity for researchers and policy analysts to address questions about the health care services, and this research can contribute to improving quality and cost of care over time.

Future research opportunities can be categorized into (1) factors approach, (2) process approach, and (3) political approach. The results of research may help managers broaden their search for new strategic options. The results may also provide health care organization designs by helping them to understand their long-term consequences, particularly in terms of impacts of HCIS with which the organizational system must interact.

Factors approach attempts to identify static forces which lead to successful HCIS development. Potential factors include: characteristics of task (task complexity, task variety, task uncertainty, and autonomy), top management support of HCIS, good HCIS design, and appropriate user interaction and understanding. Process approach focuses on the dynamics of development of HCIS, examining social change activities and suggesting when implementation success occurs. Potential forces include: commitment to change, top management efforts, and extensive planning. HCIS is continuing to evolve, and forthcoming technologies will have a significant impact on the process of the near future. The HCIS infrastructure must be robust enough to support the new applications appropriate to the health care process. The political approach recognizes that the diverse vested interests of stakeholders affect implementation efforts and that successful implementation depends on recognizing and managing this diversity (Markus, 1983). In order to maximize the benefits from HCIS, hospitals must understand and manage their design and development processes. HCIS, when viewed in the appropriate context, is a company-wide program that can yield significant improvements in health care service value.

Summary and Conclusion

This chapter has attempted to provide a brief overview of an important management perspective to HCIS. It has addressed the impacts of HCIS on the health care organization and individuals. The purpose of this chapter was to provide a basic framework for understanding the management issues of HCIS. It is important for hospital managers to understand a number of basic concepts related to diverse areas of information technology in order to more fully appreciate the design, implementation, and execution of a useful HCIS in the hospital. Today's hospital managers must begin to anticipate future applications of information technology, and to develop an application for the many and varied behavioral and organizational consequences of the use of HCIS for the support of managerial work in the health care environment.

References

Ackoff, R.L., & Emery, F.E. (1972). *On Purposeful Systems*. London: Tavistock.

Alderfer, C. (1972). *Existence, Relatedness, and Growth: Human Needs in Organizational Settings*. New York: The Free Press.

Alexander, J.A., & Morrisey, M.A. (1988). Hospital-physician integration and hospital costs. *Inquiry, 24*(3), 388-401.

Banta, H.D. (1990). Future health care technology and the hospital. *Health Policy, 14*, 61-73.

Berwick, D.M. (1987). Monitoring quality in HMOs. *Business and Health, 1*, 9-12.

Chakravarthy, B.S. (1987). On tailoring a strategic planning system to its context: Some empirical evidence. *Strategic Management Journal, 8*(6), 517-534.

Chandler, A. (1962). *Strategy and Structure*. MA: MIT Press, 3-16.

Check, W.A. (1984). New drugs and drug delivery systems in the year 2000. *American Journal of Hospital Pharmacy, 41*, 1536-1547.

Clegg, C.W. (1988). Appropriate technology for manufacturing: Some management issues.

Applied Ergonomics, 19(1), 25-34.

Clegg, C.W., & Corbett, J.M. (1987). Research and development into humanising advanced manufacturing technology. In Wall, T.D. Clegg, C.W. and Kemp, N.J. (Eds.), *The Human Side of Advanced Manufacturing Technology*. Chichester: Wiley.

Daft, R.L., & Steers, R. (1985). *Organizations: A Micro/Macro Approach.* IL: Foresman.

Dale, A. (1986). *Matching Information Technology to Your Organization.* UK, Sheffield: Manpower Services Commission.

Duncan, W.J. (1981). *Organizational Behaviour.* Boston: Houghton Mifflin.

Feigenbaum, A.V. (1983). *Total Quality Control.* New York: McGraw-Hill.

Fera, M., & Finnegan, G. (1986). Building a productivity improvement team through MIS leadership. *Hospital & Health Services Administration, 31*(4), 7-17.

Galbraith, J. (1973). *Designing Complex Organizations.* Reading, MA: Addison-Wesley.

Gardner, R.M., West, B.J., Pryor, T.A., Larsen, K.G., Warner,H.R., Clemmer, T.P., & Orme, J.F. (1982). Computer-based ICU data acquisition as an aid to clinical decision-making. *Crit. Care Medicine, 10*, 823-828.

Glanz, K., & Rudd, J. (1988). *Effects of Quality of Care Information on Consumer Choices of Physicians and Hospitals.* Washington, DC: U.S. Congress Office of Technology Assessment Health Program.

Goldberg, J.P., & Martin, H.J. (1990). Control and support: What physicians want from hospitals. *Hospitals & Health Services Administration, 35*(1), 27-37.

Helmer, F.T., & Suver, J.D. (1988). Pictures of performance: The key to improved nursing productivity. *Health Care Management Review, 13*(4), 65-70.

Hickson, P.J. Hinnings, C.R. Lee, C.A. Schneck, R.R., & Penning, J.M. (1971). Strategic contingencies theory of interorganizational power. *Administrative Science Quarterly, 16*(2), 216-219.

Johnson, D., Ranzenberger, J., & Pryor, T.A. (1984). Nursing applications on the HELP system. *Proceedings of the 8th SCAMC, 8*, 703-719.

Keen, P.G.W., & Scott Morton, M.S. (1978). *Decision Support Systems: An Organizational Perspective.* Reading, MA: Addison-Wesley.

Keen, P.G.W. (1981). Information systems and organizational change. *Communications of the ACM, 24*(1), 24-33.

Kim, K.K., & Michelman, J.E. (1990). An examination of factors for the strategic use of information systems in the health care industry. *MIS Quarterly, 14*(2), 201-215.

Kraut, R.E. (1987). *Technology and the Transformation of White-Collar Work.* Hillsdale, NJ: Erlbaum.

Kulikowski, C.A. (1983). Expert medical consultation systems. *Journal of Medical Systems, 7*, 229-234.

Leavitt, H.J., & Whisler, T.L. (1958). Management in the 1980's. *Harvard Business Review*, November-December, 41-48.

Leavitt, H. (1965). Applied organizational change in industry. In *Handbook of Organization*. March, J.G. (Ed.), IL: Rand McNally, Ch 27.

Lee, S., & Treacy, M.B. (1987). *Information technology impact on power and influence*. Working paper, MA: MIT.

Lewin, K. (1952). Group decision and social change. In *Readings in Social Psychology*, Newcombe & Hartley (Eds.), NY: Henry Holt and Company, 459-473.

Linder, J.C. (1985). Computers, corporate culture and change. *Personnel Journal*, 49-55.

Lucas, H.C. (1984). Organization power and the information services department. *Communications of the ACM, 27*(1), 58-65.

Lucas, H.C. (1990). *Information Systems Concepts for Management*. New York: McGraw-Hill.

Madnick, S.E. (1977). Trends in computers and computing: The information utility. *Science, 185*, 1191-1199.

Madnick, S.E., & Wang, Y.R. (1988). A framework of composite information systems for strategic advantage. *Proceedings of the Twenty-First Annual Hawaii International Conferences on System Sciences*.

March, J., & Simon, H. (1958). *Organizations*. New York: Wiley.

Markus, M.L. (1983). Power, politics, and MIS implementation. *Communications of the ACM, 26*(6), 430-444.

Markus, M.L. (1984). *Systems in Organizations: Bugs and Features*. MA: Pitman.

Markus, M.L., & Robey, D. (1988). Information technology and organizational change: Causal structure in theory and research. *Management Science, 34*(5), 583-598.

McFarlan, F.W., & McKenney, J.L. (1983). *Corporate information systems Management: The Issues Facing Senior Executives*. IL: Irwin.

Miles, I. (1987). *Information technology and the services economy*. Oxford *Surveys in Information Technology, 4*, 25-55.

Moham, R., & Caley, R. (1983). Standardization of therapy machine interfact for treatment monitoring. *International Journal of Radiation Oncology Biology and Physics, 9*, 1225-1229.

Mumford, E., Land, F., & Hawgood, J. (1978). A participative approach to the design of computer systems. *Impact of Science on Society,28*, 234-253.

Osborn, C.S., Madnick, S.E., & Wang, Y.R. (1989-1990). Motivating strategic alliance for composite information systems: The case of a major regional hospital. *Journal of Management*

Information Systems, 6(3), 99-117.

Ottaway, R.N. (1979). *Changes Agents at Work.* Associated Business Press.

Pettigrew, A.M. (1972). Implementation control as a power resource. *Sociology, 6*(2), 187-204.

Pryor, T.A., Gardner, R.M. Clayton, P.D., & Warner, H.R. (1983). The HELP system. *Journal of Medicine Systems, 7*, 87.

Rockart, J.F., & Flannery, L.S. (1983). The Management of end user computing. *Communications of the ACM, 26*(10), 776-784.

Rockart, J.F., & Scott Morton, M.S. (1984). Implications of changes in information technology for corporate strategy. *Interfaces, 14*(1), 84-95.

Schein, E.H. (1982). Mechanisms of change. In *Interpersonal Dynamics*, Bennis, W. (Ed.), IL: The Dorsey Press, 199-213.

Spyker, D.A., Stier, D.M., O'Dell, R.W., Anne, A., & Edlich, R.F. (1984). A user-oriented information system for emergency medicine. *Comprehensive Therapy, 10*, 42-47.

Swasey, L. (1990). Technology is changing the face of key industries. *MIS WEEK*, May 21, 23.

Tornatzsky, L.G., & Klein, K. (1982). Innovation characteristics and innovation implementation: A meta-analysis of findings. *IEEE Trans. Engineering Management, 29*(1), 28-45.

Van Beekum, T., & Banta, H.D. (1989). Possibilities and problems in the development of home care technology. *Health Policy, 12*,301-307.

Van Bemmel, J.H. (1985). *Man and computer in the hospital of tomorrow*. Department of Medical Informatics, Working paper, Amsterdam: Free University of Amsterdam.

Venkatraman, N. The concept of fit in strategy research: Toward verbal and statistical correspondence. *The Academy of Management Review, 14*(3), 423-444.

Part V
Information Technology and Organizational Strategy

Chapter 14

Online Databases in Environmental Scanning:
Usage and Organizational Impact

Abdul M.A. Rasheed
University of Texas at Arlington

Deepak K. Datta
University of Kansas

In recent years the business environment has been characterized by increasing complexity, uncertainty and turbulence and in order to grow and be profitable, firms need to correctly analyze the environment and respond to the strategic opportunities and threats that it offers. In this context, environmental information has become a critical resource. While firms use a wide variety of methods for the collection of such information (Diffenbach, 1983), the advent and increasing availability of online databases promise to transform radically the practice of environmental scanning. However, in spite of its obvious importance, very little attention has been paid in both academic and practitioner literature to the impact of such databases on the task of environmental scanning and its effectiveness.

In this chapter we examine the role of online databases in environmental scanning by identifying the external (environmental) and organizational factors that potentially influence the degree to which online databases are used by organizations. In addition, we identify the possible impact of the use of online databases on the process of environmental scanning, as well as its impact on the effectiveness of scanning. Also provided is a brief description of some of the key online databases which are currently available, including

The authors would like to thank Anthony J. Daboub, Gregory G. Dess, and James T.C. Teng for their helpful comments on earlier versions of the paper.

their relevance for the task of environmental scanning.

The paper is structured as follows. First, we provide a brief description of the task of environmental scanning in the development of strategic and operational plans. After a brief discussion of some of the more widely used databases, we identify the key external and organizational factors that are likely to influence the extent to which a firm uses such databases in its scanning activity. Next, we provide our assessment of the possible impact that the increased use of online databases might have on both the process of scanning and its effectiveness. Finally, we discuss the limited empirical research on the use of online databases and provide directions towards possible future research.

The Role of Environmental Scanning

Strategic environmental scanning relates to the acquisition of information about events, trends, and relationships in an organization's environment, which can assist top management in the identification of strategic threats and opportunities (Aguilar, 1967). Stated differently, it represents an organization's efforts at systematically dealing with uncertainty and, hence, plays a key role in the strategic planning process.

The organizational environment can be conceptualized as consisting of a variety of components, which can be meaningfully classified into two broad categories: the 'general' (or remote) and the 'task' (or industry) environment (Dill, 1958). The former (i.e., general environment) consists of environmental areas over which a single firm has very limited influence (e.g., political, economic, social and technological forces), whereas variables influencing a firm's immediate competitive situation constitute the 'task' environment. Environmental analysis involves the collection and assessment of pertinent information towards the development of effective and proactive strategies. Figure 1 depicts a typical environmental scanning and assessment system and mentions the sources of environmental information along with the scanning approaches used by organizations.

Sources of Information and Scanning Approaches

The activity of environmental scanning is typically undertaken at various organizational levels. While there is some evidence to suggest that top executives devote considerable time to scanning (Feldman & March, 1981; Mintzberg, 1973), the task of gathering environmental information and its preliminary analysis is often left to staff analysts in the planning department. Also, the sources of environmental information are many and varied. Authors (Aguilar, 1967; Stabell, 1978) have classified these sources into two broad categories, namely, 'personal' and 'impersonal', based on the directional specificity of information transmitted. While 'impersonal' or secondary

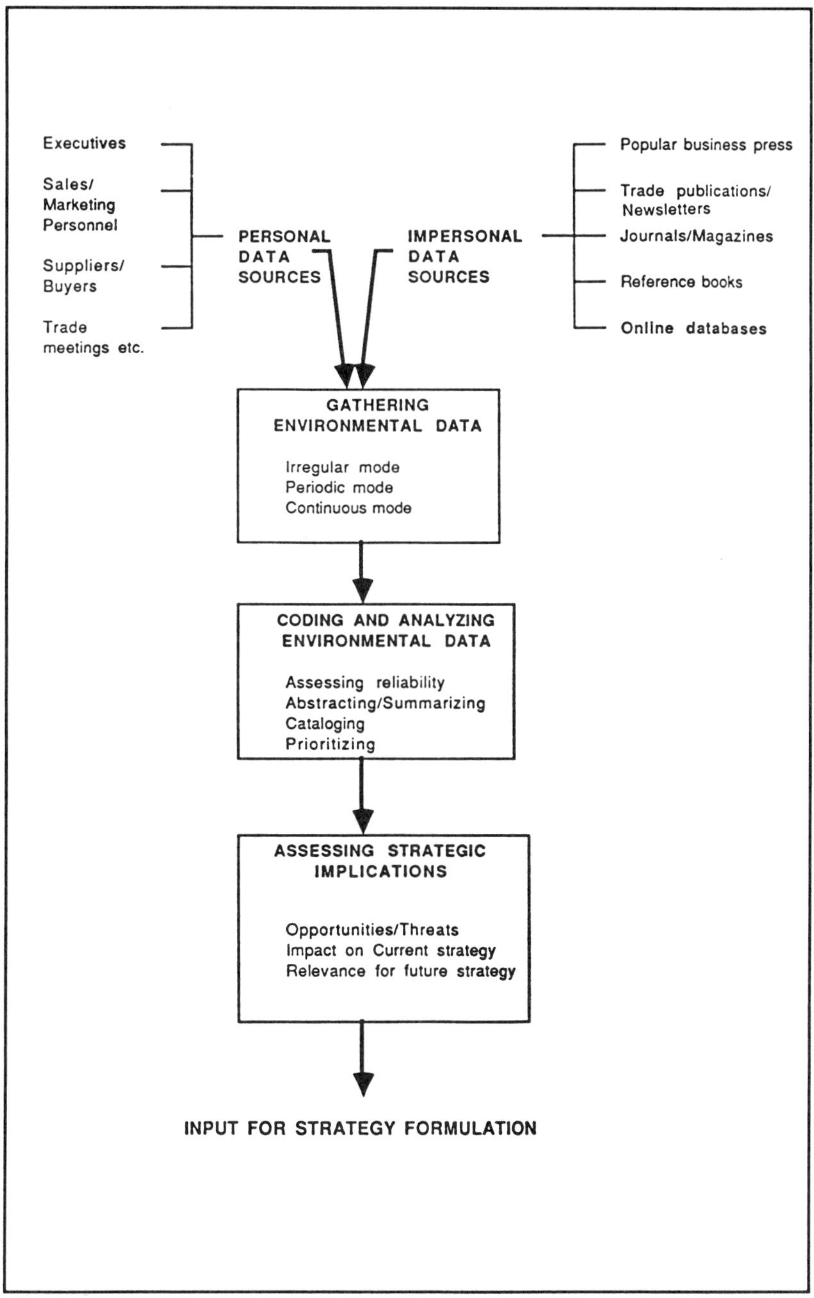

Figure 1. Environmental Scanning and Assessment System

sources of information are relatively free of personal biases and, therefore preferable, managers often rely extensively on personal or informal information sources (El Sawy, 1984; Higgins, 1983; Keegan, 1974). This might, in part, be due to "relevant" secondary information not being readily available, or alternatively, a lack of awareness of the availability of accessible secondary sources.

Organizations also differ in the scanning approach used to gather environmental information. Fahey, King and Narayanan (1981) identify three scanning approaches widely used by organizations: irregular, periodic, and continuous. The irregular mode is reactive (generally crisis initiated) and is very similar to the problemistic search suggested by Cyert and March (1963). On the other hand, scanning in the continuous mode involves a continuous and proactive search for relevant environmental information which is generally accomplished through a formal environmental scanning unit. As the name suggests, the periodic approach refers to environmental scanning being done only at specific intervals (generally coinciding with the frequency of the planning exercise). The choice of the scanning mode and the extent of scanning undertaken by organizations obviously depends on a wide variety of factors including the level of environmental uncertainty, the likely impact that environmental changes might have on organizational effectiveness, and the availability of resources for scanning (Datta & Chinta, 1986). Studies (Blandin & Brown, 1977; Culnan, 1983a) suggest that higher levels of environmental uncertainty and volatility are generally accompanied by a more intensive search for information and that executives in high-performing companies scan more frequently and more broadly than their counterparts in low-performing companies (Daft, Sormunen, & Parks, 1988). Not surprisingly, there has been a significant shift in scanning practices, with an increasing number of organizations adopting the continuous mode.

The above phenomenon has also resulted in an increased demand for environmental information by most organizations, and it was inevitable that computers would come to play an increasingly important role in the collection, dissemination and analysis of such information. What is surprising is that only recently organizations have begun to consider seriously the use of computers and computerized databases in environmental scanning. However, the growing availability of online databases in recent years suggests that their use has been rapidly increasing and will continue to do so in the future. In the following sections, we discuss some representative databases which are used for scanning organizational environments.

Online Databases

According to recent estimates, there are over 4,000 online databases currently available in the market and this number is likely to grow even further.

In the early 1980's, Horton (1983) had predicted that "the real mega industry of the late 80's and beyond will be that sector of the broad electronic marketplace that adds value to raw data products and service." His contention has been supported by the phenomenal growth of online databases. Marsh (1984), Miller (1987), and Spanbauer (1988) provide brief descriptions of some of the more widely used databases and also the types of information available in such databases. Other published sources which provide information on available online databases include the *Directory of Online Databases, The Complete Handbook of Personal Computer Communication,* and *How to Look it up Online.*

While the data contained in online databases often pertain more to operational rather than strategic decisions, the following paragraphs describe how some of them can be meaningfully used in the formulation of strategic decisions through the scanning of both competitive and general environments.

Competitive Environment

Online databases can be particularly helpful in identifying and analyzing competitors, suppliers, financial organizations, stockholders, regulatory bodies, interest groups, etc. Suppliers and customers can be easily located through EIS databases, Electronic Yellow Pages, Disclosure II, Million Dollar directory, etc. All these databases can be sorted according to size or location and many provide information at the plant level. Special subject databases like RAPRA, Surface Coating Abstracts etc. provide information on specific industries. In addition, Dunsprint Service which provides credit reports on companies can be a valuable source of information on the financial stability of suppliers and customers. Similarly, Spectrum databases provide information on stockholders and financial institutions.

On the other hand, PTS databases such as MARS and PROMT (provided by Predicast Inc.) could be very useful in conducting an industry analysis by providing relevant information on factors such as market size, market growth, segmentation, customer profitability, concentration, capital and labor intensity, and degree of unionization. Study of demand and price levels at different points in the value chain can be effectively done through time series analysis, data for which is available through the I.P. Sharp system as well as Citibank databases. Most of the software necessary for such analysis comes with these databases. Engineering databases like INSPEC, ISMEC, and COMPENDEX can provide information on technological developments within an industry. Likewise, databases on trademarks such as Trademarkscan and World Patent Index provide useful information in the assessment of technological entry barriers in an industry.

As an example, online databases can provide particularly useful information in decisions involving mergers and acquisitions. Such decisions, however, require extensive analysis of the potential target firm and its industry. In this context, databases such as D&B Million Dollar Directory and Disclosure Online provide a wealth of financial information required in such analysis and, when combined with assessments of industry attractiveness that can be obtained from databases like Arthur D. Little Online, Find/SVP and Investext, can certainly help organizations make better acquisition decisions. In addition, information available online can be invaluable to banks which desire an international portfolio of assets or any firm which seeks to expand its operations internationally. International Dun's Market Identifiers, Textline, and ICC Datasheets are some of the online databases that provide information on foreign companies and markets.

General Environment

Forces in the general environment typically have a significant influence on the corporate and business level strategies of any organization and, therefore, need to be carefully monitored. Several online databases are currently available which can help planners in their monitoring of the general environment, both at the macro and micro level. Examples include 'BLS Labor Statistics' database containing demographic data, Citibank and Economic Abstracts International databases, 'Environline' on environmental issues, Pollution Abstracts, etc. Others can be useful in scanning the regulatory environment. These are Federal Index Databases and Federal Register Abstracts which can help track regulatory bodies. Similarly, the LegiSlate database provides the current status of any bill or pending regulation and also provides information on judicial opinions which can be extremely valuable in examining possible legal implications of contemplated strategic actions.

It is apparent from the above discussion (which mentions only a few of the more commonly used databases) that there is a wide variety of online databases which can be used for the purpose of environmental scanning. Obviously, there are considerable differences across databases, not only in terms of their content but also in terms of accessibility, ease of use and cost. Some online services provide easy to use, menu-driven interfaces while others have command-driven interfaces and are often more difficult to use. Also, online services generally charge a nominal, one-time sign-up fee and then bill users every month based on their total connect time. Connect time charges vary greatly across databases and are based on time of day, day of week, and the bps rate. They range from a low of $5 to over $200 an hour. In addition to primary services, many on-line services also provide gateways to one or more sister services for a small premium.

In the following sections, we present a discussion of the organizational and external factors that influence the use of online databases as well as the impact online databases may potentially have on the process and effective-

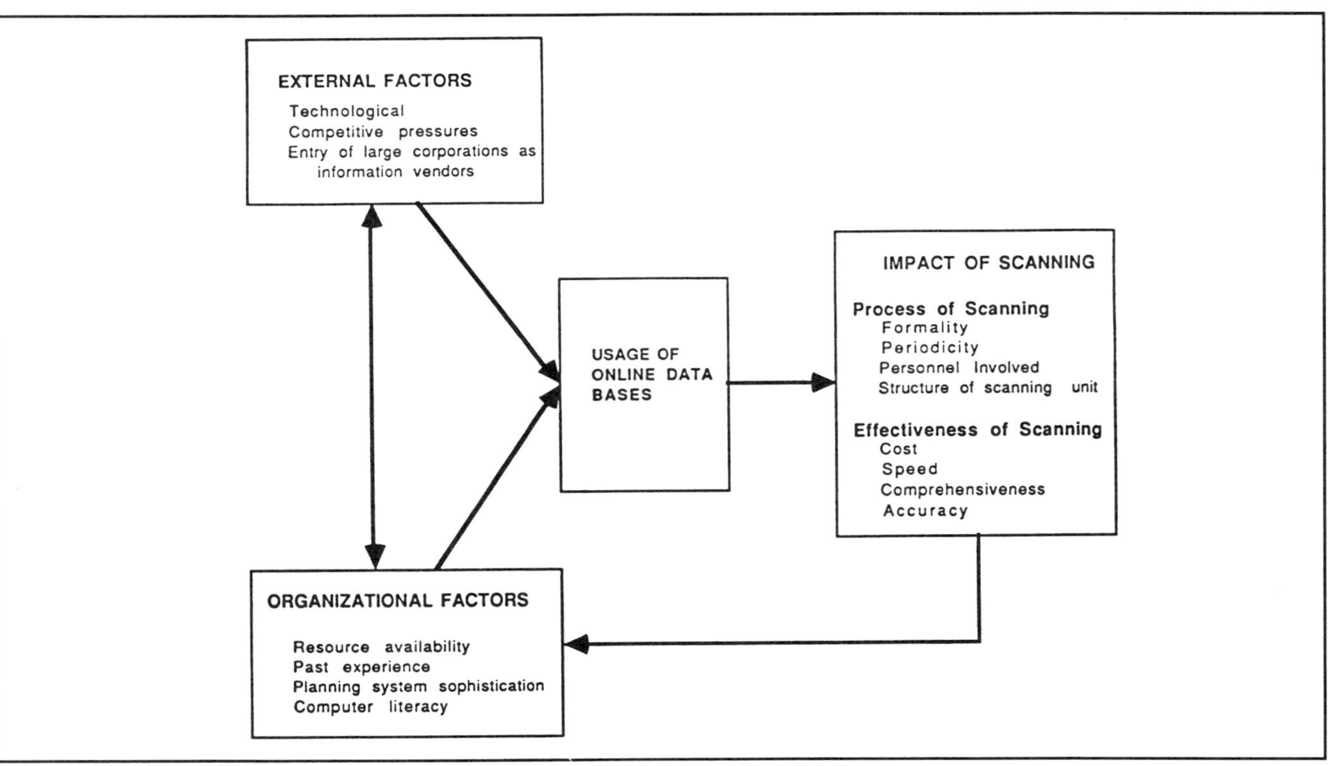

Figure 2. Use of Online Databases for Environmental Scanning

ness of scanning. This discussion is organized around the conceptual framework of Figure 2.

Factors Influencing the Use of Online Databases

Both external and organizational factors influence the use of online data bases for environmental scanning purposes. In this section, we discuss some of these factors.

External Factors

Environmental or external factors obviously have an important impact on the use of online databases by organizations. Such factors include technological developments, growing competitive pressures and increased awareness of information as a competitive resource. In addition, as discussed below, the entry of several major corporations as vendors of online databases has resulted in a wider set of choices among databases.

Technological developments. Technological developments in the last decade have played a very important role in increasing the popularity of online databases. For example, the availability of more *user friendly terminals* coupled with the declining prices of hardware (especially personal/ microcomputers) has undoubtedly been an important contributory factor in the increasing use of online databases (Seligman, 1985). Even if hardware prices stabilize, it is unlikely that we will see a slackening of demand since increased sophistication will open up new and more efficient uses. Moreover, the availability of *'intelligent modems'* has made it much easier for organizations and individuals to access online databases, resulting in a greater number of users. This, in turn, has helped bring down unit costs.

In addition to the above, other technological developments have also contributed to the growing popularity of online databases. These include the introduction of *front end* software packages which permit users to enter search items off line — the software then performs the search automatically, downloads the data and 'logs off.' In addition, the development and introduction of *gateway systems* have enabled users to access a variety of databases through a menu driven system instead of a regular subscription system. Also contributing to the increasing use of online databases is the progress that has been made in the last few years in the development of *relational databases*. The resultant ability to manipulate data in these databases has made them extremely versatile and useful for a variety of tasks, including environmental scanning.

Increased competitive pressures. The increased competitiveness of markets brought about by industry restructuring, accelerating technological change, and the globalization of markets has created a greater corporate appetite for relevant external information (McGrane, 1987). For example, a company could find its products suddenly obsolete due to technological developments in industries which might normally be considered as unrelated. Also, increased globalization of markets means that actions taken by foreign competitors or changes in the trade policies of foreign governments might critically impact the operating decisions of U.S. firms. In short, the volatility of the competitive environment along with the growing interdependence of markets has made the timely acquisition and analysis of external information particularly crucial for firms if they are to survive, grow and remain profitable.

The above developments have made organizations realize that information can be a major competitive weapon (Ives & Learmonth, 1984; Parsons, 1983; Porter & Millar, 1985; Wiseman & MacMillan, 1984). There is also a growing feeling among firms that the gathering of competitor intelligence is an essential element for survival in today's competitive environment (Fuld, 1988; Vella & McGonagle, 1987). Given the enormous possibilities for efficient information gathering through online databases, one can reasonably expect that increased competitive pressures will contribute to greater usage of such databases.

Entry of large corporations into the information market. As firms increasingly realize the value of information as a strategic resource, we have also witnessed a greater involvement of large companies in the marketing of 'information'. Information has become 'big' business and, currently, large companies like Lockheed, American Airlines, Mead, Dow Jones, Reader's Digest, H & R Block, and Citicorp, among others, have become actively involved in the selling of information, often using online databases. The resource availability of these organizations, their capacity to absorb short term losses and their marketing capabilities have certainly been major factors in the growth of online databases. However, the development and marketing of online databases has obviously not been limited to just large corporations. As Gross (1988) argues, the growth in the information vending business provides ample opportunities for entrepreneurs in the field of information delivery. The phenomenal success enjoyed by companies such as Sheshunoff & Co. (which specializes in information relating to commercial banks) testifies to the above statement (Angrist, 1987).

However, as in other industries in the early stages of their life cycle, the computerized information business has had its fair share of failures (Connely, 1985). Prodigy Services Co., a joint venture formed by IBM and Sears, Roebuck & Co., with the objective of providing home computer users with information, entertainment, electronic shopping and communication at a price lower than that of a newspaper subscription, has spent over $600 million and is still experiencing both technical and marketing difficulties (Schwadel,

1989). A major failure in the video text industry has been Covidea, jointly formed by Chemical Banking Corporation and AT&T. In contrast, video text efforts in Britain (Prestel), West Germany (Bildschiimtext) and France (Minitel) have enjoyed considerable success (Hall & Terran, 1987).

Organizational Factors

Along with external factors, the extent to which organizations are likely to use online databases in their environmental scanning activity depends on a number of organizational factors. These are briefly described in the following paragraphs.

Resource availability. Traditionally, environmental scanning has been viewed as a resource intensive activity because it often involved the laborious collection of primary data. As a result, firms faced with a situation of resource scarcity, often scaled back on their scanning activity. However, as previously discussed, the increased use of online databases should make the task of environmental scanning easier and should significantly lower the cost of collecting information. As a result, in most medium or large firms, lack of resources is unlikely to be a major impediment to engaging in environmental scanning. In fact, online databases will probably permit even smaller and relatively resource poor firms to scan relevant environmental areas routinely as a part of their strategic planning efforts.

Past experience. Previous experience that a firm and its members have in using online databases can be an important determinant of the extent to which such databases are likely to be used for environmental scanning. Allen (1977) and Culnan (1983a) suggest that use of an information source reinforces or improves perceptions of the sources' accessibility, in turn resulting in greater usage. Also, as organizational members become more experienced and develop greater expertise in using online databases, they would be able to "search" databases more efficiently, reducing connect times and making the scanning process more cost-effective.

Planning system sophistication. While formal strategic planning has become a common feature in organizations, there is considerable variation in levels of sophistication. A planning system may be considered to be more sophisticated when there is an explicit commitment on the part of top management to formalized and systematic strategic planning, and when the planning staff has expertise in the use of various analytical techniques in the planning process. In such cases it is likely that the organization is also involved in the comprehensive and continuous evaluation of environmental information and may even have a formal scanning unit as a part of the planning department. Such organizations are likely to find online databases particularly useful and cost-effective and we can reasonably expect them to make greater use of online databases.

Computer literacy among corporate executives. Bralove (1983) estimated that nearly ninety percent of the approximately ten million U.S. managers in the early 1980s were computer illiterates. In another research study, Wright (1984) found that a large percentage of managers develop cyberphobia, i.e., a fear of computers. However, the situation seems to have changed rapidly in the last few years, aided in part by greater computer oriented education. A recent survey by Control Data Corporation found that computer usage among top managers went up from 8% to 35% during the years 1984-86, with roughly one-third of the users actually using online databases. As computer literacy increases we should expect online systems to be increasingly used by all levels of managers. However, whether the use of online databases in environmental scanning actually results in competitive advantage also depends on the management's ability to integrate, analyze and use the data so collected. Superior ability to make such analysis obviously provides an impetus to the greater use of online databases.

In addition to the above, there are other organizational factors which may influence the use of online databases for the purpose of scanning. These include factors which hinder their use, for example, low credibility associated with the task of environmental analysis among management in an organization, or scanning units located far away from organizational centers, or perceptions of low competence on the part of individuals engaged in environmental analysis.

Impact of Online Databases on Environmental Scanning

Online databases with enormous capabilities to store, sort and retrieve (even analyze) large volumes of information should have a significant impact on the practice of environmental scanning. This section, which discusses the nature of such an impact, is divided into two parts. The first discusses the likely influence of online databases on the process of environmental scanning and the second analyzes the possible impact that online databases will have on the overall effectiveness of the scanning process.

Impact on Process

The increased use of online databases could have a pervasive influence on many aspects of the process of environmental scanning. Online databases have allowed the effective mass marketing of data on a hitherto unprecedented scale. Before the advent of online databases, the collection of such information often required expensive and time consuming searches through primary and secondary data sources. Online databases, with their increasingly universal availability and accessibility, can significantly reduce information asymmetries across companies by allowing even relatively resource poor companies access to required environmental information. On the other hand, they may also contribute towards greater information asymmetries and increased centralization of information gathering and processing within organizations. This is because users of online databases are charged on the basis of connect time and therefore organizations are likely to authorize only a few individuals to access online databases for monitored periods, especially when connect time is expensive.

Conversely, it is likely that in some organizations the use of online databases may actually lead to greater decentralization and less formalization of all aspects of environmental scanning. First, additional reduction in the cost of online access in the future may mean that such expenditures will fall well within the discretionary level of unit managers. Also, the current trend towards getting line managers increasingly involved in strategic planning activities is likely to result in divisional and functional managers being more involved in environmental information gathering than they traditionally have been. In such cases it is quite likely that they will become active users of online databases.

The shifts in the centralization and formality of environmental scanning should have significant structural consequences. Changes in technology often lead to changes in organizational structures by altering institutionalized roles and patterns of interaction (Barley, 1986). In organizations where the use of online databases leads to multiple decentralized scanning units, there will be a need for formalized integrating mechanisms.

The increased use of online databases in organizations also means that a different set of skills will be required of people associated with the scanning process. Traditionally, a majority of the tasks related to environmental scanning has been carried out by staff groups (e.g., members of the planning department). Because of significant differences in instructions and syntaxes across databases, it is unlikely that managers will find it cost-effective to do the scanning themselves. In all likelihood they will be forced to rely on the help of technically competent corporate librarians to undertake specific issue-related online searches. These issue-related or event triggered searches would be supplemented by general, continuous scanning undertaken by the corporate scanning unit. In either case, scanning through 'chauffeured' use of online databases would most likely be the dominant mode in the future.

With greater availability of online databases, we should also see a shift from 'informal' to 'formal' or secondary sources of information. Studies (El Sawy, 1984; Higgins, 1983; Keegan, 1974) have found that, in the past, top management relied extensively on informal sources for environmental information, due in part to the effort involved in doing a comprehensive search of formal sources, and also the time lag between the occurrence of an event and its dissemination through print media. Given the search capabilities of online databases and the ease with which they can be continuously updated, it is conceivable that online databases will result in reducing the reliance on informal sources for external information. Also, given the decreasing cost, increasing comprehensiveness and the ease with which scanning can be done using online databases, the frequency of scanning should increase. In other words, firms doing 'irregular' or 'periodic' scanning might shift to a more continuous mode.

So far we have identified likely changes in the practice of environmental scanning as online databases are increasingly used. However, the impact of online databases is not a function of just their technical characteristics. As studies (Danziger, 1977; Downs, 1967) have suggested, computer-based systems increase the influence of those who have access to technology and can organize data to their advantage. Access to information via online databases has the potential of affecting the power distribution among individuals and subunits within the organization. However, the social and political setting within the organization would also play an equally important role in determining the organizational consequences.

Impact on Effectiveness

With increased environmental volatility, the success of organizations is closely tied to their ability to understand, interpret, and adapt to changes in the environment. Environmental scanning is vitally important in responding intelligently to developments in the external environment and must be efficient and cost-effective. One of the major advantages of online databases over traditional sources of information is the speed with which vast amounts of data can be accessed and analyzed. Another area where online databases can have a positive impact is the comprehensiveness of the planning process. Organizational limitations in collecting and analyzing vast quantities of information have been a major constraint in the ability to achieve a reasonable degree of comprehensiveness, a limitation that can be partially overcome through the use of online databases. Also, it might often be possible to verify the accuracy of environmental information from one source by searching through different databases for similar data and ensuring that there is some degree of convergence in the information reported. This should result in more accurate forecasts and better decisions and strategies.

In addition to improving the quality and quantity of informational input, online databases have the potential for improving the process of planning itself. One of the recent developments in strategic planning is the emergence of Strategic Issues Management System (SIMS) (Camillus & Datta, in press). SIMS is often recommended as a substitute for strategic planning systems and is designed to act on weak signals picked up from the environment. It requires the identification of a selected number of important variables which are then continuously monitored to achieve the desired degree of sensitivity to environmental changes. Online databases will greatly enhance organizational capability to continuously monitor chosen environmental segments, thereby making SIMS more effective.

While there are undoubtedly a number of benefits, the use of online databases for environmental scanning might also be associated with potential problems. Improperly used, they can result in information overload for a planner, hindering rather than helping in the making of strategic decisions. Also, the tendency to engage in comprehensive scanning could result in the consideration of significantly greater numbers of alternatives. While beneficial in some respects, a thorough analysis of these alternatives might consume more time and effort than is desirable. In such situations, organizations unfortunately become more preoccupied with analyzing the vast amount of information gathered than in developing strategies and tactics. Similarly, the inappropriate use of online databases could result in planners being swamped with irrelevant data, or even in relevant data getting lost due to improper presentation formats used by vendors of such databases.

In summary, properly used, online databases can significantly improve the timeliness, accuracy, and comprehensiveness of environmental information gathering while reducing the costs involved in doing so. Improperly used, databases may create problems which outweigh the benefits. Moreover, whether they result in better strategic decisions depends on a number of other factors, including the characteristics of the planning system itself.

Research on the Use of Online Databases

While there has been a significant increase in the use of online databases, empirical research on their use has been rather limited. In an interesting field study, Culnan (1983b) examined the use of online databases and the characteristics of professionals who utilize them, either directly as end users or through an intermediary. She found that 'chauffeured' access was preferred when the individual has only a one-time need for information. On the other hand, individuals found direct access to be more appropriate when a database is used on a regular basis. However, the study found only partial support for the hypotheses that 'direct' users are likely to be better educated, belong to more professional organizations and read more professional literature than 'chauffeured' users. One explanation might be that the desired mode of access (direct versus chauffeured) is not necessarily just a function of an individual's characteristics but also of the characteristics of the database being accessed, a factor which the author did not examine in her study.

In another study on the use of online databases by 258 Norwegian companies, Methlie and Tverstol (1982) found that usage of external information services, including online databases, is determined by company size, industry category, awareness and attitudes. Attitudes, in turn, are formed by prior experiences. They also found that industries in turbulent environments make greater use of such services. The frequency of use was also found to be positively correlated to the size of the company. The users were generally satisfied with the level of service they received and valued the timeliness, precision and relevance of the information obtained through external information services. In a related study, Kibirige (1983) investigated how users reacted to the cost aspect of online database services in libraries and found that fees were not a deterrent to their usage. He did not find enough support for his hypothesis that industrial users accept fees more readily than academic users.

McGrane (1987) describes how companies such as Polaroid, Travelers, General Dynamics and Mead Corporation increasingly use online databases for a variety of planning tasks. Also, a study by the editors of Knowledge Industry Publications Inc. found that, in general, the greater the information used for an operating decision, the more suitable it is for electronic delivery (Sigel, 1983).

It is evident from the above discussion that we not only have a limited number of empirical studies, but also that the research domain has been rather narrow. For example, studies typically relate to factors influencing the use of online databases and not on the impact of online databases on the effectiveness of scanning and related activities. Clearly, there is a need for additional research towards obtaining a better understanding not only of the factors that influence the use of such databases but also their impact on the effectiveness of scanning.

The framework of Figure 1 can serve as a useful starting point for further empirical research in this area. The framework incorporates both external and organizational factors that lead to usage of online databases. Further, it focuses on their organizational impact, both in terms of the process of scanning and its effectiveness. It is possible to derive several testable propositions from this model. Careful empirical testing of propositions suggested by the model should lead to a richer understanding of the profound impact that online databases have on the practice of environmental scanning.

Conclusion

The use of online databases for environmental scanning is a relatively new phenomenon. However, with hardware and software technologies evolving so rapidly, online databases will almost certainly bring about major changes in the task of environmental scanning. Developments that can elevate the status of online databases from file drawers to counselors, include the development of 'intelligent' scanning systems attached to databases which monitor predefined parts of the environment and transmit information to appropriate levels of the organizational hierarchy as well as 'personalized' scanning systems that enable the user to analyze data according to his/her personal preferences and assumptions. Additional benefits can accrue from an integration of DSS and AI capabilities (Huber, 1984).

The rapid evolution of communication and computer technologies is leading to a profound transformation in the way organizations are managed, something that strategic planners cannot afford to ignore if they are to remain effective. Undoubtedly, one of the important areas of strategic planning where developments in information technology are likely to play a significant role is the area of environmental scanning. As discussed in this paper, environmental and organizational factors should result in online databases being increasingly used in environmental scanning in the future. However, online databases are already facing competition from a new data-delivery technology called Compact-Disc Read Only Memory (CD-ROM). Each of the laser optical compact discs can currently carry the equivalent of 150,000 single-spaced typed pages (Briggs & Coleman, 1987) and can be reproduced in volume for about $5 to $10 each. They are not only portable, but also eliminate the need for costly connect times to online databases. Online databases, however, still have the advantage of being able to provide the most current information because of the ease with which such databases can be continually updated. It is, therefore, likely that online databases will continue to play a critical role in the gathering and analysis of environmental information. It should also provide opportunities for meaningful future research along some of the issues discussed in this paper — research which should prove to be of considerable practical and theoretical significance.

References

Aguilar, F.J. (1967). *Scanning the business environment.* New York: MacMillan.

Allen, T.J. (1977). *Managing the flow of technology.* Cambridge, MA: MIT Press.

Angrist, S.W. (1987, November 2). Smiling all the way to the bank. *Forbes*, 48-51.

Barley, S.R. (1986). Technology as an occasion for structuring: Evidence from observations of CT scanners and the social order of radiology departments. *Administrative Science Quarterly, 31*, 78-108.

Bralove, M. (1983, March 7). Computer anxiety hits middle management. *Wall Street Journal*, 4.

Briggs, W. & Coleman, J. (1987). Compact discs: New tool for competitive analysis. *Planning Review, 15*(6), 32-37, 48.

Blandin, J.S., & Brown, W.B. (1977). Uncertainty and management's search for information. *IEE Transactions on Engineering Management, 24*(4), 114-119.

Camillus, J.C., & Datta, D.K. (in press). Designing sensitive systems: Integrating strategic planning and issues management. *Long Range Planning*.

Connely, M. (1985, March 28). Ailing videotex ventures haven't slowed plans to market the information services. *Wall Street Journal*, 1.

Culnan, M.J. (1983a). Environmental scanning: The effects of task complexity and source accessibility on information gathering behavior. *Decision Sciences, 14*, 195-205.

Culnan, M.J. (1983b). Chauffeured versus end user access to commercial databases: The effects of task and individual differences. *MIS Quarterly, 7*, 55-67.

Cyert, R.M., & March, J.G. (1963). *A behavioral theory of the firm*. Englewood Cliffs, N J: Prentice Hall.

Daft, R.L., Sormunen, J., & Parks, D. (1988). Chief executive scanning, Environmental characteristics, and company performance: An empirical study. *Strategic Management Journal, 9*, 123-139.

Danziger, J. (1977). Computers and the litany to EDP. *Public Administration Review, 37*, 28-37.

Datta, D.K., & Chinta, R.R. (1986). Environmental scanning: A Three dimensional contingency framework. *Eastern Academy of Management Proceedings*, 44-47.

Diffenbach, J. (1983). Corporate environmental analysis in large U.S. corporations. *Long Range Planning, 16*(3), 107-116.

Dill, W.R. (1958). Environment as an influence on managerial autonomy. *Administrative Science Quarterly, 2*, 409-443.

Downs, A. (1967). A realistic look at the final payoffs from urban data systems. *Public Administration Review, 27*, 204-209.

El Sawy, O.A. (1984). Understanding the process by which chief executives identify strategic threats and opportunities. *Academy of Management Proceedings*, 37-41.

El Sawy, O.A. (1985). Personal information systems for strategic scanning in turbulent environments: Can the CEO go online? *MIS Quarterly, 9*(1), 53-60.

Fahey, L., King, W.R., & Narayanan, V.K. (1981). Environmental scanning and forecasting in strategic planning. *Long Range Planning, 14*(1), 32-39.

Feldman, M., & March, J.G. (1981). Information in organizations as signal and symbol. *Administrative Science Quarterly, 26*, 171-186.

Fuld, L., (1985). *Competitive intelligence: How to get it; How to use it*. New York: John Wiley & Sons.

Gross, A.C. (1988). The information vending machine. *Business Horizons, 31*(1), 24-33.

Hall, A., & Terran, J. (1987). An effective strategy for information delivery. *Journal of Business Strategy, 8*(1), 21-27.

Higgins, J.M. (1983). *Organizational policy and strategic management: Text and cases.* Dryden Press.

Horton, F.W. (1983). Tapping external data sources. *Computer World, 17*(33), 15.

Huber, G.P. (1984). The nature and design of post industrial organizations. *Management Science, 30*, 928-951.

Ives, B., & Learmonth, G.P. (1984). The information system as a competitive weapon. *Communications of the ACM, 27*, 1193-1201.

Keegan, W.J. (1974). Multinational scanning: A study of the information sources utilized by headquarters executives in multinational companies. *Administrative Science Quarterly, 19*, 411-421.

Kibirige, H.M. (1983). *The information dilemma: A critical analysis of information pricing and the fees controversy.* Connecticut: Greenwood Press.

Marsh, S.G. (1984, June 20-25). Strategic planning information online. *Database*, pp. 20-25.

McGrane, J. (1987). Using on-line information for strategic advantage. *Planning Review, 15*(6), 27-30.

Methlie, L.B., & Tverstol, A.M. (1982). External information services: A survey of behavioral aspects of demand. *Information and Management, 5*, 269-277.

Miller, M. (1987, March-April). Competitive intelligence: Staying alive in the jungle. *Online Access Guide*, pp. 43-57.

Mintzberg, H.E. (1973). *The nature of managerial work.* New York: Harper & Row.

Parson, G.L. (1983). Information technology: A new competitive weapon. *Sloan Management Review, 25*(1), 3-14.

Porter, M.E., & Millar, V.E. (1985). How information gives you competitive advantage. *Harvard Business Review, 63*(4), 149-160.

Schwadel, F. (1989, May 11). IBM-Sears venture in videotex is hit by service outage. *Wall Street Journal*, B4.

Seligman, D. (1985, February 4). Life will be different when we're all online. *Fortune*, pp. 68-72.

Sigel, E. (1983, July). Is videotext vendible? *Datamation*, 209-222.

Spanbauer, C. (1988, October). On-line services and data bases. *PC World*, 196-201.

Stabell, C.B. (1978). Integrative complexity of information environment perception and information use: An empirical investigation. *Organizational and Human Performance, 17*, 116-141.

Vella, C.M., & McGonagle, Jr., J.J. (1987). *Competitive intelligence in the computer age - An introduction.* Westport, CT: Quorum Books.

Wiseman, C., & MacMillan, I.C. (1984). Creating competitive weapons from information systems. *Journal of Business Strategy, 5*(2), 42-49.

Wright, P. (1984). Computers, change and fear: An unholy trio. *Management Decisions, 22*(2), 31-35.

Chapter 15

The Organizational Impact of Strategic Systems

Efrem G. Mallach
University of Lowell

The organizational impact of computers and computing has been studied since computers moved out of closed computer rooms into the corporate mainstream. This outward movement first took place on a large scale in the late 1960s. It resulted from three factors:

- Refinement of multiprogramming software, including the ability to coordinate shared access to a database. (The word "database" is used in its generic sense and does not necessarily imply the use of database management software.)
- Improved hardware cost-performance, which reduced the barriers imposed by the resource requirements of on-line computing.
- Development of low-cost CRT display devices, which could display alphanumeric information quickly and quietly enough to be useful in an office environment.

As computers moved into organizations they began to affect the people in those organizations. Research was accordingly directed to studying this impact. This research had two focal points. At the level of the individual worker, are jobs changing? Are communication patterns changing? At the level of the organization, does the use of computers affect the span of control? The number of levels in an organization? The organizational structure itself?

Typical of the resulting studies were by Whisler (1970a, 1970b). In (1970b), Whisler studied the impact of computers on twenty life insurance companies. He found that firms which used computers widely (as the term "widely" was understood in the late 1960s) reduced the size of their clerical staffs, but tended to add supervisory and managerial personnel. Consistent with this finding, the average skill level of workers increased. There were changes in decision making: it was rationalized, was made less flexible, was moved upward in the organization and was centralized. The content of about half the jobs changed, with comparable movement up and down. Communication patterns changed, with the total amount of communication increasing.

These and similar findings had little impact on the use of computers because computer usage was changing too quickly for them to affect it. By the mid-1970s, central corporate computers were a standard fixture of the business landscape. Their organizational impact was a fait accompli, no longer of research interest. Were there a question, it would be "what would be the organizational impact of taking the mainframe away?"

Meanwhile, decreasing hardware costs were moving computers further into the organizational mainstream. Office Automation arrived. It changed the rules of the game. OA moved the use of computers beyond "terminal workers" such as data entry clerks. The entire clerical and managerial staff of an organization became "automated." This had an organizational impact far beyond that of on-line terminal usage alone.

As the impact of OA was larger, so was the number of people who studied it. Hirschheim (1986) lists 181 references. The great majority of these are studies of the social and organizational impact of office automation. Their collective conclusion is, in a word, inconclusive. In fact, Hirschheim shows that the a priori views of an investigator appear to be the single most important factor in determining a study's conclusions. According to Hirschheim, both optimists and pessimists have specific, deterministic, but opposed views of this impact, while relativists believe that the impact varies from one situation to another and can be controlled. As he puts it, "...our collective understanding of the social impact of information technology—whether office automation or other—is highly deficient. The implications written about are largely hunches. There is a pressing need for further research."

We did not have time to sort this situation out before the rules changed again. Technology again moved faster than those who study its impact. Departmental computing came hard on the heels of office automation. While early computer applications, including office automation, focused on efficiency in automating existing tasks, departmental computing focused on effectiveness in accomplishing the overall purpose of a work group. Since departmental computing can restructure work in a fundamental way, its impact on organizations could be far greater than even that of office automation. Yet the step beyond departmental computing will in turn eclipse it. That step is strategic systems.

Strategic Systems

Strategic information systems (Canning, 1984) can be defined as information systems which address the fundamental strategic factors in an industry. Their objective is to give their users a competitive advantage through the use of information (Porter & Millar, 1985). Strategic systems affect the way organizations are structured, the way work is accomplished, and the way information is used. They connect the islands of information usage that McFarlan (1982) calls the "information archipelago." They change the focus of information usage, which has evolved from efficiency to effectiveness, to innovation. Here, their proponents claim—and there is considerable, albeit largely anecdotal, evidence to support this viewpoint—is where the real payoff is to be found. The potential value of information has been recognized for decades, e.g., by McDonough, (1963) and by Gregory & Van Horn (1963, Chap. 15). The first computer system to have been recognized, in retrospect, to fall into the strategic system category, is generally considered to be American Airlines' Sabre automated reservation system, originally implemented on second-generation IBM 7000 series hardware. Sabre offered unprecedented convenience in making reservations and quickly won the allegiance of travel agents. By displaying American Airlines' flights more prominently than those of American's competitors, Sabre increased its sponsor's market share dramatically on routes that American served. Competitors responded by developing their own similar systems and by legal action to prevent a system's owner from biasing travel agent displays to favor its own flights. They were eventually successful on both counts. Yet American's market share gains remained.

Automated reservation systems did more than streamline the work of departments. They eliminated departments. Acres of sneaker-shod clerks in stadium-sized rooms with blackboards covering every inch of wall space, binoculars trained on blackboard listings of available seats, rushing to the blackboards to adjust seat counts, then rushing back to telephones to confirm reservations, became unnecessary overnight. Computers made the reservation process quicker and less prone to human error.

Another classic example is that of American Hospital Supply. They gave customers the ability to order supplies directly from terminals in the customer's own offices. Customers found this service far more convenient than earlier ordering methods and responded by making AHS a preferred supplier. Interestingly, AHS happened on this system by accident as a result of a sales representative's over-enthusiastic promise to a customer. Once it was in place, they did not capitalize on their advantage by enhancing it vigorously to maintain their lead. AHS was eventually acquired, in large part on the strength of their customer order entry system's reputation. Once the purchasers studied the situation from the inside, they discovered that the AHS system was behind the times and behind its competitors, but by then it was too late to rescind the merger. The moral: strategic information systems are similar to any other source of strategic advantage. They must be carefully monitored to ensure their continued usefulness.

Strategic systems have been studied by several authors. McFarlan (1984) related the competitive advantage provided by a strategic system to the five sources of competitive impact defined by Porter (1980). These are the threats of new entrants and of substitute products, the bargaining power of suppliers and of customers, and the rivalry among existing firms. Not surprisingly, he found that Porter's framework is as valid for strategic information systems as it is for any other source of strategic advantage. Ives and Learmonth (1984) studied several dozen strategic systems. They related these systems to the thirteen stages a customer goes through in using a product or service, using a list adapted from Burnstine (1980). These stages begin with establishing requirements and continue through accounting for an asset and disposing of it. They found that strategic systems can apply to each of the thirteen stages. Ives and Learmonth also related strategic systems to the concept of Inter-Organizational Systems, which is elaborated on by Cash and Konsynski (1985). Bakos and Treacy (1986) have more recently supplied an overall perspective on the work in the field of strategic systems.

The Issues

The fundamental issues that arise in studying the organizational impact of strategic information systems are the same as those that arise in studying the organizational impact of computing in general. Hirschheim (1986) lists twelve factors that researchers have identified in studying the organizational impact of office automation (OA). His list follows. The letters (o, p, r) in parentheses after each factor relate to Hirschheim's finding that people who hold different overall opinions on the organizational impact of OA tend to consider different factors in their assessments of this impact. The letter "o" indicates that the factor in question is considered important by researchers holding optimistic views. The letters "p" and "r" indicate that a factor is considered important by researchers holding pessimistic or relativistic views, respectively.

1. Productivity (o, p)
2. Employment (o, p)
3. Work (arrangement, variety) (o, p)
4. Communication (o, p)

5. Quality of working life (o, p)
6. Power (p)
7. Personal privacy (p)
8. Bureaucratic rationalism (p)
9. The underground economy (r)
10. Participation (r)
11. Ethical and social acceptability (r)
12. Education (r)

These twelve points cover the areas in which computers (or, for that matter, anything else) can affect an organization. They are as inherently applicable to strategic systems as they are to OA. Since this list has already been published and widely referenced, and the organizational impact of these items on one important class of computing has already been assessed, we should strive to use the same list wherever it is appropriate. Accordingly, these points serve below as a framework for studying the organizational impact of strategic systems.

Hypotheses

We formulate hypotheses below for the impact of strategic systems on eleven of these twelve potential impact areas. Support for the hypotheses is limited at this point, though the author is not aware (through research, the trade press, or his own personal experience) of a situation which contradicts any of them.

The value of these hypotheses to a practitioner is in suggesting areas of concern that must be monitored carefully in the implementation of a strategic system, beyond those areas that must be monitored carefully in the implementation of any information system. They also suggest areas in which benefits can be anticipated and areas which can safely be given less attention.

To a researcher, these hypotheses provide an agenda for further work. Work published thus far on strategic systems by Bakos & Treacy (1986), Cash & Konsynski (1985), Canning (1984), Ives & Learmonth (1984), McFarlan (1982, 1984), and Porter & Millar, (1985) is devoted to justifying them and putting them in organizational, industry and economic contexts. It does not address their organizational impact because research on the impact of any phenomenon must await its widespread appearance. This is happening only now with strategic systems. It is possible in 1990 to address the hypotheses presented here and either accept or reject them.

Key points for managers or researchers to keep in mind, beyond the general need to be aware of or to study an area, are noted following the letters M or R respectively after each item is discussed below.

1. Productivity. Strategic systems are not justified on the basis of increased labor productivity. They are justified on the basis of enabling a firm to do a better job as perceived by customers. If successful, they will eventually result in higher profits. But this does not necessarily mean increased productivity.

For example, suppose a firm could increase its profits by $1,000,000 if its sales revenue increased 5% with its current labor productivity. Suppose, too, that a firm is unable to obtain this sales increase with its available products, services, and marketing methods. Finally, suppose that a strategic system can yield the 5% increase with an expenditure of $400,000 for the system. Productivity at the corporate level has gone down: the firm is spending $400,000 more than it otherwise would have spent at the same sales volume. Part of this amount pays for additional staff to develop and operate the new information system. Yet profits have gone up by $600,000 as a result of developing the system, justifying management's decision to do so.

Different definitions of productivity are possible. If we define productivity as a return on assets, it may or may not go up in the previous example. Which it does depends on the value of capitalized assets that are reflected in the above $400,000 figure and on what the firm's ROA was previously. Nor do other definitions of company-wide productivity lead to a clear conclusion about the likely impact of strategic systems on that figure.

Increased volume can spread fixed costs over a larger number of units. If management is vigilant not to increase fixed costs unnecessarily, large firms are potentially more productive than small. To the degree that the use of expensive strategic systems (which are available only to organizations able to invest substantial resources in developing them) increases supplier concentration in an industry, overall productivity might increase through elimination of smaller, relatively less productive, producers. However, the history of the U.S. economy does not suggest that large firms are more productive than smaller ones after a size threshhold which varies by industry has been reached. We, therefore, do not see a significant contribution to the overall productivity of the economy resulting from the use of strategic systems either.

Second, strategic systems do not, as a rule, affect a large fraction of an organization's work force directly. They multiply the effectiveness of a small number of people manyfold. To revisit the classical airline reservation system example: such systems multiply the effectiveness of reservation clerks dramatically, but reservation clerks represent a tiny fraction of an airline's payroll. The same is true of strategic systems that streamline other aspects of a firm's interactions with its customers. The number of order entry clerks may go down at a firm, but most firms don't have that many order entry clerks to begin with. Strategic systems that make it possible to treat each customer as a niche market of one may result in increased production costs due to smaller runs and more variety, but will be worth their cost many times over at the bottom line.

Interestingly, the company that pioneered strategic systems—American Airlines—is, in 1990, installing a $150,000,000 corporate-wide office automation system called InterAAct. According to Davis (1990), this system is intended to boost management and clerical productivity by 7 to 9 percent. However, despite the cost and corporate scope of this system, American is careful not to call it "strategic."

We hypothesize that strategic systems will usually have little effect, or a slight negative effect, on productivity. However, because of the wide variations among these systems—wider by far than the variations in operational systems from one organization to the next—we also expect to see a wide variance in this effect. A given system may increase or decrease productivity dramatically.

R: Researchers must guard against generalizing from too small a sample in this respect. This caution will constrain the validity of reported results.

2. Employment. It is a matter of ongoing debate whether or not the jobs created exceed the jobs eliminated by new technologies. It is generally accepted that dislocation does take place. Manual functions, and those who perform them, are eliminated. At the same time, jobs are created to develop and support the new technology. Strategic systems accentuate this dislocation. Airlines no longer need "acres of sneaker-shod clerks in stadium-sized rooms, binoculars trained on blackboard listings..." Hospital supply firms no longer require as many salespeople. However, as discussed under the previous point, strategic systems are not necessarily justified on the basis of net cost savings to their users. There is, therefore, no a priori reason to believe that they result in a net drop in employment. Since the jury is still out on whether systems that are intended to yield cost reductions indeed have the desired effect, it is even more doubtful that strategic systems do. Our hypothesis on this point is that strategic systems are essentially neutral regarding employment.

M: Dislocation should be expected and planned for.

R: If an effect on employment is small, finding it will require a diligent search.

3. Work (arrangement, variety). Strategic systems integrate many steps of an existing information flow into a coherent whole. In a manufacturing environment, such a system may respond to a customer order by determining that necessary parts are not in stock and submitting an order for them. This eliminates the steps of moving order information from the sales information system to the production scheduling system, and then to MRP. The entire process is presumably under the purview of one or more individuals, rather than each step being handled individually. This increased scope leads to job enrichment for these people. We hypothesize that strategic systems lead to an increased variety of work and to an increased flexibility in the arrangement of work. This hypothesis echoes the arguments of Giuliano (1982) and Spinrad (1982) that computerized work is more fulfilling than conventional work.

M: Increased scope of work can be presented to employees as a benefit of the new system.

4. Communication. Office automation is generally agreed to increase communication. (It is not as generally agreed that the increased communication leads to increased communication effectiveness.) Strategic systems, however, do not necessarily increase communication. They change communication patterns, often by reducing or eliminating the need for people to communicate. Communication may still be required to deal with exception situations and to sort out problems, but much day-to-day contact will be eliminated. Lower work-related communication volume takes place in the context of fewer people communicating. We cannot hypothesize that communication volume will go up or down on a per-person basis. However, we do not envisage strategic systems leading to vast percentages of the work force having their faces glued to CRT screens with no human contact.

5. Quality of working life. Here, Hirschheim (1986) found the most dramatic differences of opinion between the optimistic and pessimistic positions. This results from the difficulty of defining such an ephemeral concept in terms of anything that can be measured or even quantified. To the degree that the quality of working life, or job satisfaction, relates to the nature of the work performed (item 3 above), strategic systems should enhance it. Furthermore, the dislocation effect (if we assume the quantity of workers remains the same) results in some workers having an increased scope in their current positions while others move into technical positions. To the extent that self-selection operates, the job satisfaction of both groups should increase. We, therefore, hypothesize that strategic systems will have a positive effect on the quality of working life. This hypothesis differs from some researchers' conclusions about the impact of other types of information systems. Kraut, Dumais and Koch (1989) found that several measures of the quality of employment suffered when a computerized system was installed to automate record retrieval for customer service representatives in a large utility. The system in question made it unnecessary for people to leave their desks, thus reducing human interaction. The system was not strategic in the sense of this paper. We mention it here as a caution not to generalize overmuch to strategic systems from studies of information systems of a more limited scope.

M: This improvement can be presented to employees as a benefit of the new system.

6. Power. Information is power, as noted by Bariff & Galbraith (1978), Mintzberg (1983, esp. pp. 184-185) and Pettigrew (1972). It is widely felt that senior managers will support systems that increase their power. Strategic systems concentrate more information in one place. This, in turn, increases the potential for a concentration and an abuse of power. As Markus (1983) points out, "Access to information is probably less important as a basis of power than is the ability to control access to information or to define what information will be kept and manipulated in what ways." Strategic systems give many people, including people outside the organization, access to information, but they concentrate the control over that information. They thus concentrate that aspect of information that is most important from a power point of view.

The role of the Chief Information Officer as conceived by Synnott and Gruber (1981), and as previously foreseen by Kanter (1970) (who called this person a Top Computer Executive, or TCE) is an important safeguard against this abuse. A CIO is a top-level executive responsible for an organization's information base wherever this information may exist. As organizations recognize the importance of their information to their strategic position, they tend to create the position of CIO in fact if not in name.

Our hypothesis on this point reflects human frailty. Whatever can go wrong will, at least some of the time. We, therefore, hypothesize that strategic systems will lead to increased concentration of power in the hands of management, and that this concentration of power will at times be abused.

M: Vigilance is required to prevent abuses. A strong CIO can help in this regard.

R: Managers may not see this happening when they themselves are the cause.

7. Personal privacy. Concern for personal privacy invasion by computers results largely from the collection of a great deal of information about individuals in one place. Strategic systems tend to integrate a great deal of information in one place. Some of this information is about individuals. For example, one airline can find out a great deal about this author through his membership in their frequent flyer plan, combined with his use of a credit card that earns him mileage credits under that plan. Coordinating these two systems was a brilliant marketing move on this airline's part. At the same time, it provides the potential for greater invasion of my privacy. While the specifics of this situation do not cause me a great deal of lost sleep, the principle applies to more worrisome situations. Strategic systems increase the potential for privacy abuse. The same safeguards that deal with other potential invasions of privacy by computer, some of which involve legislation, apply to strategic systems. Organizations must be vigilant in the use of these safeguards. Here, as with the previous point, the Chief Information Officer has a key part to play. Our hypothesis on this point parallels that on the concentration of power. Strategic systems will lead to increased concentration of information in one place. This concentration of information will, at times, be abused for the invasion of personal privacy.

M: Again, vigilance is required to prevent abuses and a strong CIO can help.

R: A researcher who is not an expert on privacy issues may need assistance in this area.

8. Bureaucratic rationalism. "Bureacratic rationalism refers to the movement of organizations toward more bureaucratic structures in their drive for higher degrees of rationalism" (Hirschheim, 1986). Strategic systems, by replacing human judgment with automated decision making (or automated recommendations which are usually accepted by human approvers) do lead to higher degrees of rationalism. However, they do so via automated, not bureaucratic, structures. Concern that strategic systems may increase the bureaucratic nature of an organization is not well founded. Our hypothesis: strategic systems will not increase bureaucracies.

9. The underground economy. One author reported by Hirschheim (Darhendorf, 1982 as cited in Hirschheim, 1986) sees increased use of office automation and related techologies as facilitating unofficial forms of work and as creating a more positive climate for it. Be that as it may, we hypothesize that strategic systems will have a negative effect on the underground economy. It is not practical for an isolated individual or small group of individuals to develop anything resembling a strategic system. To the degree that such systems increase their sponsors' attractiveness to customers, they will move customers toward the larger firms which have such systems and away from "underground" operations.

M: This may be an opportunity for "mainstream" businesses to recapture lost customers.

R: The impact of a strategic system may extend beyond a firm's visible competitors.

10. Participation. It has been a truism for decades (for example, (Gregory & Van Horn, 1963)) that non-technical end users should participate in defining the computer systems with which they will work. Such advice is more difficult to solicit when those affected are from many parts of an organization. The perspective of each potential user is limited to his or her own sphere of experience. The change in the nature of the work that will result from the successful implementation of a strategic system makes it more difficult to make meaningful suggestions in advance. While the arguments for user participation remain valid, the difficulty of achieving it increases. Greater difficulty in reaching any goal whose desirability has not changed, will result in the goal being reached less often. Accordingly, we hypothesize that user participation in the definition of strategic systems will be relatively less common, in any one organization, than it is in the definition of other types of systems.

M: Top management must push strongly for participation to counter this effect.

11. Ethical and social acceptability. If the above concerns (primarily employment and privacy) are satisfactorily addressed, there is no reason strategic systems should not be ethically and socially acceptable. We have no further hypothesis on this point. (This item is included here only to retain Hirschheim's list from (1986) unchanged.)

12. Education. People need to be educated in the way strategic systems affect their lives. The impact of a strategic system is broader than the impact of several individual systems, even if the individual systems collectively cover the same ground. This has the potential of creating fear which must be countered through education. Education is also necessary to enable people to move into the new, information-related jobs that will be created in the work dislocation process. Finally, corporate education is necessary to make people aware of the dangers to privacy and the danger of concentration of power that strategic systems make increasingly possible. All in all, the emergence of strategic systems is a powerful argument for more and better education in information management. We hypothesize that, in response to this need, education in information management will become more common. (We note in passing that information literacy is not the same as computer literacy, as pointed out by Kanter (1986) and Mallach (1986).)

M: The necessary education programs must be put in place.

Summary

There is no denying that strategic information systems will come into increasingly widespread use. They will have a correspondingly widespread impact on the organizations which use them and on the customers of these organizations.

Some of this impact will be positive. If our hypotheses are correct, these systems will provide job enrichment and flexibility without reducing employment. They will thus improve the quality of working life. They will also force increased education in information literacy, with corresponding benefits to all concerned.

On the negative side:

• Strategic systems may be developed with less user involvement than is desirable. This may in turn lead to resistance to their use. MIS managers must make an extra effort to ensure sufficient user involvement in strategic system's definition and development.

• The increased concentration of information in one place that strategic systems create will increase the potential for abuses of power. Abuses of power are difficult to prevent. A CEO's "Because I say so!" can justify almost any action that is within the bounds of legality and gross morality. The major concern, though, is not with CEOs. It is with lower-level managers within an organization overstepping their bounds. Top management must be aware of the potential problems that strategic systems can create and must monitor the actions of their subordinates to ensure that abuses of power do not arise.

• The potential for abuses of privacy also increases for the same reason. Abuses of privacy must be prevented. The same safeguards that apply to other systems should be used. The need for such safeguards may arise for the first time in an organization when strategic systems are implemented. Even if such safeguards have existed in the past, they may be needed to a greater degree or with more careful supervision. MIS management and EDP auditors have key roles to play here.

It is within the power of senior management in any organization to prevent these problems. For their organization to achieve the full potential benefit of strategic systems, it is important that they do so.

References

Bakos, Y. J., & Treacy, M. E. (1986). Information technology and corporate strategy: A research perspective. *MIS Quarterly, 10,* 107-119.

Bariff, M. L., & Galbraith, J. R. (1978). Intraorganizational power considerations for designing information systems. *Accounting, Organizations and Society, 3,* 15-27.

Burnstine, D. C. (1980). *BIAIT: An emerging management engineering discipline.* (Working Paper). Petersburg, N.Y.: BIAIT International Inc.

Cash, J. I., & Konsynski, B. R. (1885). IS redraws competitive boundaries. *Harvard Business Review, 64* (2), 134-142.

Canning, R. G. (1984). Developing strategic information systems. *EDP Analyzer, 22* (5), 1-8.

Darhendorf, R. (1982). *On Britain.* Chicago: University of Chicago Press.

Davis, S. G. (1990, July 9). Office automation takes off—again. *Computer Systems News,* 1.

Giuliano, V. E. (1982). The mechanization of office work. *Scientific American, 247* (2), 66-76.

Gregory, R. H., & Van Horn, , R. L. (1963). *Automatic Data Processing Systems* (2nd ed.). Belmont, Calif.: Wadsworth Publishing.

Hirschheim, R. A. (1986). The effect of a priori views on the social implications of office computing: The case of office automation. *ACM Computing Surveys, 18,* 165-196.

Ives, B., & Learmonth, G. P. (1984). The information system as a competitive weapon. *Comm. of the ACM, 27,* 1193-1201.

Kanter, J. (1970). *Management Guide to Computer System Selection and Use*. Englewood Cliffs, N.J.: Prentice-Hall.

Kanter, J. (1986). *Information literacy for the CEO* (Working Paper 86-14-P). Wellesley, Mass.: Babson College Center for Information Management Studies.

Kraut, R., Dumais, S., & Koch, S. (1989). Computerization, productivity, and quality of work life. *Comm. of the ACM, 32*, 220-238.

Mallach, E. G. (1986, March 10). The case against computer literacy. *Computerworld, 20*, (10), 17.

Markus, M. L. (1983). Power, politics, and MIS implementation. *Comm. of the ACM, 26*, 430-444.

McDonough, A. M. (1963). *Information Economics and Management Systems*. New York: McGraw-Hill.

McFarlan, W. F. (1982). The information archipelago—maps and bridges. Harvard Business Review, 60, (5), 109-119.

McFarlan, W. F (1984). Information technology changes the way you compete. *Harvard Business Review, 62*, (3), 98-103.

Mintzberg, H. (1983). *Power In and Around Organizations*. Englewood Cliffs, N.J.: Prentice-Hall.

Pettigrew, A. M. (1972). Information as a power resource. *Sociology, 6*, (2), 187-204.

Porter, M. E. (1980). *Competitive Strategy*. New York: Free Press.

Porter, M. E., & Millar, V. E. (1985). How information gives you competitive advantage. *Harvard Business Review, 63*, (4), 149-160.

Spinrad, R. J. (1982). Office automation. *Science, 215*, 808-813.

Synnott, W. R., & Gruber, W. (1981). *Information Resource Management*. New York: Wiley.

Whisler, T. L (1970a). *Information Technology and Organizational Change*. Belmont, Calif.: Wadsworth.

Whisler, T. L (1970b). *The Impact of Computers on Organizations*. New York: Praeger.

Chapter 16

Controlling the Introduction of Strategic Information Technologies[1]

Macedonio Alanis[2]
University of Detroit

Using computers to automate existing functions can increase speed and accuracy of operations, increase processing capacity, and occasionally save money. However, there is a point where there is not much to gain from just automating existing systems. In the early 1980's, declines in the cost of technology, changes in the economy due to global competition, and deregulation of certain industries, forced organizations to search for new ways to use technology (Ives & Learmonth, 1984).

That search for opportunities revealed that information systems can be used in several ways to improve the strategic position of a company. Some companies used information technology to develop new products and services. Others used it to create closer ties with customers or suppliers, or to prevent competitors or substitute products from entering the market. Information systems can radically change the balance of power in buyer-supplier relationships, and in most instances shift the competitive position of players (Cash, McFarlan, McKenney & Vitale, 1988).

Overlooking this potential of information systems may be dangerous. New products and services can change the balance of the market permanently. If a competitor uses new technology to gain market share by locking in customers, to mention one form, it may be difficult to recover the lost share. On

[1] This chapter is based in part on the results of the meetings of the discussion group on managing technological change, sponsored by the Management Information Systems Research Center of the University of Minnesota. Parts of this article are based on the article "Successful Introduction of Information Technologies," by J. Gray and M. Alanis, under review for publication by the Journal of Management Studies.

[2] Most of the work on this project was completed while the author was affiliated with the University of Minnesota.

the other hand, if the impacts of the change are not considered carefully, a new technology may create opportunities which the company is not ready to take and leave open doors to new, or existing, competitors. Innovations may become strategic necessities, i. e., systems that must be developed to match a threat from competitors, but these seldom convey competitive advantage since every player has those systems (Clemons & Kimbrough, 1986; Clemons & Row, 1987). If a competitor is better suited to absorb the costs of such strategically necessary systems, those necessities may leave the company in a weaker position to compete.

The stakes are so high, that the decision to introduce a new technology must be explicit and well planned (Cash et al., 1988). Few companies can control the evolution of technology. What can be controlled is the speed at which innovations enter the corporation and the impact of these innovations on their markets. To address the issues related to technology implementation, a discussion group formed by researchers from the Management Information Systems Research Center of the University of Minnesota and managers in charge of implementing new information technologies in large organizations met for over two years. The group was sponsored by the Management Information Systems Research Center of the University of Minnesota. Members meet monthly in the facilities provided by the participating organizations. The group produced a set of guidelines to reduce the risk of failure of the introduction of information technologies in an organization.

Based on the recommendations of the group, and current research on strategic information systems (McFarlan, 1984; Ives & Learmonth, 1984; Porter & Millar, 1985), innovation diffusion theory (Rogers, 1983; Brancheau, 1987), and agency theory (Ross, 1973; Fama & Jensen, 1983; Eisenhardt, 1989), this paper presents three steps organizations can use to select the right innovations, implement them properly and fully realize their potential. It also introduces two policies that can be adjusted to control the speed at which innovations enter the corporation.

Issues

Cincinnati Milacron, Inc. sells nearly 65,000 different grinding wheels and hundreds of cutting fluids through distributors. Faced with this variety, distributors found it difficult to recommend the appropriate product for a particular application. Armed with portable computers, they now enter answers to a series of questions and the computer program recommends the appropriate product. Cincinnati Milacron has nearly doubled their five percent market share with this program. Milacron has made this improvement without adding any employees (Business Week, 1987).

Dean Witter Financial Services takes part in the competitive market of originating mortgages. They recognize that mortgages are a commodity provided by many suppliers. They decided to differentiate their product by providing superior customer service. They developed a computer-based system which analyzes the borrower's income and price of the home. Based on this analysis the system provides a complete menu of all fixed, variable, conventional and government backed mortgages available in every maturity. Their system provides for an on-the-spot commitment coupled with the ability to close within fourteen days (Deering, 1987).

Perhaps the most publicized case of information systems reshaping an industry occurred in the airlines industry. American Airlines' reservation system (SABRE) changed the work of travel agents by making up to date information always available. The system was so effective in boosting the company's sales that competing airlines sued to force changes in how it displayed their flights (Smith, 1987). The system has been so successful that American started a new company, AMR Information Services, Inc., a major diversification effort involving the marketing of new kinds of data processing applications. Moreover, American maintains a competitive edge by being able to use the information collected by the system to make changes in schedules and fares to best take advantage of available traffic.

Not every aspect of information technologies is positive. While one company increases its market share, others have to lose it. Using information technology may sound good if you are the one taking advantage of it. Some times, the impact of an innovation is such that the whole market is affected. For example, while American Airlines's introduction of the Sabre system provided a competitive advantage, computerized reservation systems played a role in reducing margins in the industry. Sabre, and similar systems which followed, gave customers more power by providing them with information about cheaper fares and better schedules.

Information technologies may also become strategic necessities and force every player in the industry to invest in them to remain competitive. An example of this technology is the automatic teller machine (ATMs). Analysts agree that the universal adoption of ATMs benefited the substantial portion of retail customers who use them. However, since ATMs are offered by almost all banks, they confer neither margin nor market share advantage. Moreover, they do not appear to have reduced bank's expenses (Clemons & Kimbrough, 1986). Those strategic necessities may distract needed funds from other more profitable areas. They may give the advantage to the competitor with a better technological base.

Being aware of the potential of information technologies is important. Managers must be ready to react to changes introduced by competitors, or to take advantage of opportunities available. Most managers cannot control changes in the technology, but they can speed up or reduce the rate of change and the amount of technology that enters their organization and market.

This readiness requires understanding the information technology phenomenon, how this information can be used strategically, and how it can be properly implemented. The literature on strategic information systems helps explain the importance, impacts and search process for these opportunities. To understand the information technology diffusion process, we will use two theories: innovation diffusion theory and agency theory. The former helps explain how innovations are spread through time in an environment. The latter describes the political and social factors involved at the individual and departmental level.

Significant Related Research

Strategic Information Systems

Traditionally, competition was viewed too narrowly because it was primarily based on projections of market share and market growth (Porter, 1980). Porter's work on competitive strategy asserted that the economic and competitive forces in an industry are the product of a broader set of factors. He presented a framework of five basic forces that shape competition: Bargaining power of suppliers, bargaining power of customers, potential new entrants, threat of substitute products, and the intensity of rivalry among existing competitors (Porter, 1980). Although his work did not initially include information systems, it provided a good framework for investigating the role that information systems can play.

McFarlan (1984) mapped the competitive applications of information technology into Porter's framework and produced a list of five questions to help determine whether information technology is likely to be of strategic importance.

1. Can information technology build barriers to entry to new competitors?

2. Can we use information technology to provide services which make it difficult, or costly, for our customers to switch to our competitors?

3. Can the technology differentiate our company and change the basis for competition?

4. Can information improve our bargaining power with suppliers?

5. Can information technology create new products?

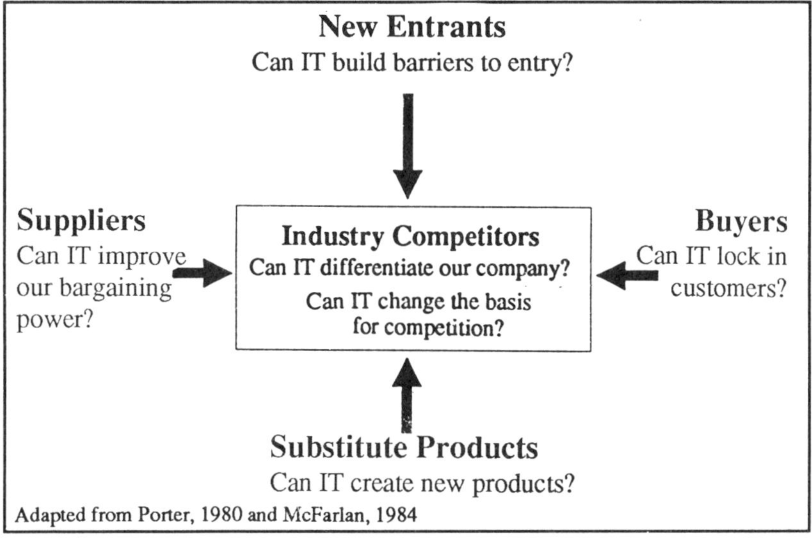
New Entrants
Can IT build barriers to entry?
Suppliers
Can IT improve our bargaining power?
Industry Competitors
Can IT differentiate our company?
Can IT change the basis for competition?
Buyers
Can IT lock in customers?
Substitute Products
Can IT create new products?
Adapted from Porter, 1980 and McFarlan, 1984

Figure 1. Questions to assess the impact of information technology on the competitive forces

If the answer to one or more of these questions is yes, information technology is a strategic resource. It can change the company's strategic position, its relative position in the market, or the market itself.

For information technology to become a viable competitive weapon, management must understand its effects on the competitive environment. Taking a broader perspective than McFarlan, Parsons (1983) views the impacts of information technology at three levels: the industry level, the firm level, and the strategy level.

At the industry level, information technology can change the fundamental nature of the industry in which the firm competes. These changes may modify the basic assumptions under which the firm is operating. Not understanding those potential changes may have devastating effects on an organization.

At the firm level, information technology can directly change the shape of the competitive forces. Analyzing Porter's five competitive forces can help in understanding the potential for change.

At the strategic level, Parsons indicated that successful firms position themselves relative to industry forces by implementing one or more generic strategies:

- Cost leadership,
- Product differentiation, or
- Creating a niche for their products.

Information technology can impact the firm's ability to execute a particular strategy.

Parson concludes that by understanding when, where, and how information technology will impact a firm, management can recognize opportunities and threats, and develop a strategy that directs resources to take advantage of opportunities and mitigate threats. Without such understanding, firms will continue to ride a speeding technological roller coaster.

Innovation Diffusion Theory

To control the rate at which information technology enters an organization, it is necessary to understand how innovations are spread and accepted. Contingent factors such as the technology, the individuals involved, the structural and cultural characteristics of the organization, and the environment in which the organization operates may affect the diffusion process (Brancheau & Davis, 1986). The first step in determining how it happens in a specific case is understanding the general process of diffusion.

Innovation diffusion has been studied in a variety of contexts from many perspectives. Each study indicates that these innovations follow markedly similar patterns (Brancheau & Davis, 1986). In a field study and historical analysis of the diffusion of spreadsheet software in organizations, Brancheau (1987) found substantial support for innovation diffusion theory in the context of the use of information systems in organizations.

Much of the theory building work in innovation diffusion theory has been done by Evert Rogers. He used meta-research methods to analyze and classify over one thousand diffusion studies (Rogers, 1983)

Rogers (1983) defines diffusion as the process by which an innovation is communicated through certain channels over time among the members of a social system. An innovation is an idea, practice, or object that is perceived as new to an individual. Most new information technologies can be considered innovations. These innovations can vary in scope from the enhancement of an existing software package to the installation of a multipurpose workstation (Niederman & Brancheau, 1987).

Innovation decision making is the process by which an individual, or organization, passes from the first knowledge about an innovation to its permanent adoption. Rogers' diffusion model suggests that innovation decision is not an instantaneous act, but a process which occurs over time (Rogers, 1983). The process includes five steps: Knowledge, persuasion, decision, implementation and confirmation. In the organizational context, it is useful to view the process as occurring in two phases: initiation (including knowledge, persuasion and decision) and implementation (composed of implementation and confirmation). In an organizational context, each phase can be carried out by different units (Brancheau & Davis, 1986).

The initiation phase occurs first. During this phase the innovation is identified and studied, a favorable or unfavorable attitude is formed towards that innovation, and a decision is made to adopt or reject it.

The implementation stage begins after an innovation has been accepted. This stage starts when the innovation is acquired or begins to be used. Originally, the innovation is put to use in the way it is received and for the purposes intended. However, depending on the type of innovation, it may be modified, or used for other purposes. This phase concludes when the innovation either becomes routine, is modified or rejected.

Agency Theory

The diffusion of technological innovations takes place at different levels. Innovations may be adopted by individuals, departments or organizations. At every level it may take a different amount of time and require different types of support. The reason is the different set of goals and needs at every level, and the technical, political and behavioral dimensions of the innovation. Controlling the rate at which an innovation is developed or adopted requires understanding these interactions as well. Agency theory can help managers understand those interactions and prepare policies to encourage or deter change.

Agency theory attempts to explain the relationship between two people or two units. An agency relationship is any cooperative effort where one person (the agent) is working for another (the principal) (Kirsch & Beath,

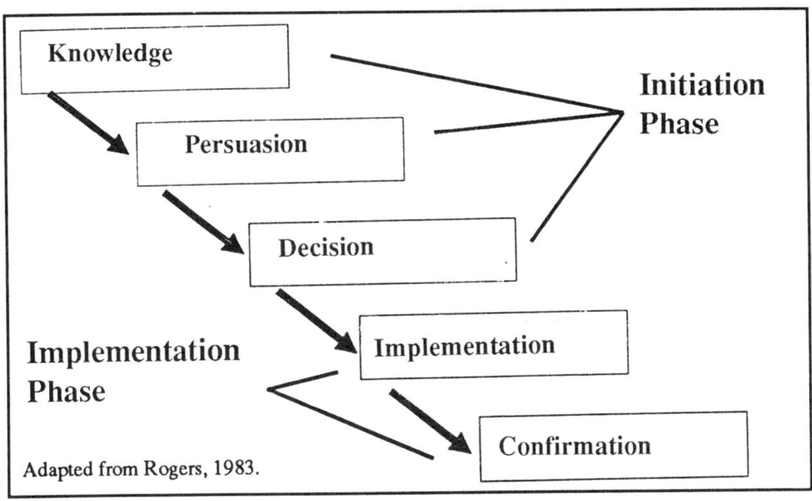

Figure 2. The Innovation-Decision Process

1989). This relationship is a contract between a principal and an agent, where the principal engages the agent to perform a service on the principal's behalf and delegates some decision-making authority to the agent (Jensen & Meckling, 1976). Agency theory helps incorporate the political and behavioral dimensions of decision making (Fox, 1984).

Agency theory research generally shares the following assumptions about humans, organizations and information (Eisenhardt, 1989).

- Humans are assumed to be self-interested, risk averse, and subject to bounded rationality.
- Principals are less risk averse than agents, due to the principal's greater ability to diversify risk.
- Organizations are assumed to use efficiency as the effectiveness criterion.
- Partial goal conflict and information asymmetry are assumed to exist within an organization.
- Information is assumed to be a purchasable commodity.

The relation is defined through a contract. Agency problems arise because there are differences in goals, different attitudes towards risk and information asymmetries. Agents and principals have different utility functions, where utility is defined as a personal measure of satisfaction (Fox, 1984). Both agent and principal act to maximize their own utilities. This produces suboptimal decisions. To overcome that, contracts can be based on behavior or outcome (Eisenhardt, 1989). Behavior oriented contracts assume a monitoring capacity of the principal, and reduce the premium required for risk by the agent. Outcome oriented contracts pass some of the risks to the agent, who requires a premium, but may not require strong monitoring.

Although agency theory, like any other economic theory, is subject to criticism, it is useful for studying the role of information systems in organizations (Kirsch & Beath, 1989). The organization, and its management, can be viewed as the principal; the agent are the employees or departments. Managers can offer incentives to encourage employees, or departments, to discover new opportunities or adopt new technologies. On the same lines, policies can be adjusted to reduce the rate of change in an organization.

Solutions

Steps to Follow

The identification and correct implementation of information technology is a problem common to many organizations. Those issues were addressed by the discussion group on managing technological change, sponsored by the Management Information Systems Research Center of the University of Minnesota. The group, formed by practitioners and academicians, reviewed the theories available and the experiences and cases of every organization. It concluded that management can successfully introduce information technology by paying close attention to three critical steps:

- screening and planning,
- implementation, and
- post audit.

Screening and planning. Screening and planning is a decision making process. The objective is to analyze the possible applications of information technologies and choose the most appropriate for the organization at every particular point in time. During this stage, management gathers information about the potential uses of information technology in their industry, analyzes the impacts at every level and decides whether to implement them or wait. The complexity of the information and analysis required calls for participation of members of every level of the organization. However, top management is ultimately responsible for these decisions. Because these decisions can have a significant impact on the organization and its market, they have to be made at a level high enough to be able to assess the full extent of the impacts of the new technology.

Management should constantly be on the lookout for opportunities. However, the amount of attention paid to this area depends on the value adding potential of information technology and on the quality of information systems resources available (McLaughlin, Howe & Cash, 1983). An organization with high value-adding potential but weak systems resources is vulnerable to competitors with better information systems resources. On the other hand, if the value adding potential is high, and the systems group is strong, the organization may be in a position to compensate for the technological weaknesses (Ives & Learmonth, 1984).

One way to assess the potential of information technology is by using Porter's framework (Porter, 1980). As suggested by McFarlan (1983), if information systems can have an impact on any of Porter's competitive forces, then they are likely to be of strategic importance. Management should look for information technologies that can:

- help build barriers to new entrants,
- change the basis for competition,
- reduce costs,
- create new products,
- or create closer ties with customers or suppliers.

Management can look for opportunities by studying and keeping current of changes in four areas:

- the available information technologies,
- their product's market,
- interorganizational relations,
- and their own organizational environment.

Changes of prices and capabilities of the available technologies can create opportunities where none existed before. Only a few years ago, organizations needed justifications for major investments in equipment every time they wanted to use a computer in one of their processes. Today, it is not uncommon to have a computer dedicated to a specific application or project. The depreciation terms for that equipment is generally the duration of the project. Sometimes the expected duration is up to seven years, but when technology changes fast, equipment is depreciated in only one year (Martin, 1990). Having a dedicated piece of hardware for a project is not only feasible, but economic.

A close study of the target market for the organization's products may suggest ways in which information can improve current products or create new ones. The case of Dean Witter Financial Services, discussed above, is a good example of how information systems can help differentiate a product. Those new products can help a company enter new markets, or gain market share.

Interorganizational relations refer to the links with the suppliers or customers. An analysis of the way the customers receive information about the company's products and place orders may reveal opportunities for the introduction of information technologies. The case of Cincinnati Milacron clearly illustrates this.

The internal organization environment can suggest applications that can save costs or create new products. Through automation, Ford Motor Company had planned to cut 100 of the 500 people who process payments for purchases. Management screened the process and devised a bill paying system which eliminates many steps and eventually 375 positions. The new system will enter information on parts received on a terminal at the receiving dock. A central computer will multiply parts price times the quantity received and authorize payments by a check-writing machine. Suppliers will no longer need to send an invoice to Ford. Ford will save the costly steps of matching purchase orders, invoices, and receiving reports. Purchasing agents now have to make sure the computer file of prices used to pay suppliers is up to date as well as negotiate with suppliers (Business Week, 1987).

The Ford project requires coordination between manufacturing, finance, purchasing and data processing. This cooperation is possible only because top management participated in the screening and planning of information. However, line managers must participate in these decisions. They understand the management issues and how information technology can solve problems. Early involvement of line management also helps them to buy into the plan.

According to the innovation diffusion theory, innovations take some time to spread through the organization when they are left alone. Management can play an active role in speeding up the innovation diffusion process by enlisting the help of individuals at all levels of the organization to fully assess the opportunities available. Management policies can help locate new opportunities and simplify the implementation of solutions.

Implementation. Deciding what technology to implement is only the beginning of the work. Making sure that the technology is available and used in the way it was designed is an important step. Many good ideas have failed because of poor implementation.

In a study of 110 organizations, Dray and Yelsey (1985) found that 75% of those organizations were experiencing problems implementing new technologies. In these companies, decisions about what technology to use and how to introduce it were made without sufficient consideration of organizational impacts. Most of the companies studied failed to manage the change triggered by technology. Most managers in these companies felt increasingly powerless and driven by technology.

The study concluded that:

- The impacts of information technology can not be avoided, but can be managed.

- Management is not powerless. Values and philosophy can, and must, drive the process of technology introduction

- Technology does not stand-alone. It must be integrated into organizations. Managing technology introduction means managing change.

- Technology acts as a magnifier. Well-run organizations often become smoother while those with serious problems almost always become worse.

• The human element is crucial. In the long run, ignoring the human element can be as disastrous as ignoring the technical element.

• There is no best way to introduce new technology. Different approaches are needed to reflect differences in organizations, markets and structures.

The introduction of new technology requires the balancing of technical, organizational, and behavioral issues (Niederman & Brancheau, 1987). The interaction between the organizational and technology's features is what Markus calls the interaction perspective (Markus, 1984). This perspective suggests that the greater the divergence between a system and its setting, the more likely it is to be rejected or modified, reducing its ability to achieve significant change. The solution to interaction problems consists in aligning the system and the organization by modifying the system or changing the organization. Organizations that have successfully implemented and exploited information technologies seem to exhibit double loop learning (Cash & McLeod, 1985). Double loop learning (Argyris & Schon, 1978) involves changing the norms and standards of an organization in reaction to innovations. In general, the organization not only learns to correct the problems created by the introduction of new technology but also to anticipate and avoid them (Cash & McLeod, 1985).

Careful change management can reduce implementation problems. The keys to success are the communication between the parties affected by the new technology and the organizational flexibility to accept change.

Post audit. Post audit is an important but frequently overlooked step. Management must make sure that the technology is working as expected.

After implementing the new system, management should review job design and staffing levels. Reducing the time required to do a job from 30 minutes to 10 does not benefit the company unless the employee is assigned additional work. Otherwise, the time saved provides the employee a reduced workload with little benefit to the company. Management should also question if the reduction in time required to do a job has shifted the bottleneck to another task. If so, output and productivity may not increase greatly until management resolves the new bottleneck.

As information technology changes the way of doing business, management should also reassess the assumptions underlying the company's strategic plans. Are the assumptions still appropriate under the new conditions? If not, management must revise mission, goals and strategic plans, completing, in effect, the double loop learning process cited above.

Following these steps will improve the chances of success in adopting new information technology. However, management must address the policy issues to control the speed at which innovations are introduced to the organization. Agency theory suggests that the major issues are organizing the effort, and sharing the risk.

Two Management Policies to Control the Rate of Change

Organization of the effort. Traditionally, top management has chosen and implemented information technology with the advice of the corporate management information systems staff. The advent of micro computers can shift this balance by making it possible for users to make more information systems decisions. Management must decide whether to centralize or decentralize management information systems decisions.

Centralizing decisions about new information technologies facilitates control of the innovations. Since most of the information costs appear in the budget of the management information systems group, rather than spread throughout the organization, it becomes easier to control costs. Also, a centralized group can take advantage of economies of scale, both in equipment and software purchases. Centralized groups may be better equipped to deal with technical problems. However, the major advantage of this approach comes in the flexibility to shift efforts. It becomes relatively simple for decisions to reach many areas of the organization.

On the other hand, a decentralized group may be more sensitive to the needs of users of information technology, particularly those in remote locations. More innovation may come from users working in the solution of their own problems. Decentralization helps gain user involvement in developing new approaches using information technology.

It is often better for top management to provide freedom but still provide some direction. Many managers create groups to screen new technologies, make appropriate modifications, and offer support after installing the application. This direction helps assure that new technologies match the corporate goals.

The company's policy should be the one that is most effective for their needs. Pepsico has tried to blend the advantages of both approaches. They have centralized major information group decisions, but at the division level rather than the corporate level (Deering, 1987).

Risk sharing. In most organizations, each unit has a certain degree of freedom in making its own decisions. However, management can still influence the unit's attitudes towards technology. Agency theory indicates that reducing the risk shared, by strict rules and failure to share the cost of lost time or resources from the introduction of information technology, will discourage change. Management limits risk sharing by:

1. Limiting the department's budget.

2. Penalizing attempts to introduce a technology which fails.

3. Not acknowledging successful introduction of technology.

These policies have the effect of making each department absorb the full risk of the adoption of new technologies. Since they have a limited ability to diversify risk, these policies will reduce the department's willingness to change the current procedures.

On the other hand, if top management wants to encourage experimentation it must absorb some of the risk and show supporting attitudes. A company may do this by:

1. Establishing independent budgets for research and development of information technologies.

2. Accepting that new technologies do not always work well.

3. Offering technical and organizational support once the new technology is in place.

4. Rewarding both successful and unsuccessful attempts to adopt new technologies.

Willingness to reward both successful as well as unsuccessful attempts will ultimately determine the risk shared between the company and the group introducing the new technology. Rewarding unsuccessful attempts may be dangerous, since local managers may attempt new technologies even if they are inappropriate. However, as noted earlier, competition may make attempting nothing as dangerous as occasionally trying a technology which does not work.

Summary and Conclusions

Comments for practitioners

Information technology has the power to change organizations, relationships with the customer, with competitors and even the market itself.

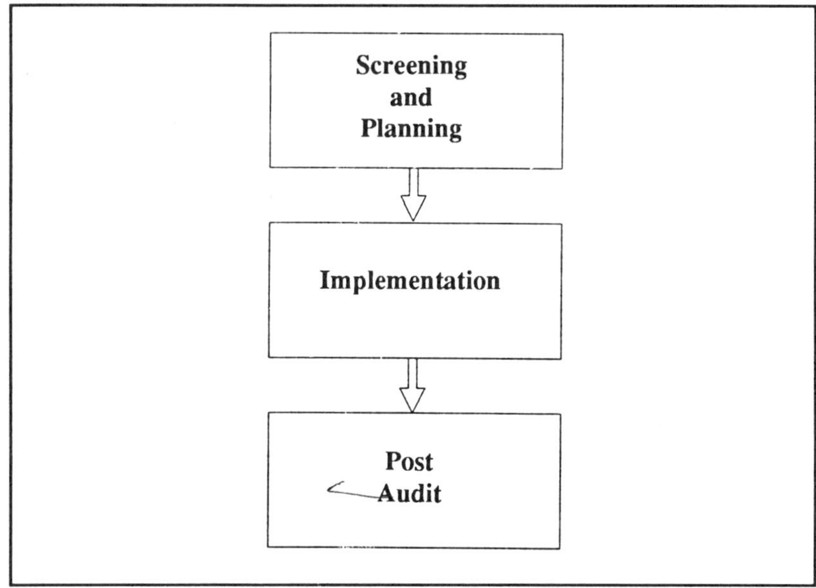

Figure 3. Three critical steps to control the introduction of strategic information technologies

Companies which successfully adopt new information technologies can expect big payoffs. This article recommends policies and steps for implementation which lead to favorable results.

New information technologies may change the way business is done so radically that revisions may be required in the company's strategic plans. It is therefore critical to carefully plan the introduction of new technologies, monitor their progress through time, and conduct post audits of new applications.

How a company organizes for innovation in information technology are critical to success. In addition, the way the company shares the risk of innovation governs the type and pace of innovation.

As information technologies reach maturity, more and more organizations are realizing the importance of careful planning and control of their information resources. Those who take the lead in innovative information technology will lead their markets.

Comments for researchers

Although the use of information systems to gain a competitive advantage is not a new topic, most of the work in this area is descriptive in nature. There is little theory based work that provides guidelines for practice. With the continuous reduction of costs of the technology, and the constant increase in competition, many organizations and industries will experience dramatic changes in the next few years. This will provide the settings and variety of environments required for research, in time to permit the results to still influence practice.

This area of research offers the opportunity to conduct theory based research of significant impact to practice. Relatively little has been published about the applicability of innovation diffusion theory to the MIS area. The work on agency theory in MIS is also just beginning. Far from being a well understood area, we are just understanding that there is a lot to learn, and gain, from research here.

References

Argyris, C. & Schon, D. A. (1978) *Organizational Learning: A Theory of Action Perspective*, Addison-Wesley Publishing.

Brancheau, J. C. (1987) "The Diffusion of Information Technology: Testing and Extending Innovation Diffusion Theory in the Context of End-User Computing," Unpublished Doctoral Dissertation, University of Minnesota, Minneapolis, Minnesota.

Brancheau, J. C., & Davis, G. B. (1986) "A Model for the Diffusion of End-User Computing" Management Information Systems Research Center Working Paper Series, MISRC-WP-87-07, Minneapolis, MN, November, 1986.

Business Week, (1987) October 12, 1987, 134-146.

Cash, J. I., McFarlan, F. W., McKenney, J. L. & Vitale, M. R. (1988) *Corporate Information Systems Management: Text and Cases*, Second Edition, Homewood, Ill, Richard D. Irwin.

Cash, J. I. Jr., & McLeod, P. L. (1985) "Managing the introduction of information systems technology in strategically dependent companies", *Journal of Management Information Systems 1*(4), 5-23.

Clemons, E. K., & Kimbrough, S. O. (1986) "Information Systems, Telecommunications, and Their Effects on Industrial Organizations," *Proceedings of the Seventh International Conference on Information Systems*, 99-108, San Diego, California.

Clemons, E. K., & Row, M. (1987) "Structural Differences Among Firms: A Potential Source of Competitive Advantage in the Application of Information Technologies," *Proceedings of the Eight International Conference on Information Systems*, 1-9, Pittsburgh, Pennsylvania.

Dray, S. & Yelsey, A. (1985) *Impact of Information and Communication Technologies on Honeywell Culture, Organizations and People*, Internal White Paper, Corporate Information Management and Corporate Human Resources, Honeywell Inc., Minneapolis, MN.

Eisenhardt, K. M., (1989) "Agency Theory: An Assessment and Review," *Academy of Management Review, 14*, 57-74.

Fama, E. F. & Jensen, M. C. (1983) "Separation of Ownership and Control," *Journal of Law and Economics, 26*, 301-325.

Fox, R. P., (1984) "Agency Theory: A New Perspective," Management Accounting, 62, 36-38.

Ives, B. & Learmonth, G. P. (1984) "The Information System as a Competitive Weapon,"

Communications of the ACM, 27(12), 1193-1201.

Jensen, M. C., & Meckling, W. H., (1976) "Theory of the Firm: Managerial Behavior, Agency Costs, and Ownership Structure," *Journal of Financial Economics, 3*(4), 305-360.

Kirsch, L. J., & Beath, C. M. (1989) "A Review of Agency Theory for IS Scholars," Management Information Systems Working Paper Series, MISRC-WP-90-03, University of Minnesota, Minneapolis, Minnesota.

Markus, M. L., (1984) *Systems in Organizations: Bugs + Features,* Cambridge, MA, Ballinger Publishing Company.

Martin J., (1990) "What's Your Strategy For Buying PC's?," *Datamation, 36*(14), 69-72.

McFarlan, F. W., (1984) "Information Technology Changes the Way You Compete", *Harvard Business Review, 62*(3), 98-103.

McLaughlin, M. H., Howe, R. & Cash, J. I., (1983) *Changing Competitive Ground Rules: The Impact of Computers and Communications in the 1980's,* Unpublished Working Paper.

Niederman, F., & Brancheau, J. C. (1987) "Planning and Managing Technology Change," Management Information Systems Research Center Working Paper Series, MISRC-WP-87-14, Minneapolis, MN.

Parsons, G. L., (1983) "Information Technology: A New Competitive Weapon," *Sloan Management Review, 25*(1), 3-14.

Porter, M. E., (1980) *Competitive Strategy: Techniques for Analyzing Industries and Competitors,* New York, NY, The Free Press.

Porter, M. E. & Millar, V. E. (1985) "How Information Gives You Competitive Advantage," *Harvard Business Review, 63*(4), 149-160.

Rogers, E., (1983) *The Diffusion of Innovations,* 3rd Edition, New York, NY, Free Press.

Ross, S. A., (1973) "The Economic Theory of Agency: The Principal's Problem," *American Economic Review, 63,* 134-139.

Smith, R., (1987) "AMR's Outlook Is Clouded as Competition Erodes Profit of Airline Unit's Sabre System", *The Wall Street Journal,* June, 1987.

Chapter 17

Information Technology in R&D: Enhancing Organizational Competitiveness via R&D Information Systems

Thomas L. Case
Georgia Southern University

John R. Pickett
Georgia Southern University

It is clear from the work of Porter (1985) that successful competition depends on a combination of factors including the exploitation of technological change. Innovations may be classified by type (e.g. product innovations versus innovations in the processes used to produce or deliver products) and also by source (internally developed versus acquired externally). Some firms are likely to be in a better position than others to improve their competitiveness through the appropriate selection of innovative activities.

R&D's Role in Contemporary Organizations

Research and Development (R&D) departments or divisions have traditionally been the subunits within organizations with the formal charge of generating or sustaining competitive advantage through a steady stream of new products, processes, and other innovations. R&D's role in the innovation and technology transfer processes is illustrated in Figure 1. Gold (1989) has done one of the best jobs of describing what R&D departments are expected to contribute to these processes and to their organizations as a whole. Gold,

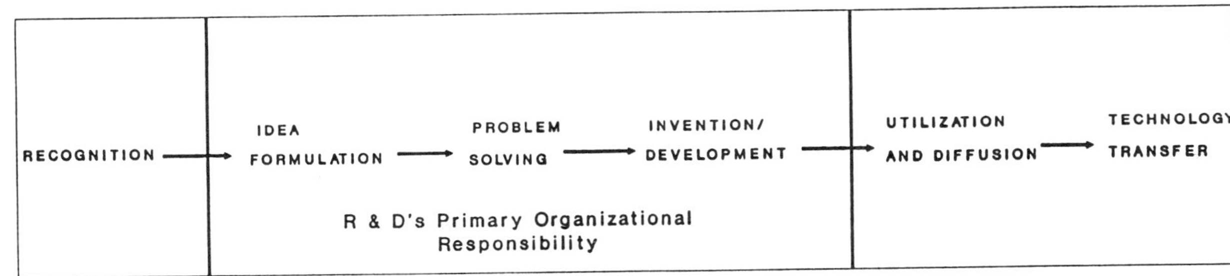

Adapted from Fischer, 1980.

Figure 1. R & Ds Role in the Process of Technological Change

Peirce, Rosseger, and Perlman (1984) have noted that R&D is expected to contribute one or more of the following:

> (a) improvements in the capabilities and quality of existing products and processes, including modifications to facilitate wider applications;
>
> (b) development of new products or processes yielding major commercial advantage over competitors; and
>
> (c) advances in knowledge likely to yield future improvements in products and processes.

Gold (1989) adds that R&D programs also often generate three additional kinds of improvements which can yield other important contributions to competitive performance:

> (a) reducing or minimizing increases in the costs of producing existing products;
>
> (b) reducing lags behind competitors' innovations in products and processes; and
>
> (c) adapting designs and processes to shifts in the supply and prices of inputs.

Oster (1990) cites a number of related approaches that may be used to enhance a firm's competitiveness through R&D.

R&D's Impacts on Organizational Competitiveness

According to AT&T Bell Laboratories' Ian M. Ross (1986) most (50 to 60 percent) of our economic growth can be attributed to technological innovation fueled by R&D. Ross also notes that technological innovation is a major force for improving productivity and industrial competitiveness and observes that:

> Industry's R&D role, to put it in the most basic terms, is to create wealth. Industry is supposed to combine new science and engineering knowledge with other resources - capital and labor - to develop and distribute new products and services. In doing this it fosters economic growth, creates jobs, and raises peoples' standard of living. If it does this successfully, it also makes a profit, much of which it should

reinvest in more R&D, plants, and people so that it can continue this creative and productive process (p. 375).

It is apparent that Gold (1989) also feels that R&D can play a critical role in the competitiveness of organizations. This view has been echoed widely. Morbey and Reithner (1990) found a direct relationship between R&D intensity and subsequent growth in sales for both large and small companies. Brenner and Rushton (1989) and Morbey (1988) also have reported direct relationships between R&D intensity and sales growth. Grilliches (1987) found that increased investment in R&D can also lead to enhanced organization productivity (output per manhour). Hence, it would seem adequate capital expenditures in R&D can lead to both more marketable products and more efficient production operations downstream.

Merrifield (1989) notes that "accelerated R&D investments, particularly in productivity enhancing innovation and automation are now essential for U.S. competitiveness" (p. 71). Smith and Treece (1990) point out that U.S. companies spent $65.2 billion on R&D in 1989, more than any other nation. However, they also note that the top 200 foreign-owned (non-U. S.) companies spent $63.4 billion on R&D in 1989. Also, R&D spending growth rates are higher in most of the countries that we consider to be our major competitors.

Foreign competition is threatening the U.S. lead in R&D. Port (1990) reports that the Sony Corporation devotes up to 25% of its R&D to "precompetitive" (basic) research; no U.S. firm spends half that much. Clark and Fujimoto (1989) found that Japanese automobile manufacturers have a 12 month product development lead time advantage over U.S. and European competitors. Merrifield (1989) notes that rapidly advancing technology has collapsed the product and process life cycles to three to five years for most consumer products and that continued shrinkage should persist throughout the remainder of the century. He also reports that during the last five years, more than 50% of the patents issued by the U.S. patent office have been awarded to foreign-owned companies. This may be one of the reasons why foreign-owned plants in the U.S. created more jobs than did domestic factories (Port, 1990). These trends, when considered as a group, suggest that many of the domestically produced consumer items currently on U.S. market shelves will be replaced by ones manufactured by foreign-owned companies. They also suggest that our foreign competitors pose a significant threat to U.S. lifestyle and economic well-being.

The revitalization of research and development seems to be a critical requirement for re-establishing the competitiveness of U.S. industry. It would seem to follow that investments within R&D which are likely to enhance the productivity and innovation potential of R&D professionals should be directly related to improvements in industrial competitiveness. The acquisition and/or development of information technology and applications which facilitate R&D performance may be one of these crucial investments. It is our contention that the strategic importance of R&D information systems has been overlooked and that such systems share many of the characteristics that have been used to describe strategic information systems. One of the purposes of this chapter will be to show that R&D information systems should be considered as having strategic value. We will also describe how information technology is currently being utilized within R&D and how it could be more fully utilized. Emerging trends will be examined with particular emphasis on networking and intelligent applications. Finally, the impacts of the use of information technology on R&D productivity and activities will be examined.

Issues

Strategic Information Systems

Wiseman (1988) has developed a framework for looking at strategic information systems which is primarily based on Porter's (1980, 1985) theories. Wiseman defines a strategic information system (SIS) as "a use of information technology intended to shape or support the competitive strategy of the enterprise" (p. 18). According to Wiseman, there are five major methods for attaining competitive advantage through a SIS: (1) developing a differentiated product, service, or image for the firm; (2) reducing costs; (3) creating innovative products or processes; (4) permitting growth in products or functions; and (5) facilitating alliances with other organizations.

It should be apparent from Gold's (1989) points about how R&D can contribute to competitive advantage (summarized previously) that an effective R&D organization may improve a firm's competitiveness in many of the same ways that a strategic information system (as described by Wiseman) would. It would seem to follow that an R&D information system which is capable of enhancing R&D performance should also be considered a strategic application of information technology. And in light of the increased number of joint ventures (and shared R&D), a R&D information system might also (necessarily) be linked with those of other organizations. This would further enhance the strategic importance of the information systems within these subunits.

However, the strategic potential of R&D and R&D information systems is overlooked by Wiseman and many U.S. companies. As Ross (1986) notes "many companies ... must learn to use R&D as a strategic resource to optimize manufacturing flow, to reduce production time, and to control costs and inventory." (p 377). While Wiseman's (1988) book is full of examples of strategic information systems, most of these are services. He overlooks the importance of developing and implementing effective information systems within strategically important subsystems. The only substantial point that

Wiseman makes about R&D is that:

> Just as a company with competitive advantage achieved through the development of innovative products and manufacturing processes might have a research and development center dedicated to discovering such products and processes, the SIS advanced guard must also invest in SIS R&D. This involves not only scouting for new information technologies but also building and testing prototype SIS applications, working closely with technology vendors to get first crack at emerging new products, and establishing close ties with strategic targets that might contribute to the development and use of an SIS. (pp. 411-412).

This quote manifests that Wiseman grasps R&D's strategic role in organizations and the importance of its contributions; however, it overlooks (or takes for granted) the crucial role that information technology could play in what he calls "SIS R&D". Surely the SIS R&D center envisioned by Wiseman would have an information system to facilitate the performance of the tasks it would be responsible for.

Reynolds (1989) notes that there are two different ways of utilizing information technology to have a strategic impact on a company. One route, which he calls the "glamorous" approach, is to develop a "new and dramatically better solution to a business problem—a solution that would not be available without information technology" (p. 87). He offers the SABRE and ASAP systems as examples of this route; most of the examples described by Wiseman (1988) would also fall into this category. The other route to achieving strategic impact is for an organization to utilize existing information technology more effectively than its competitors. It is our feeling that the effective utilization of information technology within R&D may have such a strategic impact, even if this impact is not of the glamorous variety.

Strategic Management of Information Technology

Both end-user managers and corporate executives have higher expectations for the support they receive for information systems technology (Rockart, 1982). Cash and Konsynski (1985) note that the strategic advantages offered by information systems have caused top management to covet new and creative ways of using information resources in the pursuit of strategic objectives. An increasingly important part of the information resource management function is that which is focused on the development and management of information systems that significantly improve operational efficiency, promote innovative products and services, and build a strategic information resource base that can improve the competitiveness of the organization (Porter and Millar, 1985). Because of the traditional roles that R&D has played, the time has come for top management to recognize the payoffs that are likely to result from the development of effective R&D information systems.

Role of Information and Information Technology Within R&D

A recent *Datamation* article entitled "Shaking up the Process" (1989) notes that:

> The top 10 process corporations also rely heavily on computers in their research and development operations. Computers assist in monitoring and simulating chemical interactions and help formulate manufacturing processes that generate higher quality and higher yields. Local networks connect these and other departments to one another to help ensure that development isn't carried out in isolation from manufacturing engineers or marketers (p. 73).

In this same article Henry Pfeldt, director of Kodak's information technology service states that: "In evaluating our IS operations, we concluded that managing and operating data-processing facilities is not a strategic function. Using the information is" (p. 73). Reynolds (1989) would agree that these are not "glamorous" uses of information technology for strategic advantage, but they may be just as effective in the long run.

R&D has long been recognized as an information intensive area whose success is dependent on effective communication and information flow (Marquis, 1972; Tushman, 1979). Tushman and Nadler (1980) have gone so far as to describe R&D as an information processing system. Figure 2 is an adaptation of Tushman and Nadler's description. It shows that R&D should be considered an open system which has important information and communication exchanges with the external environment and other organizational subunits.

The importance of strong information linkages between R&D and other organizational subunits has been widely recognized. Urban and Hauser (1980) maintain that an organization should encourage communication between marketing and R&D and should allocate effort to understand and monitor the process of technological transfer; however, these authors do not entertain the notion of electronic linkages between these subunits. Bright (1969) and Martin (1984) contend that there must be strong linkages not only between marketing and R&D, but also between the production and R&D subsystems. Effective information flows between these areas are essential to the achievement of R&D goals and the achievement of superordinate organizational goals. Wormell (1989) states that the strategic management of informa-

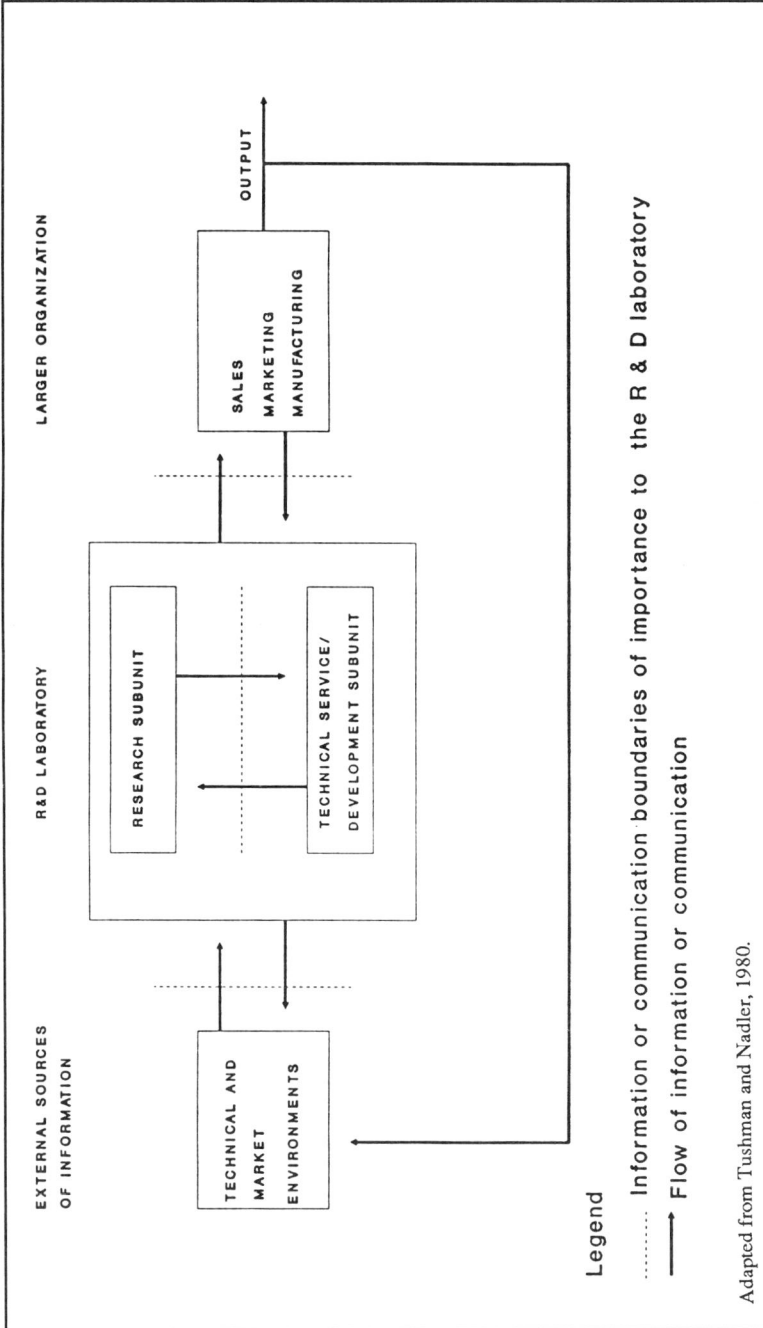

EXTERNAL SOURCES OF INFORMATION

R&D LABORATORY

LARGER ORGANIZATION

TECHNICAL AND MARKET ENVIRONMENTS

RESEARCH SUBUNIT

TECHNICAL SERVICE/ DEVELOPMENT SUBUNIT

SALES MARKETING MANUFACTURING

OUTPUT

Legend

.... Information or communication boundaries of importance to the R & D laboratory

⟶ Flow of information or communication

Adapted from Tushman and Nadler, 1980.

Figure 2. The R&D Information Processing Cycle

tion within organizations provides linkages between "the functional strategies of the business (such as finance, manufacturing, marketing, product development, R&D, human resources, etc.) and demands the information management strategies be effectively integrated into, and supportive of both the functional and business strategies of the firm" (p. 200).

Wormell (1989) also believes that the strategic management of information requires heavy dependence on the quality of information resources and sufficient integration of the firm's internal and external domains of knowledge. The importance of integrating both external and internal information by R&D is depicted in Figure 2. A consistent flow of scientific and technical information from external sources to R&D professionals has been recognized for some time as a determinant of R&D effectiveness and company performance (Allen, 1970, 1977; Opren, 1985; Utterback, 1974; Woleck, 1984). Davis and Wilkof (1988) have pointed out that many organizations need to improve the flow of scientific and technical information to and through R&D; however, they overlook the role that information technology could play in this process.

There has been much greater realization that information technology can facilitate communication and coordination within R&D. Ross (1986) notes that it is important for organizations to recognize how the quality and flow of information influences the manufacturing process as well as the designs of the products being made. He states:

> Information becomes a resource that can add value at every point in your manufacturing - but beyond that, at every point throughout the entire innovation process, from design through marketing. The key to success becomes how well you manage the flow and feedback among all the people and machines involved in the process. Information technology becomes an important tool for carrying out this approach. Such technology goes far beyond the computer. It encompasses data recognition equipment, communication technologies, factory automation, and other hardware and services (p. 377).

Merrifield (1989) also feels that establishing strong linkages between R&D, production, and marketing through flexible computer integrated manufacturing (FCIM) will be necessary to maintain the competitiveness of U.S. industries.

Kizer and Decker (1985) were among the first to recognize that information technology had the potential to enhance the effectiveness of R&D activities. Schilling and Lezark (1988) have taken the stand that R&D laboratories "are in the business of adding value to information. They are information factories and computers are the production machinery." (p. 39). These authors see several advantages to having a state-of-the-art data handling environment within R&D labs including (1) greater productivity of laboratory staff; (2) better quality information in terms of accuracy and presentation; (3) enhanced ability to attract high quality scientists and engineers; (4) computational methods which complement experimental and theoretical methods; and (5) the potential of computer generated solutions to provide insight when experimental procedures are not feasible. Ross (1986) echoes one of these concerns by stating that he feels that R&D is likely to be optimized when organizations are capable of attracting and retaining high quality scientists and engineers. He recommends that organizations should "provide them with good facilities and an environment that stimulates their creativity...make sure that they have access to a free flow of information both within the company and with the outside scientific and engineering communities" (p. 376). The desire to create such an environment and to reap the potential benefits listed by Schilling and Lezark (1988) has led to the emergence of integrated R&D information systems and what Ouchi (1986) has called "laboratory information management systems" (LIMS).

Solutions/Remedies

While awareness of R&D as a strategic resource has been increasing and many organizations have started to recognize that information technology can have a major impact on R&D activities and effectiveness, there is very little research which has directly addressed how information technology is being utilized within R&D. Rossini and Porter (1986) provided the first empirical analysis of the application of computer technology in R&D. In 1988, Rossini, Porter, Jacobs, and Abraham published the first longitudinal assessment of the utilization of information technology within R&D. The most recent and extensive assessment of the state-of-the-art of R&D information systems has been provided by Case and Pickett (1989b). Because the most comprehensive picture has been provided by this study, its findings will be the focus of the next few paragraphs.

Information Technology Applications within R&D

The Case and Pickett (1989b) survey addressed numerous issues including organizational, budgetary, major types of applications, hardware brands used, networking, access to external database, system development, areas which have been enhanced by information technology, and how current R&D systems could be improved. Their respondents were *Fortune* 500 companies that were almost evenly split between centralized and decentralized operations. The median R&D staff size was 250 and more than half of the organizations reported having an in-house information processing staff that was dedicated to meeting R&D information processing requirements. A median of 4% of the annual R&D budget was devoted to information processing activities but more than three quarters of the respondents indicated that this percentage had been increasing over the last five years.

Microcomputers are used on a routine basis by almost all R&D professionals, but minicomputers and mainframes are utilized daily by more than half of the R&D professionals in the responding organizations. IBM and DEC equipment predominated in almost two-thirds of these R&D departments; Ruby (1986) reported a similar pattern. Some type of network was reported in more than 60% of the organizations, but most were mainframe based or consisted of some type of micro-to-mainframe communications

Application	Currently Used (%)	Plan to Use in Future (%)
Word Processing	96	1
Statistical Analysis	92	1
Scientific Calculations	92	0
Database Management	89	3
Data Collection	87	6
Spreadsheets	87	3
Patent Searches	83	11
Lab Automation	80	7
Computer Aided Design	77	4
Data Telecommunications	75	6
Modeling/Simulation	73	13
Process Control	72	13
Project Management	70	11
Cost/Benefit Analysis	69	8
Financial Planning/Analysis	65	1
Software Development	54	7
Computer Aided Engineering	44	18
Prototyping	44	8
Robotics	41	3
Feasibility Analyses	38	13
Hardware Development	37	6
Expert Systems	34	28
Computer Aided Manufacturing	34	24
Artificial Intelligence	32	27
Decision Support Systems	31	21
Market Forecasting	30	4
Product Portfolio	24	10
Technological Forecasting	14	15

Table 1: Current and Projected Applications of Computers in R&D

interface; self-contained LANs were not commonly reported although Rossini et al.'s (1988) findings suggest that LANs are much more prevalent than this (it is possible that many of these were also host-based systems).

Table 1 shows a breakdown of 28 functional applications of information technology within R&D as reported by Case and Pickett (1989b). Rossini and Porter (1986) and Rossini et al. (1988) also report utilization statistics which are largely consistent with those provided by Case and Pickett. It is apparent in Table 1 that a mixture of general purpose and special purpose software is used to support the activities of R&D professionals. Over sixty percent of Case and Pickett's respondents indicated that they had standardized on commercial software; only 30% of the applications had been developed in-house. However, over 100 different products were mentioned (Lotus 1-2-3 was cited most often with dBase III Plus and RS/1 tied for second).

Special Applications of Information Technology Within R&D

There are several packages that have been developed specifically for R&D that should be mentioned. Some of these are for data acquisition, storage, retrieval and analysis; others are for activities such as patent searches, project selection, and group decision support for R&D professionals. Although many of these activities are unique to R&D, they warrant some attention because they help clarify how information technology is being used to enhance competitiveness by facilitating the performance of R&D tasks.

Data Acquisition

Case and Pickett (1989b) found that one of the most commonly mentioned packages that R&D laboratories had standardized on was RS/1. This was developed to allow laboratories to collect and analyze realtime signals and has been adopted by many top R&D labs as a general purpose R&D support package (Russell, 1985). RS/1 and competing products such as ASYST, Labtech Notebook (see Ruby, 1986), and the Hewlett-Packard Data Acquisition Manager - DACQ/300 package (see Drenkow, 1987) have made it feasible to use PCs for data acquisition. Of course, to use a PC for data acquisition, a hardware interface alternately known IEEE-488 connection or General Purpose Instrument Bus (GPIB) is needed to connect the laboratory instruments and monitors to the PC (Putnam & Rhoads, 1988). These are available as separate converter boxes which connect to an RS-232 port or as add-on boards which fit a standard expansion slot in a PC. Besides data acquistion, these packages provide an array of file management, graphics, and decision-support features that assist with many of the data management tasks associated with

R&D projects. The developers of RS/1 claim that it can reduce the time spent gathering, analyzing and graphing data by an average of 40% and provide annual savings of at least $8000 per user (Russell, 1985).

Process Control

Several packages have also been designed to facilitate the transfer of new processes from R&D to production. Labview, a Macintosh based product that provides user-friendly catalytic testing, monitoring and comparison functions, has been adopted by chemical, pharmaceutical, oil refining R&D Labs (Willcox, 1988). Omnimap RS50 and Spectramap SM200 are other PC-based process control systems with decision-support software which makes it possible to fine-tune newly developed processes as soon as they are implemented in the production flow (Perloff and Smith, 1987).

Patent Searches

LEXPAT is a computerized full-text patent retrieval service that many R&D labs routinely take advantage of (Simpson, 1985). Not only can it inform R&D professionals if they are re-inventing someone else's wheel, but this service can be a valuable source of technical information. Technological developments are often reported in patents before they appear in journal and trade literature (and sometimes they never appear in these media). The data base is current and the full-texts of the nearly 1200 patents issued each Tuesday by the U.S. Patent Office are available the following Monday. Patent trial decisions, technical publications, and other R&D relevant information can also be accessed through Mead Data Central's LEXIS and NEXIS services (Simpson, 1985).

Project Selection

The selection of R&D projects is a decision problem of vital importance in the company's long-range strategy. Several studies have demonstrated the potential usefulness of decision analysis in R&D project selection decisions (e.g. Baker and Freeland, 1975; Becker, 1980; Liberatore, 1987). Ruusunen and Hamalainen (1989) have developed two packages HIPRE (Hierarchical Preference analysis) and NETPRE (Network Preference analysis) software to assist R&D managers in this very important process. Although these are new products, they have shown considerable promise.

R&D Decision Support Systems

Scott-Morton (1971) identified decision support systems (DSSs) as interactive computer-based systems which help decision makers utilize data and models to solve structured and unstructured problems. Toda, Hiraishi,

Shintani, and Katayama (1988) remind us that R&D is a set of complex problem-solving activities that requires advanced knowledge information processing such as concept formation and all sorts of inferences including deduction, induction, and abduction. They have developed a model and working prototype of a complex Research Decision Support System (RDSS) which is capable of assisting R&D professionals with a range of activities from information structuring to information integration. The overall purpose of the RDSS is to help organizations assess the effectiveness of alternative strategic paths for corporate R&D.

R&D Networks

Both Case and Pickett (1989) and Rossini et al. (1988) found that networking was the exception rather than the rule in R&D operations. The LIMSs that Ouchi (1986) has discussed are primarily composed of networked microcomputers. Business telecommunications is commonplace in many R&D operations and is essential to data acquisition and process control applications (Hooper, 1986). The increasing incidence of multinational joint ventures, and the shared R&D that often results from these, have made electronic linkages and data transfers almost a necessity (Ziegler, 1990). Several of the international programs aimed at mobilizing researchers and engineers (e.g. EUREKA, ESPRIT, and RACE) have been forced to confront telecommunication issues such as hardware connectivity and protocols (Ferne, 1989). ARCNET has provided R&D related services and software to R&D and engineering laboratories, universities, etc. since the early 1970s (Carson, 1981). There is a large variety of applications within ARCNET's production library capable of supporting R&D activities; these may be accessed for a very reasonable timesharing charge. Hewlett-Packard, whose instruments and hardware are utilized in many R&D departments, has developed NCS (Network Computing System); this product has become the centerpiece in several R&D computer networks located on university campuses and within corporations (Marcus, 1990).

In short, there is great variety in the way that information technology is utilized within R&D. Many of the applications are unique and many others are general purpose applications that are likely to be found in other organizational subunits (see Table 1). It is quite possible that this variety will increase. In 1985, the market for R&D software was estimated to be $6 billion and market expansion was projected for the next decade (Russell, 1985). The size of this market would seem to provide an incentive to software developers to generate a wider array of R&D applications and to enhance the user-friendliness of existing products.

Future Trends and Emerging Information Technologies in R&D

Case and Pickett (1989b) asked respondents to indicate the areas in which information technology applications not currently available in R&D were planned or soon to be implemented. The five most frequently mentioned areas are shown in Figure 3. This data gives us some idea about where information technology applications within R&D are going. It is clear from Figure 3 that expert system (ES) and artificial intelligence (AI) applications are likely to be major emphasis areas; while only a little more than a third of Case and Pickett's respondents were currently utilizing these technologies, nearly another third were expecting to exploit them in the future. Computer aided manufacturing (CAM), decision support systems (DSS), and computer aided engineering (CAE) applications also seem likely to be strong growth areas within R&D.

Expert System and Artificial Intelligence Applications

The high level of interest shown in expert systems (ES) and artificial intelligence (AI) applications is probably a reflection of the general growth of these technologies. According to Feigenbaum, McCorduck, and Nii (1988) more than half of the *Fortune* 500 companies are using or developing ES/AI systems; over 2000 are being used daily and over 10,000 are being developed. Sales of AI/ES were predicted to reach $650 million in 1989 and to return $115.5 million on their investment (*Information Center*, 1989). These trends led Thomas Peters to contend that any company which is not using or considering the use of ES/AI is getting dangerously behind the game (Peters, 1987).

Because of the significant interest shown in the development of expert system and artificial intelligence applications within R&D by Case and Pickett's (1989) respondents, Pickett and Case (1990) conducted a second survey which focused on the utilization of these technologies within R&D. They re-surveyed all respondent's indicating that they were utilizing or developing and ES/AI and also sent out questionnaires to a mailing list of potential respondents provided by one of the leading ES consultants and seminar leaders. Thirty-three usable responses were received (this translates into a response rate of just over 30%); nineteen of the respondents were currently using ES/AI. Users reported having significantly larger budgets and number of R&D professionals than non-users.

The survey instrument employed by Pickett and Case (1990) asked respondents about the reasons why the decision was made to develop ES/AI

50
40
30
20
10
0

ES AI CAM DSS CAE

Application

PLAN TO USE CURRENTLY USE

Adapted from Case and Pickett, 1989.

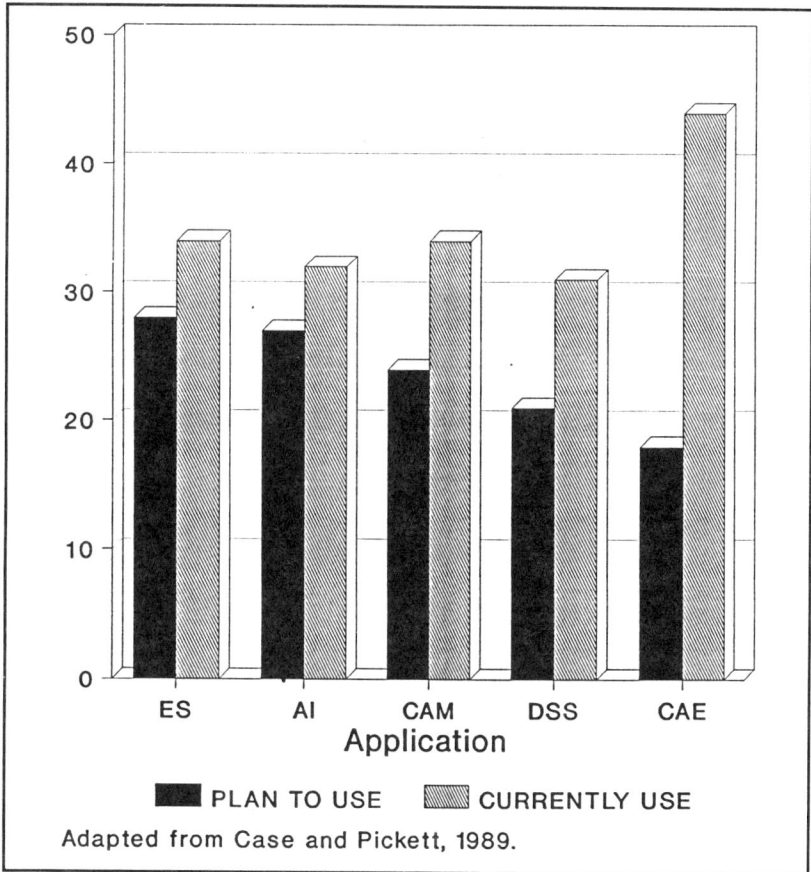

Figure 3. Current and Planned Use of Five IS Applications Within R&D: Percent of Respondents*

systems, the development methods that were used, the types of logical processes incorporated into the systems, how ES/AI was being applied, and the types of applications currently being developed. Respondents were also asked to indicate what types of benefits had been derived from utilizing ES/AI and what types of problems had been encountered. While the sample size was small, the results are probably representative of the current situation in R&D organizations in regard to the development, use, and perceptions of ES/AI.

Two major reasons were cited for developing ES/AI applications by Pickett and Case's (1990) respondents. The first was to capture the expertise of technical experts who are viewed as irreplaceable assets; some organizations also indicated that they were utilizing expert systems for training or to distribute scarce expertise to widely dispersed operating locations. The second major reason was to either better understand the results of complex systems or to improve control over complex systems. Both of these responses are very consistent with well publicized ES/AI applications including those mentioned in other chapters in this book.

The types of applications of ES/AI within R&D found by Pickett and Case (1990) are displayed in Table 2. The pattern of responses which was obtained is largely consistent with the motivation behind the development of these technologies. Jenkins (1986) describes some similar types of ES/AI applications within R&D. Researchers at Westinghouse developed a system which accurately pinpoints problems with X-Ray machines 90% of the time; the X-ray machine is a vital part of R&D activities at Westinghouse and this application has reduced its downtime considerably. In another application at Westinghouse, an expert system collects and analyses data from test probes; this assists in the development of new integrated circuits (Jenkins, 1986). These probes collect and analyze more data than human experts could possibly process and help ensure that available data is not being overlooked. A more "bread and butter" ES/AI application at Westinghouse helps the company prepare bids for large, customized contracts; it helps ensure that the contract is designed to the customer's specification, and that the proposed equipment included in the contract constitutes a system that meets federal, state, and local regulations in the most cost effective manner (Jenkins, 1986). It is quite clear from these examples that there are a range of ways in which ES/AI can be applied within R&D.

Pickett and Case (1990) asked their respondents how satisfied they were with their ES/AI systems. Over two-thirds indicated that they were satisfied with the systems which were in place. These researchers also asked their respondents to list the benefits that had been derived from utilizing ES/AI within R&D; the results of this query are listed in Table 3. Improvements in decision-making (including that of less experienced staff) were reported by two-thirds of the respondents utilizing. Improved productivity or efficiency is also reported by a solid majority of the users of these systems. According to Jenkins (1986), even the R&D experts who serve as the model for the ES application are likely to benefit from its development; formalizing the rules that they follow when applying their knowledge often has the side effect of making them apply the rules more consistently and effectively. All in all, these reported benefits are likely to further enhance the interest of R&D professionals in ES/AI applications.

Table 4 shows a summary of the most significant problems associated with using ES/AI within R&D which were identified by Pickett and Case's (1990) respondents. The most frequently reported problem seems to revolve

Application	Current Use	Current Development	Future Development
Developing machines which are operated or maintained more efficiently (e.g., which give advice rather than readings, take rule-based actions, ask for preventive maintenance or repair, warn of dangerous situations, operate in hazardous environments).	13	19	25
Developing systems which control or interpret other complex systems (e.g., which control complex systems, optimize software performance, explain observed data, infer outcomes of complex events).	16	10	14
Developing systems which lead to more efficient operations (e.g., which carry our intelligent database searches, find faults, solve equations, do technological forecasting, assist in reverse engineering, configure objects under constraints, identify vulnerabilities, correct trainee performance).	13	5	8
Capturing and distributing expertise (e.g., which capture knowledge from experts, integrate knowledge from business process/systems).	13	16	6
Developing simulations, voice recognition systems, and applied robotics (e.g., which simulate experiments or industrial processes, develop voice recognition and robotic systems).	4	3	3

Table 2. Current and Future AI/Expert System Developments: Frequency of Responses (n-33)

around managerial ignorance or inertia. Many respondents described difficulties in convincing management that these applications have potential for significant payoff. In other cases, the respondents expressed the belief that ES/AI had been oversold to top management within their organizations or that cultural and behavioral barriers were inhibiting their approval, development, and implementation. Over half of Pickett and Case's (1990) respondents who were using ES/AI added written comments citing the need for better develop-

13 Improved decision making and problems solving.
13 Automation of some decision making.
13 Development of expertise in less experienced staff.
12 Improved staff productivity.
12 Improved user efficiency or effectiveness.
10 Reduction in staff time spent on some decision making tasks.
10 Enhanced product or service quality.
10 Wider distribution of scarce expertise.
9 Faster problem diagnosis.
9 Faster response time.
7 Faster project completion.
6 Satisfaction of urgent user needs.
6 Easier operation of complex equipment.
5 Better forecasting ability.
5 Better strategic plans.
4 Enhanced awareness of potential impact of information processing technology on R&D activities.
3 Enhanced creativity or idea generation.
2 Improved morale.

Table 3. Frequency of Reported Benefits from AI/Expert Systems in R&D (n-19)

10 Selling management on the value of this approach.
6 Development of the knowledge base.
5 Lack of suitable development tools.
5 Systems need periodic update.
4 Systems only operate in a narrow domain.
4 Finding qualified personnel.

Table 4. Significant Problems in Using AI/Expert Systems in R&D: Response Frequencies (n-19)

ment tools and better integration of ES/AI into both the hardware and software structure of conventional computing systems.

Computer Integrated Manufacturing

Case and Pickett (1989b) found that computer assisted manufacturing (CAM) and computer assisted engineering (CAE) were each used by at least a third of their responding organizations; these were also among the five areas most frequently mentioned for future development. CAE within R&D seems very predictable, but the interest in CAM goes beyond R&D and involves the production subsystem. This interest may be consistent with some of Merrifield's (1989) thoughts:

> Continuous development of advanced technology, alone, will not be enough to maintain industrial competitiveness. Flexible computer integrated manufacturing (FCIM) will be required also...Moreover, given the rapid continuous evolution of new technology, the FCIM systems offer the opportunity for continual reprogramming to meet new or changing specifications...an FCIM facility can run night and day for economies of scale, while serving customers on demand as needed...Manufacturing in this sense will become essentially a service function, in which a given facility may make hundreds of products for different companies in different industries. It will be driven by continuous innovation on the one hand, and by marketing and distribution on the other...Continuous development of new components and software for the FCIM systems themselves also will offer competitive opportunities, and allow these facilities to be continually upgraded for longer lives and better paybacks.

A conceptualization of the skeleton of the FCIM model described by Merrifield is provided in Figure 4. Merrifield's comments give us a glimpse of where R&D's future may very well lie; it is apparent that its future role may call for it to be much more closely linked to the production and marketing subsystems than it is currently.

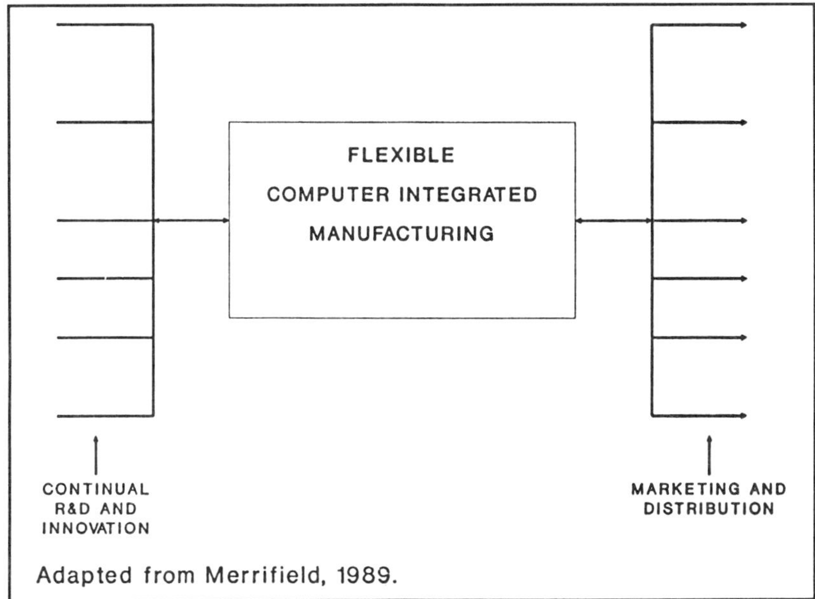

Figure 4. A Conceptual Model of R&Ds Role in a FCIM Organization

Improved Area	Percent of Respondents
Research/information	35
Statistical analyses	32
Time/error reduction	10
Report writing	7
Activity coordination	7
Automation of activities	7
Other	6

Table 5. Areas of Enhanced Productivity/Performance

Impacts of Information Technology on R&D

Handman (1986) was among the first to provide compelling evidence that increasingly available computing resources can enhance the productivity of R&D professionals. The Case and Pickett (1989b) survey also asked respondents if the utilization of information processing technologies had improved R&D performance. Ninety-six percent of the respondents indicated that it had. Case and Pickett also asked respondents to identify the specific areas in which information technology had enhanced productivity or performance. The responses to this question are summarized in Table 5. The most frequently mentioned area of improvement was in the ability to do more thorough research and/or compile more accurate and complete information. The ability to perform more powerful or sophisticated statistical analyses was also men-

Needed Improvement	Response Frequency
Better integration/networking	23
More computers/capacity	15
Increased utilization	6
Better training	4
Better security/communication/control	3
Greater user-friendliness	3
Other	4

Table 6. Areas Where R&D Information Systems Could Be Improved (n=76)

Information Processing Issue	**Percent of Respondents**
Improving productivity	76
Security of databases	76
Security of system	65
Computer literacy of staff	65
Upgrading software	64
Faster project turnaround	62
Upgrading hardware	56
Information processing budget	55
Using more applications	47
Getting more hardware/software	39

Figure 7. Organizations Indicating High Concern with Ten Information Processing Issues

tioned by nearly a third of the respondents. The automation or coordination of R&D activities (including report generation) and time/error reductions were the other most frequently mentioned areas of improvement. Pickett and Case (1990) found that most organizations utilizing expert systems and artificial intelligence reported positive impacts; these have already been summarized in a previous section of this chapter.

Case and Pickett (1989b) found that there was still room for further improvement in many areas. Indeed, these researchers found that only 10 percent of the organizations in their sample reported that they were "extremely" satisfied with their current system (however, 97% indicated that they were at least somewhat satisfied with their current R&D information system). Sixty-eight percent of the responding organizations indicated that system development programs were currently underway to improve or upgrade the current system. According to Case and Pickett's respondents, there are several specific areas in which R&D information systems could be improved. These are displayed in Table 6. It is quite apparent that better system integration and networking were the most frequently reported targets for further improvement. Increased processing capacity was another frequently mentioned area for improvement.

Table 7 displays the percentages of Case and Pickett's respondents who expressed "high" or "extreme" concern with each of ten information processing issues. The pattern of responses that was obtained provides additional insight into the areas in which R&D information systems can be improved. It might be concluded from Table 7 that further improvement of the productivity of R&D professionals via information technology is the overall concern. This may be accomplished by obtaining more or upgrading existing hardware and software and/or enhancing the computer literacy of the staff. However, a majority of the respondents also expressed concern about having the information processing budget needed to make these improvements.

Information Linkages between R&D and Other Subunits

Because networking was perceived most frequently by Case and Pickett's (1989b) respondents as being an area in which improvement was necessary, Case and Pickett (1989a) took a closer look at R&D networking issues. Although the importance of strong information linkages between R&D and the production and marketing subsystems has been stressed by many writers (e.g. Bright, 1969; Marquis, 1972; Merrifield, 1989; Martin, 1984; Urban and Hauser, 1980), Case and Pickett (1989a) found very little empirical evidence establishing the existence of such linkages. Indeed, these authors found that less than 40 percent of their R&D respondents could access production or production control databases. Less than a third had access to market research databases. Case and Pickett (1989a) report further that R&D databases are extensively used to develop marketing plans in less than 20% of the responding organizations; R&D databases are extensively use in the development of production plans in about one-fourth of the organizations. In sum, the current flow of information between R&D and these other subsystems could be much more substantial. It would seem that most organizations have considerable information linkages to build before they would be capable of implementing the FCIM model previously described by Merrifield (1989) (see Figure 4) or be considered an information processing model (see Figure 2 and Tushman and Nadler, 1980).

Case and Pickett (1989a) expected to find that organizations with decentralized (distributed) R&D activities would have more sophisticated networks (in order to facilitate coordination and control of these dispersed research endeavors) than would organizations with centralized R&D activities. The reverse was found; networking was more prevalent in organizations with centralized R&D. Organizations with centralized R&D functions were also found to be more extensively integrated with the corporate strategic planning process.

Strategic Planning and R&D: Information Linkages

R&D has long been recognized as a key subunit in both the development and implementation of business strategies. Urban and Hauser (1980) point out that R&D "strategy should reflect the overall policy and strategy of the organization" (p. 19). Glasser (1982) notes that R&D professionals have two major interactive roles with the organization's strategic plan; one is concerned with generating, monitoring, and updating the plan, while the other is concerned with interactions between the plan and R&D activities. Gibson (1981) indicates that R&D plans and activities are usually reflective of general organization strategies whether they be defensive turf-protective ones, imitative "copycat" (reactive) strategies, or proactive "first to market" strategies. Some companies may go so far as the development of "product innovation charters" which specify how R&D goals are linked to corporate goals as well as the actual programs that are to be used to achieve these goals (Crawford, 1983). From these descriptions, it would be expected that there would be strong information linkages between R&D and the strategic planning process.

Case and Pickett (1989a) did find information linkages between the corporate planning process and R&D, but most were human linkages. More than half (55%) of Case and Pickett's (1989) respondents indicated that the head of R&D was "heavily" or "extensively" involved in the corporate planning process (another 38% indicated that this person was somewhat involved). In contrast, less than one quarter of the respondents indicated that corporate planners had access to R&D databases. However, more than half of their respondents felt that R&D information should play a larger role in the strategic planning process. Apparently, the human information linkages were not having as strong an impact as others would have liked.

Research Directions

It should be apparent from the foregoing discussion that information technology is utilized in an ever increasing variety of ways to support R&D activities and to enhance the productivity of R&D professionals. It should also be apparent that there are quite a number of improvements which could be made and that there are a number of areas that would benefit from further research. Perhaps the most important question is whether or not increased capital expenditure in information technology applications within R&D translates into improved R&D productivity. If so, does improved R&D productivity translate into organizational productivity and sales growth? If a direct link between investment in R&D information systems and enhanced competitiveness were established, it would be hard not to view these as strategic information systems.

A second key area that should be addressed by future research concerns R&D project selection. Are the "best" or "right" project alternatives being pursued in order to maintain or enhance competitiveness? Are the projects which are selected consistent with corporate imperatives and long-range goals of other organizational subunits, especially those of marketing and production? How could information technology improve the project selection and project management process?

A third research area should address the organizational structure and organization culture factors which determine the degree to which R&D professionals are allowed to be innovators. While Peters (1987) and others have stressed the importance of tolerating failure and creating an environment which encourages innovation, there are often organizational and cultural factors at work which stifle rather than promote innovation. The ability of information technology to enhance creativity and innovative activities within R&D should also be explored.

International comparisons of how information technology is utilized within R&D might also provide some valuable insights. Are the improvements that have been observed in the R&D efforts of foreign-owned companies due to different patterns of application of information technology? If any differences were observed, considerable refocusing and re-allocation of information resources might occur.

The payoffs received from investing in specific information technologies (both special and general purpose) within R&D have not yet been adequately addressed. R&D managers would like to hedge their bets by investing in systems which have been shown to have significant impacts on the operations of other R&D organizations. Comparative studies on the impacts of particular applications would have tremendous value to them. Ideally, these would focus on both short and long-term impacts. For example, in emerging areas such as that of expert systems, the jury is still out. While the Pickett and Case (1990) study has provided a snapshot idea of how expert systems are being applied within R&D, it is unknown if these "pioneers" will continue to report positive results five years down the road.

The empirical literature, while sparse, also suggests that the information linkages between R&D and other subunits (and the external environment) could be much stronger. While many theorists have argued that strong internal and external information flows should exist, the reality seems to be something considerably short of this. Information and communication technologies may be mechanisms for strengthening these linkages and improving the flow of information from the external environment to R&D, within the R&D organization itself, and from R&D to other organizational subunits. Once again, comparative studies of the information technology alternatives available for establishing these linkages would be extremely valuable to R&D managers and corporate planners.

Summary and Conclusions

The traditional role of R&D in providing a steady stream of product and process innovations has caused many writers to recognize it as a subunit with strategic importance. Research has shown that capital expenditures in

R&D can translate into sales growth and productivity gains. Other writers and practicing R&D managers have indicated that the effective application of information technology within R&D can improve the productivity and performance of R&D professionals. However, most discussions of the strategic application of information technology have overlooked the potential for enhancing corporate competitiveness through the development, implementation, and maintenance of effective R&D information systems.

Only a handful of empirical investigations have addressed the application of information technology within R&D on a comparative basis. As might be expected, the trend is toward greater utilization of information technology and greater diversity in the types of applications. Networking and taking advantage of expert systems and other intelligent applications are important in many R&D organizations. A number of applications designed particularly (or primarily) for R&D (such as patent searches, data acquisition, and project selection) seem to be gaining in popularity. Still, there is evidence that considerable work needs to be done if R&D is to establish both the internal and external linkages needed for it to play the role that has been envisioned in flexible computer integrated manufacturing organizations of the future.

References

Allen, T.J. 1970. Communication networks in R&D laboratories. *R&D Management, 1*: 14-21.

Allen, T.J. 1977. *Managing the flow of technology: Technology transfer and the dissemination of technological information within the R&D organization*. Cambridge, Mass.: MIT Press.

Baker, N., & Freeland, J. 1975. Recent advances in R&D benefit measurement and project selection methods. *Management Science*, 1164-1175.

Becker, R.H. 1980. Project selection checklists for research, product development and process development. *Research Management, 23* (September-October): 34-36.

Brenner, M.S., & Rushton, B.M. 1989. A strategic analysis of sales growth and R&D in the chemical industry. *Research - Technology Management. 32* (March-April): 8-15.

Bright, J.R. 1969. Some management lessons from technological innovation research. *Long Range Planning, 2*: 36-41.

Carson, R.P. 1981. Distributed data analysis in computer networks. *Industrial Research & Development*, (May): 130-135.

Case, T.L., & Pickett, J.R. 1988. *Applications of expert systems in R&D*. Invited paper presented at the national meeting of the Operations Research Society of America/The Institute of Management Sciences, Denver, Colorado.

Case, T.L., & Pickett, J.R. 1989a. R&D information systems and their strategic linkages in Fortune 500 companies. *Proceedings of the Annual Meeting of the Southeastern Decision Sciences Institute*, Charleston, SC.

Case, T.L., & Pickett, J.R. 1989. R&D information systems: Utilization, applications, impacts, and major concerns. *Research - Technology Management, 32* (July-August): 29-33.

Cash, J., & Konsynski, B. 1985. IS redraws competitive boundaries. *Harvard Business Review,* 63 (March-April): 134-142.

Clark, K.B., & Fujimoto, T. 1989. Lead time in automobile product development: Explaining the Japanese advantage. *Journal of Engineering and Technology Management, 6*: 25-58.

Crawford, C.M. 1983. *New Products Management.* Homewood, Ill.: Richard D. Irwin.

Davis, P., & Wilkof, M. 1989. Scientific and technical information transfer for high technology: Keeping the figure in its ground. *R&D Management, 19* (January): 45-58.

Datamation. 1989. Shaking up the process. 37 (November, 15): 73.

Drenkow, G. 1987. Data acquisition software that adapts to your needs. *Research & Development, 29* (April): 84-87.

Feigenbaum, E., McCorduck, P., & Nii, H.P. 1988. *The Rise of the Expert Company.* New York: New York Times Books.

Feme, G. 1989. R&D programmes for information technology. *The OECD Observer, 159* (August-September): 10-13.

Gibson, J.E. 1981. *Managing research and development.* New York: John Wiley & Sons.

Glasser, A. 1982. *Research and development management.* Englewood Cliffs, NJ: Prentice-Hall.

Gold, B. 1989. Some key problems in evaluating R&D performance. *Journal of Engineering and Technology Management, 6*: 59-70.

Gold, B., Peirce, W.D., Rossegger, G., and Perlman, M. 1984. *Technological Progress - Industrial Leadership.* Lexington, Mass: D. C. Heath - Lexington Books.

Grilliches, Z. 1987. R&D and productivity: Measurement issues and economic results. *Science, 237* (July): 31-35.

Handman, S. 1986. Computer addition multiplies research output. *Research & Development, 28* (April): 84-85.

Hooper, W.G. 1986. Telecommunications systems are a company's information arteries. *Research & Development, 26* (September): 27-28.

Information Center. 1989. Expert systems trends. 5 (June): 17.

Jenkins, A. 1986. Artificial intelligence meets practicality. *PC Week, 3* (September, 23): 37-40.

Kizer, D.O., & Decker, C.D. 1985. Improving R&D effectiveness via computers. *Research Management, 38* (July-August): 19-21.

Liberatore, M.J. 1987. An extension of the analytic hierarchy process for industrial R&D project selection and resource allocation. *IEEE Transactions on Engineering Management,* EM-34: 12-18.

Marcus, S. 1990. Now your workstation can generate this fractal. *Research & Development, 32* (February): 144-148.

Martin, M.J.C. 1984. *Managing technological innovation and entrepreneurship.* Reston, Va.: Reston.

Marquis, D.G. 1972. The anatomy of successful innovations. *Managing advancing technology, 1*: 35-48.

Merrifield, B. 1989. The overriding importance of R&D as it relates to industrial competitiveness. *Journal of Engineering and Technology Management, 6:* 71-80.

Morbey, G.K. 1988. R&D: Its relationship to company performance. *Journal of Product Innovation Management. 5*: 191-200.

Morbey, G.K., & Reithner, R.M. 1990. How R&D affects sales growth, productivity, and profitability. *Research - Technology Management. 33* (May-June): 11-14.

Opren, C. 1985. The effect of managerial distribution of scientific and technical information of company performance. *R&D Management, 15*: 305-308.

Oster, S.M. 1990. *Modern competitive analysis.* New York: Oxford.

Ouchi, G. 1986. How much LIMS does your lab need? *Research & Development, 28* (April): 78-81.

Perloff, D.S., & Smith, A.K. 1987. Software aids transfer of processes from lab to line. *Research & Development, 29* (May): 109-114.

Peters, T.J. 1987. *Thriving on chaos: Handbook for a management revolution.* New York: Alfred A. Knopf.

Pickett, J.R. & Case, T.L. 1990. Expert systems and artificial intelligence: R&D applications. *Research - Technology Management, 33* (May-June): 35-38.

Port, O. 1990. Why the U. S. is losing its lead. *Business Week.* July 15: 35-39.

Porter, M. 1980. *Competitive strategy: Techniques for analyzing industries and competitors.* New York: Free Press.

Porter, M. 1985. *Competitive advantage: Creating and sustaining superior performance.* New York: Free Press.

Porter, M., & Millar, V.E. 1985. How information gives you competitive advantage. *Harvard Business Review, 63* (July-August): 149-160.

Putnam, F.A., & Rhoads, J.G. 1988. Use your computer to improve data acquisition. *Research & Development, 30* (June): 68-73.

Reynolds, C. 1989. The strategic IS dilemma. *Datamation, 35* (September, 15): 87.

Rockart, J.F. 1982. The changing role of the information systems executive. *Sloan Management Review, 24* (Fall): 3-13.

Ross, I.M. 1986. Successful R&D management: Catalyst for competitive advantage. *Vital Speeches of the Day. 52* (April): 374-378.

Rossini, F.A., & Porter, A.L. 1986. Who's using computers in industrial R&D and for what?

Research Management, 29 (May-June): 39-44.

Rossini, F.A., Porter, A.L., Jacobs, C.C., & Abraham, D.S. 1988. Trends in computer use in industrial R&D. *Research -Technology Management, 41* (September-October): 36-41.

Ruby, D. 1986. Laboratory micros. *PC Week, 3* (May 20): 89-90.

Russell, C.H. 1985. R&D software users identify their needs. *Research & Development, 27* (December): 46-49.

Ruusunenen, J., & Hamalainen, R.P. 1989. Project selection by an integrated decision aid. In B. Golden, E. Wasil, & P. Harkins (Eds.) *The analytic hierarchy: Applications and studies*. New York: McGraw-Hill.

Schilling, P.E., & Lezark, A.P. 1988. Planning for laboratory computing and communications. *Research - Technology Management, 41* (July-August): 39-43.

Scott-Morton, M.S. 1971. *Management decision systems: Computer-based support for decision making*. Cambridge, Mass.: Harvard University.

Simpson, J.W. 1985. Using patent data to boost R&D. *Design News,* (February 18): 89-91.

Smith, E., & Treece, J.B. 1990. Glimpsing the future in the numbers. *Business Week.* July, 15: 194-196.

Toda, M., Hiraishi, K., Shintani, T., & Katayama, Y. 1988. *Information structuring for advanced decision support systems: Proposal of a research decision support system and its implementation*. Technical Report No. 82, International Institute for Advanced Study of Social Information Science.

Tushman, M.L. 1979. Managing communication networks in R&D laboratories. *Sloan Management Review, 20*: 37-49.

Tushman, M.L., & Nadler, D.A. 1980. Communication and technical roles in R&D laboratories: An information processing approach. In B. Dean & J. Goldhar (Eds.), *Management of research and innovation:* 91-112. Amsterdam: North-Holland.

Urban, G.L., & Hauser, J.R. 1980. *Design and marketing of new products*. Englewood Cliffs, N J: Prentice Hall.

Utterback, J.M. 1974. Innovation in industry. *Science, 183* (February 15): 620-626.

Willcox, D. 1988. Catalyst testing is easier with 'virtual instruments'. *Research & Development, 30* (June): 76-79.

Wiseman, C. 1988. *Strategic Information Systems*. Homewood, Il: Irwin.

Wolek, F.W. 1984. Managers and the distribution of scientific and technical information. *R&D Management, 14:* 225-229.

Wormell, I. 1989. Strategic information management to improve competitiveness. *Information Services & Use, 9* (November): 197-204.

Ziegler, B. Today's strategic alliances link multinational companies. *Atlanta Journal-Constitution*. (June 7): D1, D9.

Part VI
Information Technology and Organizational Design

Chapter 18

Beyond the Technical Perspective: Towards an Expansion of the Systems Design Domain

Uldarico Rex Dumdum, Jr.
Marywood College

Despite huge investments made in Information Systems (IS) research, despite tremendous advances in information technology, and despite a constant flow of methods, tools and frameworks, organizations are still confronted with serious and widespread system design and implementation failures. In this paper, a design failure is broadly defined as that which results in an inappropriate "solution" to a "perceived problem" and an implementation failure is that which results in an inappropriate implementation of the design or non-acceptance of the system developed. While systems developers have attributed these failures to the technical complexity of the project, the volatility of both the organizational environment and technology, and to unrealistic user expectations of what the system can do, it is the contention of this paper that the problem lies deeper than these. This paper argues that many of these failures are due to the restricted scope of the design and implementation process - a process where far too much emphasis has been on the technical side of system development. This imbalanced emphasis constrains both the domain by which problems are recognized and characterized and the domain from which "solutions" are derived. Such constraints manifest themselves in the ways problems are perceived, in the reaction to emerging and pressing problems, and the corresponding prescriptions for initiating, developing, directing, and evaluating system development efforts. For example, technically perceived problems typically call for technical solutions.

While this paper recognizes that technical remedies are also helpful, it argues that such remedies are narrowly defined and hence do not deal with all the fundamental issues and organizational realities affecting the design, development, and the effective use of information systems. It further argues that there are many reasons and factors that lie outside the realm of technology that are equally critical to the success or failure of a system. The main purpose of this paper therefore is to establish the case for an expansion of the currently restricted scope of the design and implementation process. The paper proceeds as follows: Section 2 characterizes the technical perspective (TP) and its assumptions. Section 3 characterizes the information system failure syndrome and argues the need to broaden the design and implementation process and Section 4 proposes a framework that can systematically take into account most of the major factors and issues that affect systems success and failure. The paper concludes by discussing the implications of the proposed framework.

The Technical Perspective and Its System Development Assumptions

The technical perspective (TP) sees technology as essential to the survival of a computer society just as food, clothing, housing, education and health services (Evans, 1979). Technology is believed to be the only way to cope with an increasingly dynamic and unstable environment (Galbraith, 1973). The TP sees technology as a capital intensive resource with certain benefits, costs and skill requirements which are, in principle, easily identifiable and which provide specifiable information processing service (Lyytinen and Lehtinen, 1984).

The TP is a strongly-held belief among many information system (IS) developers. Such developers see their role to be selfless, value-free servants of the organization. They see their mission to be the determination of how best to provide the right amount of information to the right people at the right time in the right quality and form. An IS is then evaluated as successful once one gets the "right system right" (once one is able to derive the information needs of the user consistently, correctly, and completely - and once all these information requirements are met by the correct and complete implementation of the system) and once the information is conveyed in the form most usable to the user in the fastest way at the least cost (Senn, 1984, 1990; Laudon & Laudon, 1988; McLeod, 1990; Dickson, Senn, & Chervany, 1977).

From a TP, system development efforts proceed with the following underlying assumptions:

a. In principle, systems development is no different from the production of other technological products like cars and bridges. An IS is therefore seen as a technical artifact that can be designed and implemented successfully by better hardware and software engineering (Lyytinen & Lehtinen, 1984) and by better project management and control methods (Keider, 1974; Page-Jones, 1985). Systems implementation is seen primarily as a technical change. Any implementation therefore is seen merely as a mechanical process of installing technology (Hirschheim, 1985). If human problems arise, these problems are assumed to be separable from the technical problems and are seen to usually concern the border of human and technical systems - ergonomic or human factors design (Lyytinen & Lehtinen, 1984);

b. Direct Substitution and Incremental Aggregation are two major strategies employed. Direct substitution is the replacement of tasks, procedures, and/or people by newer and "better" (more integrated and more efficient) set of tasks and procedures. Incremental aggregation is the enhancement of the existing system by the accumulation of new capabilities (Kling & Scacchi, 1980). Implicit in these two strategies is a prevailing attitude called "technological dominance" where not only is technology and its application the solution to the problem but also that the application of more technology is always better (Hershauer, 1978). While nobody denies that technology has created dilemmas, TP assumes that the application of even more "benevolent" technology is the only way out of the dilemmas (Martin, 1981). In addition to this, TP further assumes that improved technical attributes directly translate into better organizational attributes, e.g., better organizational performance, greater work-efficiency, and better decision-making (Lyytinen & Lehtinen, 1984). The system is seen as "good" and hence assumed to be organizationally acceptable because it meets an economic and technical need (Kling, 1980).

The Need for an Expanded Design and Implementation Process

The efforts invested in making the TP work have not paid off. System failures are widespread and serious. Mowshowitz (1976), for example, noted that the performance of automatic data processing systems in business indicate that "20 percent are successful, 40 percent are marginal, and 40 percent are failures." Current IS studies indicate the same phenomenon (Kumar & Welke, 1984; Lucas, 1973, 1975, 1981; Zmud, 1983; Hirschheim, 1985; Bostrom & Heinen, 1977; Morgan & Soden, 1973; Gause & Weinberg, 1989). While the postulated reasons for system failures have been varied, more and more studies are suggesting that the primary cause for system failure has been the dominating emphasis on merely the technical side of system development. For example, after studying system developer values, Kumar and Welke (1984) presented empirical evidence indicating the following: (a) "that the technical and economic value concerns are highly likely to influence and govern system design choices and behavior" (p.19), hence in the design of new systems, these concerns shall have top priority and shall be the bases of major design decisions; and (b) socio-political-psychological values are found relevant only if they contribute directly to the system development project or are related to systemic goals and concerns of the client; and (c) job satisfaction and quality of working life related value concerns are likely to be ignored by system developers concerned with bringing the system development project on schedule and within budget (p.19). While Mumford, Land, and Hagwood (1978) attribute system failures directly to a lack of knowledge of human needs and motivation on the part of technically oriented system analysts and designers, Noble (1984) argues that the politics of technology development and its underlying reward (funding, approval and support) structure within organizations systematically leads to a conscious disregard for human and social dimensions. Furthermore, this disregard is typically translated into a machine-centered design (Kling, 1977) with mechanistic orientations (Zmud, 1983) and a limited goal orientation - that of optimizing only the technical aspects of the system (Pettigrew, 1973, 1980). Such disregard may lead into the generation of inappropriate "solutions" primarily because it was the wrong problem to begin with - an error of the third kind (Mitroff, 1980, 1987) and also into the inappropriate implementation of systems (Lucas, 1975, 1981; Doktor, Schultz & Slevin, 1979; Weinberg, 1982a). Still, other implications of this ignorance and/or disregard of social and human dimensions are as follows: research and development efforts are still heavily leaning toward the development of tools and methodologies (a) that are insensitive to people, politics and influence (Noble, 1984); (b) that are uncritical in their outlook towards the rational and political use of IS and in their consequences (systems are as much a problem solver as they are problem generators) (Hirschheim, 1985; Klein & Hirschheim, 1985; Kling, 1980; Kling & Scacchi, 1980; Gause & Weinberg, 1982; Franz & Robey, 1984); and (c) that continue to reinforce myopic and restrictive criteria in initiating system development efforts and evaluating system success (Klein, 1984; Bjorn-Andersen, 1984; Linstone, et al, 1981; Linstone, 1984).

Due to the profession's rather dismal level of success in system development, Weinberg (1982a) proposes that we rethink the way we do systems analysis and design, that we reevaluate why we succeed and why we fail, and that we begin to extract lessons that can be vital to doing systems analysis and design better. In addition to this, he suggests that "we outgrow our adolescent fascination with toys and develop an adult interest in people [so that] we will begin, as other technologies have done, to master the social and psychological forces that are the real power behind successful technology - and the real reason for technology in the first place."

While there have been some suggestions to improve our current ways of doing systems analysis and design (cf. Linstone et al, 1981; Linstone, 1984;

Kling, 1980; Mitroff, 1980, 1985; Allison, 1971; Weinberg, 1982a, 1982b; Gause & Weinberg, 1989; Wood-Harper, Antill & Avison, 1985), there still seems to be a lack of a broad framework that can be used to incorporate both the technical and human/social aspects and subject these aspects to systematic analyses. The next section proposes such a framework.

The Proposed Framework

The proposed framework is an expansion of the behavioral perspective of the implementation of management science by Schultz and Slevin (1979). It also uses the concept of concentric systems (Weinberg, 1982a) where one system encapsulates another system into a successively larger system and that "the system" is really a series of nested capsules. The proposed framework is as shown in Figure 1. The framework consists of six categories which define the critical factors that should be applied in evaluating the desirability of information systems. The next sections will discuss each of the six categories.

Technical Validity

This category is most familiar to systems professionals. It refers to the technical aspects of the design, development and usability (production of and

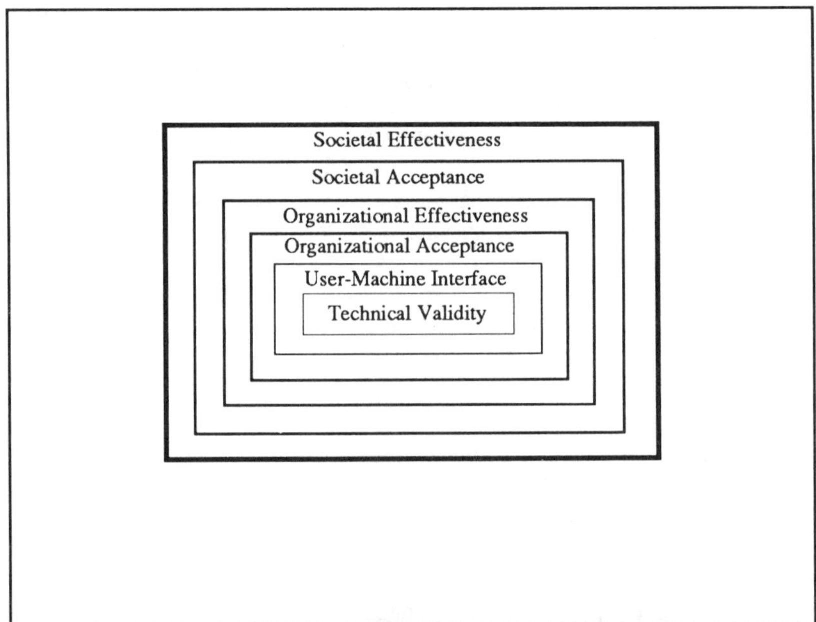

Figure 1. The Proposed Framework

delivery of information), expandability and maintainability of information systems. The IS in this category is seen as a model. Hence, from this perspective, the job of systems professionals is that of model building - the heart of their training and traditionally, their most significant contribution to the organization. Once the model is developed and "works" (produces the right outputs when the right instructions are input), the model is evaluated as "valid" and the involvement of analysts and designers is seen to be terminated until further maintenance (either corrective, adaptive, or perfective) is required.

Issues relevant to this category include "getting the right requirements" and "getting the requirements right" (Nosek & Sherr, 1984; Davis, 1982; Bostrom & Heinen, 1982). Some techniques used revolve around data acquisition and data representation. Some of the data acquisition techniques include interviews, questionnaires, observation, and document reviews (Senn, 1984). Some of the data representation techniques include decomposition diagrams, dependency diagrams, data flow diagrams, HIPO charts, Warnier-Orr and Jackson diagrams, Action diagrams, Decision Trees, and decision tables (Martin & McClure, 1985). Since most of these diagramming techniques are manually driven, and since the applications are getting more complex (increasing complexity of time-dependent behavior, increasing complexity of data, and increasing complexity of functions), many of these diagramming techniques are getting automated (Yourdon, 1986).

The major concern of this category is to derive and implement the correct, complete, and consistent requirements of the model to be built. Since it is difficult to extract the complete requirements from the abstract, techniques such as prototyping and user involvement have been recommended (Boar, 1984; Naumann & Jenkins, 1982; Ives & Olson, 1984; Mumford, 1984). In addition to this concern, this category is also involved with the issues that the field of Software Engineering is concerned with, e.g., system building quality assurance -walkthroughs, inspections, reviews (Freedman & Weinberg, 1982), software reliability (Myers, 1976), management and control of software projects (DeMarco, 1982; Brill, 1984; Keider, 1974), and software maintainability and system expandability (Leintz & Swanson, 1980; Glass & Noiseux, 1981; Martin & McClure, 1983).

User-Machine Interface

This category focuses on the system's "user friendliness" and "understandability". According to Kling (1977), the key element in interface design is its "user-centeredness" rather than "machine-centeredness". This category has become a critical issue in systems design for many reasons. First, because the user community has expanded beyond data processing (DP) professionals to include novices with little or no technical training (Schneiderman, 1982). Second, because of the many unnecessary details and complexities traditional design imposes upon users, Wilcox (1982) and Kling (1977) claim that many systems designs have been produced according to the designers' assumptions, values, experience, myths, great interest and heavy exposure to information handling, and insensitive misconceptions about the users. And third, because user behavior patterns (acceptance and use) is claimed to determine whether the system is ultimately successful or unsuccessful, regardless of the system's technical sophistication and validity (Wilcox, 1982). Schneiderman (1982) substantiates this by indicating that the amount of user acceptance and use is very much affected by the amount of human factors that are considered in the design of the system.

The primary concerns of this category are in tailoring the system to the level of sophistication of the user, in building a level of tolerance into the system that enables the user to cope with complexity and permits minor error and variable performance on the part of the user, and on making the system understandable (Moreland, 1982; Gause, 1983 Class Lecture Notes; Gilb, 1976). Gause suggests that systems are easier to use when systems are understandable. He proposed 5 "hip-pocket" measures of understandability which are as shown below:

- A system is understandable if it is easy to memorize.
- A system is understandable if it is easy to recognize errors in design.
- A system is understandable if one can predict, by the pattern in the existing design, what an extension of the design would look and act like.
- A system is understandable if the system performs as we expect it to.
- A system is understandable if the concepts inherent in the design are familiar to the users.

Expanding on the work of Moreland (1982), Meyer, Harper and Gas (1984) proposed other aspects of user-friendliness. They are as follows:

- Clarity of purpose, functional flexibility, unity of concept, completeness, and error-acceptability of the designed system.
- Clear, complete and understandable documentation, training, and aids.
- Easy and non-distracting access to the system.
- Understandable user-system dialogues and intelligible error messages.
- Kind and reassuring system response to user tasks and inquiries.

	MACHINE-CENTERED DESIGN	USER-CENTERED DESIGN
1.	Efficiency is emphasized.	Systems are valued that increase personal competence and pride in work
2.	Systems are designed in purely functional terms	People are accepted as non-rational & error prone
3.	Human error is not tolerated. Systems are not error-tolerant and provide limited diagnostic information.	Jobs are designed to be personally satisfying. Automated procedures are designed to fit job needs.
4.	Users are forced to match the precision required by the machine.	The burden of precision is placed on the machine. Systems are forgiving.
5.	Jobs and procedures are designed to simplify machine processing.	Users easily obtain and create systems to meet their needs.
6.	Human relations are ignored as long as the the job gets done.	Users can initiate, veto, and collaborate in systems design.
7.	System design is imposed upon users. They initiate but never veto system design.	Designs and assumptions are transparent and intelligible to users through appropriate techniques and clear documentation.

Adapted from Kling, 1977

Table 1. Some Characteristics of Machine and User-Oriented Designs

In addition to these, Kling (1977) presents an expanded framework in which the personal characteristics of the user, the tasks, and the organization can be incorporated into the design of better user-machine interface. He does this by contrasting machine-oriented design from a user-centered design. Shown in Table 1 are some of the design implications of each orientation:

Organizational Acceptance

Organizational acceptance is the result of the dynamics of information technology diffusion within the organization. It is a function of how an IS is viewed within the organization (perceptions, confidence, and attitudes toward the IS), the "needs" the IS is meeting (actual and perceived) and at whose expense and benefit (the interests it serves and does not), the way the IS is introduced (participation and implementation issues) within the organization, and the "arena" where the various organizational "games" are played. Organizational acceptance can be actual or symbolic. The "arena" in a complex organization can be seen either from a "rationalist" perspective where the organization is composed of groups working in harmony toward a common goal or from a "segmented institutionalist" perspective where the organization is seen to consist of federations of groups that build a shifting set of alliances and in the process serve several selected parochial interests, legitimate or illegitimate (Kling, 1980; Kling & Scacchi, 1980). The importance of the two perspectives is twofold: first, it provides the designer with a viewing lens to see or understand how action within an organization is taking place - who the groups are, how they interact, and what parochial interests they represent and serve; and second, it contrasts the inherent assumptions of each view of the organization and each view's corresponding system goals and solution space. The two views are as seen in Table 2.

For example, one way of viewing organizational acceptance is as follows: All an IS needs to be, in order to be organizationally accepted, is to be effective in meeting tangible goals, e.g., producing the right output when right input is made; is to be efficient, and is to be, for most times, reliable and correct. If the IS satisfies these, it is then assumed that an IS is "automatically" accepted because it, by economic substitution, "meets the needs" of the organization. To enhance acceptability, or at least minimize resistance to the IS, all one needs to do is to involve users in the design of the system and educate them on how to use the system. Being "wonderful and functional," the resulting IS will sell itself (Kling & Scacchi, 1980).

Another view of organizational acceptance however can be as follows: In an organization, the actors involved: (a) have conflicting needs, motivations, and goals; (b) are not always obedient to organizational goals and are, in essence, political agents representing different parochial interests (Mintzberg, 1983; Markus, 1983, 1984); and (c) may not necessarily be

ASPECTS OF THE VIEW	RATIONALIST	SEGMENTED INSTITUTIONALIST
Consequences of computing to social life.	Positive	Positive and negative, legitimate or illegitimate
Consensus on Social Goals relevant to computing use.	Marked consensus. Harmony.	Intergroup conflicts are as likely as cooperation and reciprocation. Each group has parochial interests to defend.
Predominant goals.	Economic and organizational efficiency.	Revolves around sovereignty of individuals and groups over critical aspects of their lives, integrity and social equity
Main interests	Managers' interests are more legitimate than those of subordinates .	Each group's interests and goals are just as important as those of others.
Solution space	To satisfy, motivate, and make workers productive is by finding the best computing arrangements through proper design and implementation.	Serve the interests of all parties involved.

Adapted from Kling & Scacchi, 1980

Table 2. Characteristics of Two Views of an Organization

working together in harmony (Kling, 1980). These actors are therefore involved in some sorts of "games" and may even use illegitimate and/or clandestine means to maintain or extend their "turf" or for some other purpose. Acceptance or rejection of an IS then is seen not to be a homogeneous response. The response may be covert or overt depending on the perceived "benefits" and "penalties" or "costs." For example, one may "accept" and "use" an IS (and yet derive his or her information from other sources and means) because the "boss" - the rewarder of those who diligently "obey" him - requires the use of an IS. Other reasons could be for reciprocity - alliance formation purposes - "support my project and I'll support your upcoming project" or as a means to build or maintain powerbases or defeat rivals. Some may also reject the IS as a symbolic or real defiance to authority. For example, a worker who will be replaced by a system will naturally oppose or even sabotage the system, no matter how technically well designed the system is or how well the IS contributes to the efficient operation of the organization. From this view of the organization, one can therefore make the conjecture that the acceptance or rejection of an IS can be beyond the technical and user-machine interface aspects of the IS. This view asserts that it will be a faulty assumption to immediately evaluate an IS as successful and accepted just because it produces and delivers information reliably and in the form desired.

Organizational Effectiveness

This category is a function of: (a) the evaluator's criteria - the way the organization is perceived and the aspects of the operation of the organization being measured against predetermined goals (it must be noted that how an organization is perceived influences one's expectations of how an organization should be operated and hence evaluated (Cameron & Whetten, 1983); (b) the values and perceived reward structures in place within the organization (people do what gets measured and the things that get rewarded get done (LeBoeuf, 1985; Hedberg, 1978); (c) the organization's goals; (d) the environment the organization is in; and (e) the resources, means, and strategies at the insider's disposal to assure the organization's survival, maintenance, or growth. From these, one will note that an IS becomes but one of the many aspects that go into the evaluation of organizational effectiveness. Hence, one way of evaluating an IS is to look at the actual or perceived contribution of the IS towards the achievement of organizational goals. While the IS may be technically valid, while it may possess good user-machine interface, and while it may be organizationally accepted and used, the IS may not necessarily contribute towards, say, helping workers perform better jobs while and at the same time upgrading the quality of their working lives and reinforcing their integrity and sense of belonging.

For example, one view of assessing the contribution of an IS towards organizational effectiveness is by posing some basic management questions like: "What is the IS worth to the business?", "What is the business value of the proposed system?" or "What is the strategic importance of the IS to the business?" It is noted by Robinson (1982) that these questions often remain unasked by analysts during the analysis and design phases of IS. Hence, if one is involved in a market-oriented business, then an IS may be of value if the IS enhances the business' market position by providing competitive advantage or by bringing the business to par with its competitors. If, however, the current orientation of the business is improving its operations, then the value of the IS may be measured by its contribution toward measurable cost savings through operational efficiencies, greater throughput and elimination of bottlenecks, and by intangible benefits such as better service, faster turnaround, or better recordkeeping. And finally, if the current system necessitates replacement due to obsolescence or due to regulatory changes (e.g., new tax laws, new reporting schemes), even though the proposed IS may add little to the competitiveness of the business, a new system may have to be developed, the value of which is in keeping the system up to date or in its regulatory compliance (Robinson, 1982).

Another view of assessing the contribution of an IS toward organizational effectiveness is the degree to which the IS serves the interest of all legitimate parties, instead of just the dominant party or parties. This view sees organizations as individual and group need-meeting cooperatives. Hence, organizational effectiveness evaluation goes beyond the typical "what are the economic costs and operational benefits of the system?" to include issues such as:

- Who are likely to benefit and suffer and in what forms? Is it temporary or permanent?
- Who will the IS alienate, replace, or deskill? Are provisions made for them in the design and use of the proposed system?
- What new forms of threats, pressures, and aggravations are introduced?
- What structures, social meanings and alliances are going to be replaced or modified?
- What is the effect of the IS on the overall quality of working life and employee satisfaction?
- Who participated in the project and who did not?
- Whose views and goals are achieved or maintained by the development of the system? Whose views and goals were not incorporated?

Societal Acceptance

Societal acceptance is the result of the dynamics of IS and information technology diffusion on a global scale. It is a function of how various segments of the global population perceive, proact and react to how an IS and its related technologies contribute to meeting of needs, the way it is introduced and used, and at whose benefit and expense. As one will note, many of the concerns within the organizational domain are similar to the concerns within a societal domain. The major difference is in its much greater complexity and unmanageability due to scale. For example, on a global scale, many rivers of change are converging on the individual, families, businesses, nations and societies: the marketplace is now global (Pescarella, 1989; Diebold, 1990; Ohmae, 1989; Wenk, 1989), technology is advancing, converging and feeding and growing at each other at an accelerating pace (Hirschheim, 1985; Wenk, 1989; Pescarella, 1989), and there has been an explosion of new values and lifestyles in the last 25 years (Pescarella, 1989). Ohmae (1989) claims that while on a political map, the boundaries between nations are as clear as possible, such boundaries fast disappear when one traces the real flows of industrial, financial and information activity. He perceives this global market to be a "geography without borders" where people and nations become global producers and consumers. Diebold (1989) sees information technology as increasingly part of the societal infrastructure linking consumers, businesses, government agencies and other institutions. He claims that this rapidly evolving environment will necessitate changes in the nature of businesses and how they will operate and interrelate. Due to these, he exhorts leaders and managers to expand their focus beyond the traditional internal uses of technology within the corporation to include and avail themselves of the significant changes external to the corporation. Wenk (1989) sees that the future will entail more technology, not less, and that technology will have such a powerful influence in human affairs and will affect every dimension of life - positively and negatively. He sees technology as a social, economic, political and ecological amplifier. He sees that technology:

> may expand or diminish the human spirit, equality of opportunity, quality of life, freedom and self-esteem. It amplifies wealth and the imperative for its acquisition; it spurs material appetites for comforts and conveniences; and it has increased the disparity everywhere between rich and poor. It stretches the boundaries and the volume of communication networks. It expands the number of options. It has injected new demands on our institutions, perhaps faster than they can respond. Most significantly, in considering risk, its sources and perceptions, technology amplifies problems in its social management (p. 173).

In assessing the implications of these seemingly inevitable changes, it is important to consider a broad spectrum of impact. A consequentialist perspective (Hirschheim, 1985; Klein & Hirschheim, 1985) can assist us in this. This perspective contends that "the desirability of a proposed change [in this case, the design and application of IS and related technologies] is to be judged by the consequences it brings about regardless of whether these were intended or unintended" (Hirschheim, 1985). Underlying these perceptions are value positions which influence expectations, the desirability or repulsiveness of technological change and how society should deal with it. Within this perspective are three broad viewpoints about the nature of change and its effects: optimistic, pessimistic, and pluralistic. They are as summarized below (Hirschheim, 1985, pp. 205-237):

The optimistic viewpoint asserts that information systems and their related technology will increase productivity, create at least as many jobs as they destroy, increase organizational effectiveness and enhance communication. It further asserts that the quality of working life will improve with fewer menial tasks and with the creation of new and interesting varieties of work. The society at large then is seen to experience a better quality of life.

The pessimistic viewpoint contends that information systems and their related technologies will do little to increase productivity, will create widespread unemployment, will deskill the remaining work thus deleteriously affecting job satisfaction and the overall quality of working life. It also sees that technology will amplify the current foul-ups of today (Dodswell, 1983), be a source of enslavement rather than freedom (Braverman, 1985), will confine the worker to a blind round of servile duties instead of broadening their horizons of work (Braverman, 1974), will reinforce a social system based on the exploitation of human labor (Briefs, 1980), and will lead to a greater concentration of power and a lessening of personal privacy and power.

The pluralist viewpoint is neither optimistic nor pessimistic, but a compromise between the two. It argues that technology is neutral and that there are possibilities created by technology which are positive and negative depending on the way and context in which technology is applied. Burns (1981) asserts technology is "not apolitical ... [and] may be compared to the building block, which is itself neutral but when part of a working structure ... reflects the desire and wishes of whoever designs or controls the system." Jones (1982) claims that technological change has "an equal capacity for the enhancement and degradation of life, depending on how it is used." Hirschheim (1985) captures this viewpoint cogently:

> Whereas the optimist and pessimist seek to define a technology as good or bad somewhat independently of a specific application, the pluralist attempts to define what criteria would need to be satisfied in

AREA OF CONCERN	OPTIMIST	PLURALIST	PESSIMIST
Productivity	Marked improvement	<------------------------------------->	Little increase
Employment	Increase or no change	<------------------------------------->	Steady decline
Work	Better	<------------------------------------->	Worse
Communication	Better	<------------------------------------->	Mixed
Quality of Working Life	Improved	<------------------------------------->	Deteriorated
Power	Implied better power sharing	<------------------------------------->	Further centralized
Personal Privacy	No change	<------------------------------------->	Decreased
Neutrality of Technology	Neutral	<------------------------------------->	Not neutral

Adapted from Hirschheim, 1985

Table 3. Comparison of the Optimist, Pessimist & Pluralist Positions

> a given application for that technology to be considered positive. The pluralist, therefore, devotes attention to the development of criteria for social and technological acceptance and then the application of "acceptable" technologies in appropriate circumstances. It follows that the pluralist does not believe that ... technology ... is a fait accompli ...[Any technology] has to be judged in the context of its setting and use (p.234).

An adaptation of a comparison of the three perspectives are as shown in Table 3.

The importance of these viewpoints is in trying to anticipate the concerns and possible reactions of various segments of this global society to the introduction and intensive use of IS and related technologies. It must be noted that the value position underlying each viewpoint influences how such segments will react to technological change. For example, while those bearing the optimistic view will welcome change and those with a pessimistic view will oppose and reject change, the pluralist view will hold a "neutral" position and focus its attention into choices of technological options and then opt for an "appropriate technology."

Societal Effectiveness

The focus and challenge of this category is how to steer information technology to help produce satisfactory outcomes. Satisfactory outcomes, however, imply goals and futures segments of the population collectively want. Steering assumes the existence of goals and requires strategies for their achievement. This is no small feat, since the complexity of delivery systems, the diversity of institutions and stakeholders involved and the requisite linkages require greater demands for obtaining coherence (Wenk, 1989). Underlying this category is the notion of appropriate technology. According to Willoughby (1990) the key to appropriate technology "is the interrelationships between the technologies themselves, or the artifacts, and the society in which they are employed ... [and] unless the process of tailoring technologies to fit their contexts has been pursued, technologies should not be deemed to be 'appropriate'." For example, he states that:

> A solar energy device cannot be called "appropriate" simply because it may be compatible with the principle of environmental conservation; it must also address the real needs of the community.

In this category, we shall take a look at the following broad areas where information systems can be applied: national competitiveness, amelioration of global problems and achievement of human ideals. In addition to these, the potential impact of information system failure shall be discussed.

National competitiveness (NC), according to Michael Porter (1990), has become one of the central preoccupations of government and industry in every nation. He claims that while the notion of a competitive company is clear, the notion of national competitiveness is not. He offers six popular notions on national competitiveness with each notion evoking its own set of strategies (Porter, 1990, p.84): (a) As a macroeconomic phenomenon, driven by variables such as exchange rates, interest rates, and government deficits; (b) As a function of cheap and abundant labor; (c) As a function of bountiful natural resources; (d) As a function of government policy: targeting, protection, import promotion, and subsidies; (e) As a function of management practices, including management-labor relations; and (f) As a function of productivity and a relentless process of ongoing improvement in existing and anticipated industries by raising product quality, adding desirable features, improving product technology, or boosting production efficiency. The issue therefore, from an NC point of view, is the perceived and actual value and contribution of an IS towards national competitiveness. For example, since Porter claims that it is a myth that a nation can be competitive in all industries and that national competitiveness can only be fully understood at the level of particular industries, how then can an IS help in understanding the determinants of productivity, the viability of existing and projected skills and technologies, the identification of those decisive characteristics - those specific industries and industry segments - so as to create and sustain competitive advantage in particular fields?

From an amelioration of problems and achievement of human ideals point of view, some of the ways the contribution of an IS may be assessed are as follows: On the notion that the quality of the future depends on the quality of governance, how can IS be used to enhance the role of government in the areas of military and economic security, health and safety, human rights, quality of life, public process and environmental protection (Wenk, 1990)? On the notion of the social and moral obligation of businesses and governments in addressing environmental problems, how can IS help businesses and governments decide what it can and ought to do about the ecological crisis and what its relationship and levels of cooperation ought to be (Frederick, 1989)? On the notion of achieving human ideals (e.g., freedom, autonomy, self-determination, affection, recognition, meaning and continuity), how can an IS enhance the quality of dialogues, debates and participation of all stakeholders in the ranking of social and environmental priorities, in the allocation of resources, in the development of the criteria for which economic and political activities must achieve, and in the resolution of conflicts? These notions and evoked strategies are primarily geared to answer four prevailing questions: Can we do it?, Ought we do it?, Can we afford it? and Can we manage it (Wenk, 1989)?

In addition to all these, there is yet another area which is critical in assessing the value and contribution of information systems: the impact of information system failures. For example, many studies indicate that information system failures have led to serious consequences, e.g., the accident at Three Mile Island, the chemical spills at Bhopal, India, and at Institute, West Virginia, and the loss of the space shuttle Challenger (Wise & Debons, 1987). Disaster warning systems are also of interest because if they fail to operate as anticipated, serious life loss, injury, and property damage almost inevitably occur (Foster, 1987). For example, Foster (1987) states that 225,000 people were drowned when a major cyclone hit Bangladesh on November 12, 1970 and 22,000 people were killed by volcanic landslides in Central Columbia on November 13, 1985 despite, rather than without, warning. In our growing dependence on highly integrated, multifunction, massive information systems and in our world of wars, strikes, sabotage, embargoes, corruption, mental illness, drug abuse, interagency competition, shifting social values, and dramatic technological change (Foster, 1987), how can we better understand information system failures, how can we prevent or minimize the impact of their failures, and how can we take appropriate steps to design to accommodate the erratic stresses as mentioned above?

Conclusion

The proposed framework uses the concept of concentric systems (Weinberg, 1982a). This concept looks upon a system as a series of nested capsules (in the proposed framework, as a series of nested categories). System failures can occur due to factors (a) within a particular category, and/or (b) by the lack of provision of a proper continuity from one category to the next category(ies) which encapsulate it. The value of this concept is explained by Weinberg:

> If we understand that the system is a series of nested capsules, we can accomplish system changes that have exactly the proper continuity with the existing systems that encapsulates them. If we don't understand this concept, we're likely to design systems that fail when interacting because of some aspect of an "outer system." We'll alibi our failure by saying "that part doesn't count - it was outside the system so I didn't have to account for it.
>
> Proper continuity means confessing that every new development is actually MAINTENANCE of some larger system. But continuity also means we can use these outer systems to make the inner system succeed. Rather than counting them as potential threats

to our lifeblood, we can count on them to support our efforts (p.35).

From the above, one will note that the ultimate value of the proposed framework is that it serves as a broader base upon which systems, the related development efforts, and their consequences can be understood and evaluated. It also requires that we explicitly consider the wider implications of our actions. It further offers a broader base upon which problems can be characterized and a larger domain from which solutions can be derived. Specifically, some of the key implications are as follows:

a. That there are categories beyond the technical and user-machine interface aspects of a system that are just as important to system success. If we continue to merely focus on these two innermost categories and continue to disregard the categories that encapsulate them, our system development efforts shall continue to experience widespread and serious failures. It therefore requires that the current design and implementation process be broadened to allow for the integration of organizational and social factors. While we do not claim that the explicit consideration of these factors in the design and implementation process will GUARANTEE system success, we however believe that it will provide a BETTER CHANCE at success.

b. That design must proceed with both an understanding and healthy respect for the existing organizational and social structure so that any change introduced will either reinforce, maintain, or at least offer a viable alternative to the existing structure. Merton (1968), for example, states that "any attempt to eliminate an existing social structure without providing adequate alternative structures for fulfilling the functions previously fulfilled by the abolished organizations is doomed to failure."

c. That a good understanding of the organization - the key players, the dominant and the dominated parties, the games they play, their motivations, goals, and aspirations, the organization's culture and reward structure - is necessary if one is to channel these valuable resources (opportunities and hindrances) to help develop and support (or effectively oppose) system development efforts.

d. That while there can easily be a unanimous assent to the need for integrating organizational and social factors into the design and implementation process, there is an undeniable lack of know-how of how one can actually do the integrating. For example, while there seems to be a growing mistrust for developing systems driven by the technical perspective, there is yet to be seen an equal amount of proposals for actual and workable techniques and methods that incorporate the categories beyond the innermost two categories to aid practitioners in their development efforts. While this dilemma may be due to the IS field's rather youthful existence, it still mandates that more efforts and resources be allocated toward an expanded yet integrative design.

e. That if we are to achieve an expanded yet integrative design - a design that caters to both the technical and human/organizational/social factors systematically on an equal priority basis - we need to establish some fundamental mechanisms. These mechanisms must include:

- a mindshift which entails an explicit reexamination and rethinking of the way we typically develop systems and the types of tools and methodologies we have typically preoccupied ourselves with in developing and using systems (Mitroff, 1985).

- a vision expansion to allow us to shift from a single perspective to multiple perspectives so that the typical blindspots can be seen, recognized, and accounted for in the design and implementation process. (It must be noted that whereas a mindshift seeks to expose an approach's assumptions and examine their validity, a vision expansion seeks to unearth alternatives, radical or otherwise, to complement or replace existing approaches.) (Linstone, 1984).

- an environment shift to allow us to shift from an environment that restricts integrative design to where integrative design is not only accepted in principle, but also practiced and allowed to flourish (Kanter, 1983).

ACKNOWLEDGMENT

The author wishes to thank Dr. Heinz K. Klein for helpful comments on earlier drafts of this paper.

References

Allison, G. (1971). *Essence of Decision: Explaining the Cuban Missile Crisis*. Little, Brown and Company.

Baber, R.L. (1982). *Software Reflected: The Socially Responsible Programming of Our Computers*. North-Holland Publishing Company.

Bjorn-Andersen, N. (1984). "Challenge to Certainty." In Bemelmans, Th. (ed), *Beyond*

Productivity: Information Systems Development for Organizational Effectiveness.

Boar, B. (1984). *Application Prototyping*. Addison-Wesley.

Bostrom, R., & Heinen, S. (1977, Sept & Dec). "MIS Failures: A Socio Technical Perspective: Parts I and II." *MIS Quarterly.*

Bostrom, R., & Thomas, B. (1982, August). *Getting the Requirements Right: A Communications Perspective.* Working Paper.

Braverman, H. (1974). Labor and Monopoly Capital: The Degradation of Work in the Twentieth Century. *Monthly Review Press.*

Briefs, V. (1980). "The Effects of Computerization on Human Work: New Directions for Computer Use in the Work-Place." *In Human choice and Computers* (Ed. A. Mowshowitz), Vol 2, North-Holland.

Burns, A. (1981) *The Microchip: Appropriate or Inappropriate Technology?* Ellis Horwood.

Cameron, K.S., & Whetten, D. (eds). (1983). *Organizational Effectiveness: A Comparison of Multiple Models.* Academic Press.

Davis, G.B. (1982). "Strategies for Information Requirements Determination." *IBM Systems Journal, 21,* No. 1.

DeMarco, T. (1982). *Controlling Software Projects: Management, Measurement and Estimation.* Yourdon Press.

Dickson, G., Senn, J., & Chervany, N. (1977, May) "Research in MIS: The Minnesota Experiments. "*Management Science, 13,* No. 9.

Diebold, J. (1990). "How Computers and Communications are Boosting Productivity: An Analysis." *International Journal of Technology Management, 5*(2).

Dodswell, A. (1983). *Office Automation*, Wiley.

Doktor, R., Schultz, R.L. & Slevin, D. L. (1979). "The Implementation of Management Science." *TIMS Studies in the Management Sciences, 13,* North-Holland.

Evans, C. (1979). *The Mighty Micro.* Gollancz Publishers.

Foster, H.D. (1987). "Disaster Warning Systems: Learning From Failure." In *Information Systems: Failure Analysis* (Wise & Debon: Ed). Springer-Verlag NATO ASI Series.

Franz, C., & Robey, D. (1984, Dec). "An Investigation of User-Led System Design: Rational and Political Perspectives." *Communications of the ACM, 27,* No. 12.

Frederick, R.E. (1989). "Introduction: Is Business Obligated to Protect the Environment?" *Business Ethics Report: Highlights of Bentley College's Eighth National Conference on Business Ethics.*

Freedman, D.P., & Weinberg, G. (1982). *Handbook of Walkthroughs, Inspections, and Technical Reviews: Evaluating Programs, Projects and Products.* Little and BrownComputer Systems

Series.

Galbraith, J. (1973). *Organizational Design: An Information Processing* View. Addison-Wesley.

Gause, D. (1983). *Lecture Notes: Watson School of Engineering*, SUNY-Binghamton.

Gause, D., & Weinberg, G. (1982). *Are Your Lights On?: How to Figure Out What the Problem Really Is*. Winthrop Publishers, Inc.

Gause, D., & Weinberg, G. (1989). *Exploring Requirements: Quality Before Design*. Dorset House.

Gilb, T. (1976). "Some Principles of System Design." *Datamatics, 5*(1).

Glass, R.L., & Ronald A. Noiseux. (1981). *Software Maintenance Guidebook*. Prentice-Hall.

Hedberg, B. (1978). "Using Computerized Information Systems to Design Better Organizations and Jobs." In N. Bjorn-Andersen (ed), *The Human Side of Information Processing*, North-Holland.

Hershauer, J.C. (1978, Apr). "What's Wrong With Systems Design Methods? It's Our Assumptions." *Journal of Systems Management*.

Hirschheim, R.A. (1985). *Office Automation: A Social and Organizational Perspective*. Wiley Series in Information Systems.

Ives, B., & Olson, M.H. (1984, May). "User Involvement and MIS Success: A Review of Research." *Management Science, 30*(5).

Kanter, R.M. (1983). *The Change Masters: Innovations for Productivity in the American Corporation*. Simon and Schuster.

Keider, S.P. (1974, Dec). "Why Projects Fail." *Datamation, 20*(12).

Klein, H., & Lyytinen, K. (1984, Sept). "The Poverty of Scientism in Information Systems." *Proceedings of IFIP WG 8.2 Conference, Information Systems Research Methodologies - A Doubtful Science*, Manchester, England.

Klein, H., & Hirschheim, R. (1985). "Fundamental Issues of Decision Support Systems: A Consequentialist Perspective." *Decision Support Systems, 1*, North-Holland.

Kling, R. (1977, Dec). "The Organizational Context of User-Centered Software Designs." *MIS Quarterly*.

Kling, R. (1980, Mar). "Social Analyses of Computing: Theoretical Perspectives in Recent Empirical Research." *Computing Surveys, 12*(1).

Kling, R., & Scacchi, W. (1980). "Computing as Social Action: The Social Dynamics of Computing in Complex Organizations." *Advances in Computers, 19*, Academic Press, Inc.

Kumar, K., & Welke, R. (1984, June). *Implementation Failure and System Developer Values: Assumptions, Truisms and Empirical Evidence*. Working Paper, No. 009, Center for Accounting Research and Education, University of Waterloo, Ontario.

LeBoeuf, M. (1985). *The Greatest Management Principle in the World.* Putnam's Sons.

Laudon, K.C., & Laudon, J.P. (1988). *Management Information Systems: A Contemporary Perspective.* MacMillan.

Leintz, B.P., & Swanson, E.B. (1980). *Software Maintenance Management.* Addison-Wesley.

Linstone, H.A. (1984). *The Multiple Perspectives for Decision Making: Bridging the Gap Between Analysis and Action.* North-Holland.

Linstone, H.A. et al. (1981). "The Multiple Perspective Concept: With Applications to Technology Assessment and Other Decision Areas." *Technological Forecasting and Social Change,* 20.

Lucas, H.C. (1975). "User Reactions and the Management of Information Services." *Management Informatics, 2*(4).

Lucas, H.C. (1981). *Why Information Systems Fail.* Columbia University Press.

Lyytinen, K., & Lehtinen, E. (1984). *Seven Mortal Sins of Systems Work.* Working Paper, University of Jyvaskyla, Finland.

Markus, M.L. (1983, June). "Power, Politics and MIS Implementation." *Communications of the ACM, 26*(6).

Markus, M.L. (1984). *Systems in Organizations: Bugs + Features.* Pitman Publishing.

Martin, J. (1981). *The Telematic Society.* Prentice-Hall.

Martin, J., & McClure, C. (1983). *Software Maintenance: The Problems and Its Solutions.* Prentice-Hall.

Martin, J., & McClure, C. (1985). *Diagramming Techniques.* Prentice-Hall.

McLeod, R. (1990). *Management Information Systems: A Study of Computer-Based Information Systems.* Macmillan.

Merton, R.K. (1968). *Social Theory and Social Structure.* Free Press.

Meyer, Harper, & Gas. (1984, Mar). "Issues and Opinions." *MIS Quarterly.*

Mintzberg, H. (1983). *Power in and Around Organizations.* Prentice-Hall.

Mitroff, I. (1980). "Towards a Logic and Methodology for 'Real-World' Problems." In N. Bjorn-Andersen (ed), *The Human side of Information Processing,* North-Holland.

Mitroff, I. (1985). "Mindshift: The Mental Revolution We are In." *Technological Forecasting and Social Change,* 28.

Morgan, H., & Soden, J. (1973, Winter). "Understanding MIS Failures." *Database, 5*(2,3,4).

Moreland. (1982, Feb). "Aspects of User Friendliness." *Datamation.*

Mowshowitz, A. (1976). *The Conquest of Will: Information Processing in Human Affairs.* Addison-Wesley.

Mumford, E., Land, F., & Hagwood, J. (1978). "A Participative Approach to the Design of ComputerSystems." *Impact of Science on Society, 28*(3).

Mumford, E. (1984). "Participation - From Aristotle to Today." In Bemelmans, Th. (ed), *Beyond Productivity*,North-Holland.

Myers, G. (1976). *Software Reliability: Principles and Practices.* Wiley and Sons.

Naumann, J., & Jenkins, A.M. (1982, Sept). "Prototyping: The New Paradigm for Systems Development." *MIS Quarterly.*

Noble, D. (1984). *Forces of Production: A Social History of Industrial Automation.* Knopf.

Noble, D. (1982). "Is Progress What It Seems to Be?" *Datamation.*

Nosek, J., & Sherr, D. (1984). "Getting the Requirements Right Vs. Getting the System Working: Evolutionary Development." In Bemelmans (ed), *Beyond Productivity*, North-Holland.

Ohmae, K. (1989, May-June). "Managing in a Borderless World." *Harvard Business Review.*

Page-Jones, M. (1985). *Practical Project Management: Restoring Quality to DP Projects and Systems.* Dorset House.

Pettigrew, A.M. (1973). *The Politics of Organizational Decision Making.* Tavistock Publications.

Pettigrew, A.M. (1980). "The Politics of Organizational Change." In N. Bjorn-Andersen (ed), *The Human Side of Information Processing*, North Holland.

Pescarella, P. (1989). *The Purpose-Driven Organization: Unleashing the Power of Direction and Commitment.* Jossey-Bass Publishers.

Porter, M. (1990, Mar-Apr). "The Competitive Advantage of Nations." *Harvard Business Review.*

Robinson, D. (1982, April). "Synchronizing Systems Development with Business Values." *Datamation.*

Schultz, R.L., & Slevin, D.P. (1979). "Introduction: The Implementation Problem." In Doktor, Schultz & Slevin (eds), *The Implementation of Management Science, TIMS Studies in the Management Sciences*, 13.

Senn, J.A. (1984). *Analysis and Design of Information Systems.* Wiley and Sons.

Senn, J.A. (1990). *Information Systems in Management.* Wadsworth.

Schneiderman, B. (1982). "How to Design With the User in Mind." *Datamation.*

Weinberg, G.M. (1982a). *Rethinking Systems Analysis and Design.* Little and Brown Computer Systems Series.

Weinberg, G.M. (1982b). *Understanding the Professional Programmer.* Little and Brown Computer Systems Series.

Wenk, E. Jr. (1989). "The Politics of Technology and the Technology of Politics: Issues for the U.S. Congress' Third Century." *Technological Forecasting and Social Change,* 36.

Wilcox, R. (1982). "Behavioral Misconceptions Facing the Software Engineer." *Datamation.*

Willoughby, K. (1990). *Technology Choice: A Critique of the Appropriate Technology Movement.* Westview Press.

Wise, J.A., & Debons, A. (1987). *Information Systems: Failure Analysis.* Springer-Verlag NATO ASI Series.

Wood-Harper, A.T., Antill, L., & Avison, D.E. (1985). *Information Systems Definition: The Multiview Approach. Blackwell Scientific Publications.*

Yourdon, E. (1986, June). "What Ever Happened to Structured Analysis." *Datamation.*

Zmud, R.W. (1983). "CBIS Failure and Success." In *Information Systems in Organizations.*

Chapter 19

Information Technology and the Humanization of Work

Zdravko Kaltnekar
University of Maribor

The development of information technology has been rapidly accelerating, thus giving rise to new possibilities in designing information systems. As a result true relationships in work processes are changing, and so are numerous other fields of man's private and public life. Both in turn considerably modify the role of man in all areas. All too often the role of technology is being overestimated and the significance of man in the overall development diminished. In this so called "hardware orientation" one can pose the question of whether after all it is informatics that exists for the sake of man or man for the sake of informatics. It appears that modern information systems somehow enslave man, making him dependent on purely technical matters. Primacy is given to those who have these technical facilities at their disposal and who control them. Many computer experts have therefore made this technology a "fetish", thus being able to control the crowd of the "unacquainted". Luckily, modern technology is getting cheaper and is becoming available to the public at large. In addition, it is getting less complicated and therefore more easily controllable. Both in turn reduce the chances of monopoly over it.

Even today experts are sometimes confronted with a dilemma: whether informatics is one of the organization functions or, perhaps, vice versa. At one extreme, the organization is completely subordinated to informatics. At the other, informatics is defined as a major means of organization (Tarman, 1989).

This issue is becoming a pressing one due to the fact that organization theory and, above all, practice do not quickly follow developments in both informatics hardware and software. Therefore purely technical solutions outstrip requirements to establish appropriate relations among people linked by organization structures. The need for technological development and the impact exerted on human relations can by no means be dismissed. However, this should not blur the basic requirement that all work processes are designed because of man and for man. The second challenge posed by modern technology, and particularly by the development of information technology, to organization theory and practice and thus to the definition of man's role in processes is the growing rate of automation and robotization. The latter is increasingly penetrating into the industrial production and at a slower pace into day to day life. Again one faces a half-serious yet worrying question: will there be any place for man in the future automated world?

Technical possibilities are in progress bearing hope as well as fear. On the one hand robots take over dangerous and difficult work, enabling man to devote his abilities and strength to creative pursuits. On the other hand information technology increasingly spreads to creative fields (thinking machines, expert systems, intelligent knowledge based systems, etc.), replacing man here too. A higher rate of mechanization and automation may make man dependent on technical means, thus making him a mere link in the chain. Economy makes its own contribution by calling for best possible use of expensive devices. The overall technological development, information technology, automation and robotization represent hope for mankind coupled with the challenge of its best use and exploitation.

Man is the most vital element in each work process. His role in the work process varies depending on his position in the social structure as well as on his role in the work process. All his relations to the work process boil down to three basic roles:

- Each work process is basically organized because of man and for man, he is the user of process operation results.

- Man designs, directs and manages work processes, thus being their designer, planner and organizer.

- He works directly during the work process itself, performing operational tasks, thus being an active element of the process as its cooperator.

Of course this is not always the same person. Different people are involved in the overall social work division and distribution of results created in common. It is above all social order that essentially defines the relations by means of which an individual takes part in each particular role. This is the reason why his influence on individual fields varies considerably.

The role of man in information systems also varies. In this context one can single out systems professionals responsible for the variety of tasks related to information processing and distribution, and system users (end users) who actually make use of information (Khosrowpour, 1990). Various categories of end users can be singled out, e.g. top management, functional management and operating personnel (Bahl, 1990). There are even systematic lists of them (Amoroso, 1990). Further development will make this classification immaterial. The information technology development makes informatics available to its users. Computers in general and personal computers in particular have already gone out of data processing centers (Kashitza, 1989). Individual cooperator's participation in the process is growing and each one is becoming in a way his own manager. This also augments the responsibility for decision making and transfers it on to a larger number of cooperators. It is no longer some particular service for informatics that is responsible for the entire information systems, but all those in positions where information arises, is processed, transmitted and/or used. This in turn enhances the role of man and each individual taking part in these processes.

All decision making rests on necessary information, and adequacy of decision making depends on adequacy and sufficiency of such information. The bulk of data and information circulating within the information system is ever growing and therefore one can refer to it as informational throng (Liptak, 1990). It is essential to select from this throng only the information relevant to a particular decision. Ever more sophisticated information technology (systems of artificial intelligence, systems of natural like intelligence) enables the automatic selection of information and also its direct application for individual decisions. The computer with its possibilities is taking over operational decisions and is penetrating into the field of tactical decision making, thus somehow reducing the role of man in direct decision making. These are of course primarily routine decisions not requiring any major involvement of man's mental potentials. Actually this relieves man, enabling him to apply his abilities to more important tactical and above all strategic decision making. In this context information technology can be used as decision support in that adequate data processing can make successful and efficient decision making possible. As concerns the scope of work, man's role in information systems diminishes whereas its qualitative level increases.

Participation in the work process involves different stages of man's obligation and freedom. The form of obligation varies from thoroughly physical obligation to economic obligation. This of course defines the stage of freedom for each individual in decision making, as well as in his work. Both are conditioned by social development, technical possibilities and even more by social relations and social consciousness. The information technology development bounds man within a defined framework of individual procedures, but it also relieves man of non-creative routine work thus offering him greater possibilities for creativity.

No matter how the envisaged topic is dealt with, one can always arrive at the conclusion that the development of the information technology exerts all sorts of positive and negative impacts on man and his position in processes. The purpose of this chapter is an in-depth analysis of numerous impacts and effects in the interaction between information technology and requirements for more humane futures of man's work and life. Definitions of individual notions, contradictions and problems concerning this relationship, methods to achieve greater humanization and the influence of the information technology on these developments will be tackled.

Definition of Terms

The discussion of different terms is possible only if all participants speak the same language. It is common knowledge, however, that different people mean different things using the same terms. Different authors often tend to define the same terms differently. These definitions differ above all in length. Furthermore, definitions require a more in-depth clarification of the topic tackled in order to introduce the reader to the overall discussion. Following are the basic terms of the fields directly relevant to our discussion.

Much has been said and written about alienation in general, and in particular work alienation. It is essential to talk about it, since alienation is a special feature of our time. This can be reflected in day-to-day life, the disappearance of free time, in an ever smaller scope of communication, friendship, etc. Alienation is particularly manifested in production processes. Traditional work in agriculture with many customs and habits is giving way to mechanistic work in industrial production, where machines dictate a completely defined rhythm and man constitutes only a small part of the overall mechanism. Such an industrial way of production is also getting established in other fields of economy.

Alienation largely depends on the quality of life at work. This is the first point to be dealt with. The requirements for the quality of life at work are growing. The development of possibilities of higher living standards is coupled by the changing attitude of workers to different goods. The worker no longer takes part in the working process only to provide for himself means to meet his basic physiological needs. This process also provides for him possibilities to meet other needs and wishes including the wish for the socialization and evaluation of his individuality. In order to make his wishes come true, it is necessary to create conditions which will make this possible.

There are many factors that can help the worker find true meaning in his work, ranging from the social systems and organizational measures safeguarding fair relations among employees, to purely technical conditions of work. They all involve a whole range of measures designed to guarantee these conditions. The process of guaranteeing itself should include the whole range of bodies and services as well as each individual worker, i.e. the entire working organization.

It is very difficult to define what the quality of life at work really is. Different people have different requirements concerning work and these requirements do change. Meeting of particular needs gives rise to new needs - on a higher level (Maslow, 1970).

The worker is looking for different things at different levels of the work process, consequently his attitude towards the quality of life at work is also modified accordingly. The expectations of each individual worker depend on his mentality, his physiological and psychological properties, education, and opportunities to satisfy the demands, etc. In this particular process each worker is a separate individual with his own particular demands. He may not be aware of all his needs and wishes all the time. Therefore it is practically impossible to create such conditions that would fully satisfy each individual. All these conditions and the related quality of life at work are therefore designed for the average worker.

Actually the quality of life at work has not yet been defined in this paper. It most certainly involves the worker's attitude towards environment,as well as his attitude toward the physiological conditions at work and opportunities offered by the work. This feeling most certainly depends on the ratio between his expectations and possibilities for their realization. The quality of work life is expressed as the possibility to be effective at work by making the least possible effort.

The alienation of work is undoubtedly the consequence of poor quality of life at work at large. This confirms numerous definitions of the term alienation (many of them were quoted by Gaugler, Kolb & Ling, 1977). Alienation is thus defined as mismatch between the required expectations, values and goals on the one hand and difficulties and inabilities to implement these values or goals on the other. In such a context alienation can be referred to as an inhuman phenomenon.

One must stress alienation's multifold meaning and therefore varied use of the term. Different authors tackle this term differently, since they come from different environments. A detailed analysis of all these views and opinions exceeds the purpose of this paper. Our further discussion will center around an interesting breakdown (Tietze, 1974; Israel, 1972; Weil, 1973) into

• subjective alienation (in psychic state) and
• objective alienation (in social processes).

Subjective alienation is a psychological term defining conscience and adjustment structures of a situation in which the individual finds the working environment to be unclear, hostile and inhumane. Here alienation is a synonym for the exclusion from decision making processes directly concerning the worker as a person, for social disorganization, for tailorization of work. Objective alienation stems from political consequences entailed by the contradiction between work and capital. The division between physical and mental work should not be tackled as primarily a psychological problem, but as the repercussion of the political and economic features of the system.

The contradictions arise with the question of where subjective alienation must necessarily stem from objective alienation. Would the elimination of objective alienation automatically cancel subjective alienation, or vice versa: can the measures taken to reduce the subjective also influence the objective alienation? This contradiction is coupled with three fundamental positions, briefly dealt with below:

a) The elimination of objective alienation also removes the subjective one. This is the position acquired by most authors coming from the countries of "real socialism", where alienation is not really a topic, unless referred to as the alienation in capitalism.

b) The elimination of objective alienation is not sufficient for the development of humane forms of life. This position is most certainly the closest to the actual state of affairs. The alienation is conditioned by a variety of reasons, and will probably never be completely removed. One can only seek a less alienated work and a less alienated life. It is not only total elimination, but also the alleviation of alienation that counts.

c) It is not sufficient to eliminate private property in order to remove alienation of work. Theoretically, public ownership of means of production does do away with the exploitation of workers, however, the question arises to what extent each individual feels this public property is his and actually shares in the management of it. This fact is now being corroborated by the current difficulties in the so called socialist countries. Naturally, subjective alienation cannot be removed.

The field of work offers the worker a possibility of meeting most of his needs, if it is adequately regulated and organized. The worker's needs and requirements are most certainly higher than the simple substitution of effect by money and the related possibility to meet material and non-material needs outside his field of work. By work and also directly in the work man can meet all his needs, from the lowest physiological and security needs that can be covered by his salary, to much higher ones. This field of work provides the worker with a possibility of establishing social contacts with fellow workers coupled with the feeling of belonging to the working community and the working organization at large. It also provides him with an opportunity for self-assertion and recognition, as well as for the revealing of his creative potential and thus for the development of his own personality. It is namely these latter needs that are becoming ever more important, after the lowest needs have been satisfied, that call for the adequate organization of work. Through the adequate design of work it is essential to enable the worker the access to the satisfaction of the higher needs. More varied contents of work, wider choice of work and working time, greater democratization of working processes in general constitute the basis for such possibilities.

It is very difficult to find the right answer to what humanization of work really is, since there are many possible answers to this question. It depends on a variety of factors and conditions. Different requests also influence the accuracy of this answer. In general terms it can be established, however, that humanization of work is the opposite of alienation of work. The former makes it possible for us to remove or to alleviate respectively the impact of conditions giving rise to alienation.

According to Webster's dictionary humane means "having the feelings and dispositions proper to man; kind; benevolent; tender; merciful ...", one could also say something humandecent or even humancreating. As humanization one can characterize the creation of humandecent conditions of work, the removal of inhumane conditions in the working process, especially in industry and on conveyer belts. Such use of this word can be too narrow, because it could also mean that the humanization of work should not be applied unless the conditions are inhumane. The matter is that it is very difficult to segregate humane and inhumane conditions, since they depend on too many factors (personal, economic, political, technical, technological, etc.). Therefore the humanization of work is the quest of more humane and better conditions of work.

This is also our aim. There is probably no wholly humane and also no wholly unhumane work. The humanization of work is therefore a mobile strategy, which cannot have an end aim. This means it is an endless process, a continuous pursuit of something new, something better, thus more humane. Man's demands are changing and will change in the future too. Consequently one cannot set goals today for tomorrow or forever for that matter.

It should be stressed once again, there is no final term and humanization does not represent only one possibility, one method or one procedure. Therefore it is not right to pretend (as many authors from eastern European countries do) that humanization is impossible unless private ownership of means of production is done away with. It is the matter of the goal that we want to attain. It is true that the collective ownership of the means of production can contribute to the humanization of work. But it surely is not the only way that can make this possible. First the abolition of private ownership does not necessarily lead to humanization (extreme concentration camps in Stalin's time) and second, it is not a "conditio sine qua non" and other measures can also bring about humanization of work without the abolition of private ownership (e.g. regulation of working conditions, mechanization, automation, job enrichment, etc.). It is the matter of how and how much. There are many possibilities ("many ways lead to Rome"), different approaches, methods, procedures, etc. The final goal is unclear and far away, one can only try to come closer to it.

If people are to develop into humane, free and creative personalities, it is essential to do away with all types and forms of alienation, particularly alienation of work. This however cannot be achieved by proclamations and conjuring, but by a meaningful design of working processes. There is a way to arrive at this end, namely:

- It is possible to overcome alienation in the working processes by enabling the worker to plan and prepare his work independently without anyone else telling him what exactly he is supposed to do. It is necessary to make it possible for the worker-performer to "see" his own work within the framework of the overall associated labor and to accept it voluntarily and consciously as his own duty and not only as a task imposed on him by someone else. At the same time he must be able to participate in designing his own work.

- The alienation of the results of work can be eliminated by enabling each particular worker to take sovereign decisions about the results of his work. Of course one is aware of the fact that market only rarely defines the value of each individual directly, but only the results of the associated labor. Therefore it is necessary to establish ways and means according to which each individual can establish his own contribution to the common result.

Both can be materialized only if one gets at one's disposal all data on the process and results of work in due time. The elimination of alienation is thus directly linked with proper design of information trends. The present technology of data processing enables us to form and maintain extremely complex and

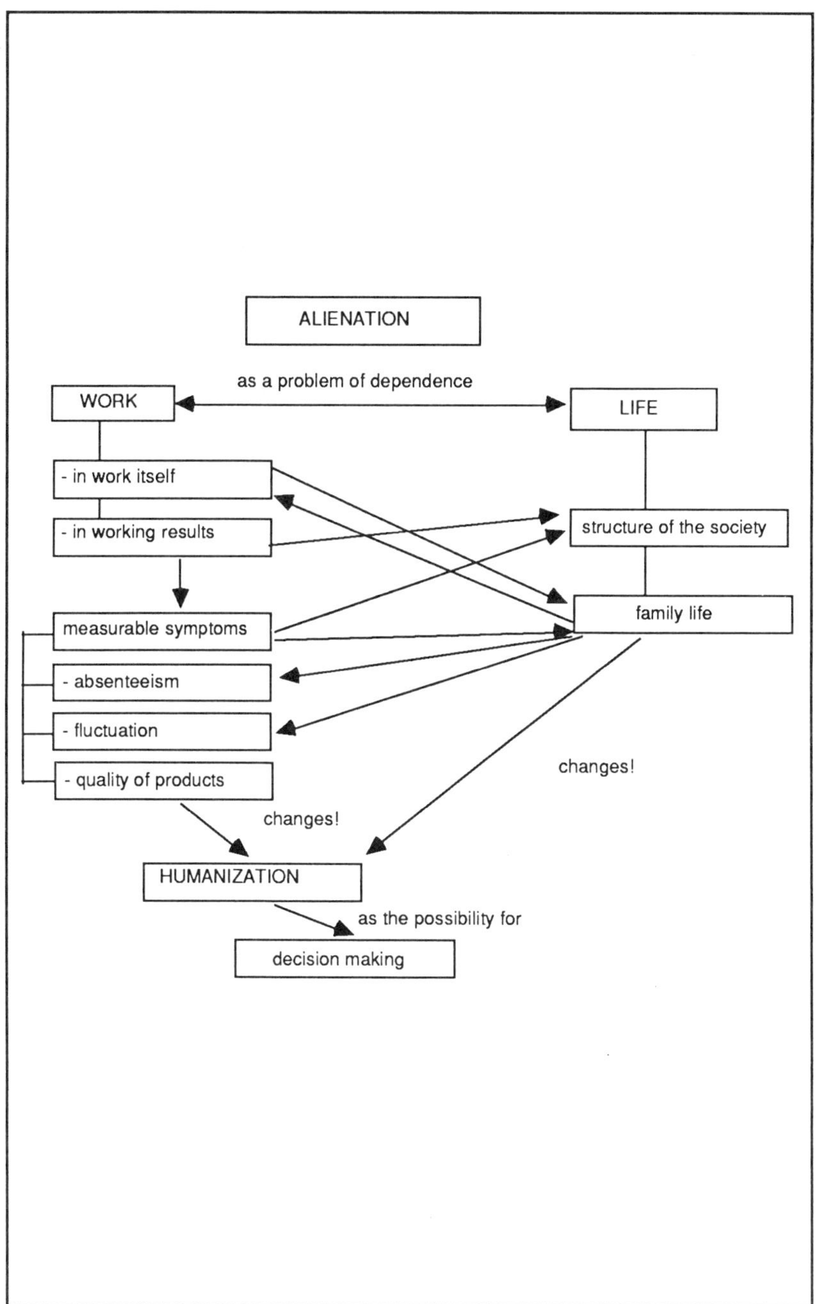

Figure 1. Problems of Alienation and Humanization of Work

extensive information systems. Their technical implementation is therefore not difficult. The requirements to include into these information processes all workers-performers of tasks causes more concern. It is possible to conclude that most information flows are most frequently interrupted due to the insufficient training of the latter. Consequently, it is the working process and the product that dominate the worker, instead of just the opposite.

The adequate training of workers for the successful participation in processes and their appropriate distribution can be guaranteed only by an appropriate organization of work that can comply with the principle: the right person for the right job. It is probably not of vital importance to search for the most appropriate among various definitions of the organization concept, yet it should be taken into account that organization always involves the definition of certain relations among people, which enable the achievement of set objectives. Organization is defined by its structure and the processes running through it.

Organization as function defines relations in the open socio-technological systems having dynamic and mutually dependent relations with its environment. It can and it must change according to changes in this environment. It can change its objectives, policy, tasks, transforming at the same time its structure and processes in it and adapting them to the modified requirements. External influences and internal changes give rise to a permanent process of adapting its components and relations, therefore organizing should be understood as a dynamic process of searching for the most appropriate relations. The successfully operating system is characterized by its synergy and synthesis of the parts in an entity so as to enable all parts to be more efficient than if they operated independently.

Organization structure comprises institutionalized arrangement of individual elements of process (people and resources) for individual tasks or groups of tasks. By means of this, certain space and time relations among all factors are established, which enable the relational process flow, and the development of hierarchy, adherence and communication structure, the establishment of numerous units with various rights of decision making, participation, planning control, etc. The smallest basic units - basic systems - combine with lower and higher levels of the structure, and produce cooperation in higher level systems linking individual work processes.

Organization processes determine dynamics of activity within systems. They determine the space and time of single element actions vital for the accomplishment of common tasks. They depend on organization structure and work conditions as well, requiring the change of organization structure.

Organization processes are being run mostly in communication structure that links individual elements of organization structure. The vital part of communication structure is information flowing within it. Therefore the interlink among organization processes and information systems is very important: processes determine the information needs, and availability of information to a great extent enables the flow of these processes, assuring their efficiency.

Automation and informatization of processes change their technical structure which, of course, to a great extent changes qualification requirements for workers by transforming their relations as well. The principal limit of the decisional approach is to be essentially "cybernetic". According to the cybernetic view of the organization, "man is not considered a 'machine' any more but a 'regulator' who makes decisions and takes corrective actions on work processes ..." (Cariati, Ciborra & Maggiolini, 1989). As has already been mentioned this, increases man's role in the process and organization.

The main purpose of this paper is not discussion of information technology and its development, but of its impacts on man and his relationships within the working process. Nonetheless, the former calls for a brief definition and above all the definition of its development fields that will be of interest for further discussion.

It is not easy to give a precise definition of the term information technology. It has been tackled differently by different authors, and this applies in particular to the scope of the term. Quite often the terms Management Information Systems (MIS), Information Resources Management (IRM) and Information Technology (IT) are being used as synonyms. The existing literature indicates that the terms MIS, IRM and IT represent organization information in information processing technology that is used to collect, process and disseminate information for managerial use (Khosrowpour, 1990). In this context, the information technology proper is to include the technologies of computers, telecommunications and office automation utilized by an organization in the management of information. Information technology can also be defined as a system whose parts are: knowledge, actions and regulations which make possible all the intensive information activities (Tunje&Savi, 1989). The information technology elements that are applied and used are already contained in man's knowledge and experience and can be incorporated in physical devices and equipment. This needs to be specially singled out, since it brings about the requirement for a meaningful segregation between abilities, limitations and tasks imposed on man on the one hand and technological ways and means on the other.

The products of information technology can be divided into two basic groups (Kobe, 1986): teleinformatics and actorics systems. Teleinformatics systems are used for information transmission and processing (computers and the like), whereas actorics systems are the technical systems driven by pieces of information received (robots and the like). Both are of interest for further discussion, since they influence the position of man in the process.

It is also fitting to note the division of information technology as concerns the degree of intelligence of each type of product (Tunje & Savi, 1989):

• technology of "non-intelligent" systems with software numeric base, precisely defined algorithms of the solutions transcribed into special program. It has only few possibilities for solving some unpredictable situations;

• technology of "artificial intelligence" systems includes the capability of either teleinformatic or actoric systems to independently create an internal model of a (limited) problem and to change it according to changeable evaluation criteria; its essential strength is represented by a symbolic data processing based upon some analogies with man's way of thinking;

•technology of "natural-like intelligence" systems is based upon the completely new approach with the task to copy the process in man's brain and nervous system. It consists of a number of mutually interconnected processors with their own memory units. These systems are not programmed but still have to learn and to perfect themselves; they must be able to react in individual situations, and their reaction could be based upon contradictory or incomplete data.

Varied development stages of information technology also define the position of man in the information systems itself as well as in other processes. Technological systems are assuming ever more demanding tasks including decision making, thus somehow pushing man aside. This, however, should not mean the exclusion of man from processes, but man's assumption of ever more demanding tasks in higher levels of the decision making hierarchy. The issue of further meaningful development of the overall technology and information technology in particular boils down to: how far should man go in giving way to the development of technological means and where are the limits of this development? And, are the limits necessary at all?

The most extensive of all technical systems in information technology is the computerization of data processing. Therefore this chapter will further focus on it. The considerations in this area can find parallels in the overall information technology.

Issues, Controversies and Problems

As the process of a computerization of a firm expands, it is increasingly evident that its dynamics, depth and width depend on more harmonious interaction between human and technical information components (Tulak, 1989). It can be stated almost everywhere that the unity of economic, technical, technological and social viewpoints is more important today than at any other time for the sake of developing successful and competitive macro and micro-spheres (Trethon, 1989). Man is and remains to be a very important element of every process - and so also in information systems - in all of its three basic roles: as a designer and developer of the systems, as a direct participant in them and as the user of the results. Of course they can all be completely different people referred to as information technology managers, information system personnel or staff and information users. Quite often these functions are personally interlinked. In this context both developers and participants are mostly included into IS personnel. This usually applies to all the staff in different computer centers and services. It is expected that further development can bring about the merger of the other two functions: the final user is becoming to an ever larger degree the participant in the application of information technology. The latter enables him direct participation in information processing, above all into input and output.

The chapter will now focus on the user and his relationships, opportunities and problems in the utilization of information technology. Nonetheless, it is essential to draw attention to some problems stemming from the development of information technology, of course above all from the point of view of humanization of work. Despite all the decentralization in the application of information technology and its availability to users, the other two categories will still need designers and developers of these systems, experts who will be involved with direct introduction of the new technology into individual processes, i.e. information technology managers. It is essential to discuss the matter of the impact of information technology on technical staff involved in the process of information processing, since many problems affecting them (e.g. the problem of work with screen) will also affect the users - though perhaps to a smaller extent.

At the heart of most information technology management problems are "people" problems. The lack of managerial orientation of information technology managers has been considered to be a major contributing factor in many information technology deficiencies (Khosrowpour, 1990). Many information technology managers are too much in love with technical solutions provided by this technology, forgetting at the same time its true role in the overall system. Information technology should be only the tool supposed to enable an efficient and successful economy. The "not understood" shut themselves into the ivory tower of their knowledge expressing their frustration by a superior attitude towards the ones who do not master all this technology the way they do. This in turn makes it impossible or at least aggravates the relationship with other managers and users in the organization. Technologists managing these information technology resources with their continuous flow mentality and sheer love of the computer tend to ignore the business requirements (Khosrowpour, 1990). All this in turn gives rise to psycho-sociological difficulties for both the users and the managers themselves.

The organization principle "the right man for the right job" comes here especially to the foreground. It was probably erroneous to look for technical experts in the first place for this paricular profession. The information technology itself was rather intricate in the beginning and called for purely technical knowledge. Consequently, the person having such knowledge at its disposal or being able to absorb it easier as a technician became the expert in this field. All of a sudden the important thing was to master the overall hardware and to be able to deal with all this "machinery". Thus many information system managers have not become acquainted in detail with the very nature of information and its role in the process, let alone with the need to organize the overall process of data collection, processing and dissemination as concerns the actual needs of the users. Only the manager with all his properties required for the manager can be considered the true expert in this field. This profession becomes even more demanding in view of the fact that all these managerial properties must be applied to a specific field of information technology.

Without claims to perfection we shall try to list major fields of his tasks, particularly future ones, defining also his own profile:

- He must first organize the structure of the process of collecting, processing and dissemination of all necessary information, establish the needs of the users and capabilities of information systems, set up adequate communication channels and links with and among other organization units as well as make sure that they are properly supplied with the necessary information.

- The information technology manager must also study new technical possibilities by experimenting, and must above all follow the new technology available at the market and introduce it into his organization. He acts as some sort of a change agent and takes care of the state of the art and integration of the overall data processing.

- The decentralization of the overall information technology consolidates the role of man as advisor in different structures and on different levels of the users, and calls for adequate training and understanding of the opportunities offered by information technology.

- His task must also involve the elaboration of expertise in issues such as information planning, information budgets, reliability of information sources, information security measures, acquisition of new information technologies, etc. (Khosrowpour, 1990).

• Last but not least he must also participate in different economic analyses both within information systems themselves and otherwise, therefore he must be well acquainted with such terms as profit and loss, return of investment, forecasting, etc.

• It is essential that he takes full responsibility for all these developments or at least co-resposibility with the general management.

The work of the information technology manager is very demanding and the demands considerably exceed the usual knowledge of technical experts, as well as their willingness to absorb additional knowledge. The latter is probably the core of the problem, since technical knowledge is undoubtedly necessary for mastering this field. Therefore, it will be extremely difficult to find the adequate educated profile of the worker for this particular profession. Specialists with highly varied basic education will have to be trained for it by the acquisition of additional knowledge.

The appropriate training will considerably reduce many frustrations that the present day information technology managers are exposed to. It will make it possible for them to participate on equal footing with managers in other fields, as well as to be involved in top management. Thus, a lot of tensions and friction will be done away with, also concerning other participants of the information process.

As the information system staff we can count above all the technical and administrative workers of different computer centers. The difficulties they are exposed to in their work roughly boil down to two different groups: ergonomic issues relative to job design, working tasks and environment, and psycho-sociological issues linked to their work. Although both groups are interlinked, they can still be tackled separately.

A more in-depth analysis by ergonomic issues exceeds the purpose of this paper. They can be divided into two groups: hardware and software problems. First comprise the purely technical structure of the workplace itself, means of work and working environment. The problems concerning work at screen and the related sight troubles as well as lighting and air-conditioning related problems come to the foreground.

Software ergonomics involves, in general terms, adjustment of working trends, work content and psychic means of work to the regulation of man's behavior, and in more specific terms, the application of knowledge of work to computer processing software design.

Psycho-sociological problems crop up above all due to the loneliness at work. Man communicates mostly with computer, and has therefore very poor interpersonal relations. This can considerably affect possibilities for socialization of these workers in the organization at large, and can bring about their alienation in this context.

In the extreme it can even give rise to some sort of obsession with the computer and man relinquishing all social contacts with his environment. This is particularly evident when the work with the computer is interesting and poses some sort of challenge. However, many computer operations are routine and rather tedious. They often do not differ from typing and the entailing monotony gives rise to some uneasiness. The pressure is sometimes due to time constraints involved in individual operations or demands for immediate results respectively.

Technical personnel problems in information systems are ever abating, since the number of this personnel is diminishing. Information technology makes all this work available to the users. Some of these problems, particularly ergonomic, will certainly affect the latter too, though to a much lesser degree, since these users will not be involved in information technology all their working time. Moreover, they will have to use varied technology, which in turn will reduce the monotony of their work.

Decision Making Process and its Informational Basis

The right to make decisions is the most important humanization measure. The man who is pushed away from the decision making process is undoubtedly alienated both at his job (since he has to follow the decisions made by others) and at the distribution of the results of his work. All other measures for greater degree of humanization can, no doubt, facilitate this work, reduce its pressure and to some extent even alienation. However, without the right to decision making these measures are imposed upon man, which in its turn considerably diminishes their impact. It will be possible to do away with the overall alienation of work only when (and if) the work becomes man's hobby. Until then only the increasing right of every individual to make decisions about his or her work will truly limit the alienation.

The most important component of organizational links is the decision making requirement - on various levels, with various degree of impact and various implications of individual decisions. Abstracting some other relations (like informal organizational structure, human relations), organization can be defined as a decision making structure. It is mainly information on decision making and the information essential for decision making that is part and parcel of communication. Thus information provides a very important basis - we dare say "conditio sine qua non" of decision. The efficiency of decision making is conditioned by the quality of information.

Decision making is a highly complicated process and information constitutes just one of the bases for making particular decisions. As a matter of fact, information (in particular the computer processed and prepared information) constitutes the basis for the numerical and logical part of the decision making process. Many decisions and above all various operation decisions require this level only, therefore they can be simply passed on to computer. Most decisions require much more than merely logical combination of various numerical data. Perhaps, at first glimpse at least, it might seem illogical to argue that intuition is gaining momentum in the decision making process. Let's try to prove it by a brief insight into the development of work processes and in particular the decision making process:

- in the first period there was a shortfall of appropriate information systems, therefore nobody had enough information and the success of decision making mainly depended on intuition.

- in the following period those who provided themselves with more advanced information bases could as a result become more efficient.

- lately the problems arising from the lack of information have become less and less frequent (available information is ever increasing), therefore decision making efficiency again depends on something else-let's call it intuition.

Two questions arise in such a context: First, what does intuition mean after all and what does it imply? And second, what does organization have to do with it?

More in-depth analysis of intuition would much exceed the scope and purpose of this paper even if applied to decision making only. Therefore, we will restrict ourselves to mentioning some of the basic items. First we must pick out from the bulk of information only the pieces relevant to the problem. This is very often not an entirely logical process. In fact, problem definition and its selection requires more than just an appropriate combination of information. The same can be applied to the selection from various scenarios of solutions which a computer can offer on the basis of processed information. In the same or similar economic conditions some are more and some less efficient than others. In our intricate world every manager should be well acquainted with logical thinking as well as intuition. It is only in the first case that he can be aided by computer systems.

Information is increasingly becoming one of the vital bases of the decision making process. Lately we have been reading much about the so called decision support system. DSS has been conceived as a class of applications designed to help middle rank and senior managers dealing with the semistructured type of problem. In its broadest sense the term Decision Support can be applied to any information system within an organization designed to help the manager improve his or her ability to make decisions (solve problems) (Fisher, 1989). In broader sense we dare say there is no information which would not be usable to such or other, current and future decision making. Therefore, each information system can be defined as Decision Support System. In its narrow sense DSS is usually defined as "computer aid" to:

- assist managers in facing semistructured problems of decision making;
- support rather than replace managerial judgment;
- improve the effectiveness of the decision making process rather than efficiency (Bajgori}, 1989).

Decision making process, however, still remains within the domain of those who determine and set particular tasks and the producers, respectively. The growing perfection of computer techniques and technology (thinking machines, expert systems, intelligent knowledge base systems) increasingly enables large number of decisions to be transfered to the computer. So far these have been numerous routine decisions, managed by previously defined rules. Computers can entirely take over the numerical level of decisions where rules are easy to form. These are mainly decisions on the operation level. There are, however, many decisions, therefore the computer role at the direct decision making should not be neglected. The logical part of decision making is pretty much available to the computer (which has its own logic). Thus more advanced computers will be able to take over the tactical level of decision making, though here too one should not neglect the role of intuition. On the strategic level, decisions are so complex that besides the numerically logical part they also require a significant portion of intuition. Therefore, let's hope, they will still remain in the domain of man. Individual stages of decision making process are shown on Figure 2. It should be noted that the computer is increasingly moving up the scheme, first as decision support and now to an ever larger extent as making decisions.

The information technology is increasingly becoming a tactical and even strategic tool. Its impacts on man are manifested in both of his functions:

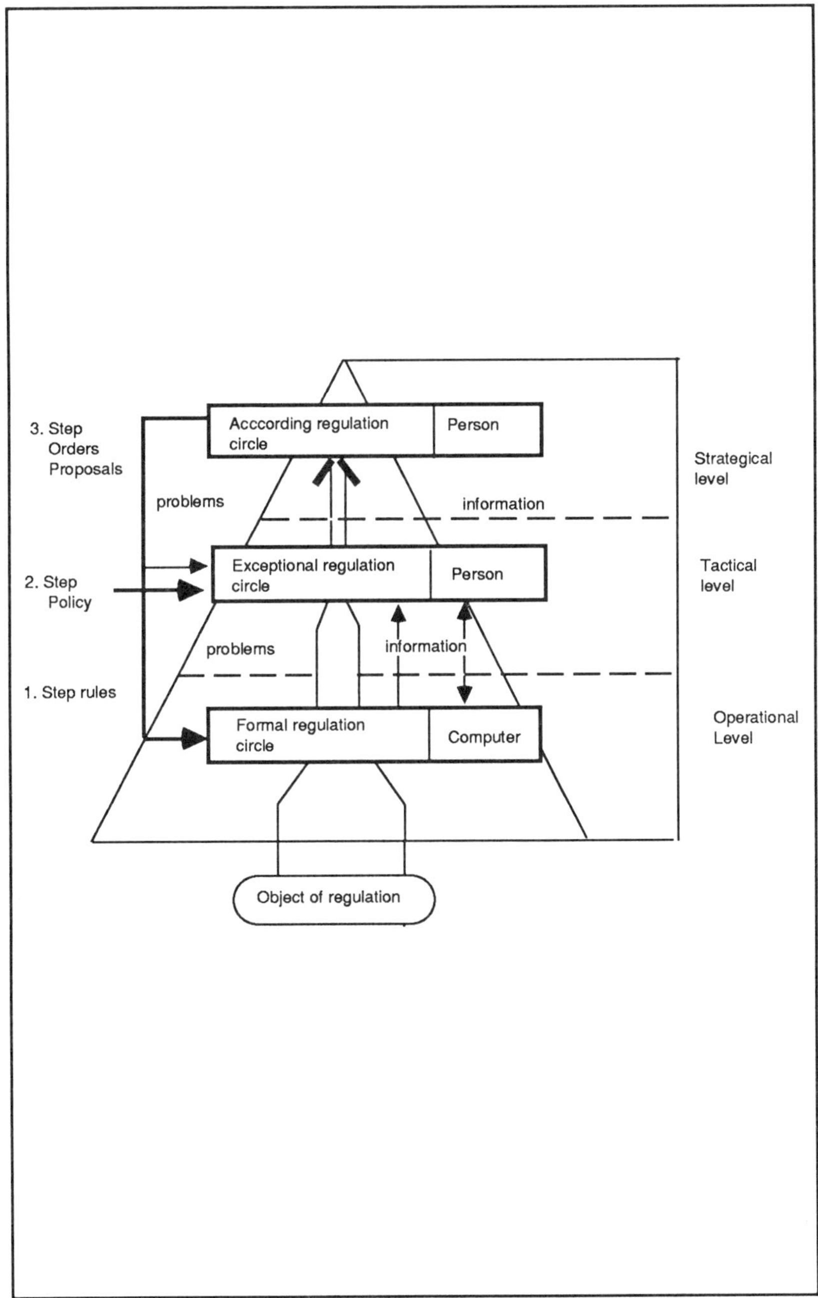

Figure 2. Different Levels and Manners of Decision Making

in direct decision making on a lower level and as decision support in tactical and strategic decisions. In both cases, however. the penetration of information technology has its favorable and adverse implications.

Information technology is becoming ever more important as decision support. Strategic and tactical decisions are very often quite intricate and they may also entail tangible consequences. Therefore, man must have at his disposal extensive and detailed information in order to make appropriate decisions. Quality of information comes here before quantity. It is the computer that can take care of quality by evaluation of data, problem diagnosis, problem analysis, the integration of solution techniques, analyses of alternatives, etc., preparation of the most appropriate decisions in the given situation.

Unfortunately, the much heralded DSS has, to some extent, been a disappointment. Many executives are still fearful of direct interaction with the system, are uncomfortable with the commands, or do not understand the relationship between their objective and the stored information. Additionally, even though a growing number of managers are computer literate, they do not have an in-depth knowledge of many activities associated with problem solving and decision making (Breslavsky & Yaverbaum, 1990). Due to insufficient education and lack of experience they are not aware of all the capacity that these tools can provide for them, they do not trust them and consequently do not use them as much as they should. This in turn puts them under stress, since in no way do they want to show their resistance to new technologies, which can easily bring about confusion in decision making.

Decision making relative to strategic affairs of an organization requires a lot of knowledge. One cannot expect that a person would have all the necessary knowledge, therefore, decision making is usually shared by several experts. As a result the problem of integration of all the participants of decision making process crops up as well as the evaluation of their often conflicting views and suggestions. Information technology can help evaluate different variants by different intelligent decision support systems and even replace the need for the involvement of a large number of experts by expert systems.

Quite often one is confronted with the lack of an appropriate framework that incorporates the simultaneous adaptation of the system and the users in the context of problem identification, formulation, analysis and choice (Breslawski & Yaverbaum, 1990), some sort of inability of information to meet the real needs of the users.

Participative Design and Implementation of Information Systems

Many problems arising from the use of information technology and the operation of information systems as well as the psycho-sociological problems of the adoption of these technologies by people are due to the fact that in a way these systems have been imposed upon users. Quite often they do not participate in the choice of information technology and the shaping of information systems and therefore both are alien to them. User involvement and participation in MIS planning, system development and system implementation has been often cited as a critical factor determining the success or failure of computer-based MIS in organizations (Bahl, 1990).

Possible dimensions for both the information system development process and the information system itself (the system development results) can be: functional effectiveness, economic efficiency, technical reliability and social acceptability (Oppelland, 1989). The implementation of these requirements can be evaluated by the users only. The only criterion can be the satisfaction of their expectations that, quite naturally, stem from their need for information.

The circle of the information users and consequently people involved in information systems is ever expanding. The use of information technology is expanding from top management and functional management to operating personnel in various fields of activity within the organization. The involvement of a greater number of users in the shaping of information systems and their introduction makes the process more complicated, which frequently results in compromises, extends the time for the adoption of the systems and increases the cost. Therefore it is essential to think well where and when the user should be included into the process. This should undoubtedly apply to the decisions relative to the users.

The involvement of users in shaping and introducing information systems can essentially improve their operation. There is no doubt that the committed user will accept and use the system. Social acceptability of the system will surely enhance its functional effectiveness and therefore its economic efficiency.

Such an involvement can produce advantages for the user himself. In the first place, the very fact that he is involved in the decision making process on information systems and information technology will do away with many stress problems and alienation problems. Furthermore, the very process will make it possible for him to get acquainted with the demands and possibilities offered by the new technology. Consequently it will be easier for him to understand and thus his resistance to the novelty will be quelled. At further use he will consider it in a way as result of his own effort and will use it accordingly.

Automation of Processes and the Role of Man

In a broader sense automation and robotization can also be included into information technology (the already mentioned actorics system). The essence of the operation of automated devices is information that needs to be transmitted to the machine by means of an appropriate program. Of course automation in turn exerts a significant impact on man's position in the process. Many of its impacts are similar to the impacts of the computer, of course in their own respective fields of activity. Therefore it is essential that position of man in automated processes be appropriately tackled.

In the near future one can obviously expect integrated processes where people, computers, robots and automated lines will work harmoniously. For such computer integrated manufacturing (CIM) directing and connecting of the entire process with a computer is required, as well as automation and computerization of various professional fields. It depends on the facilities and possibilities as well as on the level of particular procedure processing, the involvement of the computer and different robots, numerically controlled machines, transfer lines, etc. as to which fields will be sooner or later involved into the technical processing. The central role of the computer and the integration of these processes is essential to this system.

In each process double connection man - machine is established and as result, the interdependence of these basic actors of the process. It is conditioned by numerous factors, in many cases by the technical development of the process. Irrespective of the accentuated importance given to man in the process, one must admit that automation changes the role by bringing him many advantages and drawbacks. Automation, by all means, can contribute to humanization of work. Robots can take over dangerous and difficult tasks, as well as many routine ones. It is such tasks in particular that cause alienation of the worker in the process, since they make him suffer physically or mentally as he cannot participate in them as a proper generic being. Contrary to this, automated processes can contribute to the alienation if the worker is not skilled enough and does not know their nature.

Less exacting requirements for routine work can bring about much more demanding tasks. Their content is getting increasingly complex, requiring creative participation by the worker. On the one hand, this enables him to use all his (especially mental) abilities and on the other hand brings about fear whether he will be able to implement all these tasks. The tasks which are too demanding can produce numerous frustrations and alienation. Thus the problem of control of the processes arises, a new possibility proving that after all the machine does enslave the man.

Improved techniques often produce the requirement for the entirely defined way of tasks implementation. Taylor's "the one best way" again gains in importance. This is especially true for those workers who remain directly in the process (e.g., supervisors of various control devices) and partly even for designers and maintenance service of the systems (e.g., work according to the programs purchased.)

Automation reduces the overall required fund of human work and enables shortening of the working time up to unimaginable limits. This will bring to people much liberty at designing their family and social life. Of course this also brings danger that work will be reserved for a narrowly selected group of creative people. Although one should not neglect the problem of unemployment, it is not by far the worst danger. "Man does not live from bread only" (Dudincev) but work is one of its vital parts and creativity is connected with it. Should we then sentence a part of mankind to idleness?

Though man in automated processes is taking care of an ever smaller amount of tasks, his role is consolidating. These tasks are getting more complex and hence more demanding. Higher required skill of workers also enables a more direct decision making concerning the implementation of these tasks. At the same time, this increases worker's participation in management process. Each worker separately participates in the system where the responsibility and power depend on modern qualifications (use of computer techniques) and an ability to make quick decisions in continuously changing conditions (Durlik, 1989).

All these tasks give man a completely new role in the process and they distribute the tasks to others that have so far been reserved for top managers only. With the changed role the requirements change too: the unskilled and retrained worker has no place in these processes. Changed and changeable tasks demand a highly trained worker, who is willing to put his own intellectual potential into his work.

Automation is taking up new fields of activity and in particular those where the work is still difficult for man or where he is ineffective due to lower efficiency. It facilitates a significant rise in labor productivity, liberates man from routine work and enables him to use his creative abilities. It boosts production, thus increasing the living standards of the population at large. It also promotes general education and the cultural level of the population.

On the other hand, automation brings about many dangers by a pronounced differentiation of people. The difference between the capable and less capable is increasing, and despite the growing number of skilled and educated people there still remains a stratum of the "non participating". Even greater danger exists in that knowledge remains within the domain of some people only, who thus acquire an unproportional part of authority. This in turn can cause greater alienation of other participants in the process as well as co-citizens.

Automation thus generates both heaven and hell, probably always both of them simultaneously. The share of the former and the latter then depends on our ability to introduce and shape it properly. Therefore this should by no means be a purely technical process of replacing some machines by more improved ones. It should neither be an economic ratio of comparing between cost and effect. It should also take into account man, the one directly involved in the process, as well as people and mankind at large. Therefore, society with its regulations has to take a vital part in the process. Automation has to be carefully and thoroughly introduced and should be coupled with overall consideration of wishes, requirements and needs of people within the process itself and in the society at large.

Information Technology and Organization

Organization is defined as a system of relations among people. All the different relations of cooperation, superiority, and inferiority determined by organizational structure can only be set on the basis of communication links. We can prove that organizational relations are defined by information and vice versa. Information acts as a link enabling the very functioning of an organization. In fact, organization processes mainly represent various activities relative to information: input, processing, transmission and above all numerous information flows connected with it. On the other hand this very organizational structure defines information flows — their contents as well as their direction of movement.

Organization connects human structure thus representing the framework for decision making. By superior - subordinate systems and cooperation, it institutionalizes the interrelations, defines places of decision making and the one who makes decisions, as well as the responsibility for the decisions themselves and their implementation - all this regardless of the degree of rigidity of its structure. Organization also determines conditions for the selection of people for carrying out specific tasks (and making decisions concerning them). The inference drawn from the aforementioned is that the more demanding the decisions brought by information system development the greater the significance of the organization.

A greater degree of democratization in decision making requires good organization and appropriate distribution of rights and responsibilities in decision making, otherwise it can lead to anarchy. As Kent says, freedom means responsibility, which should not be general and moral responsibility only but exactly defined and in some ways institutionalized. Basic framework is offered by the organization structure. The more complex the work tasks are, the more complex should be the way to define relations and forms of cooperation in the process. The same applies to different procedures and techniques designed to safeguard the successful functioning of the system.

As a matter of fact there is no need to say anything else about the changing of the work organization demanded by the introduction of automation and robotization into the process. As the role of man in the process and the requirements for his qualification change and will further change in the future, the organization of work will have to change too. All these changes will differ from one another and will often be completely different in different parts of the process operation and its environment. Only they can make the organization successful, only they can assure proper achievement. That is why general rules cannot be set forth, and one cannot claim in advance that automation will bring about more or less rigid organization of the processes, that it facilitates more decision making democracy, humanizes or dehumanizes man's work, etc. All this can happen under different conditions.

A part of routine work still exists in automated processes. However this part decreases, but it will still be vital for a long time; despite their routine these jobs can be of great importance. Let's take an example of all control activity which is actually very monotonous, professionally often non-demanding, yet extremely important. In all such cases a very rigid organization is to be set up and all the procedures must be thoroughly defined. No flexibility is allowed, the individual must not decide on the way of work, work methods, his working time, etc. Numerous organizational regulations must precisely stipulate all the relations and the worker must also precisely consider them.

Automation, however, also decreases the scope of the routine work, providing the possibility to get involved in creative work for a growing number of people. It will be impossible to limit this trend, so there will be a need for a flexible organization with a lot of independent decision making about individual's work methods, flexible working time, with only roughly defined interrelations, project tasks of particular professional teams, etc. All this will provide greater freedom for man.

This variety of different forms makes it impossible for us to further discuss the changes of organization due to automation. One may only add that it is essential to know all numerous forms in detail and to use them properly too.

One of the most significant organizational changes has been the emergence of a separate information management function with a chief information officer (Davis, 1989). The aforementioned dispersion of the information technology across the organization gives rise to the question about the need and the way to organize the function. The enumerated tasks do show the need for the information technology manager as well as the demand for his relatively high position in the hierarchy of the organization. This is the only way in which the overall information system can be logically organized. Whether that will be an individual directly linked to the top management or a special information department will of course depend on the conditions of a specific organization.

It is the decentralization of all information as well as the dispersion of the information technology that can lead us to the thought that organizing an information system does not mean organizing one of its subsystems (i.e., the information one), but organizing the entire business system from the information viewpoint. Herein the information viewpoint can in no way be isolated either from the basis or from the management process or from their link with the environment. Thus it is evidently the question of the process of information (Kajzer & Marn, 1989). As a result a demand crops up that the information system must be directly built into the organizational structure and must also follow the processes underway in this structure.

The information system thus depends on the organization. Information technology constitutes a tool enabling the successful operation of the organizational processes. This technology enables more and better information as well as its quicker circulation. This means an easier and more meaningful provision of all users with the necessary information. Usually the number of communication channels is also reduced as a result. All this no doubt has a profound impact on the organization and calls for its change. It is above all this possibility of quicker circulation of information and its less complicated dispersion to a greater number of users that in a way reduces the number of hierarchical levels. Each manager can thus communicate with a much bigger number of the subordinate employees and the significance of long-established rules has considerably dropped.

Finally one can mention computer aid at organization system design itself. We have already discussed the influence of information and its flow on organization structures definition. It is also noteworthy that computers enable man to acquire more information in a faster way. This in turn exerts an impact on the organization of processes; more information, on the other hand, also enables a more appropriate design of organization structures and processes. Good organization cannot be calculated and its design cannot be automated either. The basis of organization is man and his relations in a determined association with all his biological, psychological and sociological features, a specific man as an individual but linked with many others (also unique items) into a determined association. The multidimensional relations which are established here cannot be calculated. The number and sort of information, in any case, help in searching for most appropriate relations. Computerization of processes, growing importance of information system design and growing capability of computers by all means affect the organization. They both enable and require the rational design of numerous relations. Therefore the influence of organization science and the entire organization activity even increase. The computer can impact all relations, yet it is merely a tool - very efficient, successful and powerful, but nothing else. The effectiveness of various associations joined in the organization depends on the conformity of its application.

It is also essential to stress the mutual interaction of all the topics and methods dealt with. This only increases the impact of organizational structures and processes setting. All the aforementioned considerably transforms approaches to work organization, work processes, production as well as the entire social life. Organization however is a permanent process of searching for the most appropriate possibilities. Environment, society, technical facilities, economy, and objectives change, and so must organization and evaluation of its effectiveness.

Methods of Humanization of Work and the Role of Information Technology

It has already been stressed that humanization comprises a wide field of activity with the basic goal: to form the work in such a way that best suits the worker carrying out the work. Such a wide strategy undoubtedly brings about different methods and measures. That is why it is difficult to devise a precise systematization of the necessary and possible measures, particularly because they are mutually interlaced and interdependent. Roughly, the following main fields of measures can be listed:

- psychological
- socioeconomic and
- technical and technological measures for the humanization of work.

An in-depth analysis of numerous methods for the humanization of work exceeds the purpose of this chapter. A schematic presentation in Figure 3 shows the variety in this field and draws attention to all the possible moves. The scheme defines the levels, fields, measures and instruments for the humanization of work (Bernhard in Oberhof, 1980).

Information plays an important role in all these measures. First it is essential to compile all the necessary information on the existing state of affairs and establish the role of man in the process, the degree of his alienation, inhuman measures and methods, the extent of their effects, etc., and on the other hand the possibilities for changing of conditions. Modern information technology with its dispersion, dissemination and availability to various levels can help a lot. It can also link workers in a direct way in the process of decision making, thus incorporating them in the process of taking these measures. This in turn can make the measures taken appropriate and meaningful and can contribute to the very implementation of these measures in practice, whereas the very incorporation of workers in the process of decision making has a humanizing effect in itself.

Humanization, however, shouldn't merely serve to eliminate unsuitable conditions, relations, procedures, methods, etc. It should bring about new possibilities and reshape them so as to enable the worker to perform his job as

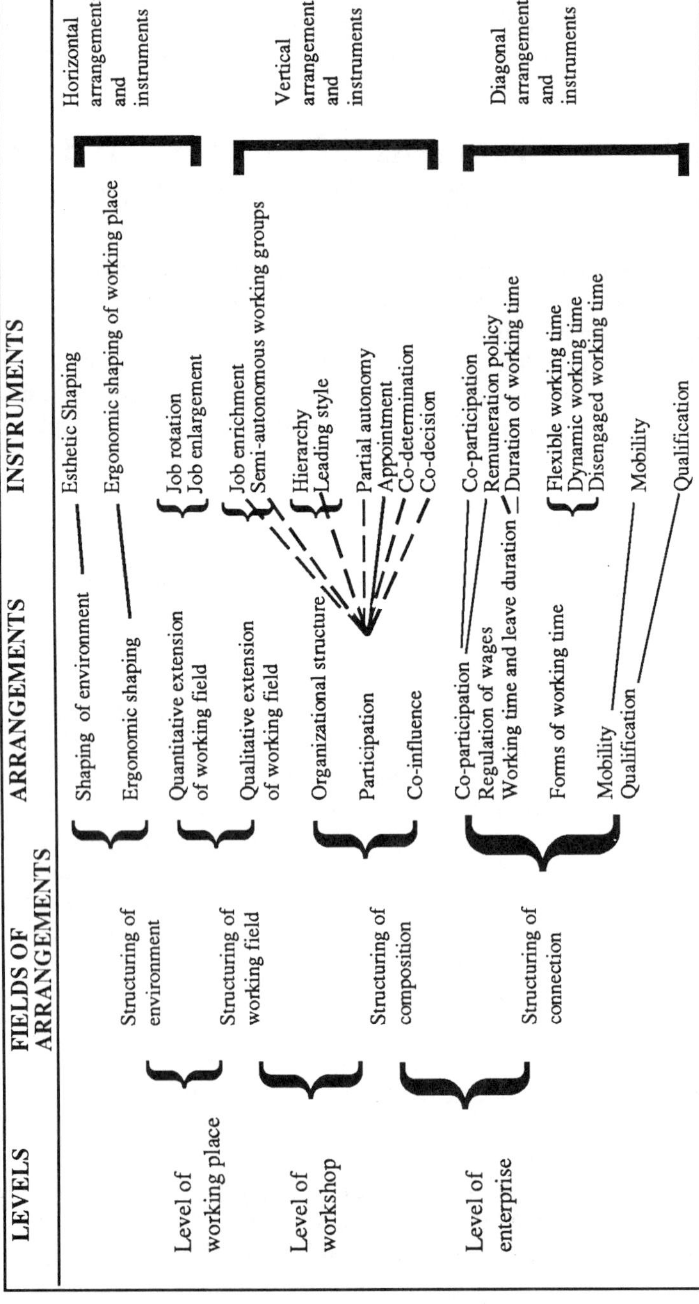

Figure 3. Arrangements and Instruments for Humanization of Work (Bernhard in Oberhof, 1980)

a complete generic being and take advantage of all possibilities for his own development and self-assertion. These are the two poles of the same process: the bad, non-humane, and difficult should be eliminated in the work and the new, better, more humane should be created. It is above all to stress the second direction, too often one tends to forget about it (e.g., work with computers, new relations in automated processes).

The possibilities for humanization of work give rise to new organizational forms, defining mutual relations among people in a completely new way. In general all these aspirations boil down to two basic directions:

- various forms of organization, participation, and workers' self-management at work and their results, and

- work structuring as a method of shaping work and its environment.

Both have been comprehensively dealt with in literature (Kaltnekar, 1976, 1979, 1980, 1980a, 1981, 1983, 1985, 1986, 1987, 1988; Kardelj, 1972; Novak, 1983; Mikolayczyk, 1983, 1986; ILO, 1979; Gaugler, Kolb & Ling, 1977, and many other books and articles), therefore it can by no means be summarized. One must admit that various forms of workers' participation at management do constitute important democratization in the working process as well as in the entire public life. Hence, this phenomenon is characteristic of both capitalist and socialist countries as well as the developing world. The question "yes or no" is no longer topical, all that matters is "how". Of course, there are numerous and various possibilities, but the main trend certainly is to strengthen the role of the worker in management.

As numerous and various are also the possibilities for work structuring. This should be the kind of work organization, work situation and work conditions that in case of maintaining or increasing of effectiveness will enable the coordination among the content of work and capabilities as well as aspirations of the individual worker. It should not matter whether this is achieved by division or extension of work, by concise or more comprehensive definition of job tasks, e.g., by already known fragmentation of job tasks or the dominating principles of job task enlargement and job enrichment (the theory often defines the latter merely as humanization of work). The required approach depends on the situation and mainly on the existing work division. Sometimes the work will have to be divided further on so as to ease access to workers abilities. On the other hand, too fragmented work will have to be concentrated. Workers' abilities can be larger or smaller than the requirements posed by the job he has.

Methods and approaches to structuring are common knowledge. One should only mention trends in enlargement and enrichment of work:

- job rotation,
- job enlargement,
- job enrichment, and
- autonomous (or semiautonomous) working groups.

All these forms are interlaced and are being further complemented. Their advantages and drawbacks must be continuously studied and as a result the most fitting among them selected. The introduction of these methods of work structuring can largely contribute to the diffusing of work alienation and to strengthening of its humanization. Thus all the methods represent a new challenge to the future design of organization. Both self-management (participation) and work structuring, and above all work in autonomous working groups, requires lots of information. Actually, it is the very development of the information technology that enables true self-management. In fact management is decision making and the latter is made possible only by the sufficient and timely supply of information about all facts that may influence decision making. The availability of the information technology to the user, the ever greater number of people trained for the work in information systems can expand the circle of employees taking part in the decision making process and can also provide all of them with necessary information. The latter in turn constitutes the prerequisite for the successful establishment of the both aforementioned approaches to the improvement of work humanization: self-management and work organization in autonomous working groups. Both approaches require the direct integration of the worker into the process of decision making.

Autonomous working groups actually constitute some kind of self-management: workers distribute the work and divide it among themselves on the basis of mutual contract. It is namely the decision regarding some specific work that is or should be the essential requirement of self-management. Consequently, both development trends are intertwined and it is this link that the paper will further focus on. The link is shown in Figure 4.

The setting up of autonomous groups means a quantitative and qualitative enlargement of the working field. A group assumes an enlarged job for some time together with partial planning and control. The tasks define the working field of the group, however, the working fields of each separate worker are not stipulated in advance and from "outside". A group forms an autonomous area and distributes the tasks among its individual workers by itself. In theory every member of the group chooses his own job within a certain enlarged task. This choice is in fact still limited, the reason being different qualifications needed for separate jobs. It is also limited due to the need for agreement with other workers in the group. Decisions made in the group are therefore

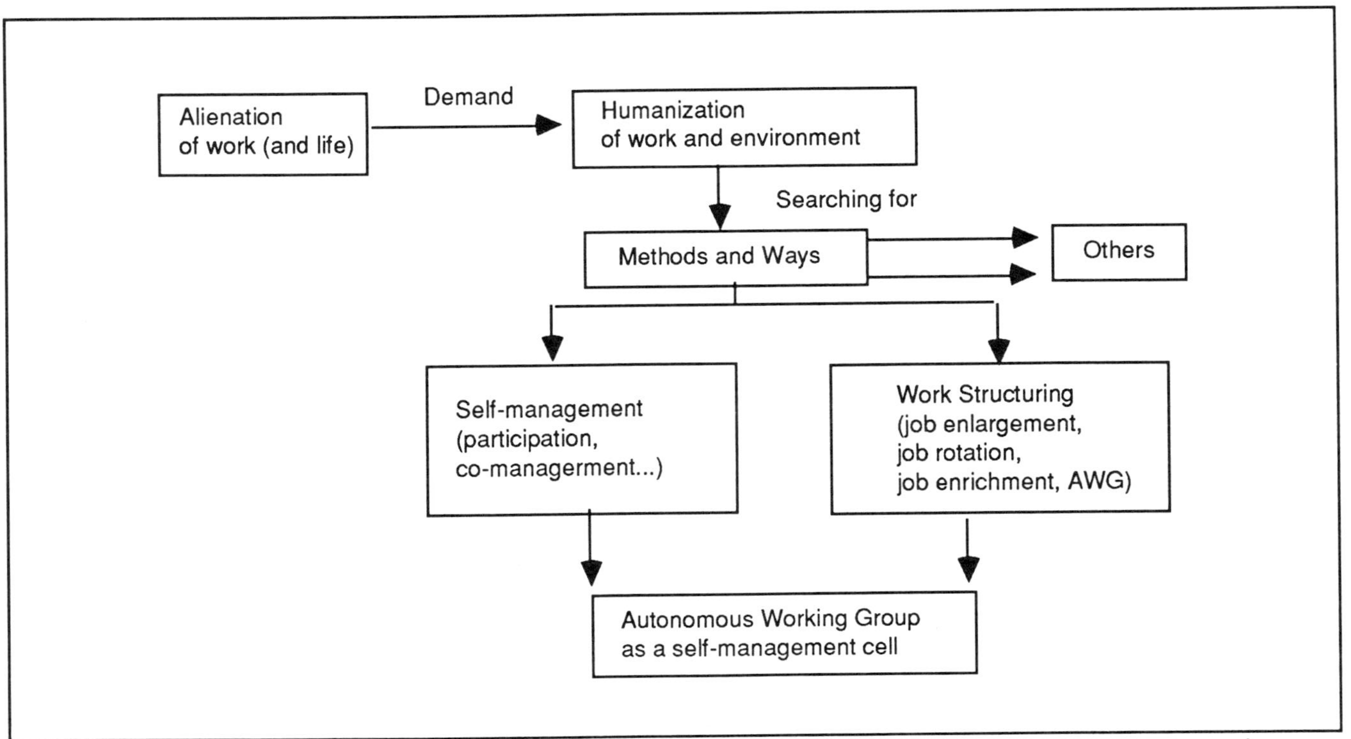

Figure 4. Methods for Humanization of Work

multipersonal and require constant negotiation among workers in the group.

The setting up of such groups can vary to a large extent. The working field of usual workshops or separate conveyer belts must, on the basis of various criteria, split into smaller units which can form a rounded off entirety. Technological criteria are most commonly used in work shaping and groups are set up according to a certain final part or product in the production process. In this way the results of the group's work can be established.

Organization of such working groups depends on the field of work and on the technology of work. A group must carry out a rounded off and interrelated part of a working process. In such a case the group must have - both in production and nonproduction field of work - a specific working program stemming from the working program of a large organizational unit. The group must mark out precisely its own working area in relation to other groups. It must have a certain autonomy within a larger collective body so as to distribute working tasks among employees independently as well as to be able to partly plan and control the tasks. A group thus assumes responsibility for the overall work and its results. Therefore the group remuneration as well as the salaries and wages of workers depend on the results of the whole group.

Having referred to the organization of autonomous working groups, we have also touched upon their competences. It must be noted, however, that we didn't have in mind the establishment of some new economic units with overall statements of account and bureaucracy that goes with it. This would of course be unfounded and irrational. They should only assume the function of direct organization and distribution of work, and should to a great extent exert impact on the distribution of income.

Autonomous groups would assume their working tasks on the basis of self-management agreements with other autonomous groups and thus with a larger organizational unit. In these agreements they will be, of course, limited a great deal, especially with technology of their work. A self-management agreement should above all retain the elements of the already mentioned work program: task, time, extent, quality and expenses. The workers would be arranging their work within working units on the basis of such agreements, too.

Such an agreement should not be a complicated document which would require a complicated procedure. As regards their tasks, the workers will most frequently discuss them without concluding specific agreements, in view of working conditions, and different qualifications.

Autonomous groups could contribute a great deal to the elimination of alienation of work. They would also make it possible for the holders of self-management rights to have more say in the management itself. Agreements between the participants of the self-management process could be more straightforward and direct, since workers can communicate easier in smaller groups than in large organizations. It is of course necessary to carry out a detailed examination and establish what rights groups should have in actual circumstances, i.e., what direct decisions they could take and what relations they would have with other groups. Experiences with other countries could be useful, but there are many problems which are typical and must be solved in accordance with the actual conditions. Even in each particular enterprise it will be necessary to examine forms of such units and search for the most appropriate ones.

The setting up of autonomous working groups is dealt with in literature only in the production field, predominantly in connection with conveyer belts and routine work. Too little emphasis is laid on the founding of such groups in other fields and areas involving different production conditions, as well as in nonindustrial branches of economy. However conditions and needs for change do exist in these areas too. Work can be very monotonous here, too, and this can also lead to alienation. Such jobs are particularly common in administrative areas.

Work in nonproduction areas also calls for some fresh motivation. As regards administrative jobs it is very difficult, sometimes even impossible, to establish the direct effect of each particular worker. White collar workers often have less say in management than blue collar workers. The issue at stake is not the number of representatives in various self-management bodies, which often exceeds the number of blue collar workers; the issue involved is the possibility for real participation in managing their own work.

The autonomous working groups in nonproduction areas should also be organized according to the "technological" principle. Workers carrying out an all-around part of a process, e.g., business functions, should also be included in such groups. These groups would thus be even more heterogeneous - as to their number and contents - than in the production area.

In conclusion: such or similar autonomous groups can be established in any enterprise irrespective of the actual social system - socialist or capitalist. In any case, the problems of alienation of workers at their work are common and should be resolved. In both cases such groups must also be able to exercise some self-management rights, no matter what this arrangement is called, self-management, participation or otherwise. There is obviously no need for further discussion on the role of the information technology in the organization of such autonomous working groups. Reaching an accord among the employees within a working group as well as within individual groups may become a highly complicated process and can thus reduce the efficiency of work. Both modes of bargaining require comprehensive information and data and, above all, their appropriate and timely combination. It is a must that should be guaranteed by the well selected information technology. The latter can also standardize individual procedures and program the conduct of bargaining as well as the mode of information compilation, processing and application. This in turn enables successful, rational and speedy bargaining. Therefore, it is fitting to note that the successful organization of autonomous working groups as basic self-management units is in a way conditioned by development of the information technology, and above all by its dispersion and availability to the user.

Future Trends and Opportunities

The development dictates future trends. Therefore, further development of the information technology can be easily predicted. The trends now underway also testify to the great pace of this development giving rise to the expectations of significant changes in the technology itself coupled with the requirements for changes in the organization of information systems and the modified impacts exerted on man in the production process and the society at large. Actually, one can claim that this would only further accentuate the aforementioned relations and contradictions. The introduction of information into all parts of the society is an undeniable fact that one should become reconciled to accepting both its advantages and drawbacks.

On the other hand, man assumes an ever more significant role, both in the society and in the working processes. Developments involving greater democracy are now underway all over the world, and man as individual is becoming ever more important. His role of a link in the chain in some electoral bulk is diminishing, and man is becoming a creative element of social changes. The significance of his participation in individual processes of social production is also increasing. Routine work is being taken over by machines, i.e., computers and robots, hence, he can deal with more or less creative tasks that he can successfully carry out only if he is appropriately qualified. Greater educational and qualification requirements in turn strengthen his importance. Furthermore, it is worthy to note the extraordinary tempo and dynamism of this development (Odegov, 1990), posing ever greater requirements and thus consolidating the role of man in the process.

The development of technology must match the development of human potential. The development of the human being must be matched with the development of technology. Man must be trained to accept and use the new technology. The reverse relationship calls for an even greater emphasis; the technological development must take into account the needs and requirements as well as the possibilities and abilities of man. This should be made possible by the involvement of users in the process of selection and application of information technology, which should also guarantee their greater responsibility for the operation of the information systems. The fast development of technology brings about new possibilities and problems affecting the relationship between production systems and their environment, technical solutions and human needs, working and social relations, production and human relations, etc. This development cannot be easily predicted, let alone dealt with in detail. Therefore only individual trends manifested as possibilities and requirements of the next decade (Bhabuta & Veryard, 1989; Liptak, 1990; Meyer, 1989) shall be referred to in this chapter:

- Organizational decision making is becoming ever more complex, decisions are based on the information decision support, the computer is assuming an ever greater part of decision making, both operative and tactical, and an ever greater number of people are taking part in the process of decision making (multipersonal decisions).

- All this calls for the establishment of appropriate relationships and interaction between the numerous factors involved in each process, and above all between its technological and human factors.

- Fast technological development exceeds the development of the abilities and the qualification of employees, and brings about some sort of "technophobia" on the part of the users and therefore calls for an appropriate training.

- New relationships are being established among the technological means themselves. The technology in turn exerts an important impact on the relationship among people, thus further alienating them. Humans need human contact and are more motivated by the rewards associated with interpersonal interaction than by the paycheck. The lack of interpersonal relations can have a demotivating effect.

- The exclusion of man from some processes (automated production, decision making processes) does liberate man from routine work, but on the other hand brings about the fear that man will be "liberated from every work". The problem of the lack of work and unemployment has already been referred to.

- The changing circumstances also call for the changes in organization. That is why it is essential to provide fairly flexible organizational structures. The participative management supported by the development of the information technology, makes possible the involvement of a greater number of employees, thus reducing the power of traditional managers.

- The process of decision making requires an appropriate amount of relevant information. The problem boils down to the shortfall and even over-abundance of information and to the related problem of selection for application.

• The process of overall informatization makes a wide range of information available to everyone. Apart from the problem of overabundance, the reverse problem of anonymity and the right for privacy also crops up.

• The increased environmental turbulence with numerous and frequent technological innovations, short life span of various solutions and products, changing circumstances of supply and demand on the market, etc., also come to the foreground. This in turn calls for further adjustment of organization systems as well as the meaningful development and application of the technology. All these and many other already applied and foreseen changes call for continuous and current adjustment of numerous and various elements taking part in these processes, in each particular case and in each particular organization.

Information technology can be used either to deliver or to enslave man. Therefore, future research work should focus on the quest for links between technological development and the potential for making man's life more humane and meaningful. Consequently, further development of the information technology should stem from the fact that we are working for man and because of man.

Conclusions

The accelerated development of information technology brings about fresh possibilities, thus changing various relationships at work and in the society at large. One can often read that we live in the information age. Information is becoming and is remaining the core of every process. The essential part of work in the information system is being assumed by computers, while many other processes are being automated and robotized.

In this context a question arises: Is man prepared and trained to cope with all these changes? The crux of the matter is that these changes are inevitably coupled with changes in both man's role in the production of tangible and intangible good and in man's private and social life. New technologies can liberate man and partly do away with his alienation. But they can also jeopardize the rights and possibilities of many people, actually of the majority of people taking an active part in these processes, thus provoking fear and anxiety.

An attempt has been made in this paper to outline basic requirements and problems posed by the development of information technology and the repercussions as they are manifested in the interaction between the development of technology and man using it. A brief definition of the ideas tackled and of the scope of the fields dealt with has been followed by a quest of the possibilities for a greater humanization of work coupled with the impact of the information technology. One can establish that the right decision making can most profoundly contribute to the humanization of work (and life at large). Man can be free only if he can freely make decisions relative to the conditions of his work and life. Therefore, he needs relevant and timely information. Information technology can supply man with such information, on the condition that it is appropriately selected and applied. The development of the possibilities for decision making and the development of the information technology are mutually supportive. Therefore it is essential to study both in their interaction and to establish effects of this interaction.

One cannot indulge in an in-depth analysis of numerous possibilities and methods for the humanization of work in such a brief chapter. Therefore we focused on the possibility to make decision making available to the large number of people involved in particular processes. Consequently, we have dealt with the autonomous working group as a basic self-management unit to present a specific case or method respectively, since it can make management and decision making available to the performer of working tasks to the greatest degree. It is information technology with all its new possibilities and above all its decentralization that should enable this direct management.

The topic of work humanization calls for further and profound study in view of the incessant development of information techology. This is the only way to take advantage and to avoid drawbacks posed by this development. It is also essential to carry out continuous and current analyses of the actual circumstances in each particular organization, since development will continue to bring about new possibilities and new problems.

References

Amoroso, D.L. (1990). Understanding the end user: The key to managing end-user computing. In *Proceedings of the International Conference on Managing Information Resources in the 1990s*, Hershey, Pennsylvania, p. 10-18.

Bahl, H.C. (1990). Strategies for user Involvement and Participation. In Khosrowpour M. & Yaverbaum G. (Eds.), *Information technology resources utilization and management*, Harrisburg, Idea Group Publishing.

Bajgori], N. (1989). Decision support systems versus expert systems: the systems approach. In *Proceedings of the International Conference on Organization an Information Systems*, Bled Yugoslavia, p. 398-403.

Bernhard in Oberhof (1980). Human strukturiertes Arbeitssystem (Human structured work-system). *Zeitschrift fuer Arbeitswissenschaft, 34*(3), p. 161-166.

Bhabuta, L., & Veryard, R. (1989). Relating IT/IS planning and strategic management of organizational effectiveness. In *Proceedings of the International Conference on Organization an Information Systems*, Bled-Yugoslavia, p. 191-216.

BMFT (1984). *Programm Forschung fuer Humanisierung des Arbeitslebens, Foerderschwerpunk Produktion* (Programm research for humanization of working life, focus production), Bonn-Germany.

Breslavski, S. & Yaverbaum, G. (1990). Expert systems applied to management information systems: Inteligent decision support. In Khosrowpour M. & Yaverbaum G. (Eds.), *Information technology resources utilization and management*, Harrisburg, Idea Group Publishing.

Cariati, T., Ciborra, C., & Maggiolini, P. (1989). Office information system planning: A transactional perspective. In *Proceedings of the International Conference on Organization an Information Systems*, Bled-Yugoslavia, p. 828-837.

Cziria, L. (Ed.) (1989). *^o ma vediet organizator prace (What an organizator of work has to know)*, Bratislava - Czechoslovakia, Praca.

Davis, G.B. (1989). Management information systems: Looking ahead ten years. In *Proceedings of the International Conference on Organization an Information Systems*, Bled-Yugoslavia, p. 9-25.

Durlik, I. (1989). Some problems concerning the managemnt of flexible manufacturing systems. In *Proceedings of the International Conference on Organization an Information Systems*, Bled Yugoslavia, p. 896-906.

Fisher, R. (1989). Decision support systems and managerial effectiveness. In *Proceedings of the International Conference on Organization an Information Systems*, Bled-Yugoslavia, p. 346-355.

Huska, A.M. (1984) *Metody a techniky rozvoja organizacie (Methods and technics of the development of organization*) Bratislava Czechoslovakia, Ustav ekonomiky a organizacie stavbnictva.

ILO (1979). *New forms of work organization*, Geneve - Switzerland, International Labor Office.

Israel, J. (1972). *Der Begriff Entfremdung* (Conception alienation), Reinbek - Germany.

Kajzer, [& Mam, F. (1989). Information system conception from the viewpoint of business system organization. In *Proceedings of the International Conference on Organization an Information Systems*, Bled Yugoslavia, p. 282-290.

Kaltnekar, Z. (1976). Nove oblike organizacije proizvodnega dela (New forms of organization of production work) *Organizacija in kadri, 5*(8), p. 629-660.

Kaltnekar, Z. (1979). *Uvodjenje promenljivog radnog vremena* (Introduction of flexible working time) Zagreb - Yugoslavia, Informator.

Kaltnekar, Z. (1980). *Lay-out proizvodnih sistema i kvalitet radnog 'ivota* (Lay-out of production systems and the quality of working life). Poslovna politika, 9(3), 52-58.

Kaltnekar, Z. & Juran, I.I. (1980a). Die Motivation und materielle Stimulation in autonomen Gruppen der administrativen und fuehren den Arbeiter. (Motivation and material stimulation in autonomous groups of administrative and managing workers). In *Proceedings of the Fifth EFPS/ EAPM International Work and Pay Conference*, Amsterdam - The Netherlands.

Kaltnekar, Z. (1981) Autonomous working group as the basic cell of selfmanagement. In *Proceedings of the Sixth International Conference on Production Research*, Novi Sad - Yugoslavia, p. 201-205.

Kaltnekar, Z. (1983). *Autonomne radne grupe kao na~in pribli'avanja odlu~ivanja proizvo/a~u* (Autonomous working groups as the way to approach the decision making to the producer), Produktivnost, 25(5), p. 19-28.

Kaltnekar, Z. (1985). Rigidity or elasticity of production planning. In *Proceedings of the Eight International Conference on Productions Research*, Stuttgart - Germany.

Kaltnekar, Z. (1986). Robotisierung und Humanisierung der Prozesse - zwei antagonische Herausforderungen der Zukunft (Robotisation and humanization of processes - two antagonistic challenges of future). In *Proceedings of the Sixten International Colloquy on Scientific Organization of Work*, Ostrava - Czehoslovakia, p.33-38.

Kaltnekar, Z. (1987). *Proces samorzadowy w obrebie autonomicznej grupy roboczej jako sposob humanizacji pracy* (Self management process in the autonomous working groups as the way for humanization of work). Folia Oeconomica, Acta Universitatis Lodziensis, No. 67, p. 77-87.

Kaltnekar, Z. (1988). *Organizacija delovnih procesov* (Organization of working processes). Kranj, Moderna organizacija.

Kardelj, E. (1972). *Protislovja dru'bene lastnine v sodobni socialisti~ni praksi* (Contradictions of societal ownership in contemporary socialistic praxis). Ljubljana - Yugoslavia, Dr'avna zalo'ba Slovenije.

Kashitza, O. (1989). Some design concepts for computerized functioning environment of work stations at VEF production association. In *Proceedings of the International Conference on Organization an Information Systems*, Bled-Yugoslavia, p. 907-911.

Khosrowpour, M. (1990). Managing information technology resources as a major organization assets. In Khosrowpour, M. & Yaverbaum, G. (Eds.), *Information technology resources utilization and management*, Harrisburg, Idea Group Publishing.

Kobe, M. (1986). *Svet i Jugoslavija u visokim tehnologijama elektronike* (The world and Yugoslavia in high technologies of electronics), Pozadina No. 4, p. 26-42.

Kurth, R. (1985). Die Auswirkungen technisher Aenderungen auf die Arbeit. (The influences of technical changes on the work) *Angewandte Arbeitswissenschaft No. 105*, p. 24-35.

Liptak, F. (1977). *Systemova organizacia prace* (System organization of work), Bratislava - Czechoslovakia.

Liptak, F. (1990). Zmeni v praci riadiacich pracovnikov na prelome tisicro~i a ~initele, ktore ih vyvolavaju (The changes in the work of managing workers on the break of millenium and factors which bring them about). In *Proceedings of the International Conference on Man and Work on the Treshold of the Third* Millenium, Bratislava - Czechoslovakia, p. 254-257.

Maslow, H.A. (1970). *Motivation and Personality*. New York, Harper - Row Publishers.

Matthoefer, H. (1980). *Humanisierung der Arbeit und Produktiviteat in der Industriegesellschaft* (Humanization of work and productivity in the industrialized society), Koeln - Germany, Bund

Verlag Gmbh.

Meyer, J. (1989). International & ethical issues for the future: The impact of technology on humans, interaction & productivity in organizations. In *Proceedings of the International Conference on Organization an Information Systems,* Bled-Yugoslavia, p. 483-487.

Mikolajczyk, Z. (1983). *Metody doskonalenia organizacji procesov i stanowisk pracy* (Methods for improvement of organization of processes and working places), Warszawa - Poland, Institut wydawniczy zwiazkow zawodowych.

Mikolajczyk, Z. (1986). *Kierunki humanizowania organizacji w przemysle* (Tendencies for humanization of organization in the enterprises). Lodz - Poland, Universytet Lodzki.

Novak, M. (1974). *Organizacija rada u socijalizmu* (The organization of work in socialism). Zagreb - Yugoslavia, Informator.

Odegov, J. (1990). Trudovoj potencijal predprijatija: Tendencii razvitija v uslovijah uskorenija nau~no-tehni~eskoga progresa (Working potential of the enterprise: tendencies of the development in the conditions of the acceleration of scientific - technical progress). In *Proceedings of teh International Conference on Man and Work in the Treshold of the Third Millenium,* Bratislava - Czechoslovakia, p. 315-319.

Oppeland, H.J. (1989). *The impact of participation and participation attitudes on information system implementation success.* Paper on the International Conference on Organization an Information Systems, Bled-Yugoslavia.

Sucharevskij, B.M. (Ed.) (1983). *Sistema upravlenija trudom* (System of management of work). USSR, Ekonomika, Moskva.

Tarman, Z. (1989). Informatics in function of organization or opposite. In *Proceedings of the International Conference on Organization an Information Systems,* Bled-Yugoslavia, p. 71-77.

Tulak, F. (1989). Resistance to computerization - some psychological and communications problems. In *Proceedings of the International Conference on Organization an Information Systems,* Bled Yugoslavia, p. 488-497.

Tunje, A. & Savi, A. (1989). *Informational organization. In Proceedings of the International Conference on Organization an Information Systems,* Bled-Yugoslavia, p. 154-162.

Tietze, B. (1974). *Humanisierung der Arbeitswelt: Theoretisches Programm un politische Praxis* (Humanization of the working world. Theoretical program and political praxis) Arbeit und Leistung No. 28(12), p. 309-315.

Trethon, F. (1989). Technical development with the human for the human. In *Proceedings of the International Conference on Organization an Information Systems,* Bled-Yugoslavia, p. 438-441.

Weil, R. (1975). Humanisierung der Arbeit - Antwort auf Entfremdung (Humanization of work - answer on the alienation) *Schriftenreihe des IfaA, No 4,* p. 2-92.

Weil, R. (1987). *Arbeitim Wandel: Auswirkungen auf Anforderungen, Leistungen und Entgelt* (Work in change: influences on the requests, effects and pay) Koeln - Germany, Institut fuer angewandte Arbeitswisseschaft.

Zink, K.J. & Ritter, A. (1989). *Verbesserung der Arbeitsqualitaet durch Problemsloesungsgruppen* (Improvement of the quality of work through the problem solving groups). Bonn - Germany, Wirtschaftsverlag NW.

CHAPTER CONTRIBUTORS

Macedonio Alanis is an Assistant Professor of Management Sciences and Information Systems at the University of Detroit. He obtained his Ph.D. in Business Administration from the University of Minnesota. His concentration area is Management Information Systems. He has an Sc.M. in Computer Sciences from Brown University and a B.S. in Computer Sciences from the Instituto Tecnologico y de Estudios Superiores de Monterrey. His current research interests include information requirements determination; software development tools and techniques, including CASE technology; and the diffusion of information technology.

Allan Bird is Assistant Professor of Management and International Business at the Stern School of Business, New York University.

William I. Bullers, Jr. is a Professor of Management Information Systems at the Anderson Schools of Management, University of New Mexico, Albuquerque, NM. He received a B.S. in Mathematics from Carnegie-Mellon University, an MBA from Duquesne University, and a Ph.D. in Management Information Systems from Purdue University. He has previously held positions as software systems analyst, computer operations shift supervisor, and application systems supervisor. His research interests are in information systems design and modeling representations, data base management systems, and knowledge representation and decision support for computerized manufacturing systems.

Thomas L. Case is an Associate Professor and Acting Head of the Department of Management at Georgia Southern University. He holds a PhD in Social Psychology from the University of Georgia. He teaches graduate and undergraduate information systems and general management courses including data communications, systems analysis and design, organization development, and research and development management. In addition to R&D information systems, he is involved with programs of research concerning managerial influence processes and supported employment services. He is currently Treasurer and Proceedings Editor for the International Academy for Information Management.

Deepak K. Datta is Assistant Professor of Management at the School of Business, University of Kansas. He holds a Ph.D. in Strategic Planning and Policy from the University of Pittsburgh, an MBA from the Indian Institute of Management, Calcutta, and a B.Tech in Mechanical Engineering from the Indian Institute of Technology. Prior to his Ph.D., he worked as a manager in multinational corporations in India. Professor Datta has published in the Journal of Management, Journal of Management Studies, Long Range Planning, and the Journal of General Management. He has also presented a number of papers at various international and national meetings. His current research interests are in the areas of mergers and acquisitions, joint ventures, and strategic planning systems.

Uldarico Rex Dumdum, Jr. is Assistant Professor in the Department of Business and Managerial Science at Marywood College. He is currently writing his Ph.D dissertation on problem formulation of ill-structured situations in the context of information systems development. He was a Fulbright scholar and holds a BS in Civil Engineering from the University of Mindanao, Philippines, Masters in Engineering from the Asian Institute of Technology, Bangkok, Thailand, and a Masters in Advanced Technology/Information Systems from SUNY-Binghamton. He also did graduate studies in development anthropology concentrating on the social impact analyses of development projects. He was head of social systems planning at the Cotabato-Agusan River Basin Development Project in Southern Philippines. His research interests lie in the area of problem formulation, requirements determination and the social and organizational consequences of technology.

William L. Gardner is the Hearin-Hess Assistant Professor of Management at the University of Mississippi where he teaches undergraduate and graduate courses in management, organizational behavior, and research methods. He received a B.S. in business administration from Susquehanna University, and an M.B.A. and a D.B.A. from Florida State University. Gardner served as the research associate for a major grant commissioned by the State of Florida to identify the key competencies of educational managers; this coupled with his experience in the wholesale hardware industry, serves as a basis for his research and teaching efforts. His research interests include managerial work, impression management, leadership, motivation, and management education. He has published articles in Academy of Management Journal, Academy of Management Review, Journal of Management, and Organizational Dynamics.

John E. Gessford received his doctorate in industrial engineering from Stanford University in 1957. He worked in industry for 14 years in computer system planning, development, and management assignments, first at International Paper Company and then as a consultant at SRI International. In 1973, he become an Associate Professor at California State University, Los Angeles, and developed an information systems program in the School of Business. In 1985, he joined the faculty of a newly formed information science program at The Claremont Graduate School and developed courses in database management and artificial intelligence, as well as a practicum course to give students practical experience in information systems development. In 1990, he joined the information systems faculty at California State University, Long Beach.

P.J. Guinon is an Assistant Professor at the School of Management, Boston University.

E. Alan Hartman received a Ph.D. from Michigan State University in 1972. He was a faculty member at Texas Christian University (1972-1976) and is now a faculty member in the College of Business Administration at the University of Wisconsin Oshkosh. Research interests include organizational entrepreneurship, role of the ratee in performance appraisal, and effects of technology on work.

Clyde W. Holsapple is Professor of Decision Science and Information Systems at the University of Kentucky where he holds the Endowed Chair in Management Information Systems. His main fields of research are decision support systems, database management, software integration, knowledge management, and business expert systems. His books include Foundations of Decision Support Systems (Academic Press, 1981), Micro Database Management (Academic Press, 1984), Manager's Guide to Expert Systems Using Guru (Dow Jones-Irwin, 1986), Business Expert Systems, (Irwin, 1987), and The Information Jungle, (Dow Jones-Irwin, 1988). He has written numerous articles for journals such as Decision Sciences, Operations Research, Policy Sciences, Information Systems, The Computer Journal,Decision Support Systems, The Information Society, Financial Management, and Human Systems Management. He is the DSS area editor for the OSRA Journal of Computing and the director of the Kentucky Initiative for Knowledge Management.

Linda Ellis Johnson is Instructor of Finance and Management Information Systems at Western Kentucky University. She is currently completing a Ph.D. in Management Information Systems at the University of Kentucky in the Department of Decision Science and Information Systems. Her research interests are in the area of decision support systems, computerized communication systems, and model management systems. Mrs. Johnson has presented and published articles at several regional conferences. Her research in computer-assisted instruction has been funded by the University of Kentucky.

Zdravko Kaltnekar is on the faculty for organizational sciences at the University of Maribor in Kranj, Yugoslavia where he holds the rank of Professor Ordinarius. He is the author of 15 books, 77 articles and numerous national and international conference papers. His main research interests are organizational systems and methods, and the humanization of work. In 1988 he chaired two international conferences: "Reduction of Working Time, Extension of Operating Time and their effect on Work and Pay" in Ljubljana, and "Organization and Information Systems" in Bled. He is on the editorial board of Analyse de Systems (Lyon, France) and of the Information Resources Management Journal (Harrisburg PA, USA).

Laura L. Koppes is an Assistant Professor of Psychology at the University of

Wisconsin Oshkosh. She received her Ph.D. in Industrial and Organizational Psychology from The Ohio State University in 1987. Her research interests include attribution theory and job stress, the impact of computer technology on people, performance appraisal, and training.

Soonchul Lee is an Assistant Professor of MIS at the School of Management, Boston University.

William G. Lehrman is Assistant Professor of Management at the R.B. Pamplin College of Business, Virginia Polytechnic Institute and State University.

Binshan Lin is an Assistant Professor of Information Systems at Louisiana State University in Shreveport. Dr. Lin has published articles in the International Journal of Operations & Production Management, Education and Computing, Applied Statistics, Interfaces, Journal of Systems Management, and Journal of Small Business Strategy.

Efrem G. Mallach is Associate Professor in the department of Operations Management and Information Systems in the College of Management Science of the University of Lowell in Lowell, Mass. He specializes in the impact of new technologies on their users and on how users can best plan for their adoption. Prior to embarking on an academic career in 1984, Dr. Mallach enjoyed a successful fifteen-year career in the computer industry. His last business position was as division general manager of a nationwide systems and software firm. He continues to work with executives of both vendor and user firms to develop appropriate strategies for coping with new technologies. Dr. Mallach holds a B.S.E. degree from Princeton University, an M.B.A. from Boston University and a Ph.D. from the Massachusetts Institute of Technology. He has published over thirty professional papers, is a regular columnist in information systems trade publications, and speaks frequently at conferences and seminars.

David Nealon is currently a graduate student in Industrial and Organizational Psychology at the University of Wisconsin Oshkosh. His research interests include validating performance appraisal systems and identifying predictors of peoples' motivation to use computer technology.

Ramakrishnan Pakath is an Assistant Professor of Decision Science and Information Systems at the University of Kentucky. He received the Ph.D. degree in Management (MIS) from Purdue University in 1988. Dr. Pakath's research interests and experience include the application of artificial intelligence and operations research to problems in the design of centralized and distributed support systems; analysis and simulation of inventory control problems in serial and non-serial manufacturing systems; and analysis and simulation of information value in optimization modeling. Dr. Pakath has several articles that are currently under review or are forthcoming in such journals as European Journal of Operational Research, Information and Management, and Computer Science in Economics and Management. Dr. Pakath's research has been funded by IBM, Ashland Oil, and the University of Kentucky.

Joy Van Eck Peluchette is a doctoral candidate in the Department of Management at Southern Illinois University, Carbondale. Her major field is organizational behavior, supported by a minor in strategic management. Ms. Peluchette has published in the Personnel Administrator and has made several presentations at the national meetings of the Decision Sciences Institute. Her research interests include issues relating to careers, impression management, electronic networking, upward influence, women, and the work-family interface.

Baron Perlman is a Rosebush Professor at the University of Wisconsin Oshkosh. He received his Ph.D. in 1974 in Clinical Psychology from Michigan State University. Since 1975 he has been a faculty member in the Department of Psychology at the University of Wisconsin Oshkosh. He is presently interested and involved in research which studies intrapreneurship, innovation and the impact of computer technology on institutions of higher education and other public organizations. He is the co-author of two books, The Academic Intrapreneur (Praeger, 1988) and Organizational Entrepreneurship (Irwin, 1990).

John R. Pickett is an Associate Professor and former Head of the Department of Management at Georgia Southern. He holds a PhD in Statistics from the University of Georgia and teaches graduate and undergraduate information systems and quantitative methods courses. He is the author of a Pascal programming textbook. His consulting and research interests include statistical process control, expert systems and artificial intelligence applications, and the utilization of information technology in R&D. He is actively involved in a number of professional organizations and served as Vice President for Statistical Services for the Block-Kelly Company.

Abdul M.A. Rasheed is currently an Assistant Professor in the Management department at the University of Texas at Arlington. He holds a Ph.D. in Strategic Planning and Policy from the University of Pittsburgh, an MBA from the Indian Institute of Management, Calcutta, and a B.S. in Physics from Kerala University. His primary research interests are environment-organization interactions, problem structuring and diagnosis, diversification and performance, and international corporate strategy. Prior to his Ph.D., Professor Rasheed held several executive positions in the banking industry in India and the Middle East.

Richard A. Reid is a Professor of Management at the Anderson Schools of

Management, University of New Mexico, Albuquerque, NM. His teaching and research interests center on the application of management science techniques to challenging managerial problems. He is particularly interested in using concepts from general systems theory to design frameworks for modeling computer-assisted management systems. He earned his Ph.D. in systems research from Ohio State University and was a postdoctoral fellow at The Harvard Center for Community Health and Medical Care. Dr. Reid has had published over 50 refereed articles in national and international journals, including the Journal of the Operational Research Society, Journal of Systems Management, Journal of Information Systems Management, Interfaces, and Production and Inventory Management.

Holly R. Rudolph is a doctoral student in accounting at Southern Illinois University at Carbondale. A Certified Public Accountant, Ms. Rudolph has worked in industry as an internal auditor and has taught accounting for the last nine years. Her research interests include group decision support systems and their behavioral implications.

John R. Schermerhorn, Jr. is the Charles G. O'Bleness Professor of Management at Ohio University, where he teaches courses in management, organizational behavior, and organization development. He holds a Ph.D. in organizational behavior from Northwestern University, the M.B.A. in management and international business from New York University, and a B.S. in business administration from the State University of New York at Buffalo. Dr. Schermerhorn taught previously at the University of Vermont and Southern Illinois University at Carbondale, where he also served as Department Head and Associate Dean. He has been a Visiting Professor at the Chinese University of Hong Kong and a Visiting Scholar at Liaoning University in China and the Technical University of Wroclaw in Poland. Past Chair of the Management Education and Development Division of the Academy of Management, Dr. Schermerhorn is the author of Management for Productivity, Third Edition (New York: John Wiley & Sons, 1989) and senior coauthor of Managing Organizational Behavior, Fourth Edition (New York: John Wiley & Sons, 1991). His scholarly research on interorganizational cooperation, organization change, and managerial behavior and performance has been published in a variety of journals. He is an active management consultant with special expertise in management training and development, and human resource strategies for productivity improvements. His work includes assignments with organizations in Malaysia, Egypt, Thailand, and Hungary among other countries.

Tatsumi Shimada is currently Professor of Management Information Systems and Office Management at Yokohama College of Commerce. He has served as director of Yokohama College Computer Center for four years. Previously he has experienced in system analysis & design at private industry (Meidensya

Electric Mfg.Co.,Ltd.) and management consulting (Japan Productivity Center). Professor Shimada is interested in research areas including MIS in big corporations, Strategic IS, and Analysis & Design of Office IS. Currently he is focusing on comparative studies of IS between Japan and other countries. He has published several books and numerous articles in professional journals in Japan.

Edward J. Szewczak is Associate Professor of MIS at Canisius College in Buffalo, NY. He received his Ph.D. from the Graduate School of Business at the University of Pittsburgh. He had previously served on the faculties of Robert Morris College and the Southern Illinois University at Carbondale. His information systems research has been published in Data Base, European Journal of Operational Research, Information & Management, Information Resources Management Journal, the Journal of Microcomputer Systems Management, Journal of MIS, Omega, and Simulation & Games as well as in scholarly books and national and regional conference proceedings. He has also published in the Journal of Management Studies, Logique et Analyse, Long Range Planning, and the Organizational Behavior Teaching Review on subjects ranging from strategic management to philosophical ethics.

Irmtraud Streker Seeborg holds a Diplom in Mathematics from the J. W. Goethe Universitaet in Frankfurt, Germany; an MBA from the University of Utah; and an MA and Ph. D. in Administrative Science from Yale University. Dr. Seeborg is the author/coauthor of several books on computer programming. She has published papers, among others, in Journal of Management Information Systems, Journal of Human Resources, Industrial Relations, and the Journal of Applied Behavioral Science, and has presented papers at meetings of the Academy of Management and other organizations. Dr. Seeborg has worked as a systems analyst and MIS consultant both in the US and in Europe. Since 1976 she has taught in the Department of Management Science, Ball State University. During the academic year 1990/91 she is on a study leave and works in the technology center of Heckler & Koch, Oberndorf, Germany on the development of expert systems and a technology database.

Wanda A. Trahan is an Assistant Professor of Psychology at the University of Wisconsin Oshkosh. She received her Ph.D. in Industrial and Organizational Psychology from Louisiana State University in 1989. Her research interests include leadership and attribution theory, the impact of computer technology on people, organizational discipline, and training.

Joseph Williams is an Assistant Professor in the Computer Information Systems Department at Colorado State University. He received the M.A. degree in 1980 in Rhetoric from the University of Wisconsin at Madision, the M.B.A. degree in 1982 and the Ph.D. degree in 1987 in Management Science & Information Systems from the University of Texas at Austin. He has held faculty positions at Bucknell University, New Mexico State University, and the University of Montana. Joseph is a member of the Association for Computing Machinery, the Decision Sciences Institute, the Institute for Management Science, the Society of Petroleum Engineers, the International Association of Energy Economists, and the American Bar Association. His research interests are in the areas of disaster recovery planning, computer security, EDP audit and control, information economics, management of IS resources, petroleum industry applications, and side-effects of IS technology. Joseph is the author of approximately a dozen papers, including publications in Datamation, the Production and Inventory Management Journal, and the Journal of Petroleum Technology.

INDEX

- J -

- K -

- L -

- M -

- N -

- O -

- P -

- Q -

- R -

- S -

- T -

- U -

- V -

- W -

WITHDRAWN
UST
Libraries